Electric Circuit Analysis

Margaret R. Taber
Eugene M. Silgalis

Cuyahoga Community College

Electric Circuit Analysis

Houghton Mifflin Company

Dallas
Geneva, Illinois
Boston
Hopewell, New Jersey
Palo Alto
London

Printed in the U.S.A.

Library of Congress Catalog Card Number: 78-69525

ISBN: 0-395-26706-4

Contents

Contents

xii

Preface

Students in the course in electric circuits at most community and technical colleges have very diverse backgrounds. Many schools have minimal or no prerequisites for this course. So— along with students who have the proper math and science backgrounds—there are those with almost no math and science backgrounds. There are some students with previous electrical-electronic experience acquired in the service, in technical high schools, or in proprietary schools. There are also former engineering students.

With such a varied student population, it is difficult to keep everyone interested and learning. Most instructors teach for the *average* student, but what percentage of the class is average? This text is written so that students with some previous experience may move through it quickly. At the same time, material is provided for the student who needs to review arithmetic, algebra, and scientific notation. Some of the advanced topics and units may be omitted or may be used as enrichment or bonus material.

The electric circuit topics are divided into units. The units are then broken down into specific objectives. The book states each objective, then discusses it, gives one or more examples of it, and follows these by one or more problems. A person who is interested only in learning or reviewing a specific electrical topic on a concentrated basis may just turn to that specific objective in the text. Having completed that particular objective, the person may go to the test at the end of that unit. These tests are provided so that the interested person may see whether all the objectives in the unit have been successfully completed. The Appendix supplies answers to the problems presented in the tests.

We have used the materials (objectives, examples, and problems) on which much of the text are based at Cuyahoga Community College since 1973. We and our colleagues have used them in both day and night classes with great success.

The text material is intended to cover one academic year (two semesters or three quarters). You will note that the beginning of the text is very simple. It was designed this way, so that students with weak mathematics backgrounds can take an algebra course concurrently. The text topics begin with very elementary material and progressively lead up through Fourier analysis and basic instruments.

After the beginning units, the mathematics background required of the student increases. Unit 14 covers time constants, Unit 15 the sine wave, and Unit 17 vector algebra. It is a good idea for students to take trigonometry along with these units. However, the explanations in these units are very detailed and they have been successfully completed by students without prior experience in trigonometry.

We start the text with some rough definitions of current, voltage, and resistance, and then briefly discuss various electrical (and electronic) symbols. We have found that this approach especially interests our engineering technology students, who are practical and who like to have a reason for learning material. Once the students cover this brief introduction and use Ohm's law, they become interested in learning more about current, voltage, and resistance. Units 5 and 6 cover these topics in detail. However, those instructors who want their students to get a good foundation in current and voltage may have them cover Unit 5 along with Unit 1.

We inserted the unit on Problem Solving (Unit 3) for those students who are weak in mathematics. A student with a good algebra background should be able to skip this unit. However, we recommend that the student try the test at the end of the unit to be sure of his or her proficiency.

Note the instructions given at the beginning of Unit 3, the unit on Problem Solving. These instructions could also be used for any of the other beginning units. The point is that the fast student or the student with previous experience may move through the beginning units quickly.

We would like to thank Rita and Barbara Pawlik for their help in typing the manuscript. We would also like to thank the following individuals who reviewed the manuscript during its preparation: Professor Frederick F. Driscoll, of Wentworth Institute; Dr. Frank T. Duda, of the Community College of Allegheny County; Professor Louis Gross, of Columbus Technical Institute; Professor Arnold Kroeger, of Hillsborough Community College; Professor Marlowe L. Sperstad, of Arizona State University; and Professor Thomas Roach, of San Diego City College.

Margaret R. Taber
Eugene M. Silgalis
Cleveland, Ohio

The Language of Electric and Electronic Circuits

Unit 1

Most of your daily activities today bring you in contact with electric or electronic circuits: your electrical appliances, digital watch, camera, automobile (see Figure 1.1), AM and FM radio, television, tape deck, just to name a few. How many items can you add?

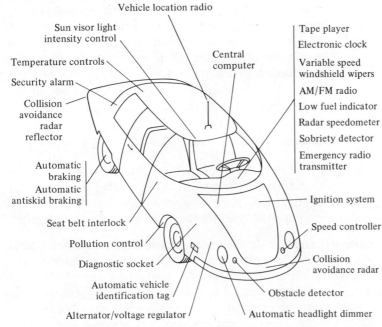

Vehicle location radio

Sun visor light intensity control

Temperature controls

Security alarm

Collision avoidance radar reflector

Automatic braking

Automatic antiskid braking

Seat belt interlock

Pollution control

Diagnostic socket

Automatic vehicle identification tag

Alternator/voltage regulator

Central computer

Tape player

Electronic clock

Variable speed windshield wipers

AM/FM radio

Low fuel indicator

Radar speedometer

Sobriety detector

Emergency radio transmitter

Ignition system

Speed controller

Collision avoidance radar

Obstacle detector

Automatic headlight dimmer

Figure 1.1 Use of solid-state electronics in a car of the future

1

If you are going to do more with these devices than just use them, you will have to learn the language of electric and electronic circuits. This book will concentrate on electric circuits. You need to understand electric circuits because they are a part of every electronic device. And, of course, all electrical appliances contain electric circuits.

Objectives

After completing all the work associated with this unit, you should be able to:

1. Recall the following for the terms voltage, current, resistance, and power: the definition, the symbol used in mathematical formulas, the basic unit of measurement, and the abbreviation or symbol used for the basic unit of measurement.
2. Identify the following electrical symbols: battery, switch, lamp, resistor, capacitor, inductor, ammeter, voltmeter, transformer.
3. Draw a schematic diagram for the electric circuit—or a circuit diagram—for a flashlight. Explain the operation of the circuit.
4. Draw the circuit diagram of a simple electric circuit that contains a battery and a resistor. Insert an ammeter into the circuit to measure current and a voltmeter to measure the voltage drop across the resistor.

We shall use certain terms frequently when we discuss electric or electronic circuits. We shall briefly introduce these terms in this unit, and you will learn about each of them in much greater detail in later units. In addition to the terms, several symbols will be introduced in this unit. However, as you can see from the objectives, you will not have to learn all the terms and symbols now. The purpose of including all of them here is just to give you an introduction. We want you to start using this new language. You will probably find yourself referring back to this unit as you progress through the text.

Electric-Circuit Terms

Objective 1   Recall the following for the terms voltage, current, resistance, and power: the definition, the symbol used in mathematical formulas, the basic unit of measurement, and the abbreviation or symbol used for the basic unit of measurement.

Current   The flow or drift of electric charge past a point in an electric circuit in a given time. The basic unit of measurement for electric current is the *ampere*. The abbreviation for ampere is A. One ampere is equal to a flow of 6.24×10^{18} electrons per second past a given point in a conductor. The symbol used for current in equations or mathematical formulas is *I*.

dc current A direct, steady, unchanging, unidirectional flow of electric charge. A battery provides the potential needed for a dc current.

ac current A current that alternates, or flows in one direction in the circuit and then in the other; it does this periodically, like a sine wave. The current that flows in lamps and in most electrical appliances in your home is ac.

Voltage The potential, force, or push that causes electric charge to move. Voltage is the term commonly used to mean electromotive force (emf, electron-moving force), electrical potential, and potential difference.

Two symbols are used for voltage in mathematical formulas: E usually represents a voltage supply, such as batteries. V usually represents a voltage drop across a circuit element.

However, sometimes these two symbols are used interchangeably. The basic unit of measurement for electrical potential, emf, and potential difference (voltage) is the *volt*. The abbreviation for volt is V.

dc voltage A direct, steady, unchanging, unidirectional potential; see Figure 1.2. A battery provides this type of potential.

ac voltage An alternating potential; see Figure 1.2. The electrical outlets in your home provide this type of potential.

Resistance The property of an electric circuit (or a circuit element) that dissipates electrical energy into heat energy. However, with your present knowledge, this definition is vague. A simpler definition that you may use in your study of *dc* circuits is this: *Resistance* is the property of a circuit element that offers *opposition* to the flow of electric current or charge. The symbol used for resistance in mathematical formulas is R. The basic unit of measurement for electrical resistance is the *ohm*. The symbol that is used for ohm is Ω, the capital Greek letter omega.

Power The rate of doing work, or electric energy per unit time. The symbol used for power in mathematical formulas is P. The basic unit of measurement for power is the *watt*, abbreviated W.

To complete the definition part of Objective 1, just remember the following rough definitions. You will improve on them as you progress through later units.

Current is the rate of flow of electric charge.

Voltage or potential is the push that is required so that current* will flow.

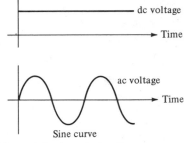

Figure 1.2 dc and ac voltage

---

* From the definition of current, you know that it is really *charge* that flows. However, the term *current flows* is commonly used to describe current in a circuit. Therefore you will see this term used throughout this text.

Resistance is the opposition to the flow of current.
Power is the rate of doing work.

Problem 1.1   Make a chart from memory, listing the following for the terms voltage, current, resistance, and power: the symbol used in mathematical formulas, the basic unit of measurement, and the abbreviation or symbol used for the basic unit of measurement.

Before you go on to Objective 2, read through the following definitions of a few more terms. You will learn more about these terms as you go along.

Resistor   A circuit element that is manufactured to have a certain resistance. It may be fixed or variable, and it comes in many sizes, shapes, and forms. It may be used to limit current, to drop a voltage, for purposes of isolation, and to obtain a voltage variation across it. Resistors are measured in units of ohms.

Capacitor   A capacitor basically consists of two conductors (metal plates) separated by a dielectric (insulator). It is a circuit element that is capable of storing an electric charge and releasing it when required. In an electronic circuit, a capacitor blocks the flow of dc current and generally passes ac current. It opposes a change in voltage across its terminals. Capacitors are measured in units of farads.

Inductor   An inductor basically consists of a coil of wire wrapped around a magnetic or nonmagnetic core. It is a circuit element that is capable of storing energy in its magnetic field and releasing it when required. It opposes a change in current through it. In an electronic circuit, an inductor allows dc current to flow and generally impedes the flow of ac current. Inductors are measured in units of henries.

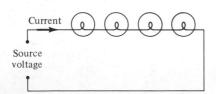

Figure 1.3   Series connection of lamps

Load   A load is anything that consumes power and is connected to a circuit or source. The load may be a lamp, TV set, air conditioner, speakers, and so forth. Switches or short copper wires (considered ideal conductors) are not considered loads because they do not consume power. In many electric and electronic circuit problems, the load is represented as a resistor, although it may not actually be one.

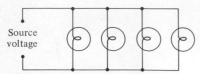

Figure 1.4   Parallel connection of lamps

Series connection   The lamps shown in Figure 1.3 are connected in series. The current *through* each lamp is the same or is common.

Parallel connection   The lamps shown in Figure 1.4 are connected in parallel. The voltage *across* each lamp is the same or is common.

Conductor   A conductor is a material that passes current easily. Copper, aluminum, silver, and gold are examples of metals that are excellent conductors.

Insulator   An insulator is a material through which it is almost impossible to conduct current. Glass, plastic, mica, and air are examples of insulators.

Semiconductor   A semiconductor is a material that is neither a good conductor nor a good insulator. Transistors, diodes, and integrated circuits are made from silicon and germanium, which are both semiconductors.

Electrical Symbols

Objective 2   Identify the following electrical symbols: battery, switch, lamp, resistor, capacitor, inductor, ammeter, voltmeter, and transformer.

The first thing you will probably come in contact with if you try to repair—or just understand—the operation of an appliance, an electronic device, or a piece of electrical or electronic equipment is a schematic. A schematic, or circuit diagram, of a phono amplifier is shown in Figure 1.5. *Don't try to understand this diagram now!* It is for illustrative purposes only. This diagram is made up of many symbols that stand for the actual elements or components.

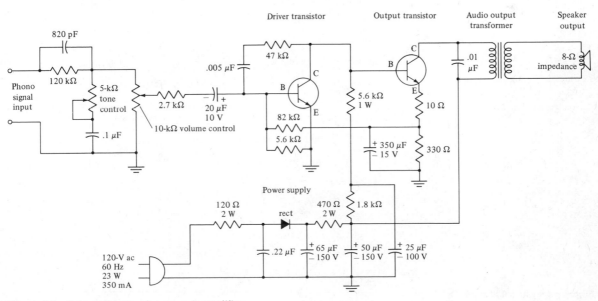

Figure 1.5   Schematic of a phonograph amplifier

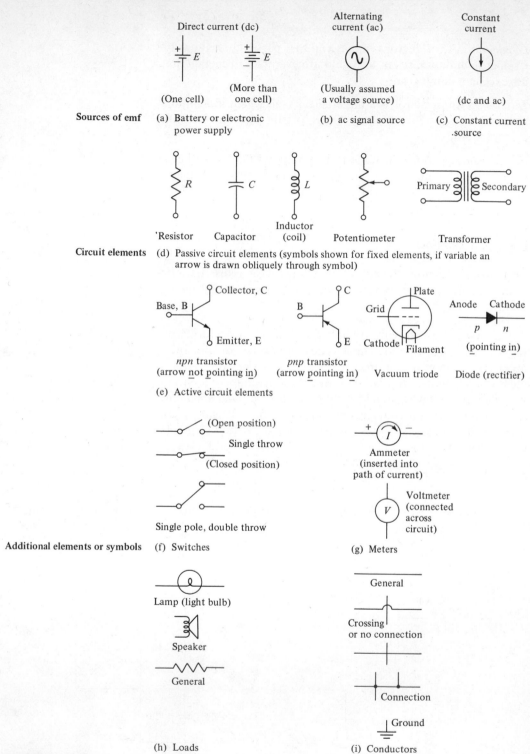

Figure 1.6   Electrical symbols

6

In Figure 1.6 you will find the symbols you need to complete Objective 2. You will also find some additional symbols that you will use in later units.

The first symbols, shown in Figure 1.6(a), are for a dc power supply, of which a battery is a very common form. You note that the long line has a plus sign placed beside it, with a minus sign beside the short line. Some dc power supplies are made up of more than one cell, as in a car battery or in most flashlights. The symbol can show this; see Figure 1.6(a). However, these symbols are usually used interchangeably. We shall usually use the letter symbol E to mean a dc supply; you see this beside the symbol. Many times, instead of E, we'll show the actual supply voltage, such as 3 V or 6 V.

The symbol shown in Figure 1.6(b) represents an ac voltage source. You are probably aware of only voltage sources, like batteries; however, with the use of transistors and integrated circuits, current sources have become more common. Figure 1.6(c) shows this symbol. In Figure 1.6(d) you see the symbols for a resistor, a capacitor, an inductor and a transformer. The letter symbol used in mathematical formulas is written next to the element. If there is more than one element of the same type, then subscripts on the letter symbol are used. Figure 1.6(e) shows some of the active circuit elements that you will use in electronics. The lamp shown in Figure 1.6(h) lights up when current flows through it.

Look at the remaining symbols in Figure 1.6. Notice the meters shown in Figure 1.6(g). An ammeter is used to measure current. Since current is the flow of electric charge per given time, to measure the flow you must insert the ammeter *into* the path of current flow. The ammeter is connected in series in the circuit. The letter symbol for current is I, so when you see an I inside the circle, you know the meter is measuring current. But any time you see a meter connected in series in a circuit, even if it doesn't have an I written in it, it is measuring current.

The voltmeter measures potential difference or voltage in a circuit. This voltage may be supplied by a dc or ac source, or it may be the voltage drop across a circuit element such as a resistor or lamp. Since a voltmeter measures the potential difference in a circuit it must be connected *across* a pair of terminals. In circuit drawings, any meter shown connected across a pair of terminals—in parallel with a circuit element or power supply— must be a voltmeter. The symbol V_{CC} is used in transistor circuit diagrams to designate a power supply connected to the collector of a transistor. Subscript letters are used so that the various resistors, capacitors, and transistors may be identified separately. If all capacitors were known by the same symbol, it would be impossible to discuss just one.

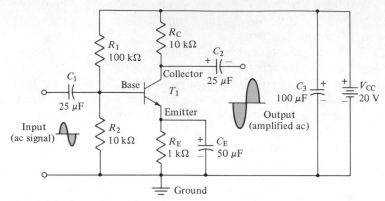

Figure 1.7   Simple transistor amplifier circuit

Example 1.1   Identify the electrical symbols for the circuit components shown in Figure 1.7.

Discussion   Before we identify the electrical symbols shown in Figure 1.7, we want to give you a little background about the circuit. It is a simple one-stage transistor amplifier circuit. Its purpose is to take a small ac signal and amplify it (make it larger). The ac signal may be the voltage from the cartridge of your record player or the output of a microphone. The transistor is a *current*-amplifying device. By this we mean that if we put a small current into the input to the transistor (*base*), we will get a large current in the output of the transistor (*collector*). Even though the purpose of this amplifier is to amplify the ac signal, the transistor needs dc voltages on its base and collector terminals to make it operate properly. This is why you see the dc supply V_{CC}.

Solution   There are actually only four different symbols in Figure 1.7. Symbol T_1 is an *npn* transistor; R_1, R_2, R_C, and R_E are resistors; C_1, C_2, C_3, and C_E are capacitors; and V_{CC} is a dc power supply. There was only one symbol for a capacitor shown in Figure 1.6. However, in Figure 1.7, did you notice that capacitors C_2, C_3, and C_E have a plus on one side and a minus on the other? Capacitors C_2, C_3, and C_E are *electrolytic* capacitors. These capacitors must be connected into the circuit as shown; if not, they will be ruined.

Problem 1.2   Identify the electrical symbols for the circuit components shown in Figures 1.8 and 1.9. That is, write the names in the boxes provided.

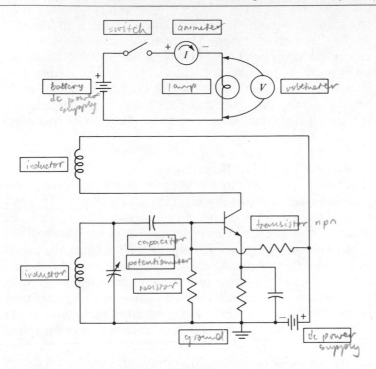

Figure 1.8 Simple electric circuit with lamp for Problem 1.2

Figure 1.9 An Armstrong transistor oscillator circuit for Problem 1.2

Flashlight Circuit Diagram

Objective 3   Draw a schematic diagram for the electric circuit—or a circuit diagram—for a flashlight. Explain the operation of the circuit.

To get you started understanding circuit diagrams or schematics, let's consider the circuit of a flashlight. The flashlight shown in Figure 1.10 consists of a power supply (batteries), a

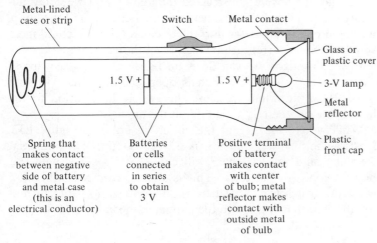

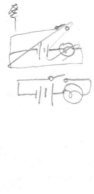

Figure 1.10   Typical flashlight

lamp, and a switch. The purpose of the switch is to interrupt the flow of current. When the switch is closed, completing the circuit, the lamp lights up. And when the switch is open, the lamp is off. If you have a flashlight handy, take it apart and examine it. Since the flashlight is such a simple device, you probably haven't thought about the circuit before. But understanding this circuit is the first step to understanding much more complicated ones.

The circuit schematic or circuit diagram for a flashlight is shown in Figure 1.11. Even though the symbol for the lamp appears to be similar to the lamp in actual appearance, the electrical symbols do not represent the appearance, but the function of the part in the circuit. In this drawing the symbols are shown in the same approximate location as they are in the flashlight. You see the electrical symbols for the lamp, the power supply, and the switch connected by straight lines. The straight lines show how these circuit elements are connected; in this case, they are all connected in series. The interconnecting lines symbolize electrical conductors. The conductors are usually copper wire, but in the case of the flashlight the conductor is the spring and the case itself.

Circuit schematics are usually drawn so that the flow of energy is from the left to the right. In this case, Figure 1.11 should be drawn as shown in Figure 1.12. Here the batteries or power supply are acting as the input to the circuit, and the lamp or light is the output from the circuit. Most circuit schematics for amplifiers, radios, and other electronic devices show the circuit input on the left and the circuit output on the right.

Before we leave the flashlight circuit, we will do one more thing with it. When you close the switch on the flashlight, the lamp lights up, showing that a current is flowing in the circuit. If the flashlight were constructed differently, and we could actually get at the connections between the batteries and the lamp, we could put an ammeter into the circuit so that we could measure how much current is flowing. Even though we can't do this with the flashlight, we can do it on paper, and you see this in Figure 1.13. When the switch is open, the ammeter will read zero. However, when we close the switch, in addition to the light being on, there is a reading on the ammeter.

While you're looking at the circuit diagram in Figure 1.13, let us again mention that this is a *series* circuit. The ammeter, the lamp, and the switch are all connected in series with the power supply, or batteries. This is a series connection because when the switch is closed, there is only one path for the current to take, and that is through the ammeter, the lamp, and then the

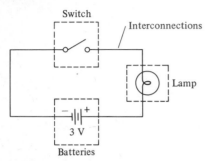

Figure 1.11   Circuit diagram or schematic for flashlight (circuit elements shown in same proximity as in flashlight)

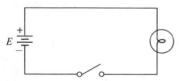

Figure 1.12   Circuit schematic for flashlight

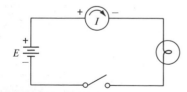

Figure 1.13   Flashlight circuit with ammeter

switch. Current is the same through each element in a series connection, or is common to all elements.

Problem 1.3 Draw the electric-circuit schematic or circuit diagram for a flashlight. Explain the operation of the circuit.

The Simple Electric-Circuit Diagram

Objective 4 Draw a circuit diagram for a simple electric circuit that contains a battery and a resistor. Insert an ammeter into the circuit to measure current and a voltmeter to measure the voltage drop across the resistor.

If you remember the definition of resistance—that is, the property of a circuit element that offers opposition to the flow of current—then you realize that the lamp in the flashlight circuit of Figure 1.12 is just a special kind of resistance, one that gives off light when current flows through it. Therefore we could make a more general electric-circuit diagram if we replaced the lamp by an actual resistor. This has been done in the circuit shown in Figure 1.14. In Unit 4 we will show you how E, I, and R relate in this simple electric circuit. But now we want to show you how to connect a voltmeter into the simple electric circuit so that you can find the voltage across the resistor R.

A *voltmeter* reads the voltage, or more correctly the potential difference, across a circuit element. The voltmeter must be connected *across* a pair of terminals. To read the battery voltage or potential difference, one connects the voltmeter across the batteries. In Figure 1.14(b), we have connected the voltmeter across the resistor terminals. In an electric circuit, when a current flows through a circuit element such as a resistor, a voltage drop occurs across that element. When the voltmeter is connected across the resistor R in the circuit, we measure a voltage drop across the resistor. The symbol that we use for a voltage drop is V. As in the flashlight circuit, we consider the wire connecting the circuit elements and the ammeter to be an *ideal conductor*. By ideal conductor, we mean that it gives no opposition to the flow of current and therefore has no voltage drop. Therefore the voltmeter that is connected across resistor R is reading the value of voltage that exists across the battery terminals. In this simple circuit, V is equal to E, the dc power supply.

Problem 1.4 Draw the circuit diagram for a simple electric circuit that contains a battery and a resistor. Insert an ammeter into the circuit to measure current and a voltmeter to measure the voltage drop across the resistor.

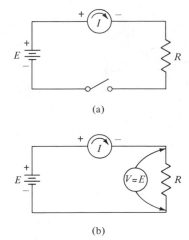

Figure 1.14 Simple electric circuit diagram

Table 1.1

Term	Potential or Voltage	Current	Resistance	Power
Symbol used in formulas	E/V	I	R	P
Basic unit of measurement	volts	ampere	ohm	watts
Abbreviation or symbol for basic unit	V	A	Ω	W

Test

1. Define current, voltage, resistance, and power.
2. Complete Table 1.1 from memory.
3. Identify the electrical symbols shown in Figure 1.15 by writing their names in the boxes provided.
4. Draw the circuit diagram for a flashlight. Insert an ammeter to measure current and a voltmeter to measure battery voltage.

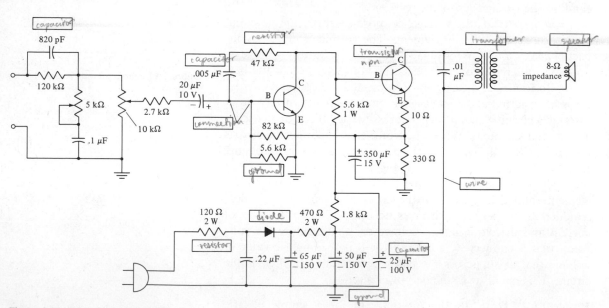

Figure 1.15   Circuit for Test Problem 3

Powers of Ten Unit 2
and Scientific
Notation,
Unit Prefixes,
and the Resistor
Color Code

Did you notice in the schematic of the phonograph amplifier, Figure 1.5, the values listed beside each component, such as 120 k (or kΩ), 820 pF, and 0.1 μF? If you didn't, go back and look at this schematic again. The k, p, and μ are *prefixes* to the units of measure. Sometimes the unit in which the basic electrical quantity is defined is too large or too small to be practical. As you will see as you go through this unit, the prefix just stands for the number 10 raised to some power.

Most of the formulas you will use in electric or electronic circuits are defined in terms of the basic units. Therefore the answers may be very large or very small numbers. Most electronic calculators give you answers that have eight places after the decimal point. But some answers may be so small that the eight places contain only zeros—no significant figures. You eliminate this problem if you use *scientific notation*, or if the calculator has provisions to change the very large or small numbers into scientific notation.

If you understand powers of ten, scientific notation, and the prefixes that represent powers of ten, you will be able to interpret the answers to electric- and electronic-circuit problems when you use a calculator. You will be able to understand the use of multipliers on meters, and you will be a step closer to understanding schematic circuit diagrams such as that in Figure 1.5.

After completing all the work associated with this unit, you should be able to:

1. Explain what is meant by the term powers-of-ten notation.
2. Change any number written in powers-of-ten notation into decimal notation, or vice versa.
3. Change any number written in decimal notation or in powers-of-ten notation into scientific notation.
4. Add, subtract, multiply, and divide numbers written in powers-of-ten notation.
5. Recall the following prefixes and their symbol or abbreviation: 10^9, 10^6, 10^3, 10^0, 10^{-3}, 10^{-6}, 10^{-9}, and 10^{-12}.
6. Convert any number using one prefix before the unit of measure to an equivalent number using another prefix or none. Or convert any number written in scientific notation to an equivalent number that uses a prefix.
7. Determine the rated value, tolerance, and/or minimum and maximum resistance values when given an actual resistor or the color bands on a resistor, and vice versa.

Powers-of-Ten Notation

Objective 1   Explain what is meant by the term powers-of-ten notation.

When a number is written using powers-of-ten notation, you see a number followed by 10 with an exponent.

$$\text{number} \times 10^{\text{exponent}}$$

Examples of numbers written in powers-of-ten notation are 56.7×10^2, 0.03×10^{-3}, and 739×10^4. But to understand what this means, you must know what decimal notation means and what 10 with an exponent is.

When you write numbers like 69.3, 5, or 0.47, you are using *decimal notation*. Even though you don't see a decimal point with the 5, it is understood that the decimal point is to the right of the 5, that is, 5.0. When a number is written in decimal notation, the value of each digit depends on its position with respect to the decimal point. (See Figure 2.1.)

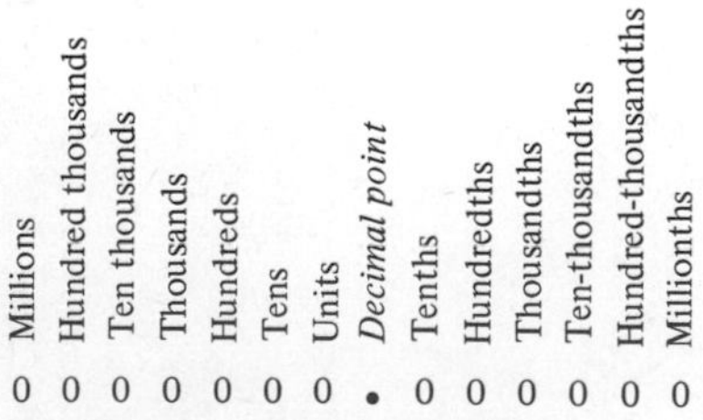

Whenever we are working with very large numbers like 6,240,000,000,000,000,000 or very small numbers like 0.000000093, we would like to have a way of writing these numbers without all the zeros. Writing all the zeros is time-consuming, and it also leads to inaccuracies. It is very easy to accidentally drop or add a zero when writing these numbers. Just try copying these num-

Figure 2.1   Value of each digit in decimal notation

bers and see how careful you have to be to get the correct number of zeros!

To work with these very large or very small numbers, you must first understand what 10 with an exponent means. The number written above and to the right of the 10 is called the *exponent* or *power*. It tells you the number of times you have to multiply the 10. See Example 2.1.

Example 2.1

$$10^1 \text{ means one } 10 \qquad \therefore \quad 10^1 = 10$$

$$\text{(the symbol } \therefore \text{ means therefore)}$$

$$10^2 \text{ means } 10 \times 10 \qquad \therefore \quad 10^2 = 100$$

$$10^3 = 10(10)(10) = 1000$$

$$10^7 = 10(10)(10)(10)(10)(10) = 10{,}000{,}000$$

Did you notice in Example 2.1 that the exponent tells you the number of zeros that follow the 1? Now you try some. See Problem 2.1.

Problem 2.1   Expand the following powers of ten.

$$10^4 =$$

$$10^5 =$$

$$10^6 =$$

$$10^1 =$$

$$10^2 =$$

So far we have discussed only positive exponents. What happens if we have a number less than 1? By this question, we mean what happens if we have a fraction? Here we run into negative exponents. See Example 2.2.

Example 2.2

$$10^{-1} = \frac{1}{10^{+1}} = \frac{1}{10} = 0.1$$

The minus sign in front of the exponent means you divide 1 by 10 raised to the power of the given exponent.

$$10^{-2} = \frac{1}{10^2} = \frac{1}{10(10)} = \frac{1}{100} = 0.01$$

$$10^{-4} = \frac{1}{10^4} = \frac{1}{10(10)(10)(10)} = \frac{1}{\underbrace{10000}_{\text{4 zeros}}} = \underbrace{0.0001}_{\text{4 places}}$$

Did you notice in this example that the number of zeros in the denominator gave you the number of places to the right of the decimal point in your answer? Try Problem 2.2.

Problem 2.2 $10^{-3} =$

$10^{-5} =$

$10^{-1} =$

$10^{-6} =$

Let's look at the movement of the decimal point a little more closely. Remember that when you write 1, the decimal point is always understood to be to the right of 1. From Example 2.2 and Problem 2.2 we saw

$$10^{-1} = 0.1_{\smile} \qquad 10^{-2} = 0.0\,1_{\smile} \qquad 10^{-3} = 0.0\,0\,1_{\smile}$$

Do you see the pattern? Since the exponent is negative, move the decimal point to the *left* the number of places indicated by the exponent. Do Problem 2.2 using this method.

So far we have discussed 10 with positive and negative exponents. Is there such a thing as a zero for an exponent? This has a special definition that you will just have to accept. That is,

$$10^0 = 1$$

What does this say? Well, if 10 is raised to the zero power, or has a zero exponent, it is the same thing as 1. Remember that, when we were discussing 10 raised to powers, we said that the exponent tells the number of zeros that follow the 1. Well, 10^0 says that *no* zeros follow the 1.

Objective 2   Change any number written in powers-of-ten notation into decimal notation, and vice versa.

What happens if we have some number other than just 10 raised to a power? You already know the answer. To check yourself out, see Example 2.3.

Example 2.3   Change the following numbers written in powers-of-ten notation into decimal notation: 2×10^1, 6×10^2, 43.97×10^3, and 2.57×10^1.

Solution $2 \times 10^1 = 2 \times 10 = 20$

$6 \times 10^2 = 6 \times 100 = 600$

Note that the number of zeros that follow the 6 is equal to the exponent on the 10.

$$43.97 \times 10^3 = 43.97 \times 1000 = 43,970$$

When we multiplied by 10^3, we moved the decimal point 3 places to the right.

$$2.57 \times 10^1 = 2.57 \times 10 = 2.5.7 = 25.7$$

The numbers originally given in Example 2.3 were written in *powers-of-ten notation*. You see a number followed by 10 raised to some power.

Problem 2.3 Change the following numbers written in powers-of-ten notation into decimal notation.

$$3 \times 10^2 =$$

$$9.28 \times 10^1 =$$

$$7.5 \times 10^3 =$$

$$4.123 \times 10^2 =$$

Let's go the other way. We will give you the number in decimal notation, and you express it in powers-of-ten notation. For the procedure to follow, see Example 2.4.

Example 2.4 Change the numbers 62,800 and 53.8, written in decimal notation, into powers-of-ten notation.

Solution

$$62,800 = 6.28 \times 10^4 = 62.8 \times 10^3 = 628 \times 10^2$$
$$53.8 = 5.38 \times 10^1 = 0.538 \times 10^2 = 0.0538 \times 10^3$$

You noticed that several answers were given in Example 2.4. There are even more correct answers that were not listed. You are probably thinking: What is the *best* way to express the answer? It all depends. If you are using scientific notation, there is only one correct answer. If you are using prefixes with units of measure, there may be only one preferred answer. We shall discuss these later in this unit.

Problem 2.4 Change the following numbers written in decimal notation into powers-of-ten notation.

$$732. =$$

$$93.5 =$$

$$132.3 =$$

Example 2.5 shows you what to do with negative exponents.

Example 2.5 Change the following numbers written in powers-of-ten notation into decimal notation: 6.53×10^{-1}, 73.5×10^{-1}, and 325×10^{-2}.

Solution $6.53 \times 10^{-1} = 6.53 \times \dfrac{1}{10} = 0.6\underset{\frown}{5}3 = 0.653$

$73.5 \times 10^{-1} = 7.\underset{\frown}{3}5 = 7.35$

$325. \times 10^{-2} = 3.2\underset{\frown}{5} = 3.25$

Therefore, when we multiply a number by 10 with a negative exponent, we move the decimal point to the left the same number of places as the exponent.

Problem 2.5   Change the following numbers written in powers-of-ten notation into decimal notation.

$$100.9 \times 10^{-1} =$$

$$576 \times 10^{-2} =$$

$$2.35 \times 10^{-3} =$$

Example 2.6 shows you how to work with 10 raised to the zero power.

Example 2.6   Change the numbers 1.2×10^{0} and 0.025×10^{0}, written in powers-of-ten notation, into decimal notation.

Solution   Remember the definition: $10^{0} = 1$.

$$1.2 \times 10^{0} = 1.2 \times 1 = 1.2$$
$$0.025 \times 10^{0} = 0.025 \times 1 = 0.025$$

Problem 2.6   Change the numbers 3.55×10^{0} and 10.62×10^{0}, written in powers-of-ten notation, into decimal notation.

Summary

Converting from powers-of-ten notation to decimal notation: When a number is multiplied by 10 raised to a *positive* exponent, move the decimal point to the *right* the same number of places as the exponent.

When a number is multiplied by 10 raised to a *negative* exponent, move the decimal point to the *left* the same number of places as the exponent.

Scientific Notation

Objective 3   Change any number written in decimal notation or in powers-of-ten notation into scientific notation.

When you write a number in *scientific notation*, you express the number using powers-of-ten notation, but with only one digit (not zero) to the left of the decimal point. Example 2.7 should clarify this statement.

Example 2.7   Write the following numbers in scientific notation.

$$87.3 = 8.73 \times 10^1$$

Note that there is one digit (not 0) to the left of the decimal point, with the correct power of ten following.

$$0.00935 = 9.35 \times 10^{-3}$$

$$7.13 = 7.13 \quad \text{or} \quad 7.13 \times 10^0$$

The 7.13 is already in scientific notation.

$$39.3 \times 10^3 = 3.93 \times 10^4$$
$$120 \times 10^{-3} = 1.20 \times 10^{-1}$$

Problem 2.7   Write the following numbers in scientific notation.

$2020 =$

$82.5 =$

$6,240,000,000,000,000,000 =$

$10,000 =$

$3.26 =$

$643 \times 10^6 =$

$47 \times 10^3 =$

$0.237 =$

$0.0735 =$

$0.0000000499 =$

$93 \times 10^{-6} =$

$1200 \times 10^{-12} =$

Objective 4   Add, subtract, multiply, and divide numbers written in powers-of-ten notation.

If we wish to add or subtract numbers written in powers-of-ten notation, the exponents must be the same. Then the number, or *coefficient*, part of the expression may be added or subtracted, just as you add or subtract numbers written in decimal notation. Write the numbers so that the decimal points line up with each other. Then add or subtract the numbers, inserting the decimal point of the answer under the other decimal points. Then copy down the powers of 10. Example 2.8 shows this procedure.

Example 2.8 $(3.25 \times 10^2) + (4.3 \times 10^1) =$

$$(4.26 \times 10^{-1}) - (2.15 \times 10^{-2}) =$$

Solution   To work the first problem, let's change the first number so that both numbers will be written with 10 raised to the first power. We could have written both numbers with 10 raised to the second power; you choose the power of 10 that you think is the easiest to use.

$$
\begin{aligned}
3.25 \times 10^2 = \quad & 32.5 \times 10^1 \\
+ \quad & 4.3 \times 10^1 \\
\hline
& 36.8 \times 10^1
\end{aligned}
$$

$\quad\quad\quad\quad\quad\quad\quad\quad\quad\quad\quad\llcorner$ Decimal point

$$
\begin{aligned}
4.26 \times 10^{-1} = \quad & 42.6 \ \times 10^{-2} \\
- \quad & 2.15 \times 10^{-2} \\
\hline
& 40.45 \times 10^{-2}
\end{aligned}
$$

Problem 2.8 $(7.56 \times 10^2) + (9.72 \times 10^3) =$

$$(1.59 \times 10^{-1}) + (6.27 \times 10^{-2}) =$$

$$(5 \times 10^4) - (8.3 \times 10^3) =$$

$$(8.15 \times 10^{-2}) - (4.934 \times 10^{-1}) =$$

To learn to *multiply* numbers written in powers of ten notation, study Example 2.9 and see if you can come up with a rule.

Example 2.9   Multiply the following numbers written in powers-of-ten notation.

$$
10^2 \times 10^3 = \underbrace{100}_{\text{2 zeros}} \times \underbrace{1000}_{\text{3 zeros}} = \underbrace{100{,}000}_{\text{5 zeros}}
$$

$$
= 10^5 \quad [Hint: 2 + 3 = 5]
$$

$$
10^3 \times 10^{-2} = \underbrace{1000}_{\text{3 zeros}} \times \underbrace{\frac{1}{100}}_{\text{2 zeros}}
$$

$$
= 10^1 \quad [Hint: 3 + (-2) = 1]
$$

$$
(2 \times 10^2)(4 \times 10^1) = 200 \times 40 = 8000
$$

$$
= 8 \times 10^3 \quad [Hint: 2 + 1 = 3]
$$

We could have multiplied the 2×4 and then handled the $10^2 \times 10^1$.

$$(4.3 \times 10^{-3})(5 \times 10^5) = (4.3)(5) \times (10^{-3})(10^5)$$
$$= 21.5 \times 10^{(-3+5)}$$
$$= 21.5 \times 10^2$$

Problem 2.9 Write the rule for multiplying numbers written in powers-of-ten notation. _____

To learn to divide numbers written in powers-of-ten notation, study Example 2.10 and see if you can come up with a rule.

Example 2.10 Divide the following numbers written in powers-of-ten notation.

4 zeros

$$\frac{10^4}{10^2} = \frac{\overbrace{10000}}{\underbrace{100}} = 10^2 \qquad [Hint: 4 - 2 = 2]$$

2 zeros

$$\frac{10^3}{10^5} = \frac{1000}{100000} = \frac{1}{100} = \frac{1}{10^2}$$
$$= 10^{-2} \qquad [Hint: 3 - 5 = -2]$$

$$\frac{10^{-2}}{10^4} = \frac{1}{100} \times \frac{1}{10000} = \frac{1}{10^6}$$
$$= 10^{-6} \qquad [Hint: -2 - 4 = -6]$$

$$\frac{4.2 \times 10^2}{2.1 \times 10} = \frac{420}{21} = 2 \times 10^1 \qquad [Hint: 2 - 1 = 1]$$

$$\frac{45 \times 10^{-3}}{3 \times 10^{-2}} = \frac{45}{3} \times \frac{10^{-3}}{10^{-2}}$$
$$= 15 \times 10^{[-3-(-2)]} = 15 \times 10^{-1}$$

Problem 2.10 Write the rule for division of numbers written in powers-of-ten notation. _____

Did you have the following answers for Problems 2.9 and 2.10?

To multiply numbers written in powers-of-ten notation: Multiply the number or coefficient part of the expression and add the exponents on the 10s algebraically, noting signs.

To divide numbers written in powers-of-ten notation: Divide the number or coefficient part of the expression and subtract the exponent on the 10 in the denominator from the exponent on the 10 in the numerator, noting signs.

Problem 2.11 Multiply the following numbers written in powers-of-ten notation.

$$(1.7 \times 10^2)(6.1 \times 10^1) =$$

$$(20 \times 10^{-3})(10 \times 10^{-2}) =$$

$$(2.93 \times 10^3)(34.1 \times 10^{-2}) =$$

Problem 2.12 Divide the following numbers written in powers-of-ten notation.

$$\frac{1.5 \times 10^3}{0.5 \times 10^1} =$$

$$\frac{27.5 \times 10^1}{2.5 \times 10^{-2}} =$$

$$\frac{20}{4 \times 10^3} =$$

Unit Prefixes

Objective 5 Recall the following prefixes and their symbol or abbreviation: 10^9, 10^6, 10^3, 10^0, 10^{-3}, 10^{-6}, 10^{-9}, and 10^{-12}.

The *basic unit* is the basic electrical quantity in which the unit is defined. For example, the basic unit of measurement for E is volts, for I is amperes, and for R is ohms. Just as we mentioned when we discussed powers-of-ten notation, sometimes quantities are so large or so small that it becomes awkward to use the basic units. If you said you read on an ammeter "decimal point zero zero zero zero five amperes," very few people would understand what you were talking about. It is much clearer to say 5 microamperes. As we will show you, the "micro" is a prefix that means 10^{-6}.

Table 2.1 Unit Prefixes

Powers of ten	Prefix	Symbol
10^{12}	tera	T
10^9	giga	G
10^6	mega	M
10^3	kilo	k
10^2	hecto	h
10^1	deka	da
10^{-1}	deci	d
10^{-2}	centi	c
10^{-3}	milli	m
10^{-6}	micro	μ*
10^{-9}	nano	n
10^{-12}	pico	p
10^{-15}	femto	f
10^{-18}	atto	a

* μ is the Greek letter mu.

You have already used something like prefixes when discussing money. You say "four thousand dollars" instead of "four zero zero zero." You say "four million" instead of "four zero zero zero zero zero zero." In electrical terms, when we refer to 4000 ohms, we say 4 kilohms; 4 million ohms is 4 megohms. Table 2.1 lists the prefixes that are used and their symbols or abbreviations.

Most of the symbols or abbreviations for the prefixes are straightforward, except for the Greek letter μ (lower-case mu), which is used to stand for micro. Except for μ, the symbol or abbreviation that is used is the first letter of the prefix name. We had already used a lower-case m and capital M for abbreviations for other prefixes, so we had to come up with another symbol, which is μ. Remember that the capital M is used for the *big* prefix, mega, whereas the lower-case m is used for the *small* prefix, milli.

Problem 2.13 Recall the prefix and the symbol or abbreviation that are used for 10^9, 10^6, 10^3, 10^0, 10^{-3}, 10^{-6}, 10^{-9}, and 10^{-12}.

Problem 2.14 Go back to the schematic of a phono amplifier (Figure 1.5) and find and list the various prefixes that are used.

Objective 6 Convert any number using one prefix before the unit of measure to an equivalent number, using another prefix, or none. Or convert any number written in scientific notation to an equivalent number, using a prefix.

The prefix scale shown in Figure 2.2 shows the prefixes most commonly used in electric and electronic circuits. Notice in the scale that all the prefixes increase or decrease in steps of 3. Also notice in the scale that when we change from a larger to a smaller prefix, for example from mega to kilo, we move to the right on the scale and move our decimal point to the right in the number. When we change from a smaller to a larger prefix, we move to the left on the scale and move the decimal point to the left in the number.

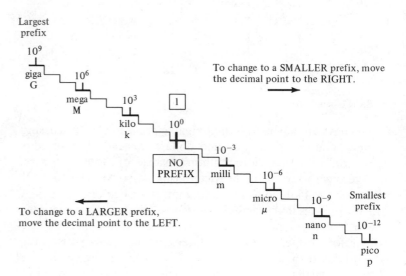

Figure 2.2   Prefix scale

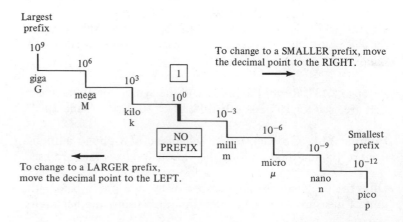

Figure 2.3   Prefix scale (simplified)

Since the prefixes used in most electric and electronic circuits increase or decrease in steps of 3, the scale of Figure 2.2 has been simplified to that shown in Figure 2.3. This prefix scale is used so much, you should memorize it. There are methods of changing from one prefix to another other than using the scale in Figure 2.3. But we have found that the scale helps you to visualize what is happening.

As you see on the scale in Figure 2.3, as you change from a larger prefix to the next-smaller one, you move the decimal point three places to the right. Example 2.11 shows you how to use the prefix scale to do this.

Example 2.11

$$0.03 \text{ amperes} = 0.030. = 30 \text{ milliamperes}$$

Change from no prefix to milli. You move to the right three places on the prefix scale, so you move the decimal point to the right three places.

No prefix

$$4 \text{ amperes (A)} = 4.000. = 4000 \text{ milliamperes (or mA)}$$

Change from no prefix to milli

$$2 \text{ megohms (M\Omega)} = 2.000. = 2000 \text{ kilohms (or k\Omega)}$$

Change from M to k

$$0.5 \text{ millisecond (ms)} = 0.500. = 500 \text{ microseconds (or } \mu s)$$

Change from m to μ

What happens if you go from the prefix μ to p? Note that the exponents change by 6. Since you are moving to the right ($\mu \rightarrow$ p) on the prefix scale, you must move the decimal point 6 places to the right. See Example 2.12.

Example 2.12

$$0.00015 \text{ microfarad (}\mu F) = 0.000150. = 150 \text{ picofarads (pF)}$$

When you change from a smaller prefix to the next-larger one, you move the decimal point three places to the left. See Example 2.13.

Example 2.13

$$4 \text{ mA} = .0\,0\,4_{\text{o}} = 0.004 \text{ A} \qquad \text{Change from m to no prefix}$$

No prefix

$$4000\,\Omega = 4.0\,0\,0_{\text{o}} = 4 \text{ k}\Omega \qquad \text{Change from no prefix to k}$$

$$600 \text{ } \mu\text{V} = .6\,0\,0_{\text{o}} = 0.6 \text{ mV} \qquad \text{Change from } \mu \text{ to m}$$

To change from a prefix of p to a prefix of μ, see Example 2.14.

Example 2.14

$$1000 \text{ pF} = .0\,0\,1\,0\,0\,0_{\text{o}} = 0.001 \text{ } \mu\text{F}$$

Summary

To change from a larger prefix to a smaller one, move the decimal point to the right by a number of steps equal to the change in powers of ten.

To change from a smaller prefix to a larger one, move the decimal point to the left by a number of steps equal to the change in powers of ten.

It also helps—when you are trying to use the more common prefixes—to express the large or small numbers in powers of ten with the exponent in some multiple of three. See Example 2.15.

Example 2.15 $4000 \text{ } \Omega = 4 \times 10^3 \text{ } \Omega$; but $10^3 = \text{k}$.   Therefore $4000 \text{ } \Omega = 4 \text{ k}\Omega$.

You will get more examples and practice doing this in Problem 2.15.

Problem 2.15   Complete Table 2.2. (To give you some examples, some of the rows have already been completed.) In the middle column, write the numbers using powers-of-ten notation. However, the whole number or numbers to the left of the decimal should be between 1 and 1000, and the powers of ten must be in multiples of 3. That is, use only 10^9, 10^6, 10^3, no tens (10^0), 10^{-3}, 10^{-6}, 10^{-9}, or 10^{-12}. In the last column replace the powers of ten by the proper prefix, or symbol for the power of ten.   Just copy the units given.

Is there a *preferred* prefix that should be used? Most of the time we try to choose the prefix so that we can express the

Table 2.2 Table for Problem 2.15

Original Number	Use Powers-of-Ten Notation (Multiples of 3)	Replace Powers of Ten by the Preferred Prefix
3900 ohms	3.9×10^3 ohms	3.9 k ohms
0.060 ampere	60×10^{-3} amperes	60 m amperes
1,500,000 ohms		
45,000,000,000 hertz	45×10^9 hertz	45 G hertz
108,000,000 hertz		
33,000 ohms		
47 ohms	47×10^0 ohms	47 ohms ↖Note no prefix
8200 ohms		
12 volts		
1,420,000 hertz		
2 amperes		
0.012 ampere		
0.00000043 ampere		
0.150 volt		
0.00000007 second		
0.000006 ampere		
0.000000000005 second		
5600 ohms		
120 volts		
0.003 volt		
390,000 ohms		
80,000,000,000 hertz		
10,000 ohms		
95,000,000 hertz		

quantity as a whole number between 1 and 1000, as you did in Problem 2.15. There are some exceptions to this; you run into one when you are using prefixes for capacitors. Capacitors are rated in either microfarads or picofarads; nanofarads are not used. So on capacitors the number may be greater than 1000 or less than 1; for example, 1200 picofarads or 0.001 microfarad.

Example 2.16   Change to the prefixes shown and list the preferred prefix.

Given Units	Change to Prefixes Shown			Preferred Prefix
1025 V	= 1.0 2 5 ↶ = 1.025 kV	= 0.001025 MV		k
25 V	=	25,000 mV =	0.025 kV	None
0.075 mA	= 0.0 0 7 5. ↷ = 75 μA	= 0.000075 A		μ

Problem 2.16   Change to the prefixes shown and list the preferred prefix.

Given Units		Change to Prefixes Shown		Preferred Prefix
9 A	=	mA =	μA	
450,000 Ω	=	kΩ =	MΩ	
8 GHz	=	MHz =	kHz	
0.5 W	=	mW =	kW	
0.00000007 s	=	μs =	ns	

Problem 2.17   The following answers given in scientific notation were determined on a calculator. Convert the answers from scientific notation to the prefix shown. Copy the units shown.

$$2.1 \times 10^{-2} \text{ A} \quad = \qquad \text{mA}$$
$$9.73 \times 10^{-4} \text{ s} \quad = \qquad \mu\text{s}$$
$$4.9999 \times 10^{-9} \text{ F} \quad = \qquad \text{pF}$$
$$7.63 \times 10^{4} \text{ Ω} \quad = \qquad \text{kΩ}$$
$$5.14 \times 10^{8} \text{ Hz} = \qquad \text{MHz}$$

Problem 2.18 The following answers given in scientific notation were determined on a calculator. Convert the answers from scientific notation into the preferred prefix. Copy the units shown.

$$1.53 \times 10^{-1} \text{ A} =$$
$$3.25 \times 10^{-5} \text{ s} =$$
$$4.82 \times 10^{-10} \text{ F} =$$
$$6.41 \times 10^{5} \text{ } \Omega =$$
$$8.07 \times 10^{7} \text{ } \Omega =$$

Resistor Color Code

Objective 7 Determine the rated value, tolerance, and/or minimum and maximum resistance values when given an actual resistor or the color bands on a resistor, and vice versa.

Most precision resistors (1%, $\frac{1}{2}\%$) and large-wattage resistors (5, 10, 20 W, and so forth) have their resistance value printed on them. Because of their small size, most carbon or molded resistors use a standard color code to designate their value and tolerance. This is known as the *resistor color code*. See Figure 2.4.

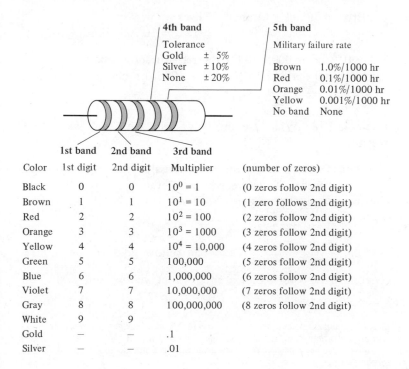

Color	1st band 1st digit	2nd band 2nd digit	3rd band Multiplier	(number of zeros)
Black	0	0	$10^0 = 1$	(0 zeros follow 2nd digit)
Brown	1	1	$10^1 = 10$	(1 zero follows 2nd digit)
Red	2	2	$10^2 = 100$	(2 zeros follow 2nd digit)
Orange	3	3	$10^3 = 1000$	(3 zeros follow 2nd digit)
Yellow	4	4	$10^4 = 10,000$	(4 zeros follow 2nd digit)
Green	5	5	100,000	(5 zeros follow 2nd digit)
Blue	6	6	1,000,000	(6 zeros follow 2nd digit)
Violet	7	7	10,000,000	(7 zeros follow 2nd digit)
Gray	8	8	100,000,000	(8 zeros follow 2nd digit)
White	9	9		
Gold	—	—	.1	
Silver	—	—	.01	

4th band

Tolerance
Gold ± 5%
Silver ±10%
None ±20%

5th band

Military failure rate

Brown 1.0%/1000 hr
Red 0.1%/1000 hr
Orange 0.01%/1000 hr
Yellow 0.001%/1000 hr
No band None

Figure 2.4 Resistor color code

The first band, which is located closest to one end of the resistor, gives the first significant figure or first digit of the resistor value. The second band gives the second significant figure or second digit of the resistor value. The third band is the multiplier, or the *number of zeros* that follow the second digit. The multiplier is determined by raising 10 to the power designated by the color band. As you learned for positive powers of ten, the power of ten gives the number of zeros. The fourth band is used for the manufacturer's tolerance; any resistor without this band has a tolerance of $\pm 20\%$. If a fifth band is given, it is used to give the military failure rate.

You can use the following mnemonic device* to help you remember the colors in order: *B*ad *B*oys *R*ape *O*ur *Y*oung *G*irls *B*ut *V*iolet *G*ives *W*illingly for *G*old and *S*ilver. Black, Brown, Red, Orange, Yellow, Green, Blue, Violet, Gray, White, 5% (gold) and 10% (silver) tolerance.

Maybe you can think up one of your own!

Example 2.17   A resistor is coded: Brown, Black, Orange, Silver. What is its nominal resistance and tolerance? Also find the minimum and maximum values possible for this resistor.

Solution

Brown	Black	Orange	Silver
1	0	000	$\pm 10\%$ = 10,000 ohms, $\pm 10\%$

Since 10% of 10,000 is 1000, the resistor should be in the range from $(10,000 - 1000)$ to $(10,000 + 1000)$ or between 9 kΩ and 11 kΩ.

Example 2.18   What are the color bands for a 270-kΩ $\pm$ 5% resistor?

Solution   Before we determine the color bands, we must get rid of the prefix. Then we can count the number of zeros, so that we can determine the third band.

$$270 \text{ k}\Omega = 2 \quad 7 \quad \underset{\text{yellow}}{\underline{0000}} \quad \pm 5\%$$

$$\text{red} \quad \text{violet} \quad \text{yellow} \quad \text{gold}$$

Therefore the color bands are red, violet, yellow, and gold.

---

* Throughout this text, we'll give you mnemonic devices, sayings, or other gimmicks to help you remember things. The origin of these is unknown. Most of these have been passed on from professor to student, and from student to professor.

Table 2.3 Table for Problem 2.19

Color Band					
FIRST	SECOND	THIRD	FOURTH	Nominal Value	Tolerance
blue	red	green	gold	6M2 Ω	±5%
gray	red	orange	none	82k Ω	±20%
white	brown	red	gold	9k1 Ω	±5%
brown	gray	brown	silver	180 Ω	±10%

Table 2.4 Table for Problem 2.10

Color Band				Nominal Value	Maximum Value	Minimum Value
FIRST	SECOND	THIRD	FOURTH			
brown	orange	green	gold	1M3		
yellow	violet	orange	none			
blue	gray	red	silver			
green	brown	brown	gold			

Table 2.5 Table for Problem 2.21

Color Band				
FIRST	SECOND	THIRD	FOURTH	Nominal Value and Tolerance
				$1.0 \text{ k}\Omega \pm 20\%$
				$110 \text{ k}\Omega \pm 5\%$
				$3900 \ \Omega \pm 10\%$
				$75 \ \Omega \pm 5\%$

Problem 2.19 Complete Table 2.3 for the given color-coded resistors.

Problem 2.20 Complete Table 2.4 for the given color-coded resistors.

Problem 2.21 Complete Table 2.5, listing the color bands for the resistors shown.

Table 2.6   Standard 10% Resistance Values

2.7 Ω	120 Ω	5600 Ω	270 kΩ
3.3 Ω	150 Ω	6800 Ω	330 kΩ
3.9 Ω	180 Ω	8200 Ω	390 kΩ
4.7 Ω	220 Ω	10 kΩ	470 kΩ
5.6 Ω	270 Ω	12 kΩ	560 kΩ
6.8 Ω	330 Ω	15 kΩ	680 kΩ
8.2 Ω	390 Ω	18 kΩ	820 kΩ
10 Ω	470 Ω	22 kΩ	1.0 MΩ
12 Ω	560 Ω	27 kΩ	1.2 MΩ
15 Ω	680 Ω	33 kΩ	1.5 MΩ
18 Ω	820 Ω	39 kΩ	1.8 MΩ
22 Ω	1000 Ω	47 kΩ	2.2 MΩ
27 Ω	1200 Ω	56 kΩ	2.7 MΩ
33 Ω	1500 Ω	68 kΩ	3.3 MΩ
39 Ω	1800 Ω	82 kΩ	3.9 MΩ
47 Ω	2200 Ω	100 kΩ	4.7 MΩ
56 Ω	2700 Ω	120 kΩ	5.6 MΩ
68 Ω	3300 Ω	150 kΩ	6.8 MΩ
82 Ω	3900 Ω	180 kΩ	10.0 MΩ
100 Ω	4700 Ω	220 kΩ	12.0 MΩ

Problem 2.22   Table 2.6 lists several 10% resistors ($\frac{1}{2}$, 1, and 2 watt) that are available commercially. We calculated that we needed the following resistor values for a circuit we were building: 35 Ω, 496 Ω, 2520 Ω, 7430 Ω, 20.5 kΩ, 59.2 kΩ, 136 kΩ, 645 kΩ, 3.1 MΩ, and 8.5 MΩ.* What resistors from Table 2.6 would you use to build the circuit?

Test

Part 1   Change the following numbers written in powers-of-ten notation into decimal notation.

1. $3.14 \times 10^2 =$ 2. $72.86 \times 10^3 =$
3. $12 \times 10^0 =$ 4. $120.7 \times 10^4 =$
5. $84 \times 10^1 =$ 6. $3.28 \times 10^{-1} =$
7. $100.8 \times 10^{-2} =$ 8. $732.5 \times 10^{-4} =$
9. $4 \times 10^{-6} =$ 10. $1000 \times 10^{-12} =$

* You may sometimes see the omega omitted; in this case the values would be written 35, 20.5 k, 3.1 M, and so forth.

Part 2 Write the following numbers in scientific notation. Just copy the units.

1. 1,500,000 Ω =
2. 0.003 A =
3. 120 V =
4. 0.000007 A =
5. 95 V =
6. 3900 Ω =
7. 0.040 V =
8. 60 Hz =
9. 10,000 Ω =
10. 9 V =

Part 3 Perform the following operations using powers-of-ten notation.

1. $(2.4 \times 10^4) + (1.3 \times 10^4) =$
2. $(1.35 \times 10^2) + (2.62 \times 10^3) =$
3. $(1.34 \times 10^3) - (1.07 \times 10^2) =$
4. $(5.62 \times 10^{-3}) - (4.2 \times 10^{-4}) =$
5. $(2.4 \times 10^4)(1.3 \times 10^3) =$
6. $(1.53 \times 10^{-2})(5.63 \times 10^2) =$
7. $(3.42 \times 10^{-1})(4.6 \times 10^{-2}) =$
8. $\dfrac{1.52 \times 10^3}{2.62 \times 10^1} =$
9. $\dfrac{4.56 \times 10^1}{8.13 \times 10^{-2}} =$
10. $\dfrac{3.43 \times 10^{-7}}{1.62 \times 10^{-3}} =$

Part 4 Change to the prefixes shown and list the preferred prefix.

Given Units	Change to Prefixes Shown		Preferred Prefix
1. 35 mV =	V =	μV	
2. 1500 kV =	MV =	V	
3. 0.00075 A =	mA =	μA	
4. 7.5 mA =	A =	μA	
5. 1,500,000 Ω =	kΩ =	MΩ	
6. 39 kΩ =	Ω =	MΩ	
7. 5000 MHz =	GHz =	kHz	
8. 2000 Hz =	kHz =	MHz	
9. 0.095 s =	ms =	μs	
10. 0.000006 ms =	μs =	ns	

Part 5   Resistor color code.

Complete the following table, for the given color-coded resistors.

Color Band				Nominal Value	Maximum Value	Minimum Value
FIRST	SECOND	THIRD	FOURTH			
1. green	blue	yellow	none			
2. red	violet	orange	silver			
3. red	yellow	red	gold			
4. yellow	orange	gold	gold			
5. orange	black	yellow	gold			
6. brown	green	orange	none			

Complete the following table, listing the color bands for the resistors shown.

Color Band				Nominal Value and Tolerance
FIRST	SECOND	THIRD	FOURTH	
7.				2000 Ω ± 5%
8.				33 Ω ± 10%
9.				910 kΩ ± 5%
10.				12 kΩ ± 20%
11.				5.6 MΩ ± 10%
12.				0.47 Ω ± 5%

Problem Solving Unit 3

To find out what resistor is required in a circuit, to solve for the required power supply or for the current, or to find the power rating required for a transistor, we use mathematics as a tool. As you will see in this unit—and as you will continue to see in future units—there is a definite problem-solving procedure.

If we are to use a tool, it must be in good working condition. So we have included this unit because some of you may have been away from school for some time. Maybe you have forgotten some of your arithmetic and basic algebra. Or maybe you had no algebra at all.

Read over the objectives for this unit. If the objectives are new to you, proceed through this unit by reading the material, studying the examples, and working the problems. When you understand all the material, check yourself by taking the test at the end of the unit. The solutions and answers for this test are given in the Appendix, so that you may grade your own test.

If you already have some of the skills required in the objectives, as you reach a known objective, go directly to the problems given after it. If you have no difficulty with the problems, then proceed to the next objective. Check your understanding of all the material in this unit by taking the test at the end of the unit.

If you already have all the skills set forth in the objectives, proceed directly to the test. If you can do all the problems on this test correctly, move on to the next unit.

Objectives

After completing all the work associated with this unit, you should be able to:

1. State the three steps required for solving a problem. The next four objectives deal with solving the problem after you have the correct form of the formula.
2. Use a formula of the form $E = IR$ to solve a problem.
3. Use a formula of the form

$$R = \frac{E}{I} \quad \text{or} \quad I = \frac{E}{R}$$

to solve a problem.
4. Use a formula of the form $I_T = I_1 + I_2$ to solve a problem.
5. Use a formula of the form

$$P = I^2 R, \quad P = \frac{E^2}{R}, \quad I = \sqrt{\frac{P}{R}}, \quad \text{or} \quad E = \sqrt{PR}$$

to solve a problem.

The next four objectives deal with solving a formula or equation for the unknown required in the problem.

6. Solve an equation of the form $E = IR$ for I and for R.
7. Rearrange an equation of the form

$$\frac{A}{B} = \frac{C}{D}$$

into $AD = BC$, and then solve for A, B, C, or D.
8. Solve an equation of the form $P = I^2 R$ for I.
9. Solve an equation of the form

$$\frac{1}{R_{eq}} = \frac{1}{R_1} + \frac{1}{R_2}$$

for R_{eq}.

Steps for Solving a Problem

Objective 1   State the three steps required for solving a problem.

At least three steps are required to solve a problem:

Step 1.   After finding the formula required for the solution to the problem, get it into the correct form for the unknown.

Step 2.   Substitute the numbers for the letters in the formula.

Step 3.   Solve the problem; that is, get an answer.

Depending on how complicated the problem is, you may have to repeat these steps several times.

Objectives 2, 3, 4, and 5 show you how to apply these three steps to solve problems. But these objectives concentrate more on steps 2 and 3 because the correct form of the formula is given. Objectives 6, 7, 8, and 9 concentrate on step 1, that is, solving a formula or equation for the unknown required in the problem.

Problem 3.1 From memory, state the three steps required to solve a problem.

Solution of Various Formulas

Objective 2 Use a formula of the form $E = IR$ to solve a problem.

As you will see in the next unit, Ohm's law states the relationship between voltage, current, and resistance. One of the forms of Ohm's law is $E = IR$. The basic unit for E is volts, for I is amperes, and for R is ohms. All the problems in this unit will use the basic units, so you don't have to worry about prefixes.

Example 3.1 Given that $I = 2$ A and $R = 100\ \Omega$, solve for E.

Solution

Step 1. Find the correct form of the formula. (It was given: $E = IR$.)

Step 2. Substitute the numbers for the letters: $E = IR = (2)(100)$

Step 3. Solve the problem.

$$\begin{array}{r} 100 \\ \times\quad 2 \\ \hline 200 \end{array}$$

Therefore $E = 200$ V.

Example 3.2 Given that $I = 0.01$ A and $R = 1500\ \Omega$, solve for E.

Solution

Step 1. Find the correct form of the formula: $E = IR$

Step 2. Substitute the numbers for the letters: $E = IR = (0.01)(1500)$

Step 3. Solve for E.

$$\begin{array}{r} 1500 \\ \times\quad 0.01 \\ \hline 15.00 \end{array}$$

There are two decimal places in the multiplier. Thus you must move the decimal point two places to the left for the correct answer. Therefore $E = 15.0$ V.

Problem 3.2   Given that $I = 3$ A and $R = 470$ Ω, solve for E.

Solution

Step 1. $E = IR$

Step 2. $E =$

Step 3. $E =$

Problem 3.3   Given that $I = 0.6$ A and $R = 50$ Ω, solve for E.

Solution

Step 1.

Step 2.

Step 3.

Problem 3.4   Given that $I = 0.15$ A and $R = 2500$ Ω, solve for E.

Solution

Step 1.

Step 2.

Step 3.

Problem 3.5   Given that $I = 0.008$ A and $R = 33,000$ Ω, solve for E.

Solution

Step 1.

Step 2.

Step 3.

Problem 3.6   Given that $I = 0.25$ A and $R = 3.3$ Ω, solve for E.

Solution

Step 1.

Step 2.

Answers: 1410 V,   30 V,   375 V,   264 V, 0.825 V

Step 3.

Objective 3 Use a formula of the form

$$R = \frac{E}{I} \quad \text{or} \quad I = \frac{E}{R}$$

to solve a problem.

The other two forms of Ohm's law are

$$R = \frac{E}{I} \quad \text{and} \quad I = \frac{E}{R}$$

Example 3.3 Given that $E = 20$ V and $I = 4$ A, solve for R.

Solution

Step 1. Find the correct form of the formula. Since we want to solve for R, we must use

$$R = \frac{E}{I}$$

Step 2. Substitute the numbers for the letters in the formula.

$$R = \frac{E}{I} = \frac{20}{4}$$

Step 3. Solve the problem.

$$\begin{array}{r} 5. \\ 4\overline{)20.} \end{array}$$

Therefore $R = 5\ \Omega$

Example 3.4 Given that $E = 6$ V and $I = 0.15$ A, solve for R.

Solution

Step 1. Find the correct form of the formula.

$$R = \frac{E}{I}$$

Step 2. Substitute the numbers into the formula.

$$R = \frac{E}{I} = \frac{6}{0.15}$$

Step 3. Solve the problem.

$$\begin{array}{r} 4\ 0. \leftarrow \text{Quotient} \\ \text{Divisor} \rightarrow 0.1\,5.\overline{)6.0\ 0.} \leftarrow \text{Dividend} \\ \underline{6\ 0} \\ 0\ 0 \end{array}$$

Since we had to move the decimal point two places to the right in the divisor, we must also move the decimal point of the dividend two places to the right. Therefore $R = 40\ \Omega$.

Problem 3.7   Given that $E = 63$ V and $I = 3$ A, solve for R.

   Solution

Step 1. $R = \dfrac{E}{I}$

Step 2. $R = $ —

Step 3. $R =$

Problem 3.8   Given that $E = 10$ V and $I = 0.05$ A, solve for R.

   Solution

Step 1.

Step 2.

Step 3.

Problem 3.9   Given that $E = 37.5$ V and $I = 0.025$ A, solve for R.

   Solution

Step 1.

Step 2.

Step 3.

Problem 3.10   Given that $E = 10$ V and $R = 1000\ \Omega$, solve for I.

   Solution

Step 1.

Step 2.

Step 3.

Problem 3.11   Given that $E = 18.15$ V and $R = 3300\ \Omega$, solve for I.

   Solution

Step 1.

Step 2.

Step 3.

Answers: 21 Ω, 200 Ω, 1500 Ω, 0.01 A, and 0.0055 A

Objective 4   Use a formula of the form $I_T = I_1 + I_2$ to solve a problem.

In the unit on parallel circuits, we shall have to solve formulas or equations of the form $I_T = I_1 + I_2$. Let's see how this is done.

Example 3.5   Given that $I_1 = 1.5$ A and $I_2 = 0.75$ A, solve for I_T.

Solution

Step 1.   Find the correct form of the formula. (It was given: $I_T = I_1 + I_2$.)

Step 2.   Substitute the numbers into the formula.

$$I_T = I_1 + I_2 = 1.5 + 0.75$$

Step 3.   Solve for I_T.

———— Line up the decimal points, and then add.

$$
\begin{array}{r}
1.5 \\
+0.75 \\
\hline
2.25
\end{array}
$$

Therefore $I_T = 2.25$ A.

Problem 3.12   Given that $I_1 = 2$ A and $I_2 = 7$ A, solve for I_T.

Solution

Step 1. $I_T = I_1 + I_2$

Step 2. $I_T =$

Step 3. $I_T =$

Problem 3.13   Given that $I_1 = 2.30$ A and $I_2 = 0.45$ A, solve for I_T.

Solution

Step 1.

Step 2.

Step 3.

Problem 3.14   Given that $I_1 = 0.008$ A and $I_2 = 0.013$ A, solve for I_T.

Solution

Step 1.

Step 2.

Step 3.

Example 3.6   Given that $I_T = 1.25$ A and $I_1 = 0.80$ A, solve for I_2.

Solution

Step 1.   Find the correct form of the formula. Since $I_T = I_1 + I_2$, we must solve this equation for I_2. This means we must get rid of the I_1 on the right-hand side of the equation. To do so, we can subtract I_1 from the right-hand side. But when we do this, we must also subtract I_1 from the left-hand side because we have an equals sign.

$$\downarrow \qquad\qquad\quad \downarrow$$
$$I_T - I_1 = I_1 + I_2 - I_1$$
$$I_T - I_1 = I_2 \qquad \text{or} \qquad I_2 = I_T - I_1$$

Step 2.   Substitute the numbers into the formula.

$$I_2 = I_T - I_1 = 1.25 - 0.80$$

Step 3.   Solve for I_2.

     —Line up the decimal points, and then subtract.

$$\begin{array}{r} 1.25 \\ -0.80 \\ \hline 0.45 \end{array}$$

Therefore $I_2 = 0.45$ A

Problem 3.15   Given that $I_T = 12$ A and $I_2 = 8$ A, solve for I_1.

Solution

Step 1.

Step 2.

Step 3.

Problem 3.16   Given that $I_T = 1.65$ A and $I_1 = 0.79$ A, solve for I_2.

Solution

Step 1.

Step 2.

Step 3.

Problem 3.17 Given that $I_T = 5.40$ A and $I_2 = 2.81$ A, solve for I_1.

Solution

Step 1.

Step 2.

Step 3.

Answers: 9 A, 2.75 A, 0.021 A, 4 A, 0.86 A, and 2.59 A

Objective 5 Use a formula of the form

$$P = I^2 R, \qquad P = \frac{E^2}{R}, \qquad I = \sqrt{\frac{P}{R}}, \qquad \text{or} \qquad E = \sqrt{PR}$$

to solve a problem.

We shall be using formulas like

$$P = I^2 R \qquad \text{and} \qquad P = \frac{E^2}{R}$$

in the unit on power. But what does the notation I^2 or E^2 mean? The I^2 is a shorthand notation for I times I. I^2 is called *I squared*. I^2 is also called I raised to the second power or I with an exponent of 2, but I squared is more common.

Example 3.7 Given that $I = 2$, what is I^2?

Solution

Step 1. $I^2 = I \times I$

Step 2. $= (2)(2)$

Step 3. $= 4$

Problem 3.18 Given that $I = 6$, what is I^2?

Solution

Step 1.

Step 2.

Step 3.

Problem 3.19   Given that $E = 13$, what is E^2?

Solution

Step 1.

Step 2.

Step 3.

Example 3.8   Given that $E = 10$ V and $R = 4000$ Ω, solve for P, using the relation

$$P = \frac{E^2}{R}$$

Solution

Step 1.

$$\text{Formula:}\quad P = \frac{E^2}{R}\quad\text{or}\quad P = \frac{E(E)}{R}$$

Step 2.   Substitute the numbers into the formula.

$$P = \frac{E^2}{R} = \frac{(10)(10)}{4000}$$

Step 3.   Solve for P.

$$P = \frac{100}{4000} = 0.025\text{ W}$$

Problem 3.20   Given that $I = 3$ A and $R = 15$ Ω, solve for P, using the relation $P = I^2R$.

Solution

Step 1.

Step 2.

Step 3.

Problem 3.21   Given that $E = 25$ V and $R = 5000$ Ω, solve for P, using

$$P = \frac{E^2}{R}$$

Solution

Step 1.

Step 2.

Step 3.

Problem 3.22   Given that $I = 0.03$ A and $R = 2000\ \Omega$, solve for P, using $P = I^2R$.

Solution

Step 1.

Step 2.

Step 3.

Answers: 36, 169, 135 W, 0.125 W, and 1.8 W

As we said before, E^2 means E squared or E times E. If $E = 5$, then $E^2 = E \times E = 5 \times 5 = 25$. Okay? Now what do we mean by $\sqrt{25}$? The symbol $\sqrt{\ }$ means *square root*. For example, $\sqrt{25}$ means the square root of 25. The square root means "What number, when squared, is equal to the number under the square-root symbol?"

Example 3.9 $\sqrt{36} = ?$

What number when squared or multiplied by itself is equal to 36? This number is equal to 6, because $6 \times 6 = 36$. Therefore $\sqrt{36} = 6$.

Example 3.10 $\sqrt{0.0001} = ?$

What number when squared (multiplied by itself) is equal to 0.0001? When we have decimals under the square-root symbol, it makes it easier to find the square root if we mark off the number in groups of two from the decimal point. If you end up with only one digit in the last group, add a zero to the right so that you have only groups of two.

$$\begin{array}{r} 0.\,0\ 1 \\ \sqrt{0.\overline{00}\,\overline{01}} \end{array}$$

When you have a square root of a fraction, like 0.0001, you should always check your answer by squaring it to see if you get the number under the square-root sign.

$$\begin{array}{r} 0.01 \\ \times\,0.01 \\ \hline 0.0\,0\,0\,1 \end{array} \quad \text{4 decimal places in the problem}$$

The answer to this problem is equal to 0.01, because 0.01×0.01 gives 0.0001.

The method given in Examples 3.9 and 3.10 is fine if the number under the square-root symbol is a perfect square. If the number is not a perfect square, then you have to use a calculator, a square-root mathematical table, or the long mathematical procedure for square roots. The numbers under the square-root symbols in the problems in this unit are all perfect squares.

Sometimes, when you are taking the square root of a number, it helps to express the number in powers-of-ten notation. But the power on the 10 must be *even*. The square root of an even power of ten is 10 raised to one-half of the original power. Study the next four examples.

Example 3.11

$$\sqrt{10^4} = 10^{4/2} = 10^2$$
$$\sqrt{10^{10}} = 10^{10/2} = 10^5$$
$$\sqrt{10^{-2}} = 10^{-2/2} = 10^{-1}$$
$$\sqrt{10^{-6}} = 10^{-6/2} = 10^{-3}$$

Example 3.12

$$\sqrt{360,000} = \sqrt{36 \times 10^4} = (\sqrt{36})(\sqrt{10^4})$$
$$= (6)(10^2) = 600$$

Example 3.13

$$\sqrt{0.0016} = \sqrt{16 \times 10^{-4}} = (\sqrt{16})(\sqrt{10^{-4}})$$
$$= (4)(10^{-2}) = 0.04$$

Example 3.14

$$\sqrt{0.00000169} = \sqrt{169 \times 10^{-8}} = 13 \times 10^{-4} = 0.0013$$

Problem 3.23 $\sqrt{16} =$

Problem 3.24 $\sqrt{196} =$

Problem 3.25 $\sqrt{0.0009} =$

Problem 3.26 $\sqrt{0.00000001} =$

Problem 3.27 $\sqrt{0.0225} =$

Answers: 4, 14, 0.03, 0.0001, 0.15, and 0.025

Problem 3.28 $\sqrt{0.000625} =$

When we are using the formula $P = I^2R$, the unknown in the problem may be I. If this is the case, then

$$I = \sqrt{\frac{P}{R}}$$

Objective 8 will show you how to obtain this. Study Example 3.15 to see how to work problems using this formula.

Example 3.15 Given that $P = 2$ W and $R = 200$ Ω, solve for I by using

$$I = \sqrt{\frac{P}{R}}$$

Solution

Step 1. Write the correct form of the formula.

$$I = \sqrt{\frac{P}{R}}$$

Step 2. Substitute the numbers into the formula.

$$I = \sqrt{\frac{P}{R}} = \sqrt{\frac{2}{200}}$$

Step 3. Solve for I.

$$I = \sqrt{0.01} = \sqrt{1 \times 10^{-2}} = 1 \times 10^{-2/2}$$
$$= 1 \times 10^{-1} = 0.1$$

Problem 3.29 Given that $P = 10$ W and $R = 40$ Ω, solve for I by using

$$I = \sqrt{\frac{P}{R}}$$

Solution

Step 1.

Step 2.

Step 3.

Problem 3.30 Given that $P = 1$ W and $R = 10,000$ Ω, solve for I by using

$$I = \sqrt{\frac{P}{R}}$$

Solution

Step 1.

Step 2.

Step 3.

When the formula

$$P = \frac{E^2}{R}$$

is used, sometimes the unknown is E. By methods given in Objective 8, $E = \sqrt{PR}$.

Example 3.16   Given that $P = 5$ W and $R = 500$ Ω, solve for E by using $E = \sqrt{PR}$.

Solution

Step 1.   Write the correct form of the formula.

$$E = \sqrt{PR}$$

Step 2.   Substitute the numbers into the formula.

$$E = \sqrt{PR} = \sqrt{(5)(500)}$$

Step 3.   Solve for E.

$$E = \sqrt{2500} = 50$$

Problem 3.31   Given that $P = 3$ W and $R = 300$ Ω, solve for E by using $E = \sqrt{PR}$.

Solution

Step 1.

Step 2.

Step 3.

Problem 3.32   Given that $P = 10$ W and $R = 4000$ Ω, solve for E by using $E = \sqrt{PR}$.

Solution

Step 1.

Step 2.

Answers: 0.5 A, 0.01 A, 30 V, and 200 V   Step 3.

In the next four objectives, you will be dealing with how to change around formulas, or equations, so that you can solve for a different unknown. To do this, you have to understand what the equals sign means in a formula or equation. An equation is a mathematical statement that two quantities are equal. The equals sign ($=$) is used to show that the two quantities are equal. For example, the statement $x = 6$ says that x is equal to 6, or the statement $E = IR$ says that E is equal to I times R.

In an equation, both sides must remain equal or in balance, like the balance or seesaw shown in Figure 3.1. For both sides

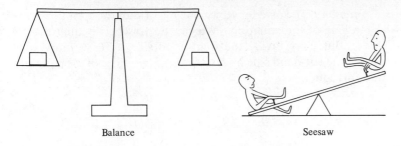

Balance   Seesaw

Figure 3.1 The balance and seesaw illustrate how an equation must be handled.

to remain equal to each other, these rules must hold:

If you subtract something from one side of the equation, you must subtract the same thing from the other side.

If you add something to one side of the equation, you must add the same thing to the other side.

If you multiply one side of the equation by something, you must multiply the other side of the equation by the same thing.

If you divide one side of the equation by something, you must divide the other side of the equation by the same thing.

Simple Equations

Objective 6 Solve an equation of the form $E = IR$ for I and for R.

This equation is Ohm's law. In the next unit, on Ohm's law, you will see the importance of being able to solve for I or R. Examples 3.17 and 3.18 will show you how to do this by using the rules given for working with equations.

Example 3.17 Given the equation $E = IR$, solve for I.

Solution To solve for I, we must get rid of the R on the right-hand side of the equation. We can do this by dividing the IR by R.

$$\frac{IR}{R} = I$$

But if we divide the right-hand side by R, we must also divide the left-hand side of the equation by R, or else the two sides of the equation are no longer equal.

$$\frac{E}{R} = \frac{IR}{R}$$

$$\therefore \quad \frac{E}{R} = I \quad \text{or} \quad I = \frac{E}{R}$$

Example 3.18 Given the equation $E = IR$, solve for R.

Solution   To solve for R, we must get rid of the I on the right-hand side of the equation. We can do this by dividing the IR by I. But if we divide the right-hand side by I, we must also divide the left-hand side by I, so that the two sides of the equation are still equal.

$$E = IR$$

$$\frac{E}{I} = \frac{\not{I}R}{\not{I}} \qquad \therefore \quad R = \frac{E}{I}$$

Problem 3.33   Given the equation $Q = It$, solve for I and for t.

Problem 3.34   Given the equation $W = QV$, solve for Q and for V.

Problem 3.35   Given the equation $P = EI$, solve for E and for I.

Problem 3.36   Given the equation $V = IR$, solve for I and for R.

Objective 7   Rearrange an equation of the form

$$\frac{A}{B} = \frac{C}{D}$$

into $AD = BC$, and then solve for A, B, C, or D.

This is just a general equation; it has no particular significance in electric circuits. However, you will use the technique you learned here on formulas that do apply to electric and electronic circuits. Example 3.19 will show you how to rearrange this equation.

Example 3.19   Rearrange the equation

$$\frac{A}{B} = \frac{C}{D}$$

into the form $AD = BC$.

Solution

Step 1.   Start with the original equation.

$$\frac{A}{B} = \frac{C}{D}$$

Step 2.   Multiply both sides of the equation by B to get rid of the B on the left-hand side.

$$\frac{A\not{B}}{\not{B}} = \frac{BC}{D}$$

Step 3.   Multiply both sides of the equation in step 2 by D to get rid of the D in the denominator of the right-hand side.

$$AD = \frac{BC\cancel{D}}{\cancel{D}}$$

$$\therefore \quad AD = BC$$

An easy way to recall this result is to remember the *crisscross*.

You must move on the criss *or* the cross. Think of this as a seesaw with the equals sign as the pivot. To get the result $AD = BC$, bring the D up with the A, and the B up with the C.

Another thing you can do with this crisscross is to interchange any two of the letters on the criss or any two letters on the cross. For example:

$$\frac{D}{B} = \frac{C}{A} \qquad \frac{A}{C} = \frac{B}{D}$$

Example 3.20   Solve the equation

$$I = \frac{E}{R}$$

for E, using the crisscross method and also the rules you learned for working with equations.

Solution   Crisscross: Remember that I divided by 1 is the same as I.

$$\frac{I}{1} = \frac{E}{R}$$

$$\therefore \quad IR = E \qquad \text{or} \qquad E = IR.$$

Also remember the rule of multiplying *both* sides of an equation by the same quantity:

$$I = \frac{E}{R}$$

$$IR = \frac{E\cancel{R}}{\cancel{R}} \qquad \therefore \quad E = IR$$

Problem 3.37　Solve the equation

$$R = \frac{E}{I} \quad \text{for } E$$

Problem 3.38　Solve the equation

$$I = \frac{Q}{t} \quad \text{for } Q$$

Example 3.21　Solve the equation

$$R = \frac{E}{I} \quad \text{for } I$$

Solution　We could make use of the criss part of the crisscross to interchange R and I:

$$R = \frac{E}{I} \qquad \therefore \quad I = \frac{E}{R}$$

Or we could first multiply both sides by I.

$$R = \frac{E}{I} \qquad IR = \frac{EI}{I}$$

Then divide both sides of this last equation by R.

$$\frac{IR}{R} = \frac{E}{R}$$

We get the same result,

$$I = \frac{E}{R}$$

Problem 3.39　Given the equation

$$P = \frac{W}{t}$$

solve for W and for t.

Problem 3.40　Solve the equation

$$G = \frac{1}{R} \quad \text{for } R$$

Problem 3.41　Given the equation

$$R = \frac{\rho l}{A}$$

solve for A and for l. (The symbol ρ is the Greek letter rho, pronounced "row.")

Problem 3.42 Solve the equation

$$I = \frac{V}{R} \quad \text{for } V \text{ and for } R$$

Problem 3.43 Solve the equation

$$P = \frac{V^2}{R} \quad \text{for } R$$

Problem 3.44 Solve the equation $P = I^2 R$ for R.

Objective 8 Solve an equation of the form $P = I^2 R$ for I.

This is the equation used to determine the power developed in a resistor R when a current I flows through it. Example 3.22 shows you how to solve this.

Example 3.22 Solve the equation $P = I^2 R$ for I.

Solution

Step 1. Start with the original equation: $P = I^2 R$

Step 2. To get I^2 all by itself, divide both sides of the equation by R.

$$\frac{P}{R} = \frac{I^2 \cancel{R}}{\cancel{R}} \qquad \therefore \quad I^2 = \frac{P}{R}$$

Step 3. Since we have I^2 (or I times I), to find I, we take the square root of the left-hand side of the equation. But this is an equation, so if we take the square root of the left-hand side, we must also take the square root of the right-hand side.

$$\sqrt{I^2} = \sqrt{\frac{P}{R}} \qquad \therefore \quad I = \sqrt{\frac{P}{R}}$$

Problem 3.45 Solve the equation

$$P = \frac{E^2}{R} \quad \text{for } E$$

Problem 3.46 Solve the equation

$$A = \frac{\pi}{4} d^2 \quad \text{for } d$$

Problem 3.47   Solve the equation

$$R = \frac{\rho l}{d^2} \qquad \text{for } d$$

Objective 9   Solve an equation of the form

$$\frac{1}{R_{eq}} = \frac{1}{R_1} + \frac{1}{R_2} \qquad \text{for } R_{eq}$$

This is the equation used to determine the equivalent resistance of two resistors when they are connected in parallel. The final form of the equation,

$$R_{eq} = \frac{R_1 R_2}{R_1 + R_2}$$

is the form you will probably memorize.

Example 3.23   Solve the equation

$$\frac{1}{R_{eq}} = \frac{1}{R_1} + \frac{1}{R_2} \qquad \text{for } R_{eq}$$

Solution   On the right-hand side of the equation, we are adding

$$\frac{1}{R_1} + \frac{1}{R_2} \qquad \text{One form of this is} \qquad \frac{1}{3} + \frac{1}{4}$$

We can't add these fractions directly because they don't have the same denominator. So first we want to get a common denominator for

$$\frac{1}{R_1} + \frac{1}{R_2}$$

The common denominator is $R_1 \times R_2$, or $R_1 R_2$. Therefore

$$\frac{1}{R_1} + \frac{1}{R_2} = \frac{R_2}{R_1 R_2} + \frac{R_1}{R_1 R_2} = \frac{R_2 + R_1}{R_1 R_2}$$

$$\therefore \quad \frac{1}{R_{eq}} = \frac{R_1 + R_2}{R_1 R_2}$$

Since we want to solve for R_{eq}, we flip both sides of the equation upside down.

$$R_{eq} = \frac{R_1 R_2}{R_1 + R_2}$$

We say that the R_{eq} is equal to the *product* of $R_1 R_2$ over the *sum* of $R_1 + R_2$.

Problem 3.48   Solve the equation

$$\frac{1}{R_{eq}} = \frac{1}{R_3} + \frac{1}{R_4} \qquad \text{for } R_{eq}$$

Problem 3.49   Solve the equation

$$\frac{1}{R_{eq}} = \frac{1}{3} + \frac{1}{4} \qquad \text{for } R_{eq}$$

Problem 3.50   Solve the equation

$$\frac{1}{R_{eq}} = \frac{1}{R} + \frac{1}{R} \qquad \text{for } R_{eq}$$

Test

1. Do Objective 1.
2. Given that $I = 0.05$ A and $R = 22,000$ Ω, solve for E, using $E = IR$.
3. Given that $E = 12$ V and $I = 0.002$ A, solve for R, using

$$R = \frac{E}{I}$$

4. Given that $E = 15$ V and $R = 5000$ Ω, solve for I, using

$$I = \frac{E}{R}$$

5. Given that $I_T = I_1 + I_2$, with $I_1 = 2.75$ A and $I_2 = 0.935$ A, solve for I_T.
6. Given that $I_T = I_1 + I_2$, with $I_T = 0.825$ A and $I_1 = 0.017$ A, solve for I_2.
7. Given that $I = 0.04$ A and $R = 5100$ Ω, solve for P, using $P = I^2 R$.
8. Given that $P = 1.8$ W and $R = 50,000$ Ω, solve for I, using

$$I = \sqrt{\frac{P}{R}}$$

9. Given that $P = 0.2$ W and $R = 720$ Ω, solve for E, using $E = \sqrt{PR}$.
10. Given the equation $P = VI$, solve for V.
11. Given the equation

$$X_C = \frac{1}{2\pi f C} \qquad \text{solve for } C.$$

12. Given the equation

$$R = \frac{\rho l}{A} \qquad \text{solve for } \rho.$$

13. Solve the equation

$$P = \frac{E^2}{R} \qquad \text{for } E$$

14. Solve the equation

$$\frac{1}{R_{eq}} = \frac{1}{R_5} + \frac{1}{R_6} \qquad \text{for } R_{eq}$$

15. Solve the equation

$$\frac{1}{R_{eq}} = \frac{1}{3300} + \frac{1}{10,000} \qquad \text{for } R_{eq}$$

Ohm's Law Unit 4

There is one law that enables you to do the following:

Determine the required supply voltage when you know the load and the load current.

Determine the size resistor required to reduce the value of a power-supply voltage to that required for a particular circuit.

Determine the amount of current through any component in a circuit or section of a circuit when you know the voltage across and the resistance of the circuit or section of the circuit.

Determine the value of the load that you may connect across a power supply.

Determine the size of resistor required to limit circuit current.

These are just a few of the unknowns that you can find using Ohm's law. You can use Ohm's law to solve for current, resistance, supply voltage, or voltage drop in simple or in very complicated circuits. If you really understand and know how to apply Ohm's law, you have progressed a good way through your electronics education.

After completing all the work associated with this unit, you should be able to:

1. Write Ohm's law in three different forms. One form should solve for current, one for voltage, and one for resistance. Include the units for each of the symbols.
2. Calculate the voltage, current, or resistance of a simple electric circuit, using Ohm's law, when any two of the three quantities are given.
3. Calculate the voltage, current, or resistance required in word problems, using Ohm's law.
4. Calculate the voltage, current, or resistance in more complicated circuits, such as those shown in Figures 4.3, 4.4, 4.5, and 4.6, using Ohm's law, when any two of the three quantities are given.

Forms of Ohm's Law

Objective 1   Write Ohm's law in three different forms. One form should solve for current, one for voltage, and one for resistance. Include the units for each of the symbols.

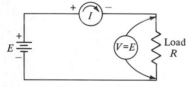

Figure 4.1   Simple electric-circuit diagram

Back in the 1800s, Georg Simon Ohm was working with the simple electric circuit shown in Figure 4.1. He found that so long as the temperature of the load did not change, the ratio of voltage across the load divided by the current through the load remained constant, or

$$\frac{V}{I} = K$$

Ohm found that, if he changed the load so as to increase the value for K without changing the power supply, the current through the load decreased. Since this property of the load opposed or resisted the flow of current in the circuit, it was called *resistance*. The letter symbol used in mathematical formulas for resistance is R.

$$R = \frac{V}{I} \qquad\qquad (4.1)$$

where V is the *voltage drop* across the circuit element; the unit of measure of V is the volt.

I is the current *through* the circuit element; the unit of measure of I is the ampere.

R is the resistance of the circuit element. To honor Ohm, the unit of measure for R is called the ohm (abbreviated Ω) instead of using volts/ampere.

[*Note*: The number to the right of the formula is a formula or equation number. We have numbered important formulas so that we may refer to them easily in other parts of this text.]

In the simple electric circuit of Figure 4.1, the load resistor is identified as R. So Ohm's law says that to find the resistance of the circuit element R, divide the voltage *across* the resistor R by the current *through* the resistor R. In Figure 4.1 the voltage across the resistor is equal to the supply or battery voltage E; therefore this gives Equation (4.1a).

$$R = \frac{E}{I} \tag{4.1a}$$

The circuit element that is known as a resistor follows Ohm's law. These are usually known as linear resistors. However, some circuit elements that do not obey Ohm's law have been purposely developed. These usually have special names. Some of these will be discussed in the unit on resistance.

By multiplying both sides of Equation 4.1 by I, we obtain another form of Ohm's law:

$$V = IR \quad \text{or} \quad E = IR \tag{4.2}$$

The first formula is probably more meaningful if the question is, "What is the voltage drop across a resistor if a certain current is flowing through the resistor?" Whereas the second formula is more meaningful if the question is, "What supply or battery voltage must be applied across a resistor to make it pass a certain amount of current?" But most technicians and engineers use both symbols for voltage interchangeably, so that either V or E could mean the voltage drop across a resistor.

By dividing both sides of Equation 4.2 by R, we can obtain a third form of Ohm's law:

$$I = \frac{V}{R} \quad \text{or} \quad I = \frac{E}{R} \tag{4.3}$$

Figure 4.2 shows you a triangle that you may use to remember the three forms of Ohm's law. To find the formula for any one of the letter symbols, just cover that letter and look at what remains in the triangle. For example, if you want the formula for I, cover I; you can see that it is equal to V divided by R. Or to find the formula for V, when you cover V, you can see that I is next to R. So V is equal to I times R.

Ohm's law is so important and is used so much that you should memorize it. We recommend that you memorize one form of it, say $E = IR$, and then use algebra to solve for the other two forms. If you need a review of how to solve an equation of the form $E = IR$ for either I or R, see Unit 3. We recommend that you learn how to solve equations of the form $E = IR$

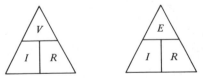

Figure 4.2 Ohm's-law triangle

by algebra, because there will be many other formulas that have this same form—formulas such as $Q = It$, $P = EI$, and $W = QV$, to mention just a few. If you learn how to get the other two forms of the formula by algebra instead of by memorization, you will have only one-third the memory work.

Problem 4.1   Take a blank sheet of paper and write Ohm's law in the three different forms. Also list the units for each letter symbol.

Simple Electric Circuits

Objective 2   Calculate the voltage, current, or resistance of a simple electric circuit, using Ohm's law, when any two of the three quantities are given.

The following examples show you how you may use Ohm's law to solve for the unknown quantity in the simple electric circuit shown in Figure 4.1.

Example 4.1   In the circuit shown in Figure 4.1, $E = 10$ V and $R = 10\ \Omega$. Solve for I.

Solution   The form of Ohm's law we memorized is $E = IR$ or $V = IR$, but we want to solve for I. Dividing both sides of this equation by R, we obtain the form we need.

Step 1.

$$E = IR \qquad \therefore \quad I = \frac{E}{R}$$

Step 2.

$$I = \frac{10\text{ V}}{10\ \Omega}$$

Step 3.

$$I = 1\text{ A}$$

Example 4.2   In the circuit shown in Figure 4.1, $E = 15$ V and $I = 3$ A. Solve for R.

Solution

$$R = \frac{E}{I} = \frac{15\text{ V}}{3\text{ A}} = 5\ \Omega$$

Example 4.3 In the circuit shown in Figure 4.1, $I = 2$ A and $R = 10\ \Omega$. Solve for E.

Solution

$$E = IR = (2\ \text{A})(10\ \Omega) = 20\ \text{V}$$

Notice that there were three steps required to solve these problems. We discussed this in the last unit.

Step 1. After finding the formula required for the solution to the problem, get it into the correct form for the unknown.

Step 2. Substitute the numbers for the letters in the formula.

Step 3. Solve the problem.

You should get into the habit of using these steps when you solve a problem. If you learn to do this now while the problems are simple, it will help you when you get to the more complicated problems and circuits. You try the next three problems.

Problem 4.2 In the circuit shown in Figure 4.1, $E = 20$ V and $R = 2\ \Omega$. Solve for I.

Problem 4.3 In the circuit shown in Figure 4.1, $E = 18$ V and $I = 2$ A. Solve for R.

Problem 4.4 In the circuit shown in Figure 4.1, $I = 3$ A and $R = 50\ \Omega$. Solve for E.

So far, all the examples and problems have had the current in amperes and small values for the resistors. In many electronic circuits the values used for R are in the kilohm or megohm range, and the values for I are in the milliampere or micro-ampere range. In other words, prefixes are used. Study Example 4.4, then work Problems 4.5, 4.6, and 4.7.

Example 4.4 In the circuit shown in Figure 4.1, $I = 2\ \mu$A and $R = 5\ \text{k}\Omega$. Solve for E.

Solution

$$E = IR$$

Before we substitute into the formula, we change the prefixes to powers-of-ten notation

$$I = 2\ \mu\text{A} = 2 \times 10^{-6}\ \text{A} \quad \text{and} \quad R = 5\ \text{k}\Omega = 5 \times 10^3\ \Omega$$
$$\therefore\ \ E = IR = (2 \times 10^{-6})(5 \times 10^3)$$

To solve for E, we work with the number or coefficient part first. Then we work with the powers of ten.

$$E = (2 \times 5)(10^{-6} \times 10^{3}) = 10 \times 10^{-3} = 10 \text{ mV}$$

Problem 4.5   In the circuit shown in Figure 4.1, $V = 10$ V and $R = 10$ kΩ. Solve for I.

Problem 4.6   In the circuit shown in Figure 4.1, $E = 18$ V and $I = 10$ mA. Solve for R.

Problem 4.7   In the circuit shown in Figure 4.1, $I = 10$ μA and $R = 200$ kΩ. Solve for E.

Word Problems

Objective 3   Calculate the voltage, current, or resistance required in word problems, using Ohm's law.

This objective is quite similar to Objective 2, but now you will not have a circuit in front of you to help you visualize what the problem is discussing. You have to visualize the problem from the description given.

Since you only have one formula, Ohm's law, to use to solve the problems, this should simplify your work now. In future units you will be learning more formulas, and you will probably have some trouble deciding which formula to use. To make word problems easier, we shall show you how to *dissect* or tear apart a word problem. First pick out the knowns and the unknown in the problem. Do this by underlining the names or units and writing the letter symbol above the underlined words. Then find the formula that relates everything. When you have the formula, this is step 1 of the three steps given previously for solving problems. This process is illustrated in Example 4.5.

Example 4.5   What current through a 2-kΩ resistor causes a voltage drop of 30 V across the resistor?

Solution

$$\begin{array}{ccc} I = ? & R & V \end{array}$$

What current through a 2-kΩ resistor causes a voltage drop of 30 V across the resistor? $V = IR$ is the formula that relates everything. The unknown is I.

$$I = \frac{V}{R} = \frac{30 \text{ V}}{2 \text{ k}\Omega} = \frac{30}{2 \times 10^{3}} = 15 \times 10^{-3} = 15 \text{ mA}$$

Note that a prefix of k in the denominator gives a prefix of m in the answer.

Example 4.6   What power-supply voltage (emf) must be applied to a 800-kΩ resistor to make it pass 15 μA of current?

Solution

$$\underset{\text{What power-supply voltage (emf) must be applied to a}}{\overset{E = ?}{}} \quad \underset{\text{800-k}\Omega}{\overset{R}{}}$$

$$\underset{\text{resistor to make it pass } \underset{I}{15\ \mu\text{A of current}}?}{}$$

The formula $E = IR$ is the formula that relates everything.

$$E = IR = (15\ \mu\text{A})(800\ \text{k}\Omega) = (15 \times 10^{-6})(800 \times 10^{3})$$
$$= 12{,}000 \times 10^{-3} = 12\ \text{V}$$

Example 4.7   What value of collector resistance in a transistor amplifier circuit is required to obtain a voltage drop of 10 V when the collector current is 10 mA?

Solution

$$\underset{\text{What value of collector } \underset{}{\text{resistance}} \text{ in a transistor amplifier}}{\overset{R = ?}{}}$$

$$\text{circuit is required to obtain a } \underset{I}{\underset{\text{voltage drop of 10 V}}{\overset{V}{}}} \text{ when the}$$

collector current is 10 mA?

$$R = \frac{V}{I} = \frac{10\ \text{V}}{10\ \text{mA}} = \frac{10}{10 \times 10^{-3}} = 1 \times 10^{3} = 1\ \text{k}\Omega$$

Note that a prefix of m in the denominator gives a prefix of k in the answer.

Example 4.8   What is the voltage drop across a 500-Ω emitter resistor in a transistor amplifier circuit when 5 mA of emitter current flows through it?

$$V = IR = (5\ \text{mA})(500\ \Omega) = (5 \times 10^{-3})(500)$$
$$= 2500 \times 10^{-3} = 2.5\ \text{V}$$

As you see from looking at the examples, when we substitute numbers into the formulas, we attach the units to the numbers. This is one way of checking whether we shall obtain the correct unit for the answer. For example, in solving for the power supply voltage E in Example 4.6, we had the units microamperes times kilohms. Then we knew that the answer couldn't be

12,000 V, because the units of amperes times ohms give the unit of volts. We are then alerted to be sure we handle the prefixes.

Sometimes you do not have to substitute the powers of ten for the prefixes as shown in the examples. As shown in Example 4.5, with volts in the numerator and kilohms in the denominator, the answer comes out in milliamperes. And in Example 4.7, with volts in the numerator and milliamperes in the denominator, the answer comes out in kilohms. A couple of additional examples: If you have milliamperes and kilohms, the answer has units of volts, because the m and the k cancel each other. If you have volts in the numerator and microamperes in the denominator, the answer comes out in megohms. You think up some additional examples of this.

Problem 4.8   What current flows through a 24-Ω resistor if a voltage of 48 V is applied across the resistor terminals?

Problem 4.9   What value of resistor draws 50 mA current when connected to a 1-kV source of electromotive force?

Problem 4.10   A fuse in the power supply for a transistor amplifier has a resistance of 0.01 Ω. What current through the fuse causes a 100-μV drop across it?

Problem 4.11   A lamp bulb, to light properly, requires 100 mA of current through it. When the lamp is on, it has 15 Ω resistance. What size battery is required to light the lamp?

Problem 4.12   What power-supply voltage must be applied to the base of a transistor amplifier if the base resistor is 1 MΩ and the base current is 10 μA?

Ohm's Law in Complicated Circuits

Objective 4   Calculate the voltage, current, or resistance in more complicated circuits, such as those shown in Figures 4.3, 4.4, 4.5, and 4.6, using Ohm's law, when any two of the three quantities are given.

This objective is included in this unit to give you some idea how useful Ohm's law really is, and to show you how to use it when you have more than one voltage, current, or resistance in a circuit. We'll give you four circuits and show you how you can apply Ohm's law. These circuits are more complicated than the simple circuit that you worked with in Objective 2.

Figure 4.3 is a series resistive circuit with two resistors.

Figure 4.4 is a series-parallel resistive circuit. The current I_1 that flows through resistor R_1 divides and flows through resistors R_2 and R_3 in proportion to their size.

Figure 4.5 is a Zener voltage-regulator circuit. This circuit holds a constant voltage across the load even though the load current or supply voltage E varies.

Figure 4.6 is a simple transistor amplifier circuit. When a small input voltage is applied at the input to the circuit, an amplified, or larger, voltage appears across the output terminals.

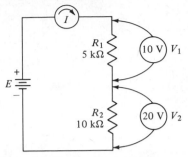

Figure 4.3 Zener voltage-regular

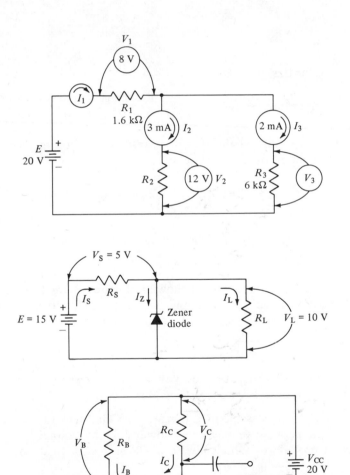

Figure 4.4 Series-parallel resistive circuit

Figure 4.5 Zener voltage-regulator circuit

Figure 4.6 Transistor amplifier circuit

Don't try to figure out how the circuits function now. Just note how you may solve for an unknown resistor, unknown current, required power-supply voltage, or voltage drop simply by using Ohm's law. You will study the circuits shown in Figures 4.3 and 4.4 in more detail in later units in this text.

The last two circuits, shown in Figures 4.5 and 4.6, you will study in more detail in a course on electronics.

Example 4.9   Solve for the current that is flowing in resistor R_1 in the circuit shown in Figure 4.3.

Solution   Ohm's law,

$$I = \frac{V}{R}$$

means for this problem:

$$I \text{ (current through resistor } R_1) = \frac{V \text{ (voltage across resistor } R_1)}{R \text{ (resistance of } R_1)}$$

To find the current through or in resistor R_1, you must know the voltage across R_1 and the resistance of R_1.

$$I = \frac{V_1}{R_1} = \frac{10 \text{ V}}{5 \text{ k}\Omega} = 2 \text{ mA}$$

Problem 4.13   Solve for the current that is flowing in resistor R_2 in the circuit shown in Figure 4.3, using Ohm's law.

Problem 4.14   In the circuit shown in Figure 4.3, the total resistance across the power supply is 15 kΩ. What power-supply voltage is required for 2 mA of current to flow through the 15-kΩ load?

Example 4.10   Solve for the voltage across resistor R_3 in the circuit shown in Figure 4.4, using Ohm's law.

Solution   It was specified that we use Ohm's law to solve this problem because we may know how to find this voltage by some other procedure. Ohm's law, $V = IR$, means for this problem:

$$V \text{ (voltage across } R_3) = I \text{ (current through } R_3)$$
$$\times R \text{ (resistance of } R_3)$$

or

$$V_3 = I_3 R_3 = (2 \text{ mA})(6 \text{ k}\Omega) = 12 \text{ V}$$

Problem 4.15 Solve for resistor R_2 in the circuit shown in Figure 4.4, using Ohm's law.

Example 4.11 In the circuit of Figure 4.5, we want to drop 5 V across resistor R_S when the source current I_S is equal to 100 mA. Solve for the required size of R_S.

Solution Ohm's law,

$$R = \frac{V}{I}$$

means for this problem:

$$R \text{ (resistance of } R_S) = \frac{V \text{ (voltage across resistor } R_S)}{I \text{ (current through resistor } R_S)}$$

or

$$R_S = \frac{V_S}{I_S} = \frac{5 \text{ V}}{100 \text{ mA}} = 50 \ \Omega$$

Problem 4.16 In the circuit of Figure 4.5, the load resistor R_L is equal to 125 Ω. Solve for the load current I_L.

Example 4.12 In the transistor amplifier circuit of Figure 4.6, 10 V should be dropped across the collector resistor when the collector current I_C is 5 mA. What size collector resistor R_C is required?

$$R_C = \frac{V_C}{I_C} = \frac{10 \text{ V}}{5 \text{ mA}} = 2 \text{ k}\Omega$$

where V_C is the voltage dropped across the collector resistor R_C.

Problem 4.17 In the transistor amplifier circuit of Figure 4.6, 19.3 V should be dropped across the base resistor R_B when the base current is 50 μA. What size base resistor R_B is required?

Test

Refer to the circuit in Figure 4.1 for Test Problems 1, 2, 3, and 4. Use Ohm's law to solve each problem.

1. Given that $E = 3$ V and $R = 10 \ \Omega$, then $I =$
2. Given that $E = 4.5$ V and $I = 300$ mA, then $R =$
3. Given that $I = 10$ mA and $R = 1$ kΩ, then $E =$
4. Given that $E = 9$ V and $R = 30$ kΩ, then $I =$
5. What is the resistance of an electric circuit that draws 2 A of current from a 100-V source?

6. What current through a 1000-Ω resistor causes a voltage drop of 10 V across the resistor?

7. What supply voltage (emf) must be applied to a 100-kΩ resistor to make it pass 1 μA of current?

8. What is the voltage drop across a 10-kΩ resistor when 2 mA of current flows through it?

9. Solve for the current I_1 through resistor R_1 in the circuit shown in Figure 4.4.

10. In the transistor amplifier circuit of Figure 4.6, the collector resistor R_C is equal to 5 kΩ. When the collector current I_C is equal to 3 mA, what is the voltage drop across the collector resistor R_C?

Current and Unit 5
Voltage

In the first units, we wanted you to start learning the language. Therefore we had you learn two rough definitions: that current is the rate of flow of electrons in a circuit and voltage is the push that is required to make the electrons flow. Before you go on and study more complicated circuits, you must have a better understanding of what we mean when we say we have a current of one ampere or a potential of one volt.

Objectives

After completing all the work associated with this unit, you should be able to:

1. Define the following terms: electron, proton, free electron, conductor, insulator, and semiconductor.
2. Define the term "quantity of electric charge." The definition must include the basic unit, the number of electrons in one unit, the letter symbol used in mathematical formulas, and the abbreviation or symbol of the unit.
3. Define the word *ampere*. The definition must include an explanation of the formula $I = Q/t$. And calculate the current, quantity of electric charge, or time, using the formula $I = Q/t$, when any two of the three quantities are given.
4. Define the term "potential difference." The definition must include an explanation of the formula V (or E) $= W/Q$. And

calculate the potential difference, energy, or quantity of electric charge, using the formula V (or E) $= W/Q$, when two of the three quantities are given or when two of the three quantities may be calculated from the given data.

5. Discuss briefly the types of electromotive force.

Definition of Terms

Objective 3 of this unit deals with current. Just what is current? You might say that you already know what current is. You learned from Ohm's law that you have a current of one ampere if you connect a one-volt battery across a one-ohm resistor. But what does a current of one ampere mean? To describe what current actually is, we have to go inside the conductor, into the inside of the atom, to the electron.

Objective 1 Define the following terms: electron, proton, free electron, conductor, insulator, and semiconductor.

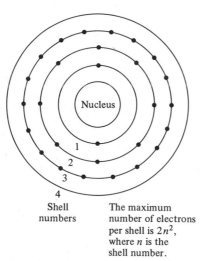

Shell numbers

The maximum number of electrons per shell is $2n^2$, where n is the shell number.

Figure 5.1 General atomic structure of an element

Figure 5.1 shows the general atomic structure of an element. There are now more than 100 elements. Some of them that you may know are hydrogen, oxygen, copper, aluminum, gold, germanium, silicon, and carbon. The center of the structure shown in Figure 5.1 is known as the *nucleus*. The nucleus contains *neutrons*, which have no charge, and positively charged particles called *protons*. The number of protons equals the number of electrons circling the nucleus. *Electrons* are negatively charged particles. Electrons contain a negative charge equal in magnitude to the positive charge on the protons. Therefore the atom as a whole is electrically neutral.

To visualize what you see in Figure 5.1, think of the solar system, with the nucleus as the sun and the electrons as the planets. The protons, which are part of the nucleus, remain immobile because they contain most of the mass of the atom. A proton has a mass 1837 times that of an electron. The elements are usually known by their atomic number, which is the number of electrons circling the nucleus or the number of protons. The electrons, as shown in Figure 5.1, are discrete particles that move in circular orbits around the nucleus. Actually electrons behave like waves; they travel in orbits only in a very loose sense. The study of the wave properties of matter is known as *quantum mechanics*. However, for the discussion in this unit, Figure 5.1 will be sufficient.

In Figure 5.1, note that the electrons exist around the nucleus in particular orbits or *shells*. The electrons in the first shell have the least amount of energy or are most under the influence of the nucleus. For an electron to move to another shell farther away from the nucleus, a certain *quantum* (amount) of energy has to be imparted to it. When enough energy is imparted to an electron, it moves so far away from the nucleus that the nucleus no longer controls it. When an electron is no longer controlled by the nucleus, it is called a *free electron*.

Each shell is able to accommodate a certain maximum number of electrons. The first shell can have a maximum of 2 electrons, the second shell 8 electrons, the third shell 18 electrons, and so on. The outermost shell of an atom is referred to as the *valence band*. Usually only the valence-band electrons are considered in the discussion of whether the element is a conductor, an insulator, or a semiconductor.

When we say conductor, insulator, or semiconductor, we are talking about what the material does electrically. That is, a material that passes current easily is called a *conductor*. It is almost impossible to pass current through an *insulator*. A *semiconductor* is neither a good conductor nor a good insulator.

To study the atomic structure in a little more detail, we have shown the simplest atoms, hydrogen (atomic number 1) and helium (atomic number 2), in Figure 5.2. Hydrogen has only one electron and one proton. The nucleus of helium contains two neutrons in addition to the two protons.

In conductors such as copper, gold, or silver, we have only one valence electron. (See Figure 5.3 for the copper atom.) This one valence electron is so far away from the nucleus of the atom that the nucleus has very little control over it. It is referred to as a *free electron*, because it is not held in a particular location in the element, but is free to move in random directions from one atom to another.

Figure 5.4 illustrates the movement of free electrons in a copper conductor at room temperature. The large circles you see represent the nuclei and all the shells of electrons except the outer shell or valence band. The small black circles with arrows attached represent the free electrons. The free electron around A may change and orbit around B. The free electron of C may orbit around A, and vice versa. These free electrons move in random directions. There is no net flow in one direction.

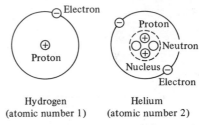

Figure 5.2 Atomic structure of hydrogen and helium

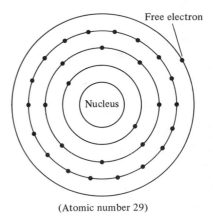

(Atomic number 29)

Figure 5.3 Atomic structure of copper

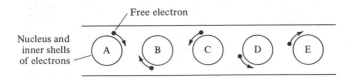

Figure 5.4 Movement of free electrons in a copper conductor (no potential applied)

Figure 5.4 tries to give you a picture of what is happening. But it is very difficult to show you an accurate picture. Figure 5.4 is drawn in two dimensions. However, a piece of copper conductor has three dimensions: length, width, and thickness. And even with an electron microscope, scientists can't see the free electron, it is so small. The mass of an electron is 9.109×10^{-31} kg. In copper at room temperature there are approximately 1.4×10^{24} free electrons per cubic inch.

If we apply an electric potential, such as a battery, across a copper conductor, electrons flow in one direction. Let us consider this in more detail. In Figure 5.5 we have the same copper conductor that was shown in Figure 5.4. But now we have attached negative and positive terminals to the conductor. The negative terminal has an excess of electrons, the positive terminal a deficiency of electrons. The positive or plus terminal attracts the free electron from A. When atom A loses its free electron, it becomes a positive ion. That is, it acts as a positive charge and attracts the free electron from B. When atom B loses its free electron, it becomes a positive ion and attracts the free electron from C.

This continues from atom to atom until we reach the negative terminal. Since the negative terminal has a supply of electrons, an electron leaves the negative terminal and orbits around E. The negative terminal gives the electrons near it a push to the left, because like charges repel each other. If the negative terminal supplies electrons at the same rate as the positive terminal accepts them, then we have a continuous flow of free electrons to the left. We have used only five atoms in a row for this description. Remember that at room temperature there are approximately 1.4×10^{24} free electrons per cubic inch of copper. So the positive terminal attracts and the negative terminal supplies many electrons. Just how many per second depends on the actual potential difference between the positive and negative terminals. This general movement of electrons in one direction is what is known as current.

Before we leave Figure 5.5, let us consider one more point. When we described the movement of the electrons in the conductor, did you note that first atom A was a *positive* ion, then B,

Figure 5.5 Movement of free electrons in a copper conductor (external electric potential applied across conductor)

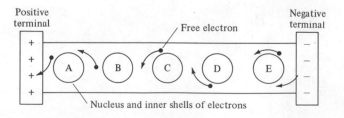

then C, and so on? As the electron or negative charge moved to the left, the movement of positive charge was to the right. No particle moved to the right, but since the positive charge did, it is "as if" a positively charged particle moved.

Problem 5.1   Define the following terms: electron, proton, free electron, conductor, insulator, and semiconductor.

Objective 2   Define the term "quantity of electric charge." The definition must include the basic unit, the number of electrons in one unit, the letter symbol used in mathematical formulas, and the abbreviation or symbol of the unit.

Since electrons are so small and so numerous in conductors, we have to have a whole bunch or quantity of electrons to have any significant electric charge. This quantity of electric charge is measured in *coulombs* (abbreviated by the letter C). One coulomb is the charge carried by 6.24×10^{18} electrons. You may think of the quantity of charge measured in coulombs somewhat as you think of a quantity of water or milk in gallons. When we talk about water, we could speak of the quantity of water in drops; but drops are very small, so we usually use the unit gallons. Therefore the formula for the quantity of charge is

$$Q = \frac{N}{6.24 \times 10^{18}} \tag{5.1}$$

where Q is the quantity of electric charge; the unit for Q is the coulomb

N is the number of electrons

Example 5.1   What is the charge carried by one electron?

Solution   When we have a quantity Q of 6.24×10^{18} electrons, we have a charge of 1 C. Or 1 C is the charge carried by 6.24×10^{18} electrons. The problem asks what charge is carried by *one* electron.

$$Q = \frac{N}{6.24 \times 10^{18}}$$

$$Q = \frac{1 \text{ electron}}{6.24 \times 10^{18} \text{ electrons/coulomb}} = 1.6 \times 10^{-19} \text{ C}$$

Problem 5.2   A coulomb represents the quantity of electric charge carried by how many electrons?

Problem 5.3   What is the charge carried by 1500 electrons?

Current

Objective 3 Define the word *ampere*. The definition must include an explanation of the formula $I = Q/t$. And calculate the current, quantity of electric charge, or time, using the formula $I = Q/t$, when any two of the three quantities are given.

In a closed or complete electric circuit, current is the flow or drift of electric charge past a point in a given time. To be specific, we have a current of one ampere when a quantity of one coulomb of charge flows past a given point in the circuit in one second. This statement is expressed mathematically in Equation (5.2).

$$I = \frac{Q}{t} \tag{5.2}$$

where Q is the quantity of electric charge; the unit for Q is the coulomb

t is the time; the unit for t is the second

I is the current; the unit for I is the ampere

To help clarify Equation (5.2), refer to Figure 5.6. Here we have shown the flashlight circuit we discussed in Unit 1. But we have left out the switch and inserted an ammeter to measure the current, or rate of flow of electrons. You also see some electrons running around the circuit. If the electrons were actually large enough to see, then you could look into the conductor as you see the person doing in Figure 5.6. Then you could count the number of electrons that move past a point in the circuit in one second. If you did this, you could determine the current. As

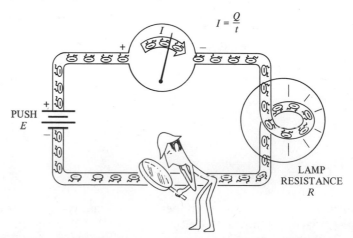

Figure 5.6 Illustration to clarify the term current (detective looking for electrons in a flashlight circuit)

you know, this would be impossible to do. Fortunately, we don't have to try to do this because the ammeter does it for us.

Again we could use a closed water system for an analogy or comparison. Current, which is a rate of flow of electrons (measured in coulombs per second), is similar to the rate of water flow (measured in gallons per second) in a hydraulic system. One coulomb per second is better known as an ampere.

Example 5.2 In an electric circuit, enough electrons are moved past a particular point in 2 min to give a total charge of 40 C. How much current is this?

Solution

$$I = \frac{Q}{t} \quad [\text{Remember that the unit for } t \text{ must be seconds.}]$$

$$= \frac{40}{2(60)} = 0.333 \text{ A} = 333 \text{ mA}$$

Example 5.3 A current of 2 A has been set up in an electric circuit. How long does it take for 4 C of charge to pass a particular point in the circuit?

Solution Using Equation (5.2), we can solve for t.

$$t = \frac{Q}{I} = \frac{4 \text{ C}}{2 \text{ A}} = \frac{4 \text{ C}}{2 \text{ C/s}} = 2 \text{ s}$$

Example 5.4 What would be the current in a wire of an electric circuit if 9.36×10^{18} electrons flowed past a point in the wire in 2 s?

Solution From Equation (5.2), the formula for current is $I = Q/t$. Recalling that 6.24×10^{18} electrons constitutes 1 C of charge, we can determine the quantity (in coulombs) by using Equation (5.1).

$$Q = \frac{N}{6.28 \times 10^{18}} = \frac{9.36 \times 10^{18}}{6.28 \times 10^{18}} = 1.49 \text{ C}$$

$$\therefore \quad I = \frac{Q}{t} = \frac{1.49 \text{ C}}{2 \text{ s}} = 0.745 \text{ A} = 745 \text{ mA}$$

In Figure 5.6, did you notice the electrons we showed in the conductor? They are running from the negative terminal of the battery to the positive terminal. This is the direction of flow of electrons, or *negative* charge carriers, in a conductor. They didn't have electron microscopes back in Ben Franklin's day, so they couldn't see what was actually happening inside conductors. But from external observable effects, they thought that electric current consisted of a flow of *positive* charge carriers.

For example, you notice the ammeters in the simple electric circuit shown in Figure 5.7 deflect to the right, and the more current, the farther to the right the pointer moves. So the direction in the conductor from the positive to the negative terminal on the battery was called the *direction of current flow*. Then all the rules for solving electric circuit problems were based on the direction of flow of positive charge carriers. When it was found that electrons were what actually moved in electric conductors, there wasn't much sense in rewriting all the laws just so they could be stated in opposite terms—especially when they were right for positive charge flow, which actually occurs in many applications.

As you can see in Figure 5.7, the flow of electrons in the circuit is from the negative terminal of the battery to the positive terminal of the battery. The conventional current—or just current—flow in the circuit is from the positive terminal of the battery to the negative terminal of the battery. Well, we have to make a decision as to which direction we want to use. We could use both. In fact, in electronics you will use both positive charge carriers (called *holes* because they are the holes left when electrons move away) and negative charge carriers (electrons) when you study semiconductors. But when you are just learning about a subject, it is best to make it as simple as possible. In circuit diagrams we shall use conventional current. The following are three reasons for using conventional current.

1. Electrical symbols are based on conventional current.
2. All mathematical circuit relationships are based on an assumed conventional current.
3. Engineers use conventional current.

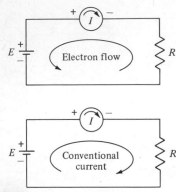

Figure 5.7 Simple electric circuit showing the direction of electron flow and conventional current flow

Problem 5.4 In a simple electric circuit, 240 C of charge passes a particular point in 80 s. What is the current?

Problem 5.5 In a short length of wire, 450 C of charge passes through in 1.5 min. Find the current in amperes.

Problem 5.6 An air conditioner draws 12 A. How many coulombs of charge must pass through it each minute?

Problem 5.7 What is the current if 12.48×10^{17} electrons pass through a wire in 100 ms?

Problem 5.8 In a flashlight circuit, 300 mA of current flows. How long will it take for 2 C of charge to pass through the switch?

Problem 5.9 A fuse is rated at 15 A. Will it blow if 840 C of charge flows through it during 1 min?

Voltage (Potential Difference)

Objective 4 Define the term "potential difference." The definition must include an explanation of the formula V (or E) $= W/Q$. And calculate the potential difference, energy, or quantity of electric charge, using the formula V (or E) $= W/Q$ when two of the three quantities are given or when two of the three quantities may be calculated from the given data.

Voltage is a term that almost everyone has heard, but it is also one of the most difficult terms to really understand. It will become clearer the more you work with electric and electronic circuits. Part of the problem is the term *voltage* itself. It is used to mean electric potential, potential difference, voltage drop, or electromotive force (emf). You might think that voltage can exist at a single point in a circuit; this is incorrect. Voltage must be the potential difference between two points. When "the voltage" for a certain point is given as 60 V, the implication is that there is a 60-V difference between this point and some unnamed zero of potential (usually ground).

To move electrons or charge through a potential difference requires work or energy. Supplying energy to move a charge through a potential difference is somewhat similar to supplying energy to move a body (mass) through a distance. When you connect a battery or other source of emf across a conductor, you are actually connecting an *energy* source.

Energy is the capacity for doing work. The work we want done is to move the electrons or charge through the conductor for a certain period of time. Energy or work is measured in *joules*. The problem is that we don't know what a joule is. But we can get a feel for the unit joule. From physics or mechanics we learn that work is done or energy is expended when a constant force on a body moves it through a certain distance or displacement. Mathematically,

$$W = Fd \qquad (5.3)$$

where W is work or energy; the unit for W is joule (abbreviated J)
F is the force in the direction of the displacement; the unit for F is newton (abbreviated N)
d is the distance or displacement; the unit for d is meter (abbreviated m)

In words, this says that one joule of energy is expended when a force of one newton acts through a distance of one meter. The units newton and meter may be new to you, but we have conversions to change the force in newtons to units of pounds and the distance in meters to units of inches.

$$1 \text{ pound} = 4.45 \text{ newtons}$$

$$1 \text{ meter} = 39.37 \text{ inches}$$

The following examples show you how to use these conversions, work with joules, and find potential energy.

Example 5.5　A person pushes on a desk with a constant force of 10 lb and moves it from position A to position B 10 ft away. How much work has the person done? Refer to Figure 5.8.

Solution

$$F \text{ (in newtons)} \times d \text{ (in meters)} = W \text{ (in joules)}$$

$$\left[10 \text{ lb} \times \frac{4.45 \text{ N}}{1 \text{ lb}} \right]\left[10 \text{ ft} \times \frac{12 \text{ in.}}{1 \text{ ft}} \right]\left[\frac{1 \text{ m}}{39.37 \text{ in.}} \right] = 44.5 \times \frac{120}{39.37}$$

$$= 135.6 \text{ J}$$

Note that the force used in this problem is the force required to keep the desk moving (the force required to overcome sliding friction). The 10 lb is *not* the weight of the desk. This problem did not consider the question of how much force was required to start the desk moving (considering static friction). To get a feel for the unit joule, think of how much work it would be for *you* to push on a desk with a force of 10 lb and move it a distance of 10 ft. Compare this with 135.6 J.

Example 5.6　The top of a building is 100 ft above the ground. A man who weighs 150 lb wants to go to the top. He takes an elevator, but it stalls 60 ft above the ground. He then climbs the steps to get to the top. How much work does the man do in climbing the last 40 ft? How much *potential energy* (energy due to position) with respect to the ground does he have when he is 100 ft above the ground?

Solution　We are assuming that the man is moving vertically with uniform motion from a point that is a distance of 60 ft

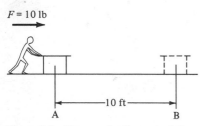

$F = 10$ lb

|← —10 ft— →|
A   B

Figure 5.8　Diagram for Example 5.5

above the reference of ground to a point 100 ft above the ground. From physics,

$$W = mg(y_2 - y_1)$$

where m (mass) times g (acceleration due to gravity) is equal to
  w (weight); the unit is newton
    y is distance from reference; the unit is meter
    W is work; the unit is joule

The man had done

$$W = \left(150 \text{ lb} \times \frac{4.45 \text{ N}}{1 \text{ lb}}\right)\left[(100 - 60) \text{ ft} \times \frac{12 \text{ in.}}{1 \text{ ft}}\right]\left[\frac{1 \text{ m}}{39.37 \text{ in.}}\right]$$

$$= 8138 \text{ J} \qquad \text{of work}$$

When the man is 100 ft above the ground, he has a potential energy of

$$W = \left(150 \text{ lb} \times \frac{4.45 \text{ N}}{1 \text{ lb}}\right)\left[100 \text{ ft} \times \frac{12 \text{ in.}}{1 \text{ ft}}\right]\left[\frac{1 \text{ m}}{39.37 \text{ in.}}\right]$$

$$= 20,345 \text{ J} \qquad \text{with respect to ground}$$

As you saw in Objective 1 and Figure 5.5, to perform the work of moving electrons through a conductor, we need a potential difference across the conductor. A *potential difference* of one volt is said to exist across a conductor when one joule of energy is required to move one coulomb of charge through the conductor.

$$V_{\text{diff}} = \frac{W}{Q} \qquad\qquad (5.4)$$

where V_{diff} is the potential difference (voltage); the unit is volt.
    The subscript diff is generally dropped, and only V
    is used. However, we have included the subscript so
    that you are aware that it is a *difference*. Again E is
    used interchangeably with V.
  W is energy; the unit for W is joule.
  Q is the quantity of charge; the unit for Q is coulomb.

When we measure a voltage in a circuit, whether it is the voltage across the power supply or the voltage across a circuit element such as a resistor, we are actually measuring a potential difference. We are actually asking: What is the potential at one point in the circuit with respect to another point? This other point in the circuit may be a reference point such as ground, but it doesn't have to be ground. Voltage (potential difference) cannot be measured with only one lead. We must always have two points in a circuit. And what we really want to know is: What is the difference between the two points? If one of the points is the circuit reference, then when a voltage is given, it should be understood that it is with respect to this reference.

Talking about voltage or potential difference in an electric circuit is similar to talking about position or heights of objects. For example, if you were standing on the ninth floor (the top floor) of a building whose base is on the ground, you would be nine floors above ground. [See Figure 5.9(a).] In the electric circuit shown in Figure 5.9(a), the plus side of the battery, or point A, has a 9-V potential with respect to ground.

Now let this building still be nine floors high, but let eight floors be below ground. [See Figure 5.9(b).] If you jump from the top floor (the ninth) of this building, how far will you fall? If you open a window and jump outside the building, you will fall only one floor to ground level. But if you jump down the elevator shaft, you will fall nine floors until you hit the bottom of the building.

Therefore, to be able to answer this question, you need to know what we are using as a reference. Is it ground or the bottom of the building? If you are jumping out the window, it makes a difference where the ground is located. But if you are jumping down the elevator shaft, it doesn't make a difference where the

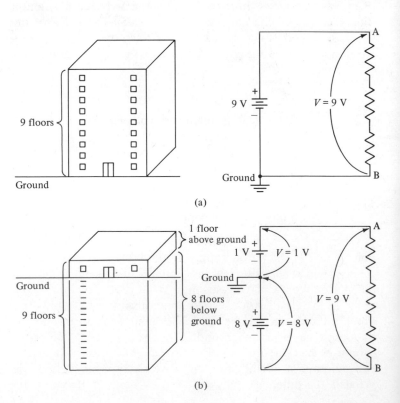

Figure 5.9 Analogy between heights in buildings and voltages in circuits

building is located with respect to the ground. It could be completely on top of the ground, completely below the ground, or somewhere in between. It's the height of the elevator shaft that determines whether you hurt yourself when you jump.

It is similar in an electric circuit. The amount of current that flows through a resistor depends on the voltage (potential difference) *across* the resistor, not just to what one end is connected. In Figure 5.9(b) the voltage (potential difference) across the resistors, point A to point B in the circuit, is still 9 V. However, the voltage at point A with respect to ground is +1 V. The voltage at point B with respect to ground is −8 V.

The following examples will clarify this discussion and the use of Equation (5.4) to solve problems.

Example 5.7   Given that it takes 40 J of energy to move 12.48 × 10^{18} electrons past a point in a circuit, what potential difference must be applied to this circuit?

Solution   The formula for potential difference, Equation (5.4), is

$$V_{\text{diff}} \text{ (in volts)} = \frac{W \text{ (in joules)}}{Q \text{ (in coulombs)}}$$

But the charge wasn't given in coulombs, so this is the first thing we must find. From Equation (5.1),

$$Q = \frac{N}{6.24 \times 10^{18}} = \frac{12.48 \times 10^{18}}{6.24 \times 10^{18}} = 2 \text{ C}$$

$$\therefore \quad V_{\text{diff}} = \frac{W}{Q} = \frac{40 \text{ J}}{2 \text{ C}} = 20 \text{ V}$$

Example 5.8   In an electric circuit, the measured voltage with respect to ground on one side of a resistor is 40 V and on the other side of the resistor is 24 V. What is the potential difference across the resistor?

Solution $V_{\text{diff}} = 40 - 24 = 16$ V. Remember, it is the potential difference across a resistor that determines the current that flows through the resistor.

Example 5.9   It takes 50 J of energy to move 10 C of charge through the resistor in the circuit shown in Figure 5.10. What is the potential difference across the resistor?

Solution

$$V_{\text{diff}} = \frac{W}{Q} = \frac{50 \text{ J}}{10 \text{ C}} = 5 \text{ V}$$

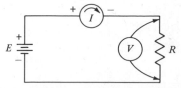

Figure 5.10   The simple electric circuit

Example 5.10 The current in the circuit of Figure 5.10 is 100 mA, and the battery provides 50 J of energy for 100 s of time. What is the potential difference across the resistor?

Solution

$$V_{\text{diff}} = \frac{W}{Q}$$

Q is not given, but information to find Q is given. From Equation (5.2),

$$I = \frac{Q}{t}$$

Multiplying both sides of this equation by t, we get

$$Q = It = (0.1 \text{ A})(100 \text{ s}) = 10 \text{ C}$$

$$\therefore \quad V_{\text{diff}} = \frac{W}{Q} = \frac{50 \text{ J}}{10 \text{ C}} = 5 \text{ V}$$

Don't forget the method of dissecting a problem given in Unit 3. It may help you solve the problems in this unit. We shall illustrate it again in the next example.

Example 5.11 In the circuit of Figure 5.10, given that R is equal to 10 Ω and that the battery E provides an emf (or applied potential) of 3 V for 1 min, how much energy has the battery expended?

                                                          R
Solution   In the circuit of Figure 5.10, given that <u>R is equal to</u>

<u>10 Ω</u> and the battery E provides an <u>emf</u> (or applied potential)
  E   *t*   *W* = ?
of <u>3 V</u> for <u>1 min</u>, <u>how much energy</u> has the battery expended?
    Looking at the information provided, it seems at first that we don't have enough to solve it. But let us write the two formulas for current and potential difference given in the unit, and see if we can find a way to start.

Equation (5.2) $I = \dfrac{Q}{t}$ Equation (5.4) $V_{\text{diff}} = \dfrac{W}{Q}$

The unknown in the problem is W, so using Equation (5.4) and multiplying both sides of the equation by Q, we find that

$$W = VQ$$

Looking again at the information given in the problem, we see that we have V, but not Q directly. However, using Equation (5.2) and multiplying both sides by t, we find that

$$Q = It$$

To find I, we may use Ohm's law:

$$I = \frac{E}{R} = \frac{3 \text{ V}}{10 \text{ }\Omega} = 0.3 \text{ A}$$

Now we can find Q:

$$Q = It = (0.3 \text{ A})(60 \text{ s}) = 18 \text{ C}$$

And finally W:

$$W = VQ = (3 \text{ V})(18 \text{ C}) = 54 \text{ J}$$

Example 5.12 The battery E in the circuit of Figure 5.10 has an emf of 3 V and is pushing 30 mA of current in the circuit. Find the rate at which chemical energy of the battery is being converted to electrical energy.

Solution The rate here is *energy per unit time* (W/t). You will learn in Unit 6 that this is *power*.

Solving Equation (5.4) for Q, we have

$$Q = \frac{W}{V}$$

Substitute this equation for Q into Equation (5.2).

$$I = \frac{Q}{t} = \frac{W/V}{t}$$

To get rid of the V on the right-hand side of the equation, multiply both sides of the equation by V:

$$VI = \frac{W}{t} \quad \text{or} \quad \frac{W}{t} = VI = (3 \text{ V})(0.03 \text{ A}) = 0.09 \text{ J/s}$$

Problem 5.10 In the circuit shown in Figure 5.10, it takes 1500 J of energy to move 500 C of charge through the resistor. What is the potential difference across the resistor?

Problem 5.11 In a circuit, it takes 36 J of energy to move 24.96×10^{18} electrons past a point. What voltage (potential difference) is applied?

Problem 5.12 For a circuit containing several resistors, we want to know the voltage drop across a particular resistor. The measured voltage with respect to the circuit reference voltage is 28 V on one side of the resistor and 19 V on the other side. What is the voltage drop across the resistor?

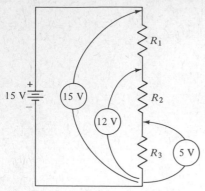

Figure 5.11   Circuit for Problem 5.13

Problem 5.13   In the circuit shown in Figure 5.11, find the potential difference across resistor R_1, the potential difference across resistor R_2, and the potential difference across both resistors R_1 and R_2.

Problem 5.14   In the simple electric circuit of Figure 5.10, E is equal to 15 V and R is equal to 1 kΩ. How much energy is required for current to flow for 5 min?

Problem 5.15   In the circuit of Figure 5.10, E is equal to 12 V. How much current flows if 600 J of energy is used for 3 min?

Objective 5   Discuss briefly the types of electromotive force.

In Objective 1 and in Figure 5.5, we talked about applying an electric potential difference across the ends of a conductor. The purpose of this potential difference is to force the electrons to move generally in one direction. This potential difference is called an *emf*, that is, an electromotive or *electron-moving force*. There are several sources of emf.

Chemical (battery or cell)   Common batteries are those used in flashlights, portable radios and televisions, appliances, toys, and automobiles. (See Figure 5.12.) In batteries a chemical reaction

Figure 5.12   Various kinds of batteries (Courtesy Union Carbide)

takes place that forces an excess of electrons on one terminal and a deficiency of electrons on the other terminal. So chemical energy is converted into electric potential energy.

Electromagnetic (generator)   Most homes and industries use electric power that is generated by rotary electromagnetic generators. So generators are machines that convert mechanical power into electric power.

Thermoelectric (heat)   An example of this is a thermocouple. Certain dissimilar metals—when joined together and then heated—generate an emf across their junction.

Photoelectric (light)   Certain materials generate an emf when exposed to light.

Piezoelectric (transducers)   Certain types of natural crystals and manufactured ceramic materials—when subjected to physical force or pressure—generate an emf.

The forms of emf we use in electric circuits are chemical (in the form of batteries), electromagnetic, and electronic power supplies. The electronic power supply changes the generated ac emf into a dc emf by means of an electronic circuit.

Problem 5.16   Name four sources of emf.

Test

1. A coulomb represents the quantity of electric charge carried by how many electrons?
2. What is the charge carried by 10 electrons?
3. What is the current in a wire of an electric circuit if 9.36×10^{16} electrons flow past a point in the wire in 1 s?
4. Given that it takes 5 s for 25 C of electric charge to flow through a switch, what is the current in the switch?
5. How many coulombs of electric charge must flow past a point in a circuit in 2 s to achieve a current of 100 mA?
6. The measured voltage with respect to ground on one side of a resistor is 10 V and on the other side of the resistor is 6 V. What is the potential difference across the resistor?
7. It takes 36 J of energy to move 18.72×10^{18} electrons past a point in a circuit. What potential difference is applied to this circuit?

8. An automobile battery is an energy source. Obviously, like any other source, it can be depleted. Suppose that 3×10^6 J of chemical energy have been stored in a 12-V battery, and that that battery is then used in a circuit to supply a current of 5 A. How long can the battery last at this rate of depletion?

9. Suppose that a 1.5-V battery is finally depleted (or exhausted) when it has caused the flow of enough charge to total 2 C. What must its initial potential energy have been?

Resistance Unit 6
and Power

In the circuit diagrams in Unit 1—particularly in Figure 1.5—there are many resistors. Some have the power rating listed next to them. If they do not, it is understood that they are $\frac{1}{2}$-W resistors. What is a resistor, and what is the property of resistance? How do you determine the resistance of an element or a circuit by making electrical measurements? How does resistance vary if physical dimensions, material, and temperature change? What is power? Why do resistors and all electric and electronic circuit elements have power ratings? Why is your electric bill determined by the power you use and the amount of time you use it?

These are just some of the questions you may have about resistance and power. You will find the answers to these questions and more as you complete this unit.

Objectives

After completing all the work associated with this unit, you should be able to:

1. State when a circuit element has a resistance of one ohm.
2. List the three physical factors that determine the resistance of metal conductors, and relate these factors in a mathematical formula. Given the resistance of a metal conductor, determine the new resistance if one of the three physical factors changes.

3. Calculate the resistance of a metal conductor of any size using the units of specific resistance given.

4. (*Advanced*) When given the conversion factor, convert the specific resistance of a conductor 1 mil in diameter and 1 ft long (in units of Ω-CM/ft) to that of a conductor of 1 m^2 cross-sectional area by 1 m long (in units of Ω-m), and vice versa.

5. Calculate the resistance of any AWG (American Wire Gauge) size conductor at 20°C, given an AWG wire table and the specific resistance of the conductor.

6. Calculate the resistance of materials at temperatures other than 20°C, given the temperature coefficient of resistance for the material.

7. Determine the conductance of a material, given its resistance.

8. Discuss the common types of resistors; that is, give the various resistors' names and general characteristics.

9. Define the term *power*. The definition must include an explanation of the equation $P = W/t$, using both joules and kilowatt-hours (kWh) to measure work W.

10. Calculate power, voltage, current, or resistance, using the appropriate form of the power equation ($P = EI$, $P = I^2R$, or $P = E^2/R$), when any two of the three quantities in the equation are given or can be found from the information given.

11. Calculate the efficiency η (Greek letter eta) of an electrical system when input and output power or energy are given. Use 1 hp = 746 W when required.

Definition of Resistance

Objective 1   State when a circuit element has a resistance of one ohm.

In Unit 4, you learned Ohm's law, which gives this definition: One ohm of resistance is that resistance which causes one ampere of current to flow when the applied potential is one volt.

$$R = \frac{V}{I} \qquad (6.1)$$

where R is the resistance of the circuit element or load in question; the unit for R is ohm

V is the voltage (potential difference) across the circuit element or load in question; the unit for V is volt

I is the current through the circuit element or the current into the load; the unit for I is ampere

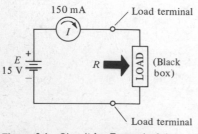

Figure 6.1   Circuit for Example 6.1

Example 6.1   In the circuit shown in Figure 6.1, what is the resistance across the load terminals? The circuit load is shown as a box (black box). This box may contain only one resistor or many circuit elements.

Solution From Equation (6.1)

$$R = \frac{V}{I}$$

The voltage across the load is the same as the power-supply voltage.

$$R = \frac{V}{I} = \frac{15\ V}{150\ mA} = 100\ \Omega$$

Problem 6.1 When a potential difference of 3 V is applied across the terminals of a lamp, 200 mA of current flows through the lamp. What is the resistance of the lamp?

Problem 6.2 A current of 5 mA flows when a 10-V power supply is connected across an unknown resistor. What is the resistance of the resistor?

Physical Factors that Determine Resistance

Objective 2 List the three physical factors that determine the resistance of metal conductors, and relate these factors in a mathematical formula. Given the resistance of a metal conductor, determine the new resistance if one of the three physical factors changes.

In Objective 1 you learned how to determine the resistance of a circuit element or circuit load electrically. That is, you measure the current that flows in the element when you apply a known potential across its terminals, and then you use Ohm's law. But to do this electrically, you must have the circuit element or conductor. If you want to make a resistor, or determine the resistance of the connecting wires in a circuit without measurement, then you must know how the physical factors of the material that you are using affect resistance.

There are three physical factors that affect the resistance of metal conductors. They are the resistance of the material itself (ρ, the Greek letter rho), the length of the material (l), and the cross-sectional area of the conductor (A).

$$R = \frac{\rho l}{A} \tag{6.2}$$

The units used for l and A in the formula depend on the units used to measure the specific resistance of the basic material. All the units except ohms must cancel, leaving the resistance in ohms. This is discussed in detail in Objective 3.

From Equation (6.2), we see that resistance is proportional to the length of the conductor. If the length of the conductor increases, then the resistance must increase in the same proportion. If only the length of a conductor changes, we may find the new resistance by multiplying the old resistance by the proportion of the *change* in length.

$$R_2 = R_1 \frac{l_2}{l_1} \tag{6.3}$$

where R_1 and l_1 are the given resistance and length and R_2 and l_2 are the new resistance and length.

We can find Equation (6.3) by taking a ratio of Equation (6.2) for the new case, call it case 2 (use subscript 2), to Equation (6.2) for the first case (use subscript 1). Do not put a subscript on the factors that do not change from case 1 to case 2.

$$\frac{R_2 = \frac{\rho l_2}{A}}{R_1 = \frac{\rho l_1}{A}} \qquad \therefore \qquad \frac{R_2}{R_1} = \frac{\frac{\rho l_2}{A}}{\frac{\rho l_1}{A}}$$

$$\frac{R_2}{R_1} = \frac{l_2}{l_1} \qquad \text{or} \qquad R_2 = R_1 \frac{l_2}{l_1}$$

Example 6.2 A copper conductor 10 ft long has a resistance of 20 Ω. What is the resistance of 100 ft of this conductor?

Solution

$$R_2 = R_1 \frac{l_2}{l_1} = 20 \ \Omega \left(\frac{100 \ \text{ft}}{10 \ \text{ft}} \right) = 200 \ \Omega$$

Problem 6.3 An aluminum conductor 100 ft long has a resistance of 60 Ω. What is the resistance of 25 ft of this conductor?

From Equation (6.2), we know that resistance is inversely proportional to the cross-sectional area of the conductor. If the cross-sectional area of the conductor increases, then the resistance must decrease in the same proportion. If only the cross-sectional area of the conductor changes, we may find the new resistance by dividing the old resistance by the proportion of the *change* in area.

$$R_2 = \frac{R_1}{A_2/A_1}$$

$$\therefore \quad R_2 = R_1 \frac{A_1}{A_2} \tag{6.4}$$

where R_1 and A_1 are the given resistance and cross-sectional area, and R_2 and A_2 are the new resistance and cross-sectional area.

If the conductor is circular in cross section, Equation (6.4) may be simplified.

$$R_2 = R_1 \frac{A_1}{A_2} = R_1 \frac{\pi \frac{d_1^2}{4}}{\pi \frac{d_2^2}{4}}$$

Circular conductors: $\quad R_2 = R_1 \left(\frac{d_1}{d_2}\right)^2 \tag{6.5}$

(Here d represents diameter.)

Example 6.3   An aluminum conductor 0.5 in. in diameter has a resistance of 30 Ω. The length of this aluminum conductor is not changed, but its diameter is decreased to 0.25 in. What is its resistance?

Solution   Since the conductor has a circular cross section, we may use Equation (6.5) to find the new resistance.

$$R_2 = R_1 \left(\frac{d_1}{d_2}\right)^2 = 30 \left(\frac{0.5}{0.25}\right)^2 = 30(2)^2 = 120 \, \Omega$$

Problem 6.4   A copper conductor with a cross-sectional area of 0.1 in^2 has a resistance of 5 Ω. The length of this copper conductor is not changed, but its cross-sectional area is increased to 0.3 in^2. What is its resistance?

Problem 6.5   A copper conductor has a resistance of 16 Ω. The length of this copper conductor is not changed, but its diameter is doubled. What is its resistance?

If both the length and the cross-sectional area of a conductor change, we may combine the results found in Equations (6.3) and (6.4).

$$R_2 = R_1 \frac{l_2}{l_1} \left(\frac{A_1}{A_2}\right) \tag{6.6}$$

For circular conductors, Equation (6.6) becomes

$$R_2 = R_1 \frac{l_2}{l_1} \left(\frac{d_1}{d_2}\right)^2 \tag{6.7}$$

Example 6.4   A copper conductor 500 ft long with a cross-sectional area of 0.002 in^2 has a resistance of 2.6 Ω. What is the resistance of 1 mi of copper conductor with a cross-sectional area of 0.005 in^2?

Solution   Use Equation (6.6) and also the conversion that 1 mi = 5280 ft.

$$R^2 = 2.6 \; \Omega \left(\frac{5280 \text{ ft}}{500 \text{ ft}}\right)\left(\frac{0.002 \text{ in}^2}{0.005 \text{ in}^2}\right) = 11 \; \Omega$$

Note that all the units except ohms cancel.

Problem 6.6   An aluminum wire 100 ft long with a diameter of 0.02 in. has a resistance of 4.25 Ω. What length of aluminum wire with a 0.04-in. diameter has a resistance of 8.0 Ω?

Resistivity or Specific Resistance

Objective 3   Calculate the resistance of a metal conductor of any size using the units of specific resistance given.

Did you notice that, in all the examples and problems under Objective 2, the resistance of a certain size conductor was given, and then you were asked, "What is the resistance of a conductor of the same material but different dimensions?" If you didn't note this, go back and look. The problem of determining the resistance of conductors would be simplified if the resistance of various conductors of *standard dimensions* were measured and tabulated. This has been done; see Table 6.1.

Table 6.1 also shows a sketch of the size conductor that was used. The second conductor, shown under the ohm-meter column, looks strange. It is very difficult to imagine a conductor with a cross-sectional area of one square meter and a length of one meter. These dimensions were chosen because the international system of units (SI)* uses meters for the unit of length. Therefore many handbooks give the specific resistance of materials in ohm-meters. In Objective 4, you will learn how to convert the units of ohm-meters into the units of ohm-CM/ft, and vice versa.

* The letters SI are short for *Système Internationale*. It means International System, and we call it the SI System because it was the French who first established it in popular use.

Table 6.1   Specific Resistance or Resistivity (ρ) at 20°C

Material	Specific Resistance	
	Ω-CM/FT	OHM-METERS
Silver	9.9	
Copper	10.37	1.724×10^{-8}
Gold	14.7	
Aluminum	17	2.83×10^{-8}
Tungsten	33	
Brass	42	
Nickel	47	
Iron	58	
Tin	69	
Manganin	271	
Constantan	295	
Nichrome II	660	

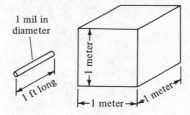

The units for the specific resistance of

$$\frac{\Omega\text{-CM}}{\text{ft}}$$

that is, ohms-circular mil per foot, may also be new to you. (We use capital CM to avoid confusion with lower-case cm, meaning centimeter.) The conductor has a circular cross section of 1 mil diameter (1 mil = 0.001 inch or 1 inch = 1000 mils) and a length of 1 foot. The specific resistance of a conductor of these dimensions is usually given as so many ohms per circular mil-foot. The circular mil-foot describes the cross section (circular), the diameter (1 mil), and the length (1 foot) of the conductor. The actual unit used for the specific resistance or *resistivity* of a material is

$$\frac{\Omega\text{-CM}}{\text{ft}}$$

so that when it is used in Equation (6.2), all units except ohms cancel.

What does the standardizing of a basic conductor size do for us? If we substitute 1 m for length and 1 m^2 for the cross-sectional area in Equation (6.6), we obtain

$$R_2 = R_1 \frac{l_2}{1}\frac{1}{A_2} \qquad \text{or} \qquad R = \frac{\rho l}{A} \qquad (6.8)$$

where ρ is the given resistance (specified as specific resistance or resistivity) of a conductor 1 m long with a cross section of 1 m^2; the unit is ohm-meter

R is the new resistance or resistance of the conductor under question; the unit is ohm

l is the length of the conductor; the unit is meter

A is the cross-sectional area of the conductor; the unit is square meter

This equation applies to all conductors. If we have a circular conductor, we want to substitute 1 ft for length and 1 mil for the diameter in Equation (6.7).

$$R_2 = R_1 \frac{l_2}{1}\frac{1}{d_2^2} \qquad \text{or} \qquad R = \frac{\rho l}{d^2} \qquad (6.9)$$

where ρ is the specific resistance of a conductor 1 ft long with a circular cross section of 1 mil diameter; the unit is Ω-CM/ft

R is the resistance of the conductor; the unit is ohm

l is the length of the conductor; the unit is foot

d is the diameter of the conductor; the unit is mil

This equation applies to circular conductors.

For circular conductors, comparing Equation (6.9) with Equation (6.2), we find the cross-sectional area by squaring the diameter, which is expressed in mils. We know that this isn't the same as finding the area of a circle. The equation for the area of a circle is

$$A = \frac{\pi d^2}{4},$$

with units of *square* mils if d is expressed in mils. So that we don't get confused with the two areas, we use units of *circular* mils for the area required in Equation (6.9).

$$A \text{ (area in circular mils)} = d^2 \qquad (6.10)$$

where d is the diameter of the conductor; the unit is mil.

The only time you may use Equation (6.10) for area is when you use Equation (6.9), where the conductor is circular in cross section and ρ is expressed in ohms for a circular mil-foot.

To see how Equations (6.8) and (6.9) can help you determine the resistance of conductors, study Examples 6.5, 6.6, 6.7, and 6.8.

Example 6.5 What is the resistance of 200 ft of copper wire 10 mils in diameter? From Table 6.1,

$$\rho = 10.37 \frac{\Omega\text{-CM}}{\text{ft}}$$

Solution The conductor is circular in cross section and ρ is given in units of Ω-CM/ft; therefore we use Equation (6.9) for resistance.

$$R = \frac{\rho l}{d^2} = \frac{(10.37)(200)}{(10)(10)} = 20.74 \ \Omega$$

Example 6.6 What is the resistance of a square aluminum conductor 0.005 m on a side and 500 m long? (Resistivity of aluminum is $2.83 \times 10^{-8} \ \Omega$-m.)

Solution We must use Equation (6.8) for resistance. The area of a square:

$$A = (0.005)(0.005) = 25 \times 10^{-6} \ \text{m}^2$$

$$R = \frac{\rho l}{A} = \frac{(2.83 \times 10^{-8})(5 \times 10^2)}{25 \times 10^{-6}}$$

$$= 5.66 \times 10^{-1} \ \Omega = 0.566 \ \Omega$$

Example 6.7 What is the resistance of a rectangular aluminum conductor $\frac{1}{4}$ in. wide by 2 in. high and 100 ft long? (Resistivity of aluminum is $2.83 \times 10^{-8} \ \Omega$-m.)

Solution Since the specific resistance or resistivity of aluminum is given in ohm-meters, the length of the conductor must be expressed in meters and the cross-sectional area in square meters.

$$l \text{ (in meters)} = 100 \text{ ft} \times \frac{1 \text{ m}}{39.37 \text{ in.}} \times \frac{12 \text{ in.}}{1 \text{ ft}} = 30.5 \text{ m}$$

$$A \text{ (in square meters): } (\tfrac{1}{4} \text{ in.})(2 \text{ in.}) = 0.5 \text{ in}^2 \left(\frac{1 \text{ m}}{39.37 \text{ in.}}\right)^2$$

$$= \frac{0.5}{(39.37)^2} \text{ m}^2$$

$$= 3.23 \times 10^{-4} \text{ m}^2$$

$$R = \frac{\rho l}{A} = \frac{(2.83 \times 10^{-8})(30.5)}{3.23 \times 10^{-4}} = 2.67 \times 10^{-3} \ \Omega$$

Example 6.8   What is the resistance of an aluminum circular conductor $\frac{1}{2}$ in. in diameter by 100 ft long? (Resistivity of aluminum is $2.83 \times 10^{-8} \ \Omega$-m.)

Solution   Even though we have a conductor of circular cross section, since ρ was specified in ohm-meters, we must use Equation (6.8).

$$A \text{ (in square meters)} = \frac{\pi d^2}{4} = \frac{\pi (0.5 \text{ in.})^2}{4} = 0.1965 \text{ in}^2 \left(\frac{1 \text{ m}}{39.37 \text{ in.}}\right)^2$$

$$= 1.27 \times 10^{-4} \text{ m}^2$$

l (in meters) from Example 6.7 is 30.5 m.

$$R = \frac{\rho l}{A} = \frac{(2.83 \times 10^{-8})(30.5)}{1.27 \times 10^{-4}} = 6.8 \times 10^{-3} \ \Omega$$

If the resistivity had not been specified and given in ohm-meters, we could have used Table 6.1 and found 17 Ω-CM/ft. With resistivity in these units and a circular conductor, we could use Equation (6.9). The diameter must be specified in mils,

$$d = \tfrac{1}{2} \text{ in.} = 0.5 \text{ in.} = 500 \text{ mils} \qquad \text{(move decimal point 3 places to right)}$$

$$R = \frac{\rho l}{d^2} = \frac{17(100)}{(500)(500)} = 6.8 \times 10^{-3} \ \Omega.$$

The solution using Equation (6.9) was much easier.

Problem 6.7   What is the resistance of 400 ft of copper wire 20 mils in diameter? Use Table 6.1 to find ρ.

Problem 6.8   What is the resistance of a square copper conductor 0.3 cm on a side and 900 m long? (100 cm = 1 m) Use Table 6.1 to find ρ.

Problem 6.9   What length of silver wire with a diameter of 10 mils has a resistance of 0.1 Ω? Use Table 6.1 to find ρ.

Problem 6.10   What diameter of tungsten wire that is 2 in. long has a resistance of 10 Ω? Use Table 6.1 for ρ.

Problem 6.11   What is the resistance of a copper rectangular conductor $\frac{1}{8}$ in. wide by 2 in. high and 200 ft long? Use Table 6.1 for ρ.

Problem 6.12   What is the resistance of 600 m of copper wire 0.002 m in diameter? Use Equation (6.8).

Conversion Between Units of Resistivity

Objective 4 (*Advanced*)   When given the conversion factor, convert specific resistance of a conductor 1 mil in diameter 1 ft long (in units of Ω-CM/ft) to that of a conductor of 1 m^2 cross-sectional area by 1 m long (in units of Ω-m), and vice-versa.

For conversion of cross-sectional areas:

$$A_{\text{square}}: \quad 1 \text{ m}^2 = (39.37)^2 \text{ in}^2$$

$$A_{\text{circle}}: \quad \frac{\pi}{4}\, d^2 = \frac{\pi}{4}\,(0.001)^2 \text{ in}^2$$

$$\text{Ratio of cross-sectional areas} = \frac{\text{area of 1 m}^2}{\text{area of circle 1 mil in dia.}}$$

$$= \frac{(39.37)^2 \text{ in}^2/\text{m}^2}{\frac{\pi}{4}\,(0.001)^2 \text{ in}^2/\text{CM}}$$

For conversion of length:

$$1 \text{ meter} = 39.37 \text{ inches}, \qquad 1 \text{ foot} = 12 \text{ inches}$$

Therefore

$$\frac{1\,\Omega\text{-CM}}{\text{ft}} \times \frac{39.37 \text{ in.}}{1 \text{ m}} \times \frac{1 \text{ ft}}{12 \text{ in.}} \times \frac{\pi\,(0.001)^2 \text{ m}^2}{4\,(39.37)^2 \text{ CM}} = 1.662 \times 10^{-9}\,\Omega\text{-m}$$

If we are given the specific resistance or resistivity in Ω-CM/ft, we can convert it to units of ohm-meters by multiplying it by 1.662×10^{-9}.

$$\rho(\text{ohm-meter}) = \rho\left(\frac{\text{ohm-CM}}{\text{ft}}\right) \times 1.662 \times 10^{-9}$$

Use the following conversion to convert from a resistivity given in ohm-meters to the units of Ω-CM/ft:

$$\rho\left(\frac{\text{ohm-CM}}{\text{ft}}\right) = \frac{\rho(\text{ohm-meter})}{1.662 \times 10^{-9}}$$

Example 6.9 The resistivity of copper is 10.37 Ω-CM/ft. What is copper's resistivity in ohm-meters?

$$\rho(\text{ohm-meter}) = \rho\left(\frac{\text{ohm-CM}}{\text{ft}}\right) \times 1.662 \times 10^{-9}$$

$$= 10.37 \times 1.662 \times 10^{-9} = 1.724 \times 10^{-8} \;\Omega\text{-m}$$

Problem 6.13 In Table 6.1, convert the resistivity in Ω-CM/ft for all the materials into ohm-meters.

Wire Table

Objective 5 Calculate the resistance of any AWG size conductor at 20°C, given an AWG wire table and the specific resistance of the conductor.

To standardize conductor sizes, North American wire manufacturers use cross-sectional dimensions that conform to the American Wire Gauge (AWG) sizes given in Table 6.2. At one end of the table you see 0000 wire (pronounced four-naught), which is almost half an inch in diameter (actually $d = 460$ mils). At the other end, you see No. 40 wire, whose diameter is 3.145 mils. Number 14 is the most common size for house wiring, with a growing trend toward Number 12.

As you look over Table 6.2, note that as the AWG number increases, the diameter of the wire decreases. Moving up the table by 3 AWG numbers decreases the area in circular mils by a factor of 2 and increases the resistance by a factor of 2. Moving up the table by 10 AWG numbers decreases the area and increases the resistance by approximately a factor of 10. For example, a No. 10 AWG wire has an area of 10,380 circular mils and a resistance at 20°C of 0.999 Ω/1000 ft, whereas a No. 20 AWG wire has an area of 1022 circular mils and a resistance at 20°C of 10.2 Ω/1000 ft. Example 6.10 will show you how to calculate the resistance values for copper wire at 20°C listed in Table 6.2.

Example 6.10 What is the resistance of 1000 ft of No. 24 AWG copper wire?

Table 6.2 American Wire Gauge (AWG) Conductor Sizes

AWG Number	Diameter d, Mils	Area d^2, Circular Mils	Ω/1000 ft (copper) @ 20°C	Ω/1000 ft (copper) @ 65°C
0000	460.0	211,600	0.049	0.058
000	409.6	167,800	0.062	0.073
00	364.8	133,100	0.078	0.092
0	324.9	105,500	0.098	0.116
1	289.3	83,690	0.124	0.146
2	257.6	66,370	0.156	0.184
3	229.4	52,640	0.197	0.232
4	204.3	41,740	0.248	0.292
5	181.9	33,100	0.313	0.369
6	162.0	26,250	0.395	0.465
7	144.3	20,820	0.498	0.586
8	128.5	16,510	0.628	0.739
9	114.4	13,090	0.792	0.932
10	101.9	10,380	0.999	1.18
11	90.74	8,234	1.26	1.48
12	80.81	6,530	1.59	1.87
13	71.96	5,178	2.00	2.36
14	64.08	4,107	2.53	2.97
15	57.07	3,257	3.18	3.75
16	50.82	2,583	4.02	4.73
17	45.26	2,048	5.06	5.96
18	40.30	1,624	6.38	7.51
19	35.89	1,288	8.05	9.48
20	31.96	1,022	10.2	11.9
21	28.46	810.1	12.8	15.1
22	25.35	642.4	16.1	19.0
23	22.57	509.5	20.4	24.0
24	20.10	404.0	25.7	30.2
25	17.90	320.4	32.4	38.1
26	15.94	254.1	40.8	48.0
27	14.20	201.5	51.5	60.6
28	12.64	159.8	64.9	76.4
29	11.26	126.7	81.8	96.3
30	10.03	100.5	103	121
31	8.928	79.7	130	153
32	7.950	63.21	164	193
33	7.080	50.13	207	243
34	6.305	39.75	261	307
35	5.615	31.52	329	387
36	5.000	25.00	415	488
37	4.453	19.83	523	616
38	3.965	15.72	660	776
39	3.531	12.47	832	979
40	3.145	9.888	1049	1230

Solution   From Table 6.2, for No. 24 wire, $d = 20.10$ mils and $d^2 = 404.0$ circular mils.   From Equation (6.9),

$$R = \frac{\rho l}{d^2} = \frac{10.37(1000)}{404} = 25.7\ \Omega$$

Problem 6.14   An electrical outlet is to be installed for an air conditioner.   Number 12 copper wire is to be used from the fuse box in the basement to the outlet on the second floor.   What is the resistance of this wire for each 10 ft of length?

Problem 6.15   What length of No. 40 AWG manganin wire has a resistance of 50 Ω?

Effect of Temperature on Resistance

Objective 6   Calculate the resistance of materials at temperatures other than 20°C, given the temperature coefficient of resistance for the material.

Up to this point we have not considered variation of temperature.   An increase in temperature is an increase in energy. The increase in energy may cause more movement of atoms and electrons in the material, with the result that there are more collisions and more interference as the free electrons move through the material.   If this is the case, the resistance of the material increases.

In some materials, the increase of energy due to an increase in temperature actually increases the number of free electrons.   In this case, the resistance decreases as temperature increases.

Most materials change their resistance with temperature in a linear manner.   Since this is true, we may determine a *temperature coefficient of resistance* for each material; see Table 6.3.

The temperature coefficient of resistance, known as α (Greek letter alpha), gives the change in resistance per degree Celsius above 20°C.   To determine the resistance at some temperature other than 20°C, we use the following formula.

$$R_2 = R_1(1 + \alpha\,\Delta T) \tag{6.11}$$

where R_2 = resistance of the material at any specified temperature

R_1 = resistance of the material at 20°C

α = temperature coefficient of resistance at 20°C

ΔT = change in temperature from 20°C

If the resistance at 20°C is not known, it may be calculated from Equation (6.8) or (6.9).

Table 6.3   Temperature Coefficient of Resistance for Various Materials at 20°C

Material	20°C
Manganin	0.000
Constantan	0.000008
Nichrome II	0.0002
Brass	0.002
Platinum	0.003
Silver	0.0038
Aluminum	0.0039
Copper	0.00393
Tin	0.0042
Tungsten	0.0045
Iron	0.0055
Nickel	0.006
Carbon	−0.0005

Example 6.11 The resistance of copper wire is 60 Ω at 20°C. What is its resistance at 40°C?

Solution

$$\Delta T = 40°C - 20°C = 20°C$$
$$R_2 = R_1(1 + \alpha \, \Delta T)$$
$$= 60[1 + (0.00393)20] = 60 + 4.7 = 64.7 \, \Omega$$

Example 6.12 What is the resistance of 200 ft of aluminum wire 10 mils in diameter at 30°C?

Solution First we must find the resistance of this wire at 20°C. Since the wire has a circular cross section, and dimensions were given in feet and mils, we may use Equation (6.9).

$$R_1 = \frac{\rho l}{d^2} = \frac{17(200)}{10(10)} = 34 \, \Omega$$

$$\Delta T = 30°C - 20°C = 10°C$$

$$R_2 = R_1(1 + \alpha \, \Delta T)$$
$$= 34[1 + (0.0039)10] = 34 + 1.3 = 35.3 \, \Omega$$

Problem 6.16 The resistance of silver wire is 4 Ω at 20°C. What is its resistance at 35°C?

Problem 6.17 The resistance of carbon wire is 50 Ω at 20°C. What is its resistance at 30°C?

Problem 6.18 What is the resistance of 150 ft of No. 10 AWG copper wire at 45°C?

Conductance

Objective 7 Determine the conductance of a material, given its resistance.

Instead of considering the resistance or opposition of a material to the flow of current, we may consider the material's ability to pass current, or the ease with which it *conducts* current. This ability to pass current is known as *conductance*. The higher the conductance, the larger the current. So conductance is the reciprocal of resistance.

$$G = \frac{1}{R} \tag{6.12}$$

The letter symbol for conductance is *G*. The SI unit for conductance is the *siemens* (abbreviated S, named after an inventor by that name). (The millisiemens is abbreviated mS; the microsiemens is abbreviated μS.)

However, many engineering personnel still use the unit that was used before the siemens was adopted. The unit that people used to use to measure conductance was the mho (ohm spelled backward); the symbol for it is the inverted Greek letter omega, $\mho$.

Example 6.13   What is the conductance of a 5-kΩ resistor?

Solution

$$G = \frac{1}{R} = \frac{1}{5 \text{ k}\Omega} = 0.2 \text{ mS} = 200 \text{ } \mu S$$

Problem 6.19   What is the conductance of a 3.9-kΩ resistor?

Being able to think in terms of conductance is helpful when we are considering parallel circuits. You will see this in Unit 8. You will be able to add the conductances of several branches, just as you will add the resistance of several resistors in a series circuit.

In Objectives 3 and 4 we discussed specific resistance or resistivity ρ. Using the concept of conductance, we may find the specific conductance or conductivity of a material.

$$\gamma = \frac{1}{\rho} \tag{6.13}$$

where ρ is the resistivity of the material; units are ohm-meters or ohms for a circular mil-foot (Ω-CM/ft)

γ (Greek lower-case letter gamma) is the symbol for conductivity; units are siemens per meter or siemens for a circular mil-foot (S-ft/CM)

Table 6.1 could be converted into a table of conductivities by using Equation (6.13).

Common Types of Resistors

Objective 8   Discuss the common types of resistors; that is, give the various resistors' names and general characteristics.

Back in Unit 1, we said that a resistor is a circuit element that is manufactured to have a certain resistance. It may be fixed or variable and comes in many sizes, shapes, and forms. Let us discuss this in more detail. Some of the resistors you will find are the following.

Molded-carbon composition resistor   These resistors are made of finely ground carbon mixed with a filler or binder. The compressed mixture is enclosed in a hard protective case of plastic or ceramic material. (See Fig. 6.2.) These resistors have wattage

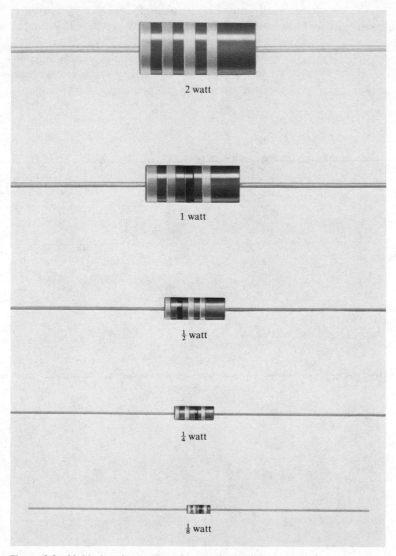

2 watt

1 watt

$\frac{1}{2}$ watt

$\frac{1}{4}$ watt

$\frac{1}{8}$ watt

Figure 6.2   Molded-carbon composition resistors (Photo courtesy Allen-Bradley Co.)

ratings of 2 W or less. They have tolerances of ± 5, 10, and 20%. The resistor color code, discussed in Unit 2, is used to identify the nominal value and tolerance of the resistor.

Wire-wound resistor Certain alloys like manganin and constantan that have near zero temperature coefficient of resistance and uniform resistance per unit length are drawn into wire. A specific length of the wire, depending on the required resistance, is wound on a core of material. It is then enclosed in some type of ceramic or high-temperature material; see Figure 6.3(a). These resistors usually have low resistance and carry high current, and so have high wattage ratings (5 W, 10 W, 20 W, etc.).

Precision resistor When we need a resistor with tolerance closer than $\pm 5\%$, we must use a precision resistor. Precision resistors are made of a metal film, like carbon vapor deposited on a glass

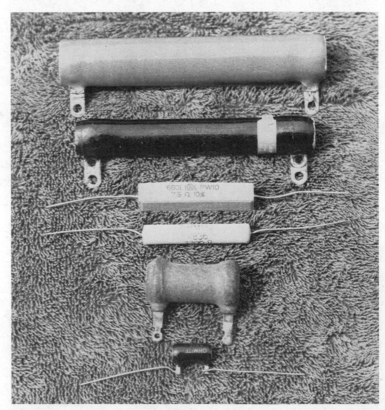

Figure 6.3 Other kinds of fixed resistors: (a) Wire-wound resistors

Figure 6.3 (b)   Precision resistors

or a ceramic form. These resistors usually have their value and tolerance written on the resistor case. [See Figure 6.3(b).] They generally have low wattage ratings. Of course, the closer the tolerance, the more expensive the resistor.

Variable resistor   These are molded-carbon composition or wire-wound resistors with a movable conductor that contacts the resistor at a particular point. (See Figure 6.4.) When the conductor or arm is moved, the resistance between it and either end of the resistor changes. Rheostats usually have large wattage ratings, and potentiometers have small wattage ratings. The variable resistors are either linear or nonlinear (called tapered or logarithmic) in resistance as the arm is moved from one end to the other end.

Integrated-circuit resistor   At least two types of resistors are used in integrated circuits. The junction or monolithic resistor

(a)

Rheostat and potentiometer

Rheostat   Potentiometer

(b)

Figure 6.4 Variable resistors:
(a) Potentiometers, (b) electrical
symbols

is used when a monolithic integrated circuit is made. (See Figure
6.5.) In a *monolithic* integrated circuit, the transistors, diodes,
resistors, and capacitors are all manufactured at the same time.
The resistor is made out of the same *p*-type (positive type) semiconductor material as the base of the integrated *npn* transistor.

Thin-film resistors are made separately and then combined
with thin-film capacitors, transistors, and diodes to make *hybrid*
integrated circuits. The thin-film resistor is fabricated by de-
positing a resistive material over a layer of silicon dioxide that
covers a *p*-type substrate or foundation. (See Figure 6.6.)

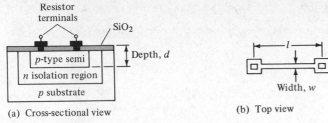

Figure 6.5 The junction or monolithic resistor

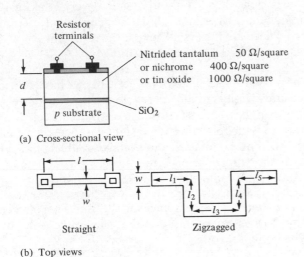

Figure 6.6 The thin-film resistor

The resistance of both of these types of resistors is still calculated by the relation given in Equation (6.2). However, since these resistors are in a very thin sheet form, the formula is modified to use sheet resistance, given in ohms per square.

$$R = \frac{\rho l}{A} = \frac{\rho l}{wd}$$

$$R = R_s \frac{l}{w} \tag{6.14}$$

where l is the length of the material (See Figures 6.5 and 6.6)

w is the width of the material (if $l = w$ then, we have a unit square of material)

R_s is ρ/d or the sheet resistance or resistance per unit square of material; the unit is $\Omega/$square

See Figure 6.6 for the sheet resistance of some of the materials used to make thin-film resistors.

Example 6.14   Design a 10-kΩ thin-film resistor using nichrome.

Solution   The length-to-width ratio required is

$$\frac{l}{w} = \frac{R}{R_s}$$

$$= \frac{10,000 \ \Omega}{400 \ \Omega/\text{square}} = 25 \text{ squares}$$

If we select a width of 2 mils, then the length must be $25(2) =$ 50 mils or 0.05 in. The length of 50 mils is very long for use in integrated circuits, where the whole IC may be only 50 mils by 50 mils. To decrease the actual length, the resistor is constructed in a zigzag pattern. [See Figure 6.6(b).]

Nonlinear resistor   All the types of resistors we have previously discussed are considered linear resistors. By linear, we mean that there is a linear or straight-line relation between E and I. This means that the resistance is constant with respect to voltage and current. For example, if the voltage doubles, the current doubles; if the voltage triples, the current triples; etc. Some common forms of nonlinear resistors are the following.

Incandescent lamp   The resistance of the tungsten filament in the lamp increases considerably as the filament heats up due to the flow of current through it.

Thermistor   The thermistor is a thermally sensitive resistor. It is made from a semiconductor and exhibits a large negative temperature coefficient of resistance. It is used in many electric and electronic circuits for protection, and for temperature measurement, control, and compensation.

Varistor   The varistor is a voltage-sensitive resistor. A small change in voltage across the device may increase the current through the device by many times. It is made from semiconductors and has a negative temperature coefficient. One of its uses is to protect circuit components from voltage surges when a magnetic or inductive circuit is interrupted. (When used for this application, it may be called a *thyrite*.)

Problem 6.20   Discuss the common types of resistors; that is, give the name and general characteristics of each.

Definition and Calculation of Power

Objective 9   Define the term *power*. The definition must include an explanation of the equation $P = W/t$, using both joules and kilowatt-hours to measure work W.

Power is defined as the *rate* of doing work. Work and energy were discussed in Unit 5. Since power is a rate, this means the amount of work or energy per unit time. The symbol used for power is *P*. Then

$$P = \frac{W}{t} \tag{6.15}$$

where *W* is the work or energy; the unit for *W* is joule
      t is the time; the unit for *t* is second

From Equation (6.15), we know that the units for power are joules per second. But in electricity the unit most commonly used is the watt. The watt is so named to honor James Watt, the inventor of the steam engine.

Using the unit of watt for power, we have a new unit for energy. Solving Equation (6.15) for *W*, we get

$$W = Pt$$

This says that a joule is equal to a watt-second. This unit of watt-second for work or energy may mean more to you than joule after you work with resistors of various power ratings in the laboratory. For larger quantities of energy, the unit kilowatt-hour (kWh) is used. The energy you use in your home is usually measured in kilowatt-hours by the electric power companies.

Example 6.15 How many joules of energy does a 20-W light dissipate in 24 h?

Solution W (J) = P (W) $\times t$ (s)
              = 20 W $\times$ 24 h $\times$ 3600 s/h
              = 1.728×10^6 W-s   or 1.728×10^6 J

Example 6.16 How many kilowatt-hours of energy does the light of Example 6.15 dissipate?

Solution W (kWh) = P (kW) $\times t$ (h)
                = 0.020 kW $\times$ 24 h
                = 0.48 kWh

Problem 6.21 A resistor converts electric energy into heat. At what rate must this conversion be made if 4500 J are produced in 5 min?

Problem 6.22 How many kilowatt-hours of energy does a 40-W light use in 1 month? Consider that a month has 30 days.

Objective 10 Calculate power, voltage, current, or resistance, using the appropriate form of the power equation ($P = EI$,

$P = I^2R$, or $P = E^2/R$), when any two of the three quantities in the equation are given or can be found from the information given.

We may obtain different forms of the formula for power [Equation (6.15)] that may be more useful by using the definitions we learned for current and voltage. Current was defined by $I = Q/t$ [Equation (5.2)]. Voltage or potential difference was defined by V (or E) = W/Q [Equation (5.4)].

$$P = \frac{W}{t}$$

Solving Equation (5.2) for t, we have $t = Q/I$.
Solving Equation (5.4) for W, we have $W = EQ$.
Substituting W and t into Equation (6.15), we obtain

$$P = \frac{EQ}{Q/I}$$

$$\therefore \quad P = EI \quad \text{or} \quad P = VI \quad\quad\quad (6.16)$$

Substituting for E in Equation (6.16), using Ohm's law, we obtain

$$P = (IR)I$$
$$\therefore \quad P = I^2R \quad\quad\quad\quad (6.17)$$

Substituting for I in Equation (6.16) using Ohm's law,

$$P = E\left(\frac{E}{R}\right)$$

$$P = \frac{E^2}{R} \quad \text{or} \quad P = \frac{V^2}{R} \quad\quad\quad (6.18)$$

The power rating on a circuit element tells how much power it can handle without changing value or being destroyed. It does not tell how much the circuit element is actually dissipating. Use Equation (6.16), (6.17), or (6.18) to do this. You will always want to make sure that the power rating on the circuit element that you are using is large enough. But occasionally, you may have to use a larger power rating than you need, because that is all that is available. You will not use more power than you would use with the exact rating. For example, a circuit diagram may call for resistors with a $\frac{1}{2}$-W power rating. If you have only 1-W or 2-W resistors, you may use these. *Don't* use $\frac{1}{2}$-W resistors when 2-W resistors are required.

Example 6.17   Given that R is a 10-kΩ, 2-W resistor. What is the maximum current it can handle without overheating?

Solution $P = I^2 R$

$$\therefore \quad I = \sqrt{\frac{P}{R}} = \sqrt{\frac{2}{10 \times 10^3}} = 14.14 \text{ mA}$$

Problem 6.23   Given that R is a 6800-Ω, 1-W resistor. What is the maximum current it can handle without overheating?

Problem 6.24   What is the maximum voltage that can be safely applied across a 15-kΩ, $\frac{1}{2}$-W resistor?

Problem 6.25   What wattage rating is required for a 3900-Ω resistor to safely pass 45 mA of current?

Problem 6.26   A transistor has a collector power dissipation rating of 350 mW. The maximum voltage across the collector is 45 V. What is the maximum collector current that the transistor can carry?

Efficiency

Objective 11   Calculate the efficiency η (Greek letter eta) of an electrical system when input and output power or energy are given. Use 1 hp = 746 W when required.

Efficiency is defined as the ratio of useful output energy to the total input energy.

$$\eta = \frac{W_{out}}{W_{in}} \tag{6.19}$$

But from Equation (6.15), W is also equal to P times t. Substituting this into Equation (6.19), we obtain

$$\eta = \frac{(Pt)_{out}}{(Pt)_{in}} = \frac{P_{out}}{P_{in}} \tag{6.20}$$

To express Equations (6.19) and (6.20) in *percent* efficiency, multiply by 100.

$$\%\eta = \frac{W_{out}}{W_{in}} \times 100 \tag{6.21}$$

$$\%\eta = \frac{P_{out}}{P_{in}} \times 100 \tag{6.22}$$

When working with power and efficiency, there are still many things, like motors, that are rated in horsepower. A conversion factor that may be useful is 1 hp = 746 watts.

Example 6.18   A useful measure of performance of transistor power amplifiers is collector efficiency. *Collector efficiency* is defined as the ratio of ac output power to dc input power. Given that we want to get 50 W output (ac power) from a class B push-pull transistor amplifier circuit, what minimum dc input power must we apply? The maximum possible efficiency of a class B push-pull transistor amplifier circuit is 78.5%. Given that a 36-V dc supply is used, what is the minimum current drain on the supply?

Solution   Solving Equation (6.22) for P_{in}, we obtain

$$P_{in} = P_{out} \frac{100}{\%\eta} = 50 \frac{100}{78.5} = 63.69 \text{ W dc}$$

$$I_{dc} = \frac{P_{dc}}{V_{dc}} = \frac{63.69 \text{ W}}{36 \text{ V}} = 1.77 \text{ A}$$

Problem 6.27   An electric motor converts electric energy into mechanical work. The motor draws 5 A from a 120-V source for 8 h, and is 80% efficient. How much work can be done?

Problem 6.28   The maximum possible collector efficiency of a class A transistor amplifier circuit using a transformer is 50%. What minimum dc input power must be applied to get 2 W output (ac power)? Given that a 6-V dc power supply is used, what is the minimum current drain on the supply?

Test

1. What is the resistance of an electric circuit, given that it takes an emf (or applied potential) of 12 V to cause 36 mA of current to flow?
2. Write a formula that relates resistance to the type of material (specific resistance), area, and length.
3. A piece of aluminum wire has a 20-Ω resistance. What is the resistance if the length of this aluminum wire is cut in half?
4. What is the resistance of 1000 ft of aluminum wire with a diameter of 10 mils? Use Table 6.1.
5. (*Advanced*) The conversion factor to change Ω-CM/ft to ohm-meters is 1 Ω-CM/ft = 1.662×10^{-9} Ω-m. The

specific resistance of aluminum is 17 Ω-CM/ft. What is the specific resistance in ohm-meters?

6. Find the resistance of 300 ft of AWG No. 14 copper wire at 30°C. Use Tables 6.1, 6.2, and 6.3.

7. What is the conductance of an 82-kΩ resistor?

8. List the following about molded-carbon composition resistors: wattage ratings, tolerance, and method of identifying nominal value and tolerance of the resistor.

9. A resistor converts 360 J of energy to heat for 6 min. What is the power to the resistor?

10. During a vacation a 40-W lamp is left on for 14 days. At 2¢ per kWh, what will this cost?

11. An iron has a wattage rating of 1000 W at 120 V. What is its resistance and current rating?

12. What voltage must be applied to a 500-W heater to pass a current of 10 A?

13. A 1000-Ω resistor has a power rating of 2 W. What is the maximum current this resistor can pass without overheating?

14. A motor has a 1-kW input and a 1-hp output. What is its efficiency?

Unit 7 *Series Resistive Circuits*

In the past few units you have been learning about the simple electric circuit, prefixes, Ohm's law, voltage, current, resistance, and power. You are now ready to put everything together and study more complicated circuits. You already know a little about the series connection. In this unit you will learn how to calculate the total resistance of several resistors connected in series. You will learn another very important law, Kirchhoff's voltage law— KVL—which you will find useful when you work with more complicated circuits. The voltage-divider principle will show you how to use resistors in series to drop a high voltage down to a low voltage. And finally, you will learn about the equivalent circuit for a power supply, because you will see that it is also a series resistive circuit.

Objectives

After completing all the work associated with this unit, you should be able to:

1. State what is meant when we say that two or more circuit elements are connected in series.
2. Calculate the total resistance when two, three, four, or more resistors are connected in series. Then use Ohm's law to solve for current and/or voltage.

3. State Kirchhoff's voltage law (KVL). And use KVL to solve for the voltage across a particular resistor or resistors and/or to solve for an unknown resistor in series resistive circuits.

4. Use the voltage-divider principle to solve for the voltage across a particular resistor or resistors in series resistive circuits.

5. Determine what happens to resistance and voltage measurements in series resistive circuits when a resistor opens or is shorted.

6. Recall that *all* power supplies have internal resistance. And calculate the power-supply terminal voltage or the load current, given the internal resistance of the power supply and the open-circuit voltage.

7. Calculate the internal resistance of a power supply from open- and closed-circuit voltage readings.

8. State the relationship for maximum power transfer in a series resistive circuit. Determine the load resistance for maximum transfer of power in a series resistive circuit.

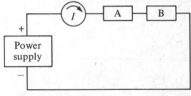

Figure 7.1 Series connection of circuit elements A and B across the power supply

Series Connection

Objective 1 State what is meant when we say that two or more circuit elements are connected in series.

Circuit elements are said to be connected *in series* when the same electric charge flows through each element. In Figure 7.1 the current I flows through component A and then through component B and back to the power supply. After the current passes through component A, it must flow through component B because there is no other path for it to follow.

Let's take the two resistors R_1 and R_2 shown in Figure 7.2 and connect them in series across the power supply E. Note that the terminals of the resistors are labeled. Resistors R_1 and R_2 are in series if the same current that flows through R_1 also flows through R_2.

In Figure 7.3 you see two possible connections. In Figure 7.3(a), resistor R_1 is placed first. However, as shown in Figure 7.3(b), resistor R_2 can be first as long as the same current flows through each resistor. In Figure 7.3(a) we connected terminal B to C. Then we must connect A to one end of the power supply and D to the other end. Once we connect the two resistors across the power supply, a current I flows. Since there is only one path through which current can flow, it must flow through R_1 and then R_2. Therefore we say that R_1 and R_2 are connected in series.

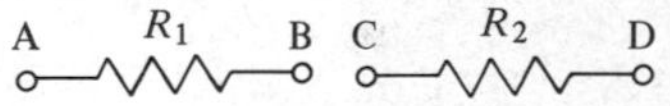

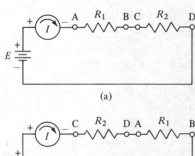

Figure 7.2 Two resistors, R_1 and R_2

(a)

(b)

Figure 7.3 Resistors R_1 and R_2 connected in series

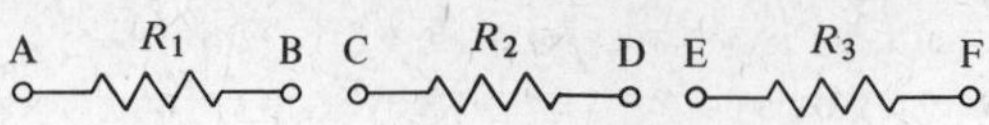

Figure 7.4 Three resistors, R_1, R_2, and R_3

Problem 7.1 Given the three resistors shown in Figure 7.4, draw a circuit diagram connecting them in series across a power supply E.

Calculation of Total Resistance

Objective 2 Calculate the total resistance when two, three, four, or more resistors are connected in series. Then use Ohm's law to solve for current and/or voltage.

To complete Objective 2, you want to be able to calculate the total resistance in a circuit with two, three, or more resistors connected in series. Before we discuss this, you should think about how we could determine this. See Problem 7.2. While you are thinking about the answer, remember the definition of resistance used for *dc* circuits: *Resistance is the property of a circuit element that offers opposition to current.*

Problem 7.2 If the electric charge has to flow through two or more equal resistors connected in series in a circuit, does the electric charge meet more opposition than it would with one resistor? Yes or no?

With the definition of resistance in mind, you correctly choose the *yes* answer. Because if we put more resistance in the circuit, there *is* more opposition to the flow of electric charge (current). Equations (7.1), (7.2), and (7.3) show how to calculate the total resistance of resistors connected in series.

$$\textit{Series} \begin{cases} \text{Two resistors } (R_1 \text{ and } R_2) & R_{\text{total}} = R_T = R_1 + R_2 & (7.1) \\ \text{Three resistors } (R_1, R_2, \text{ and } R_3) & R_{\text{total}} = R_T = R_1 + R_2 + R_3 & (7.2) \\ \text{More than three resistors} & R_{\text{total}} = R_1 + R_2 + R_3 + R_4 + \cdots & (7.3) \end{cases}$$

To determine the total resistance of any number of resistors connected in series, all we do is add up all the resistance of all the resistors connected in series. Example 7.1 illustrates this, and also shows you how to calculate the current and voltage drops in a series resistive circuit.

Example 7.1 Three resistors (1 kΩ, 3 kΩ, and 6 kΩ) are connected in series across a 20-V power supply.

(a) Draw the electric circuit diagram.
(b) Determine the total resistance in the circuit.
(c) Determine the amount of current in the circuit.
(d) Determine the voltage drop across each resistor.

Solution

(a) The electric circuit is shown in Figure 7.5.

(b) $R_{total} = R_1 + R_2 + R_3 = 1 \text{ k}\Omega + 3 \text{ k}\Omega + 6 \text{ k}\Omega = 10 \text{ k}\Omega$

(c) $I = \dfrac{E}{R_{total}} = \dfrac{20 \text{ V}}{10 \text{ k}\Omega} = 2 \text{ mA}$

(d) $V_1 = IR_1 = (2 \text{ mA})(1 \text{ k}\Omega) = 2 \text{ V}$
$V_2 = IR_2 = (2 \text{ mA})(3 \text{ k}\Omega) = 6 \text{ V}$
$V_3 = IR_3 = (2 \text{ mA})(6 \text{ k}\Omega) = 12 \text{ V}$
$\text{Total voltage drops} = \overline{20 \text{ V}}$

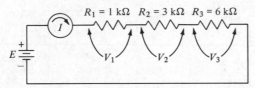

Figure 7.5   Solution to part (a) of Example 7.1

Problem 7.3   Two resistors (5 kΩ and 10 kΩ) are connected in series across a 15-V power supply.

(a) Draw the electric circuit diagram.
(b) Determine the total resistance in the circuit.
(c) Determine the amount of current in the circuit.
(d) Determine the voltage drop across each resistor.
(e) What is the sum of the voltage drops across each resistor?

Problem 7.4   Given the circuit diagram shown in Figure 7.6, determine the following:

(a) How the resistors are connected
(b) Total resistance in the circuit
(c) Amount of current in the circuit

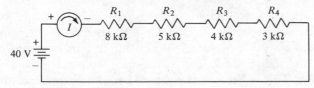

Figure 7.6   Circuit diagram for Problem 7.4

(d) Voltage drop across each resistor

(e) Sum of the voltage drops across each resistor

In part (d) of Example 7.1, look at the value you obtained for the sum of all the voltage drops in the circuit. The *sum* of the voltage drops across each resistor equals the supply voltage. This is not a coincidence! In fact, there is a law that states this. In Objective 3 you will be studying Kirchhoff's voltage law—what it is and how to use it to solve series resistive circuits.

Kirchhoff's Voltage Law

Objective 3 State Kirchhoff's voltage law (KVL). And use KVL to solve for the voltage across a particular resistor or resistors and/or to solve for an unknown resistor in series resistive circuits.

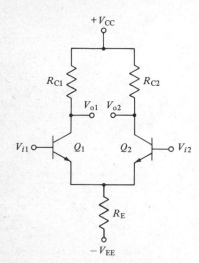

Figure 7.8 Basic circuit diagram of a differential amplifier (the negative terminal of the V_{CC} and the positive terminal of the V_{EE} power supplies are *understood* to be connected to the circuit ground)

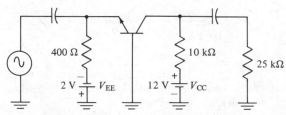

Figure 7.7 Transistor connected in a common-base amplifier configuration

Before we state *Kirchhoff's voltage law*, abbreviated KVL, you must find out what we mean when we say *voltage rise*, *voltage drop*, *positive* power supply, and *negative* power supply. Figures 7.7 and 7.8 are electronic circuits. The explanation of how these circuits operate is beyond the scope of this text. We included them to show you the use of positive and negative power supplies in circuits. The dc power supply V_{CC} shown in Figures 7.7 and 7.8 is connected as a positive supply. The dc power supply V_{EE} is connected as a negative supply.

As you see, it makes a difference how a battery or dc power supply is connected into a circuit. In most of the circuits we have shown, the power supply has been connected as shown in Figure 7.9(a). This is a positive power supply. But why is it a positive power supply? The bottom line in an electric circuit diagram such as Figure 7.9 is usually understood to be the reference for the entire circuit. The reference is usually not marked as it is in Figure 7.9(a). If the terminal marked minus is connected to the circuit reference, then we have a positive supply. This is because the top end of the circuit is connected *positively* with respect to the reference. If you connect a voltmeter

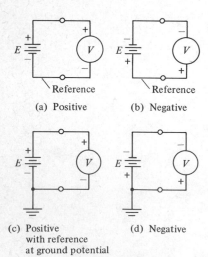

(a) Positive   (b) Negative

(c) Positive with reference at ground potential   (d) Negative

Figure 7.9 Positive and negative supplies

across the supply shown in Figure 7.9(a), you can see that you have a positive voltage.

However, some electronic devices need a negative supply to make them operate correctly—for example, the V_{EE} supply in Figures 7.7 and 7.8. In Figure 7.9(b) we have shown you how to connect the battery or power supply to obtain a negative voltage. The terminal of the power supply that is marked *plus* is connected to the circuit reference point. The voltmeter connected across the power supply in Figure 7.9(b) reads a negative potential.

In many electric and electronic circuits the circuit reference is connected to ground potential; see Figure 7.7. The water pipes in your home are at ground potential. To show that ground, or zero, potential is the circuit reference point, a ground symbol is attached to the reference line of the circuit diagram. This is shown in Figure 7.9(c) and (d).

Examine Figure 7.10. The ammeter is shown to make you aware that *current I* is common in a *series* circuit. We are assuming that this is an ideal ammeter, by which we mean that it has no resistance and therefore that no voltage is dropped across it. Then all the power-supply voltage E must be dropped across the series resistors R_1, R_2, and R_3.

Next notice some plus and minus signs on the series resistors. What these signs show, if you look at resistor R_3, is that the top of the resistor is more positive in potential than the bottom of the resistor. The top of resistor R_2 is more positive in potential than the bottom of resistor R_2. The top of resistor R_1 is more positive than the bottom of resistor R_1. In fact, the top of resistor R_1 has the most positive potential that exists in the circuit, because it is connected to the plus end of the positive power supply. Notice that we have assumed the direction of conventional current in Figure 7.10; note the ammeter. The current flows through the top of resistor R_1 down through to the bottom of resistor R_3. You read in a previous unit that when current flows through a circuit element, there is a voltage drop across that element.

In Figure 7.11 we still have the same series circuit that we had in Figure 7.10; however, there are some different notations. And we have not connected the circuit reference to ground potential.

Let's first discuss the notation used to specify the voltage drop across a particular resistor. This notation was used in Example 7.1. We use the subscript 1 on V to designate the voltage drop across resistor R_1, V_2 to designate the voltage drop across resistor R_2, and V_3 to designate the voltage drop across R_3. Now notice that we have a closed path, from the plus side of the supply to A to B to C to D to E to F, which is the same reference

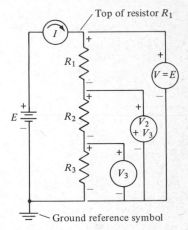

Figure 7.10   Series resistive circuit

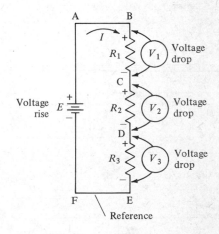

KVL: $E = V_1 + V_2 + V_3$

Figure 7.11   Series resistive circuit showing voltage drops and voltage rise

line as E, back to the minus side of the battery. Since we have a closed path, current flows. We have represented current in Figure 7.11 differently from the way we did in Figure 7.10. Many of our circuit diagrams are very complicated, and they would look even more complicated if we inserted ammeters showing all the currents. The arrow in Figure 7.11 points the direction of assumed *conventional* current flow.

Follow the current arrow around the circuit shown in Figure 7.11. You notice that, as current flows through the resistors, it is going from the side of the resistor marked plus to the side marked minus. Remember, we defined this as a *voltage drop*. Since we have a closed path, how must we picture current flowing from point F to point A in the circuit? The power supply acts as a pump. In going from point F to point A, the current encounters the minus sign first, then the plus sign of the power supply. This is just the opposite of what happened when current flowed through the resistors. Since it is opposite, we call the power supply in Figure 7.11 a *voltage rise*. All this discussion about Figure 7.11 leads us to another very important law in electric circuits, and that is Kirchhoff's voltage law.

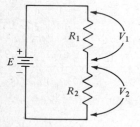

Figure 7.12   Two-resistor series circuit

Kirchhoff's Voltage Law (KVL)
The algebraic sum of the voltage drops in any closed path must equal the voltage rises (or applied voltage).

This says that in Figure 7.11, E is equal to the sum of V_1 plus V_2 plus V_3.

Example 7.2   For the circuit shown in Figure 7.12, fill in the blanks in Table 7.1.

Solution

(a)
$$E = V_1 + V_2 \qquad \therefore \quad V_2 = E - V_1 = 30 - 10 = 20 \text{ V}$$

(b)
$$E = V_1 + V_2 \qquad \therefore \quad V_1 = E - V_2 = 20 - 15 = 5 \text{ V}$$

Problem 7.5   For the two-resistor series circuit shown in Figure 7.12, complete Table 7.2.

Example 7.3   In Figure 7.11, given that $E = 30$ V and $V_1 = 14$ V and $V_2 = 6$ V, what is V_3?

Solution: KVL says $E = V_1 + V_2 + V_3$.
$$\therefore \quad V_3 = E - (V_1 + V_2) = 30 - 20 = 10 \text{ V}$$

Table 7.1

	E	V_1	V_2
(a)	30 V	10 V	
(b)	20 V		15 V

Table 7.2

E	V_1	V_2
30 V	18 V	
20 V		12 V
	6 V	9 V

Example 7.4 Find R_1 and R_2 in the two-resistor series circuit shown in Figure 7.13.

Solution From the definition of a series circuit, the current through R_2 is equal to 2 mA. Using Ohm's law, we get

$$R_2 = \frac{V_2}{I} = \frac{12}{2 \times 10^{-3}} = 6 \times 10^3 = 6 \text{ k}\Omega$$

From KVL, we have

$$V_1 = E - V_2 = 20 - 12 = 8 \text{ V}$$

And again using Ohm's law, we obtain

$$R_1 = \frac{V_1}{I} = \frac{8}{2 \times 10^{-3}} = 4 \text{ k}\Omega$$

Problem 7.6 Find R_1 and R_2 in the two-resistor series circuit shown in Figure 7.14.

Problem 7.7 Change the current of 2 A given in Problem 7.6 to 1 mA, then solve for R_1 and R_2.

Problem 7.8 Solve for I and then for R_2 in the two-resistor series circuit shown in Figure 7.15.

Example 7.5 Find R_2 and R_3 in the series resistive circuit shown in Figure 7.16.

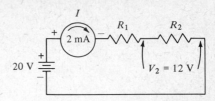

Figure 7.13 Circuit for Example 7.4

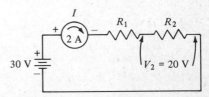

Figure 7.14 Circuit for Problem 7.6

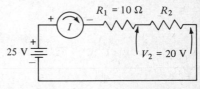

Figure 7.15 Circuit for Problem 7.8

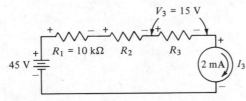

Figure 7.16 Circuit for Example 7.5

Solution Current is common or the same throughout a series circuit. Therefore $I = I_1 = I_2 = I_3$. Since we know the current *through* and the voltage *across* resistor R_3, we may find its resistance by using Ohm's law.

$$R_3 = \frac{V_3}{I} = \frac{15 \text{ V}}{2 \text{ mA}} = 7.5 \text{ k}\Omega$$

To find R_2 from Ohm's law, we need to know V_2 and I. We do know I, but V_2 is not given. However, by using KVL, we can find V_2. But before we can use KVL, we must find V_1. We find V_1 by using Ohm's law.

$$V_1 = IR_1 = (2 \text{ mA})(10 \text{ k}\Omega) = 20 \text{ V}$$

Using KVL, $E = V_1 + V_2 + V_3$. Therefore

$$V_2 = E - V_1 - V_3 = 45 - 20 - 15 = 10 \text{ V}$$

And finally,

$$R_2 = \frac{V_2}{I} = \frac{10 \text{ V}}{2 \text{ mA}} = 5 \text{ k}\Omega$$

Example 7.6 Find R_2 in the series resistive circuit shown in Figure 7.17.

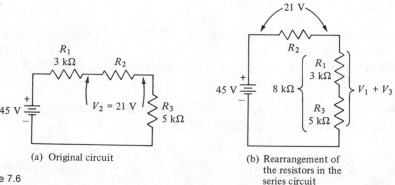

(a) Original circuit

(b) Rearrangement of the resistors in the series circuit

Figure 7.17 Circuit for Example 7.6

Solution The difficulty in this problem is that the unknown resistor R_2 is in between two known resistors. We can rearrange the resistors in the circuit as shown in Figure 7.17(b). We can do this because the circuit is a series circuit. Therefore the order of the resistors is not critical, as long as they are all in the circuit. By KVL, the voltage drop across the 3-kΩ and 5-kΩ resistors is

$$V_1 + V_3 = E - V_2 = 45 - 21 = 24 \text{ V}$$

Using Ohm's law, we find that the current in the 3-kΩ and 5-kΩ resistors is

$$I = \frac{V_1 + V_3}{R_1 + R_3} = \frac{24 \text{ V}}{8 \text{ k}\Omega} = 3 \text{ mA}$$

Since current is the same throughout a series circuit, the current in R_2 must also be 3 mA.

$$R_2 = \frac{V_2}{I} = \frac{21 \text{ V}}{3 \text{ mA}} = 7 \text{ k}\Omega$$

Problem 7.9 Find R_3 in the series resistive circuit shown in Figure 7.18.

Problem 7.10 Find R_2 in the series resistive circuit shown in Figure 7.19.

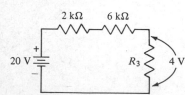

Figure 7.18 Circuit for Problem 7.9

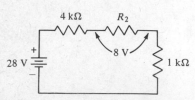

Figure 7.19 Circuit for Problem 7.10

Problem 7.11 Find R_1 and R_2 in the series resistive circuit shown in Figure 7.20.

Voltage-Divider Principle

Objective 4 Use the voltage-divider principle to solve for the voltage across a particular resistor or resistors in series resistive circuits.

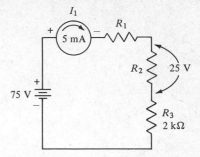

Figure 7.20 Circuit for Problem 7.11

Many times we just want the voltage across a particular resistor in a series circuit—for example, V_2 in Figure 7.21. By using the voltage-divider principle, we can find this voltage in one step.

Before we solve for the equation that we shall use, let's look at the circuit shown in Figure 7.21(b). This circuit is exactly the same as that shown in Figure 7.21(a). However, it uses a slightly different notation for the power supply E. You should become familiar with this notation, because it is very commonly used, particularly in electronics. You see the top end of resistor R_1 connected to positive E, and the bottom of R_2 connected to the reference potential. However, when you set up the circuit in the laboratory, you must connect the positive-marked lead of the power supply to the top of resistor R_1 *and* the negative-marked lead to the bottom of resistor R_2 (circuit common or reference).

To solve for the voltage across resistor R_2 in Figure 7.21 by Ohm's law, we solve for current in R_2 and then multiply this current by the resistance of R_2. Therefore $V_2 = IR_2$. Since R_1 and R_2 are connected in series across the power supply E, we can find the current by dividing the power supply E by the total resistance R_T:

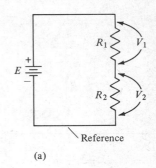

$$I = \frac{E}{R_T} = \frac{E}{R_1 + R_2}$$

and therefore

$$V_2 = IR_2 = \left(\frac{E}{R_T}\right)R_2 = \left(\frac{E}{R_1 + R_2}\right)R_2 \qquad (7.4)$$

Rearranging Equation (7.4), we obtain

$$\boxed{V_2 = \left(\frac{R_2}{R_1 + R_2}\right)E} \qquad (7.5)$$

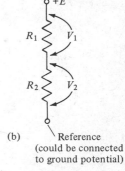

Figure 7.21 Two-resistor series circuit [circuits in (a) and (b) are the same circuit]

This form of the equation is known as the *voltage-divider* form. The circuit is known as a voltage divider as well as a series circuit, because it *divides* the total voltage E among the series resistors. So if we want to know the voltage across resistor R_2, we take the ratio of R_2 to the *total* resistance in the circuit $(R_1 + R_2)$, and then multiply this ratio by the voltage across the total resistance, which in this circuit is E.

Table 7.3

	E	R_1	R_2	V_1	V_2
(a)	20 V	1 kΩ	4 kΩ	—	
(b)	20 V	8 kΩ	12 kΩ		—
(c)	30 V	10 kΩ	5 kΩ		

In a similar manner we can obtain an equation for V_1.

$$V_1 = \left(\frac{R_1}{R_1 + R_2}\right) E \tag{7.6}$$

The following examples and problems show you how to use the voltage-divider principle.

Example 7.7 For the two-resistor series circuit shown in Figure 7.21, fill in the blanks in Table 7.3.

Solution

(a)

$$V_2 = \left(\frac{R_2}{R_1 + R_2}\right) E = \frac{4 \text{ kΩ}}{5 \text{ kΩ}} (20 \text{ V}) = 16 \text{ V}$$

(b)

$$V_1 = \left(\frac{R_1}{R_1 + R_2}\right) E = \frac{8 \text{ kΩ}}{20 \text{ kΩ}} (20 \text{ V}) = 8 \text{ V}$$

(c)

$$V_1 = \left(\frac{R_1}{R_1 + R_2}\right) E, \qquad V_2 = \left(\frac{R_2}{R_1 + R_2}\right) E$$

$$V_1 = \frac{10 \text{ kΩ}}{15 \text{ kΩ}} (30 \text{ V}) = 20 \text{ V} \qquad V_2 = \frac{5 \text{ kΩ}}{15 \text{ kΩ}} (30 \text{ V}) = 10 \text{ V}$$

[*Note:* $V_1 + V_2 = 20 \text{ V} + 10 \text{ V} = 30 \text{ V} = E$]

Problem 7.12 For the two-resistor series circuit shown in Figure 7.21, complete Table 7.4. Use the voltage-divider principle.

Table 7.4

E	R_1	R_2	V_1	V_2
40 V	120 Ω	80 Ω	—	
40 V	8 kΩ	2 kΩ		—
25 V	2 kΩ	3 kΩ		
10 V	7 kΩ	3 kΩ		

Example 7.8 Determine the V_{out} of the circuit in Figure 7.22 by the voltage-divider principle.

Solution

$$V_{out} = \frac{6 \text{ k}\Omega}{4 \text{ k}\Omega + 6 \text{ k}\Omega} (30 \text{ V}) = \frac{6}{10} (30 \text{ V}) = 18 \text{ V}$$

If we had approached the problem by starting with the voltage drop across the 4-kΩ resistor,

$$V_{(\text{drop across 4 k}\Omega)} = \frac{4 \text{ k}\Omega}{4 \text{ k}\Omega + 6 \text{ k}\Omega} (30 \text{ V}) = 12 \text{ V}$$

then we would have had to take an additional step to find the output voltage. Since the top end of the 4-kΩ resistor is connected to the 30-V supply, we find the output voltage by subtracting the drop in the 4-kΩ resistor from the supply, or 30 V − 12 V = 18 V.

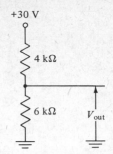

Figure 7.22 Circuit for Example 7.8

Example 7.9 Determine the voltage at point A in Figure 7.23 with respect to ground by using the voltage-divider principle.

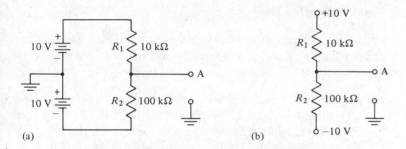

(a)

(b)

Figure 7.23 Circuit for Example 7.9 [circuits shown in (a) and (b) are the same circuit]

Solution Remember that the voltage-divider principle gives the *voltage drop across* a particular resistor. This voltage may or may not be with respect to ground, depending on the connection we are trying to find the voltage across. The voltage across the total resistance (10 kΩ + 100 kΩ) is 20 V. Therefore the voltage drop across the 10-kΩ resistor is equal to

$$V_{(\text{drop across 10 k}\Omega)} = \frac{10 \text{ k}\Omega}{10 \text{ k}\Omega + 100 \text{ k}\Omega} (20 \text{ V}) = 1.82 \text{ V}$$

The top end of the 10-kΩ resistor is connected to the positive 10-V source. Since we have a 1.82-V drop in the 10-kΩ resistor, the potential at point A with respect to ground is

$$V_{(\text{point A})} = 10 \text{ V} - 1.82 \text{ V} = 8.18 \text{ V}$$

Problem 7.13 Determine the voltage across the 3-kΩ resistor in the circuit of Figure 7.24 by the voltage-divider principle.

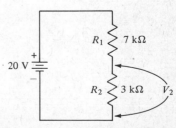

Figure 7.24 Circuit for Problem 7.13

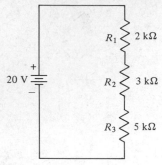

Figure 7.25 Circuit for Problem 7.15

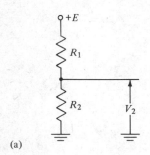

(a)

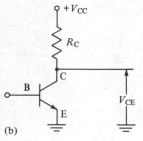

(b)

Figure 7.26 Analogy between a two-resistor series circuit and a series circuit with a load resistor and a transistor

$R = \infty \ \Omega$

(a) Open resistor

$R = 0 \ \Omega$

(b) Shorted resistor

Figure 7.27 Open and shorted resistors

The voltage-divider principle is not limited to series circuits with only two resistors. The circuits may contain three or more series resistors. You may always find the voltage drop across a particular resistor by taking the value of that particular resistor and dividing it by the total resistance in the circuit, then multiplying this ratio by the total voltage across the total resistance. For the circuit shown in Figure 7.11, the equation for the voltage drop across R_1 becomes

$$V_1 = \left(\frac{R_1}{R_1 + R_2 + R_3} \right) E \qquad (7.7)$$

Problem 7.14 List the equations that would be used to solve for V_3 and $V_1 + V_2$ in Figure 7.11, using the voltage-divider principle.

Problem 7.15 For the circuit given in Figure 7.25, find the voltage drop across the 2-kΩ resistor, across the 5-kΩ resistor, and across the 2-kΩ plus the 3-kΩ resistor. Use the voltage-divider principle.

Open or Shorted Resistor

Objective 5 Determine what happens to resistance and voltage measurements in series resistive circuits when a resistor opens or is shorted.

Many components in electric or electronic circuits act like an open circuit or a short circuit when they fail. To help you understand what happens, we can examine the series resistive circuit when one of the resistors opens or is shorted. The concepts involved also apply to circuits that contain other electric or electronic components, such as transistors and capacitors. For example, if you learn how to work with the two-resistor series circuit shown in Figure 7.26(a), then you know how to work with the transistor circuit shown Figure 7.26(b). We use the same procedures to determine the voltage V_{CE} in Figure 7.26(b) when the transistor is opened or shorted as we use to solve for V_2 in Figure 7.26(a) if resistor R_2 is opened or shorted.

Figure 7.27 illustrates what happens to a resistor when it opens or is shorted. When a resistor opens in a series circuit, there is no current in the circuit, because there is no closed path. The open resistor can be treated as a resistor that has infinite (∞) ohms resistance; see Figure 7.27(a). A resistor that has shorted has *zero* ohms resistance; see Figure 7.27(b). A shorted resistor offers no resistance to the flow of charge carriers. The following examples illustrate what happens to resistance and voltage measurements in series resistive circuits when a resistor opens or is shorted.

Example 7.10   Refer to the circuit shown in Figure 7.28. Assume that R_2, the 10-kΩ resistor, has shorted. What is the total resistance R_T in the circuit, and what is the output voltage of the circuit?

Solution

$$R_2 = 0 \ \Omega \ \text{(shorted)}$$
$$R_T = R_1 + R_2 = 5 \ \text{k}\Omega + 0 \ \Omega = 5 \ \text{k}\Omega$$

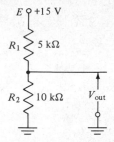

Figure 7.28   Two-resistor series circuit

We can determine the output voltage by using the voltage-divider principle.

$$V_{\text{out}} = \left(\frac{R_2}{R_1 + R_2} \right) E = \left(\frac{0}{5 \ \text{k}\Omega + 0} \right) 15 \ \text{V} = 0 \ \text{V}$$

The 15 V from the power supply is dropped across the 5-kΩ resistor.

Example 7.11   Refer to the circuit shown in Figure 7.28. Assume that R_2, the 10-kΩ resistor, has opened up. What is the total resistance in the circuit, and what is the output voltage of the circuit?

Solution

$$R_2 = \infty \ \Omega \ \text{(open)} \qquad \infty \text{ is the symbol for infinity}$$
$$R_T = R_1 + R_2 = 5 \ \text{k}\Omega + \infty = \infty \ \Omega$$

If we use the voltage-divider principle to determine the output voltage, we obtain strange results.

$$V_{\text{out}} = \left(\frac{R_2}{R_1 + R_2} \right) E = \left(\frac{\infty}{\infty} \right) 15 \ \text{V} \qquad (7.8)$$

When you see the ∞/∞, you say that it cannot be evaluated. However, we may evaluate Equation (7.8) by letting R_2 be very large compared with R_1. For example, let R_2 equal 5,000,000,000; then

$$\left(\frac{R_2}{R_1 + R_2} \right) = \frac{5,000,000,000}{5,000,005,000} \approx 1$$

Therefore V_{out} is equal to 15 V.

Let us show you another way to solve this problem. The top end of resistor R_1 is at a $+15$-V potential with respect to ground. The reason for this is that the 15-V power-supply voltage is connected to this point. No current flows through resistors R_1 and R_2 because R_2 is open.

$$I = \frac{E}{R_\mathrm{T}} = \frac{15}{\infty} = 0 \text{ A}$$

Since I is equal to 0 A, there is no voltage drop across resistor R_1.

$$V_1 = IR_1 = (0)(5 \text{ k}\Omega) = 0 \text{ V}$$

Therefore we can find the output voltage using

$$V_\mathrm{out} = E - V_1 = 15 \text{ V} - 0 \text{ V} = 15 \text{ V}$$

Problem 7.16   Refer to the circuit shown in Figure 7.28. Assume that R_1, the 5-kΩ resistor, has shorted. What is the total resistance in the circuit, and what is the output voltage of the circuit?

Problem 7.17   Refer to the circuit shown in Figure 7.28. Assume that R_1, the 5-kΩ resistor, has opened up. What is the total resistance in the circuit, and what is the output voltage of the circuit?

Be sure you study and understand Examples 7.10 and 7.11 and Problems 7.16 and 7.17. Knowing what kind of results you obtained, particularly the voltage readings, will help you trouble-shoot any circuit that may have a shorted or open component.

Internal Resistance of Power Supplies

Objective 6   Recall that *all* power supplies have internal resistance. And calculate the power-supply terminal voltage or the load current, given the internal resistance of the power supply and the open-circuit voltage.

In this objective you will learn more about power supplies. The symbol E, used in circuits such as that shown in Figure 7.29(a) or in the other previous figures shown in this unit, represents a battery or an electronic power supply. The numerical value given for E in problems represents the *terminal voltage* of the power supply; that is, the terminal voltage that exists across the resistor or series resistors when there is current in the circuit. The resistor or resistors across the power supply are called the *loads* on the power supply.

We can represent the power supply E by the equivalent circuit shown enclosed in the dashed lines in Figure 7.29(b). The equivalent circuit of a power supply is made up of an *ideal voltage source* and a *series internal resistor*. Right now we don't care how the equivalent circuit was determined. However, since the circuit shown in Figure 7.29(b) is theoretically equivalent, we can use

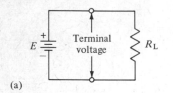

(a)

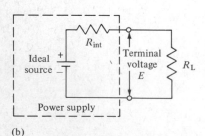

(b)

Figure 7.29   Equivalent circuit for a power supply

the theory we have been studying in this unit to solve power-supply circuits. You will learn in Unit 9 that the equivalent circuit was determined by applying Thévenin's theorem to a power supply.

When we know the internal resistance and the open-circuit voltage of a power supply, we can use the voltage-divider principle to determine the terminal voltage across the load. When we know the load voltage, we can determine the load current.

Example 7.12   The internal resistance of a power supply is 500 Ω and the open-circuit voltage is 50 V. Determine the terminal voltage of the power supply when a 2-kΩ resistor is connected as a load.

Solution   First let's make a circuit diagram for this problem; see Figure 7.30. The ideal source is equal to the open-circuit voltage of 50 V. Using the voltage-divider principle,

$$V_L = \frac{R_L}{R_T}(V_{\text{ideal source}}) = \frac{2000}{2500}(50\ V) = 40\ V$$

Here V_L is the terminal or load voltage, R_L is the load resistor, and R_T is the total resistance across the ideal voltage source.

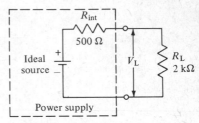

Figure 7.30   Circuit for Example 7.12

Example 7.13   Assume that, instead of the 2-kΩ resistor, a 20-kΩ load resistor is connected to the power supply given in Example 7.12. What is the new terminal voltage?

Solution   Using the voltage-divider principle, we have

$$V_L = \frac{R_L}{R_T}(V_{\text{ideal source}}) = \frac{20{,}000}{20{,}500}(50\ V) = 48.78\ V$$

Notice that we have a much smaller load current in the circuit when the 20-kΩ load is connected to the power supply.

Problem 7.18   The internal resistance of a power supply is 200 Ω, and the open-circuit voltage ($V_{\text{ideal source}}$) is 12 V. Determine the terminal voltage of the power supply when a 1-kΩ resistor is connected as a load. (See Figure 7.31.)

Problem 7.19   Assume that, instead of the 1-kΩ resistor, a 10-kΩ load resistor is connected to the power supply given in Problem 7.18. What is the new terminal voltage?

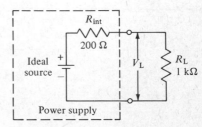

Figure 7.31   Circuit for Problem 7.18

All power supplies have internal resistance. The effect of this internal resistance is that the terminal voltage of a power supply decreases as you increase the current in the load, and vice versa. As you saw in Example 7.13, internal resistance may be so small compared with the value of the load that there is very little variation of terminal voltage with variation of load current.

Problem 7.20   As a 1.5-V flashlight battery is used, its terminal voltage drops down to 1 V.   What happened to the 0.5 V?

Problem 7.21   The size (resistance value) of the load in Figure 7.29(b) decreases.   How do the load current and the terminal voltage of the power supply change?

Since all power supplies have internal resistance, in the laboratory you always want to measure your power-supply voltage under load conditions.   That is, the load must be connected, and there must be current in the circuit.   It is a good idea to measure the power-supply voltage when no load is connected (open circuit) to make sure that you have the approximate voltage required for the circuit, but you must measure it *again* when the load is connected.   Otherwise, since you measured it only for the open-circuit condition, you are inserting a resistor (whatever the internal resistance R_{int} is equal to at the particular value of current) into the circuit in series with the circuit resistors.   If the power supply has significant internal resistance, you will have a lower terminal voltage because of the voltage drop across the internal resistance.

Objective 7   Calculate the internal resistance of a power supply from open- and closed-circuit voltage readings.

If the load can be removed from the power supply, the following is a simple procedure to determine the equivalent circuit for a power supply.

Step 1.   With the load removed from the power supply, measure the terminal or open-circuit voltage.   The open-circuit voltage is the ideal source voltage, because with the open circuit no current flows through the internal resistance.   Therefore there is no voltage drop across the internal resistance.

Step 2.   Connect a known load resistor across the terminals of the power supply, and measure the voltage across the known load resistor.

The following example illustrates this procedure.

Example 7.14   The open-circuit voltage across terminals A–B of Figure 7.32 is 13 V.   When a 20-Ω resistor is connected across terminals A–B, the voltmeter reads 10 V.   What is the internal resistance of the power supply?

Solution   The ideal supply voltage is equal to the open-circuit terminal voltage, 13 V. [See Figure 7.32(a).] With the open circuit, no voltage is dropped across the internal resistance.

The voltage across the known load resistor gives us the load current. From Ohm's law, the current in the load resistor is

$$I = \frac{V_L}{R_L} = \frac{10 \text{ V}}{20 \text{ }\Omega} = 0.5 \text{ A}$$

The circuit shown in Figure 7.32(b) is a series circuit. Therefore the current in the internal resistance is equal to the current in the load. To find the internal resistance using Ohm's law, we need the voltage *across* the internal resistance and the current in the internal resistance. We already know the current in the internal resistance. We can find the voltage across the internal resistance by using KVL.

$$V_{\text{int res}} = V_{\text{ideal source}} - V_L = 13 \text{ V} - 10 \text{ V} = 3 \text{ V}$$

Therefore

$$R_{\text{int}} = \frac{V_{\text{int res}}}{I} = \frac{3 \text{ V}}{0.5 \text{ A}} = 6 \text{ }\Omega$$

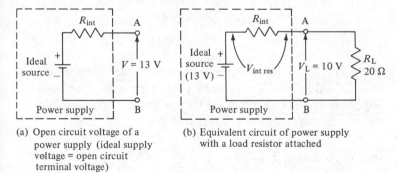

(a) Open circuit voltage of a
    power supply (ideal supply
    voltage = open circuit
    terminal voltage)

(b) Equivalent circuit of power supply
    with a load resistor attached

Figure 7.32   Circuit for Example 7.14

Problem 7.22   For the circuit in Figure 7.33, the open-circuit voltage across terminals A–B is 40 V. When a 100-Ω load is connected across A–B, the voltmeter reads 36 V. What is the internal resistance of the power supply?

When the load cannot be removed from the power supply, we can still determine the internal resistance if we can vary the load. To do this, we must know the load current at two different load values. We can measure the load current directly or determine

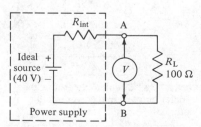

Figure 7.33   Circuit for Problem 7.22

it by measuring the terminal voltage across a known load resistance. The following example illustrates how to determine the internal resistance when two load settings are known.

Example 7.15 A power supply has a terminal voltage of 30 V when a 1-kΩ load is connected to its terminals. The terminal voltage drops to 27 V when the load is reduced to 600 Ω. What is the internal resistance of the power supply?

Solution The circuit current for the 1-kΩ load is

$$I_1 = \frac{V_{1\,k\Omega}}{R_{L_1}} = \frac{30\text{ V}}{1\text{ k}\Omega} = 30\text{ mA}$$

The circuit current for the 600 Ω load is

$$I_2 = \frac{V_{600\,\Omega}}{R_{L_2}} = \frac{27\text{ V}}{600\text{ }\Omega} = 45\text{ mA}$$

From KVL, $V_{\text{ideal source}} = IR_T = I(R_{\text{int}} + R_L)$. The ideal source or open-circuit voltage must be the same no matter what the load. Therefore

$$V_{(\text{ideal 1-k}\Omega\text{ load})} = V_{(\text{ideal 600-}\Omega\text{ load})}$$
$$I_1(R_{\text{int}} + R_{L_1}) = I_2(R_{\text{int}} + R_{L_2})$$
$$(30 \times 10^{-3})(R_{\text{int}} + 1 \times 10^3) = (45 \times 10^{-3})(R_{\text{int}} + 600)$$
$$(15 \times 10^{-3}\text{ A})R_{\text{int}} = 3\text{ V}$$
$$R_{\text{int}} = 200\text{ }\Omega$$

Problem 7.23 A power supply has a terminal voltage of 50 V when a 100-Ω load is connected to its terminals. The terminal voltage drops to 45 V when the load is changed to 50 Ω. What is the internal resistance of the power supply?

Maximum Power Transfer

Objective 8 State the relationship for maximum power transfer in a series resistive circuit. Determine the load resistance for maximum transfer of power in a series resistive circuit.

If we have the circuit shown in Figure 7.34, what size load should we attach to the terminals so that we can transfer the maximum power from the ideal voltage source V_I to the load resistor R_L? As we learned in the last section, the part of Figure 7.34 shown within the dashed lines could represent a power supply. If it does represent a power supply then $V_I = V_{\text{ideal source}}$ and $R_S = R_{\text{int}}$. However, we shall learn in Unit 9 on Thévenin's

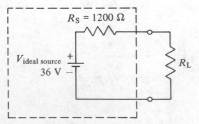

Figure 7.34 Circuit to determine the transfer of maximum power into the load

Table 7.5   Maximum Power Transfer Theorem

$$R_S = 1200 \ \Omega, \qquad V_I = 36 \text{ V}$$

$R_L,$ Ω	$R_T =$ $R_S + R_L, \Omega$	$I =$ V_I/R_T, mA	V_{R_L}, V	$P_L =$ $I^2 R_L$, mW	P_{R_S}, mW	P_T, mW	$P_L/P_T, \%$
0	1200	30	0	0	1080	1080	0
400	1600	22.5	9	202.5	607.5	810	25
600	1800	20	12	240	480	720	33.3
1200	2400	15	18	270	270	540	50
2400	3600	10	24	240	120	360	66.7
3600	4800	7.5	27	202.5	67.5	270	75
∞	∞	0	36	0	0	0	100

theorem that the part of the circuit shown within the dashed lines could represent any linear bilateral two-terminal network. So the question is an important one!

To find the answer to this question without using advanced mathematics (calculus), we have to vary the values of load resistance while we keep the series resistor R_S the same. We calculate the power in the load for each value of load resistance and see which value of R_L gives the highest power. Then we compare this value of R_L with R_S. Once this is done, we can draw conclusions for circuits with other values for the series resistor R_S. Table 7.5 shows the result of these calculations for a series resistance R_S of 1200 Ω and load resistance R_L values of 400 Ω to 3600 Ω. Table 7.5 also shows the voltage across the load resistor V_{R_L} and the percentage of total power in the load resistor, P_L/P_T.

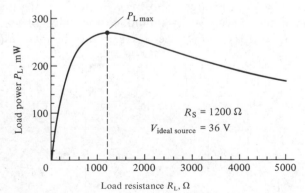

Figure 7.35   Graph of P_L versus R_L for $R_S = 1200 \ \Omega$ and $V_I = 36$ V (circuit shown in Figure 7.34)

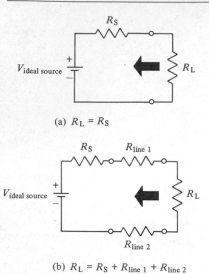

(a) $R_L = R_S$

(b) $R_L = R_S + R_{line\ 1} + R_{line\ 2}$

Figure 7.36   Maximum power transfer
theorem

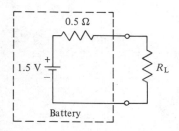

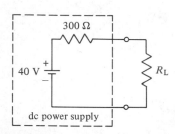

Figure 7.37   Circuits for Problem 7.25

Problem 7.24 Perform the calculations required to complete the rows in Table 7.5 for R_L equal to 1200 Ω and 2400 Ω.

Examining the results of Table 7.5, you see that the voltage across the load resistor R_L and the percentage of total power in the load resistor increase as the load resistor increases. However, the actual power in the load resistor increases up to the point at which R_L equals the series resistor R_S, then it decreases. You can see this in Table 7.5 or by studying the graph in Figure 7.35. Therefore, as a conclusion from Table 7.5 or Figure 7.35, for maximum power into the load resistor R_L, the load resistor should equal the resistance it sees when it looks back into the circuit (see the arrows in Figure 7.36). If the only other resistance in the circuit is R_S, as shown in Figure 7.36(a), then

$$R_L = R_S \tag{7.9}$$

However, if the line has some resistance, as shown in Figure 7.36(b), then

$$R_L = R_S + R_{line\ 1} + R_{line\ 2} \tag{7.10}$$

Notice that the concept of maximum power into a load resistor is different from maximum voltage across the load resistor. For maximum power into the load resistor in Figure 7.36(a), the load resistor R_L should equal the series resistor R_S. However, for maximum voltage across the load resistor, the load resistor R_L should be many times greater than the series resistor R_S.

Problem 7.25 For each of the voltage sources shown in Figure 7.37, what should be the value of the load resistor R_L for maximum power transfer into the load resistor R_L?

Problem 7.26 In the circuit shown in Figure 7.38, the line has significant resistance. What should be the value of the load resistor R_L for maximum power transfer into the load resistor?

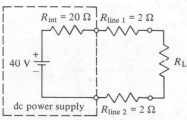

Figure 7.38   Circuit for Problem 7.26

Problem 7.27 In the circuit shown in Figure 7.39, what should be the value of the series resistor R_S for maximum power transfer into the load resistor R_L?

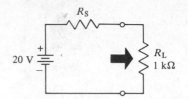

Figure 7.39 Circuit for Problem 7.27

Test

1. Three resistors (10 kΩ, 20 kΩ, and 30 kΩ) are connected in series across a 6-V power supply.
   (a) Draw the electric circuit diagram.
   (b) Determine the total resistance in the circuit.
   (c) Determine the amount of current in the circuit.
   (d) Determine the voltage drop across each resistor.
2. Given the circuit in Figure 7.40, determine the following.
   (a) How the resistors are connected
   (b) Total resistance in the circuit
   (c) Amount of current in the circuit
   (d) Voltage drop across each resistor
   (e) The sum of the voltage drops across each resistor
3. We want to run a 6-V car radio on a 12-V battery. When operating on 6 V, the radio draws 3 A. What resistor (resistance and wattage rating) must we use?
4. For the circuit diagram shown in Figure 7.12, fill in the unknown values in Table 7.6.
5. Solve for V_2 in the circuit diagram of Figure 7.41.

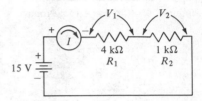

Figure 7.40 Circuit for Test Problem 2

Table 7.6

E	V_1	V_2
15 V	10 V	
21 V		9 V
	6 V	3 V

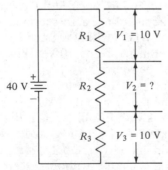

Figure 7.41 Circuit for Test Problem 5

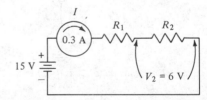

Figure 7.42 Circuit for Test Problem 6

6. Find R_1 and R_2 in the two-resistor series circuit of Figure 7.42.
7. Change the current of 0.3 A given in Test Problem 6 to 2 mA, and solve for R_1 and R_2.
8. Solve for I and then for R_1 in the two-resistor series circuit of Figure 7.43.

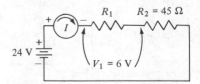

Figure 7.43 Circuit for Test Problem 8

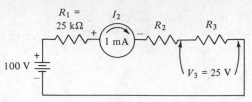

Figure 7.44 Circuit for Test Problem 9

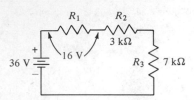

Figure 7.45 Circuit for Test Problem 10

9. Find R_2 and R_3 in the circuit shown in Figure 7.44.
10. Find R_1 in the circuit shown in Figure 7.45.
11. For the circuit diagram shown in Figure 7.12, complete Table 7.7. Use the voltage-divider principle.

Table 7.7

E	R_1	R_2	V_1	V_2
45 V	40 Ω	50 Ω	—	
20 V	5 kΩ	15 kΩ		—
20 V	6 kΩ	4 kΩ		
30 V	1 kΩ	9 kΩ		

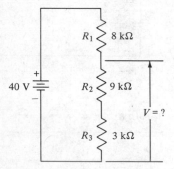

Figure 7.46 Circuit for Test Problem 12

12. Use the voltage-divider theorem to solve for V in the circuit of Figure 7.46.
13. Refer to the circuit shown in Figure. 7.47. What is V_{out} for each of the following conditions? (a) R_1 open (b) R_1 shorted   (c) R_2 open   (d) R_2 shorted
14. The open-circuit voltage across terminals A–B of Figure 7.48 is 12 V. When a 10-Ω resistor is connected across A–B, the voltmeter reads 10 V. What is the internal resistance of the source?
15. What should the load resistor R_L of Figure 7.49 equal for maximum power into the load?
16. Three resistors with the following ratings are connected in series: 2 kΩ 5 W; 1 kΩ 10 W; and 20 kΩ 18 W. What is the maximum power-supply voltage that may be applied to the circuit without exceeding the power rating of any resistor? [*Hint*: Solve for I first.]

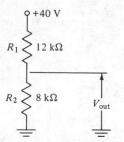

Figure 7.47 Circuit for Test Problem 13

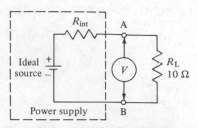

Figure 7.48 Circuit for Test Problem 14

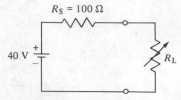

Figure 7.49 Circuit for Test Problem 15

Parallel and Unit 8
Series-Parallel
Resistive
Circuits

In Unit 7 you learned how to connect elements in series, how to calculate the total resistance of resistors connected in series, and how to change the voltage across a load resistor by inserting a resistor in series with it.

However, as you look at the complicated circuits shown in Unit 1, in particular Figure 1.5, you see that they are made up of more than just series circuits. Most of the circuit elements in Figure 1.5 are connected in parallel or in series-parallel with the other elements in the circuit. This is what this unit is all about.

In your home, your electrical outlets are connected in parallel. In your car, your lights, radio, tape deck, etc., are all connected in parallel across your battery. Why aren't they connected in series? Stop and think about this for a minute.

Objectives

After completing all the work associated with this unit, you should be able to:

1. State what is meant when we say that two or more circuit elements are connected in parallel.
2. State Kirchhoff's current law (KCL). Use KCL to solve for the current in a particular resistor in parallel resistive circuits.
3. Calculate the equivalent resistance when two, three, or more

resistors are connected in parallel. Then solve for any unknown current and/or power.

4. Calculate the current in a particular resistor in a parallel resistive circuit using the current-divider principle.

5. Define a series-parallel resistive circuit. You may use a drawing.

6. Solve for current, voltage, resistance, and/or power in any part of or in the entire series-parallel resistive circuit.

7. Determine what happens to resistance and voltage measurements in parallel and series-parallel resistive circuits when a resistor opens or is shorted.

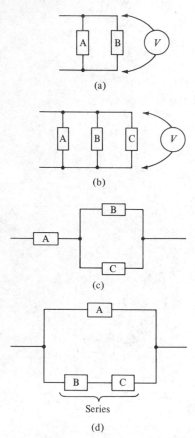

Figure 8.1 Circuit elements connected in parallel or series-parallel

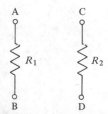

Figure 8.2 Two resistors, R_1 and R_2.

Parallel Connection

Objective 1 State what is meant when we say that two or more circuit elements are connected in parallel.

Circuit elements are said to be connected *in parallel* when the same voltage exists across each element. Since voltage is measured across a circuit element, this means that circuit elements are in parallel when they are all connected between the same two junctions or common points. In Figure 8.1 you see some examples of circuit elements connected in parallel. The circuit elements are not identified because they could be anything in parallel, such as resistors, resistors and capacitors, or a resistor in parallel with two leads of a transistor. In Figure 8.1(a), two circuit elements, A and B, are connected in parallel. Note that the voltage V across both A and B is the same.

In Figure 8.1(b), three circuit elements, A, B, and C, are connected in parallel. Again note that the voltage V across each element is the same. In Figure 8.1(c) and (d) you see some series-parallel combinations. We shall discuss Figure 8.1(c) and (d) in more detail when we get to Objectives 5 and 6. But now let's see how the circuit elements shown in Figure 8.1(c) and (d) are connected. In Figure 8.1(c), only circuit elements B and C are in parallel. When you consider all three elements, element A is in series with the parallel combination of B and C. In Figure 8.1(d), neither element B nor element C is in parallel with element A. But the *series* combination of elements B and C is in parallel with circuit element A.

Let's take the two resistors called R_1 and R_2 shown in Figure 8.2 and connect them in parallel across the power supply E. The terminals of the resistors are labeled so that you can see the connections.

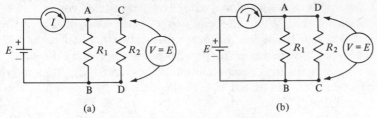

Figure 8.3 Resistors R_1 and R_2 connected in parallel

In Figure 8.3, you see two possible connections. To connect R_1 and R_2 in parallel across the power supply E, you must connect one terminal of resistor R_1 to one terminal of resistor R_2. Then connect the free terminal of resistor R_1 to the free terminal of resistor R_2. In Figure 8.3(a), you see that terminal A of resistor R_1 is connected to terminal C of resistor R_2, and terminal B is connected to terminal D.

In Figure 8.3(b), we have flipped resistor R_2 upside down and connected terminals D to A and C to B. Once the resistors are connected in parallel, then connected across the power supply E, as shown in Figure 8.3, a source current I flows. But now this current *splits* to flow through both resistors R_1 and R_2. The current split depends on the values of R_1 and R_2. When you look at Figure 8.3, you see that resistors R_1 and R_2 have the same voltage across them. This voltage is equal to the supply voltage E, because we assume that the ammeter is ideal, and so no voltage is dropped across it. Therefore, since the same voltage is applied across both resistors R_1 and R_2, we say that they are connected in parallel.

Problem 8.1 Given the three resistors shown in Figure 8.4, draw a circuit diagram connecting them in parallel across a power supply E.

Kirchhoff's Current Law

Objective 2 State Kirchhoff's current law (KCL). Use KCL to solve for the current in a particular resistor in parallel resistive circuits.

To explain a new concept in engineering, we try to make a comparison or analogy with something you already understand. An electrical system is very similar to a closed hydraulic or water system. Therefore we shall use a hydraulic or water analogy to explain what happens in an electric circuit when elements are connected in parallel.

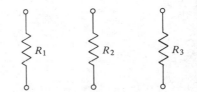

Figure 8.4 Three resistors, R_1, R_2, and R_3, for Problem 8.1

In Figure 8.5, which is a closed water system, you see two parallel paths. What happens to the water when it comes to junction A? It splits and goes into the two parallel paths. If the sizes of the tubes in the parallel paths are exactly the same, the water splits into equal amounts. The water splits differently depending on the actual size of the tube or the resistance the water meets.

The electrical system in Figure 8.6 is similar to the hydraulic system in Figure 8.5. Current is a *rate* of electron flow. When the current gets to junction A, it splits and goes through resistors R_1 and R_2. The water flow in Figure 8.5 or the current in Figure 8.6 that comes into junction A splits in the parallel path into two flows, and the two flows must total to equal the incoming flow. At junction B, the two flows must again come together and flow back to the pump. Since this is a closed system, the flow back to the pump must equal the flow out of the pump. Kirchhoff's current law states this concept that we have been discussing.

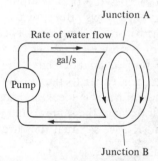

Figure 8.5 Closed water system analogy of a parallel electric circuit

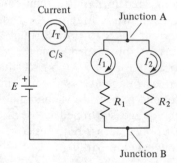

Figure 8.6 Parallel resistor electrical system

Kirchhoff's Current Law (KCL)
The algebraic sum of the currents leaving a junction in a circuit must equal the algebraic sum of the currents flowing into the junction.

$$\text{Current out} = \text{current in} \qquad (8.1)$$

KCL states that in Figure 8.6, I_T is equal to the sum of I_1 and I_2.

To illustrate KCL, the circuits are shown as in Figure 8.7, in which only the *junction* or node is shown. It is understood that power supplies must be connected to the circuit if the currents are to actually flow. We have isolated the junction that we want to work with, because the complete circuit might be so complicated that it would be difficult to find the actual junction. If we apply KCL to the junction shown in Figure 8.7, with I_1 the current that is flowing into the junction, then I_1 must be equal to the sum of I_2 plus I_3 plus I_4. Referring to Figure 8.8, you see the junction of a circuit very similar to the one shown in Figure 8.7. But now you see actual values for the three currents. Do Problem 8.2.

Problem 8.2   Refer to Figure 8.8 and solve for the current I_4.

In Figure 8.9, you see two junctions of a circuit. Example 8.1 shows you how to find currents I_1 and I_2 for this portion of the circuit.

Example 8.1   Refer to Figure 8.9. Solve for the current I_1, and also determine the value and direction (into or out of the junction) of current I_2.

Solution   Using KCL at junction A:

$$\text{Current out} = \text{current in}$$
$$I_1 = 8\text{ A} + 7\text{ A} = 15\text{ A}$$

Using KCL at junction B:

$$I_2 = \text{current in} - \text{current out}$$
$$= (I_1 + 3\text{ A}) - 5\text{ A}$$
$$= 15 + 3 - 5 = 13\text{ A } \textit{out} \text{ of the junction}$$

If we had obtained a minus sign, it would have meant that the current was flowing in the opposite direction.

Some authors state KCL in a different form. They say that the algebraic sum (represented by $\sum$, the Greek capital letter sigma) of the currents at any junction point or node must equal zero.

$$\sum I\text{'s} = 0 \qquad\qquad (8.2)$$

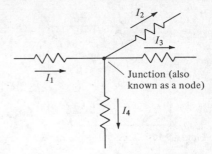

Figure 8.7   Circuit to illustrate KCL

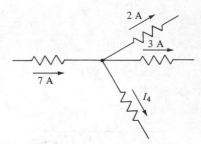

Figure 8.8   Circuit for Problem 8.2

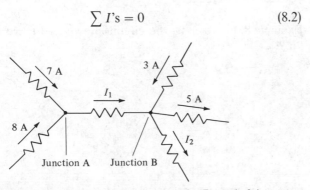

Figure 8.9   Section of circuit showing two junctions for Example 8.1

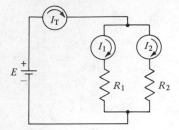

Figure 8.10   Circuit for Problem 8.3

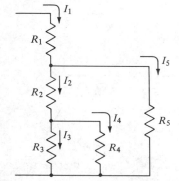

Figure 8.11   Circuit for Problem 8.4

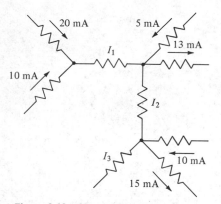

Figure 8.12   Circuit for Problem 8.5

The currents flowing into the junction are assigned a plus sign, and the currents flowing out of the junction are assigned a minus sign. Example 8.2 uses this form of KCL for the circuit in Figure 8.9.

Example 8.2   Refer to Figure 8.9. At junction B, find the $\sum I$'s.

Solution   Assuming that I_2 flows out, at junction B we have

$$+I_1 + 3\,A - 5\,A - I_2 = ?$$
$$+15\,A + 3\,A - 5\,A - 13\,A = 0$$

For the following problems, don't worry about the complicated circuits shown. For example, Figure 8.11 has series resistors, parallel resistors, and combinations of both. We are not concerned now with how the resistors are connected. We just want to see how KCL applies to the circuits.

Problem 8.3   Refer to Figure 8.10.
(a) Given that $I_T = 400$ mA and $I_1 = 300$ mA, then $I_2 =$
(b) Given that $I_T = 600$ mA and $I_2 = 400$ mA, then $I_1 =$
(c) Given that $I_1 = 20$ mA and $I_2 = 30$ mA, then $I_T =$

Problem 8.4   Refer to Figure 8.11.
(a) Given that $I_3 = 10$ mA, $I_4 = 5$ mA, and $I_5 = 20$ mA, what do I_2 and I_1 equal?
(b) Given that $I_1 = 1$ A, $I_5 = 600$ mA, and $I_4 = 300$ mA, what do I_2 and I_3 equal?

Problem 8.5   Refer to Figure 8.12. Solve for the value and direction of I_1, I_2, and I_3.

To solve various parallel-circuit problems, you may have to use both KCL and Ohm's law. The following example shows you how to do this.

Example 8.3   Refer to the circuit shown in Figure 8.13. Solve for R_2.

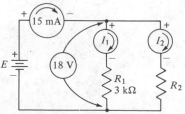

Figure 8.13   Circuit for Example 8.3

Solution   We can determine I_1 by Ohm's law.

$$I_1 = \frac{V_1}{R_1} = \frac{18 \text{ V}}{3 \text{ k}\Omega} = 6 \times 10^{-3} = 6 \text{ mA}$$

From KCL, we obtain

$$I_2 = I_T - I_1 = 15 \text{ mA} - 6 \text{ mA} = 9 \text{ mA}$$

Then we can determine R_2 by Ohm's law.

$$R_2 = \frac{V_2}{I_2} = \frac{18 \text{ V}}{9 \text{ mA}} = 2 \text{ k}\Omega$$

Problem 8.6   Refer to the circuit shown in Figure 8.14. Solve for R_1.

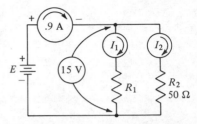

Figure 8.14   Circuit for Problem 8.6

Calculation of Equivalent Resistance

Objective 3   Calculate the equivalent resistance when two, three, or more resistors are connected in parallel. Then solve for any unknown current and/or power.

To see how we can work with parallel resistors in a circuit, let's replace the two parallel resistors R_1 and R_2 of Figure 8.15(a) by one resistor, which we call R_{eq}, as shown in Figure 8.15(b). This one resistor must act the same as the two parallel resistors in the circuit. That is, the power supply E "sees" the same load in both Figure 8.15(a) and (b), and therefore the total current, which is designated by I_T, must be the same.

You are probably wondering why we replace the two parallel resistors by one resistor. We are doing this so that we can theoretically see just what effect paralleling resistors has on an electric circuit.

Using KCL in Figure 8.15(a), we have

$$I_T = I_1 + I_2$$

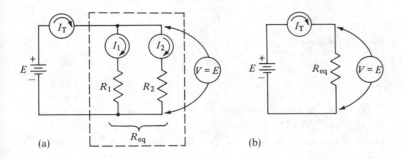

(a)   (b)

Figure 8.15   Circuit showing equivalent parallel resistance R_{eq}

Using Ohm's law and expressing these currents in terms of voltage divided by resistance, we have

$$\frac{E}{R_{eq}} = \frac{E}{R_1} + \frac{E}{R_2}$$

From Figure 8.15(b), the total current is equal to the supply voltage E divided by the equivalent resistance. This total current must also equal the current in resistor R_1, which is E divided by R_1, plus the current in resistor R_2, which is E divided by R_2. Since E appears on both sides of the equals sign, we can divide both sides of the equation by E, giving

$$\frac{1}{R_{eq}} = \frac{1}{R_1} + \frac{1}{R_2} \tag{8.3}$$

When there are only two resistors in parallel, we can use a simpler method than that indicated by Equation (8.3). Let us take Equation (8.3) and solve it for R_{eq} by finding a common denominator.

$$\frac{1}{R_{eq}} = \frac{1}{R_1} + \frac{1}{R_2}$$

$$= \frac{R_2}{R_1 R_2} + \frac{R_1}{R_1 R_2} = \frac{R_1 + R_2}{R_1 R_2}$$

Solving for R_{eq}, we find, for two resistors in parallel,

$$\boxed{R_{eq} = \frac{R_1 R_2}{R_1 + R_2}} \tag{8.4}$$

So, to find the equivalent resistance of two resistors connected in parallel, we take the *product* of the two resistors divided by the *sum* of the two resistors.

Formulas for three or more resistors are developed in a manner similar to that used to develop Equation (8.3).

For three resistors in parallel,

$$\frac{1}{R_{eq}} = \frac{1}{R_1} + \frac{1}{R_2} + \frac{1}{R_3} \tag{8.5}$$

For n resistors in parallel,

$$\frac{1}{R_{eq}} = \frac{1}{R_1} + \frac{1}{R_2} + \frac{1}{R_3} + \cdots + \frac{1}{R_n} \tag{8.6}$$

Because resistors usually have large values, such as 1 kΩ, 10 kΩ, or 500 kΩ, we may find it easier to use Equation (8.4) twice for three resistors in parallel. Referring to Figure 8.16,

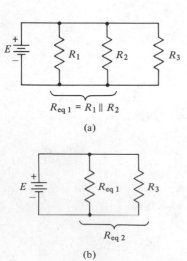

$R_{eq\,1} = R_1 \parallel R_2$

(a)

$R_{eq\,2}$

(b)

Figure 8.16 Reduction of resistors R_1 and R_2 to an equivalent resistance $R_{eq\,1}$

we do this as follows:

Step 1. $\quad R_{eq_1} = \dfrac{R_1 R_2}{R_1 + R_2}$

Step 2. $\quad R_{eq_2} = \dfrac{R_{eq_1} R_3}{R_{eq_1} + R_3}$

You first determine an equivalent resistance for two resistors. Then you take this equivalent resistance in parallel with the third resistor. And you may use the product over the sum.

When resistors of the same value are connected in parallel, we may use a shortcut to determine R_{eq}.

$$R_{eq} = \frac{R}{N} \tag{8.7}$$

where R is equal to the value of one resistor and N is the number of resistors.

The *equivalent* resistance in a parallel circuit is also the *total* resistance of the parallel circuit. We have preferred to use the term *equivalent* instead of *total* in this objective, because so many times, when asked for a total, a person is inclined to add the numbers. And in a parallel connection resistors are not added!

Here are two statements that are very helpful when you work with parallel circuits. As you read the examples and work the problems in this section, you will see why these statements are true.

The equivalent resistance of a parallel circuit is always smaller than the smallest resistor.

The equivalent resistance of two equal resistors is equal to half the resistance of one resistor.

Example 8.4 Refer to Figure 8.17. Find R_{eq} and I. Calculate the current I in two different ways.

Solution

$$R_{eq} = R_1 \| R_2$$

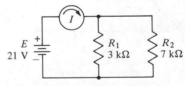

Figure 8.17 Circuit for Example 8.4

Instead of writing out the fact that R_1 is in parallel with R_2, we use the symbol $\|$ to mean "in parallel with."

Since there are only two resistors in parallel, we may use

$$R_{eq} = \frac{\text{product}}{\text{sum}} = \frac{R_1 R_2}{R_1 + R_2} = \frac{3 \text{ k}\Omega(7 \text{ k}\Omega)}{3 \text{ k}\Omega + 7 \text{ k}\Omega} = 2.1 \text{ k}\Omega$$

$$I = \frac{E}{R_{eq}} = \frac{21 \text{ V}}{2.1 \text{ k}\Omega} = 10 \text{ mA}$$

A second method of solving for I is by using KCL. I must equal the sum of the current in R_1 (called I_1) and the current in R_2 (called I_2). Since the voltage in a parallel circuit is common, 21 V is across R_1 and 21 V is across R_2. Therefore

$$I_1 = \frac{E}{R_1} = \frac{21 \text{ V}}{3 \text{ k}\Omega} = 7 \text{ mA}, \qquad I_2 = \frac{E}{R_2} = \frac{21 \text{ V}}{7 \text{ k}\Omega} = 3 \text{ mA}$$

$$\therefore \quad I = I_1 + I_2 = 7 \text{ mA} + 3 \text{ mA} = 10 \text{ mA}$$

Problem 8.7   Referring to Figure 8.15, find R_{eq} for the resistor combinations given in Table 8.1.

Table 8.1

	R_1	R_2	R_{eq}
(a)	100 Ω	150 Ω	_____
(b)	100 Ω	400 Ω	_____
(c)	500 Ω	500 Ω	_____
(d)	4 kΩ	6 kΩ	_____
(e)	20 kΩ	80 kΩ	_____
(f)	100 kΩ	100 kΩ	_____

Problem 8.8   Referring to Figure 8.15 and the supply voltage E given in Table 8.2, find I_T for each of the resistor combinations of Problem 8.7.   Use Ohm's law: $I_T = E/R_{eq}$.

Table 8.2

	R_{eq} from Problem 8.7	E, V	I_T, A
(a)		30	_____
(b)		20	_____
(c)		10	_____
(d)		12	_____
(e)		16	_____
(f)		20	_____

Problem 8.9   Refer to Figure 8.15 and Table 8.3. For each resistor combination, find I_1 and I_2 by using Ohm's law. Then find I_T by using KCL.

Table 8.3

	R_1	R_2	E, V	I_1	I_2	I_T
(a)	100 Ω	150 Ω	30	_____	_____	_____
(b)	100 Ω	400 Ω	20	_____	_____	_____
(c)	500 Ω	500 Ω	10	_____	_____	_____
(d)	4 kΩ	6 kΩ	12	_____	_____	_____
(e)	20 kΩ	80 kΩ	16	_____	_____	_____
(f)	100 kΩ	100 kΩ	20	_____	_____	_____

Example 8.5 Find the equivalent or total resistance in the circuit shown in Figure 8.18.

Solution First let us solve this problem by repeated application of Equation (8.4).

$$R_{eq_1} = \frac{R_1 R_2}{R_1 + R_2} = \frac{10 \text{ k}\Omega(15 \text{ k}\Omega)}{10 \text{ k}\Omega + 15 \text{ k}\Omega} = 6 \text{ k}\Omega$$

$$R_{eq} = R_{eq_2} = \frac{R_{eq_1} R_3}{R_{eq_1} + R_3} = \frac{6 \text{ k}\Omega(4 \text{ k}\Omega)}{6 \text{ k}\Omega + 4 \text{ k}\Omega} = 2.4 \text{ k}\Omega$$

We shall also solve this problem using Equation (8.5).

$$\frac{1}{R_{eq}} = \frac{1}{R_1} + \frac{1}{R_2} + \frac{1}{R_3}$$

$$= \frac{1}{10 \text{ k}\Omega} + \frac{1}{15 \text{ k}\Omega} + \frac{1}{4 \text{ k}\Omega}$$

$$= 0.0001 + 0.00006667 + 0.00025 = 0.00041667$$

$$\therefore \quad R_{eq} = \frac{1}{0.00041667} = 2.4 \text{ k}\Omega$$

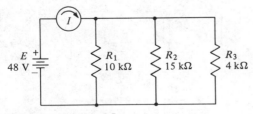

Figure 8.18 Circuit for Example 8.5

For these values of resistors, it is easy to find a common denominator instead of working with each $1/R$ individually. If we do this, then

$$\frac{1}{R_{eq}} = \frac{6 + 4 + 15}{60 \text{ k}\Omega} \qquad \therefore \quad R_{eq} = \frac{60 \text{ k}\Omega}{25} = 2.4 \text{ k}\Omega$$

You can obtain a rough approximation of the solution to this problem (Example 8.5) by using the rules given on page 145. For laboratory work the rough approximation may be all that you need. First we notice that the 4-kΩ resistor is the smallest, and so we know that R_{eq} is less than 4 kΩ. If the other two resistors are 10 times or more greater than 4 kΩ, then the answer is close to 4 kΩ. If the other two resistors are close to 4 kΩ, then the answer is near 2 kΩ. A rough approximation for the answer may be 3 kΩ.

Example 8.6   Refer to the circuit in Figure 8.19. Solve for R_2 and R_3 and for P_1, P_2, and P_3.

Solution   To solve for R_2, we need to know the voltage across R_2 and the current through R_2. Do we know these? Yes. The current as shown is 9 mA. The voltage across R_3 is shown as 18 V. Remember that voltage across parallel resistors is common, and therefore the 18 V also appears across R_1 and R_2. Thus

$$R_2 = \frac{V}{I_2} = \frac{18 \text{ V}}{9 \text{ mA}} = 2 \text{ k}\Omega$$

As always, to solve for R_3, we need to know the voltage across R_3 and the current through R_3. We know the voltage, but it seems that we don't know the current. However, we carefully examine the data given, we can find the current in resistor R_3. From KCL,

$$I = I_1 + I_2 + I_3$$

To solve this equation for I_3, we must know everything else. At first it appears that I_1 is also unknown, but if we look at the

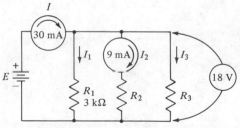

Figure 8.19   Circuit for Example 8.6

circuit carefully, we see that we can determine I_1 by using Ohm's law.

$$I_1 = \frac{V}{R_1} = \frac{18 \text{ V}}{3 \text{ k}\Omega} = 6 \text{ mA}$$

Therefore, solving KCL for I_3, we obtain

$$I_3 = I - I_1 - I_2$$
$$= 30 - 6 - 9 = 15 \text{ mA}$$

Therefore

$$R_3 = \frac{V}{I_3} = \frac{18 \text{ V}}{15 \text{ mA}} = 1.2 \text{ k}\Omega$$

To solve for the power in the resistors, you may use any one of the three power formulas. For example, for P_1, $P_1 = I_1^2 R_1$, $P_1 = V^2/R_1$, and $P_1 = VI_1$.

$$P_1 = VI_1 = (18 \text{ V})(6 \text{ mA}) = 108 \text{ mW}$$
$$P_2 = VI_2 = (18 \text{ V})(9 \text{ mA}) = 162 \text{ mW}$$
$$P_3 = VI_3 = (18 \text{ V})(15 \text{ mA}) = 270 \text{ mW}$$

Problem 8.10 Find the equivalent resistance in the circuit shown in Figure 8.20. Solve this problem two ways. (Refer to Example 8.5.)

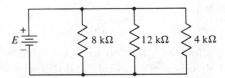

Figure 8.20 Circuit for Problem 8.10

Problem 8.11 Find the equivalent or total resistance in the circuit shown in Figure 8.21 by repeated application of Equation (8.4). Start with the two equal resistors.

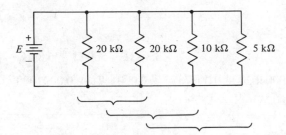

Figure 8.21 Circuit for Problem 8.11

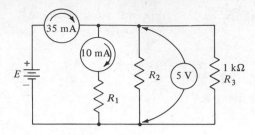

Figure 8.22   Circuit for Problem 8.12

Problem 8.12   Refer to the circuit shown in Figure 8.22. Solve for R_1, R_2, P_1, P_2, and P_3.

In the unit on resistance and power, we defined conductance G as the reciprocal of resistance: $G = 1/R$. When we use this definition for conductance, Equations (8.5) and (8.6) in terms of conductance become, for three resistors in parallel,

$$G_T = G_1 + G_2 + G_3 \qquad (8.8)$$

For n resistors in parallel,

$$G_T = G_1 + G_2 + G_3 + \cdots + G_n \qquad (8.9)$$

You may use Equations (8.8) and (8.9) to solve for the total conductance of a circuit. Once you know this total conductance, you may solve for the equivalent resistance. From the definition of conductance, the equivalent resistance is equal to

$$R_{eq} = \frac{1}{G_T} \qquad (8.10)$$

Example 8.7 shows you how to solve for the equivalent resistance of parallel resistors by using conductances. As you read this example, note that you add conductances in parallel circuits to find total conductance G_T, just as you add resistors in series circuits to find total resistance R_T.

Example 8.7   Four resistors, 3.3 kΩ, 560 Ω, 6.8 kΩ, and 10 kΩ, are connected in parallel. Determine the equivalent resistance, using conductance.

Solution

$$G_T = G_1 + G_2 + G_3 + G_4 = \frac{1}{3.3\ k\Omega} + \frac{1}{560\ \Omega} + \frac{1}{6.8\ k\Omega} + \frac{1}{10\ k\Omega}$$

$$= 0.0003030 + 0.0017857 + 0.0001471 + 0.0001000$$

$$= 0.002336\ S = 2.336\ mS$$

$$R_{eq} = \frac{1}{G_T} = \frac{1}{0.002336} = 428\ \Omega$$

Did you notice that using conductances in parallel is just the same as using the formula for resistors in series? When you use conductance, you determine the conductance of each resistor in parallel separately and then add them to determine the total conductance. Whereas, when you use Equations (8.5) and (8.6), you may find a common denominator that may make the mathematics a little simpler. This was illustrated in Example 8.5.

Problem 8.13   Four resistors, 22 kΩ, 3.9 kΩ, 100 kΩ, and 15 kΩ, are connected in parallel. Determine the equivalent resistance using conductance.

Current-Divider Principle

Objective 4   Calculate the current in a particular resistor in a parallel resistive circuit, using the current-divider principle.

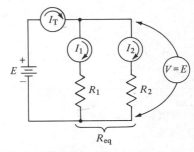

Figure 8.23   Circuit illustrating the current-divider principle

When you have a parallel circuit such as the one shown in Figure 8.23, what happens to the current I_T when it comes to the junction of the two parallel resistors R_1 and R_2? The total current I_T splits into two currents, I_1 and I_2. Therefore we are dividing the total current between the two resistors. The *current-divider principle* says that when two resistors are in parallel, to find the current in one resistor, you take the ratio of the *opposite* resistor divided by the sum of the two resistors and multiply this ratio by the total current flowing into the junction. To solve for the current I_2 in Figure 8.23 using the current-divider principle,

$$I_2 = \left(\frac{R_1}{R_1 + R_2}\right) I_T \qquad (8.11)$$

There is a similar formula for I_1:

$$I_1 = \left(\frac{R_2}{R_1 + R_2}\right) I_T \qquad (8.12)$$

However, once you have solved for I_2, it is probably easier to solve for I_1 using KCL.

We shall show you how to develop Equation (8.11). To solve for the current I_2, we use Ohm's law.

$$I_2 = \frac{E}{R_2}$$

But what is E equal to? From the simplified circuit in Figure 8.15(b), we see that

$$E = I_T R_{eq}$$

where R_{eq} is the equivalent resistance of the two resistors R_1 and R_2 in parallel.

$$E = I_T\left(\frac{R_1 R_2}{R_1 + R_2}\right)$$

Substituting this equation for E into the formula for I_2, we obtain

$$I_2 = \frac{E}{R_2} = \frac{I_T\left(\dfrac{R_1 R_2}{R_1 + R_2}\right)}{R_2} = \left(\frac{R_1}{R_1 + R_2}\right) I_T$$

Note that if you want the current in R_2, you use the ratio of the resistor *opposite* to R_2 divided by the sum of the two parallel resistors, then multiply this ratio by the total current.

Example 8.8   Refer to the circuit in Figure 8.17. Given that I is equal to 10 mA, solve for I_1 and I_2 by the current-divider principle.

Solution

$$I_1 = \left(\frac{R_2}{R_1 + R_2}\right) I_T = \left(\frac{7\text{ k}\Omega}{3\text{ k}\Omega + 7\text{ k}\Omega}\right) 10\text{ mA} = \left(\frac{7}{10}\right) 10\text{ mA}$$

$$= 7\text{ mA}$$

$$I_2 = \left(\frac{R_1}{R_1 + R_2}\right) I_T = \left(\frac{3\text{ k}\Omega}{10\text{ k}\Omega}\right) 10\text{ mA} = 3\text{ mA}$$

Problem 8.14   For the two-resistor parallel circuit shown in Figure 8.23, fill in the unknown values in Table 8.4. Use the current-divider principle.

Equations (8.11) and (8.12) are for two resistors in parallel. We may still use the current-divider principle if we have *three* resistors in parallel. But the first thing that we must do is to reduce the

Table 8.4

	I_T	R_1	R_2	I_1	I_2
(a)	5 A	100 Ω	100 Ω		
(b)	2 A	20 Ω	180 Ω		
(c)	20 mA	2 kΩ	8 kΩ		
(d)	5 mA	5 kΩ	20 kΩ		
(e)	18 mA	100 kΩ	350 kΩ		

circuit to two parallel resistors. That is, find the equivalent resistance of the two resistors opposite the one in which you want the current. Problem 8.15 illustrates this.

Problem 8.15   Refer to the circuit in Figure 8.24. Given that I_T is equal to 20 mA, solve for I_3 by the current-divider principle. To use the current-divider principle, replace the two parallel 10-kΩ resistors by their equivalent resistance.

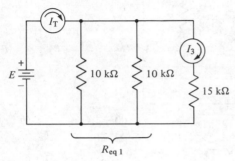

Figure 8.24   Circuit for Problem 8.15

Series-Parallel Resistive Circuits

Objective 5   Define a series-parallel resistive circuit. You may use a drawing.

A series-parallel resistive circuit is a *combination of series resistors and parallel resistors connected* in a circuit; see Figure 8.25.

In Figure 8.25, box A is in series with the parallel combination of boxes B and C. The circuit elements are shown as boxes because there may be more than one resistor making up the resistance of the box. For example, box A of Figure 8.25 may be composed of a couple of series resistors that can be reduced to one series resistor. Or box A may be made up of two parallel resistors, as shown in Figure 8.26(c). Or there may be no box A; that is, there may be no series resistance before you get to the junction of boxes B and C.

In this case, for Figure 8.26(b) and (c), you may prefer to call the circuit a *parallel-series* resistive circuit. Similar comments apply to what boxes B and C represent. They may be single resistors, as shown in Figure 8.26(a), resistors in series, as shown in Figure 8.26(b), or resistors in parallel, as shown in Figure 8.26(c). Think about what would happen if box B or C had no resistance, that is, if either one were a direct short.

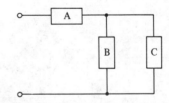

Figure 8.25   Basic series-parallel circuit

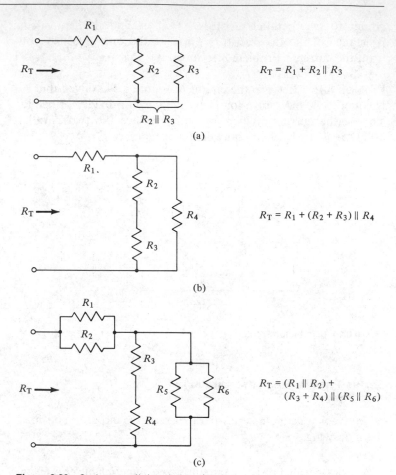

Figure 8.26　Series-parallel resistive circuits

Problem 8.16　Go through this unit and pick out all the series-parallel resistive circuits you have seen up to this point.

Objective 6　Solve for current, voltage, resistance, and/or power in any part or in the entire series-parallel resistive circuit.

You already know how to solve series-parallel resistive circuits. You know how to replace two or more series resistors by one total resistance. You know how to replace two or more parallel resistors by one equivalent resistor. Therefore, when you have a series-parallel resistive circuit, you just combine these. You look for a part of the circuit in which you can replace two or more resistors by their total or equivalent resistance. Continue to do this until you have reduced the circuit to one resistor across the power supply. Now you can find the total current. If you need to find the remaining currents, just reverse what you did to reduce the circuit. The current-divider principle should help you find the remaining currents. Example 8.10 illustrates this.

The ladder network shown in Figure 8.27 looks very complicated. However, if you start with the three series resistors (R_5, R_6, and R_7) on the extreme right, the circuit reductions are straightforward. Study the various reductions shown in Figure 8.28. If you can solve this type of circuit, you should be able to solve any series-parallel resistive circuit. Study the following examples, and especially look for the method used to start each problem. Because getting started is usually the hardest part of the problem.

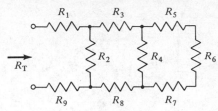

Figure 8.27 Ladder network

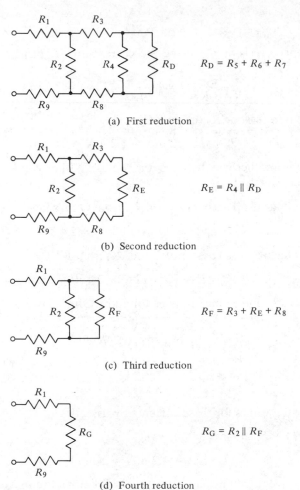

$$R_D = R_5 + R_6 + R_7$$

(a) First reduction

$$R_E = R_4 \| R_D$$

(b) Second reduction

$$R_F = R_3 + R_E + R_8$$

(c) Third reduction

$$R_G = R_2 \| R_F$$

(d) Fourth reduction

Figure 8.28 Reduction of the ladder network given in Figure 8.27

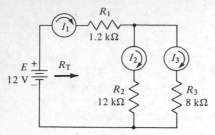

Figure 8.29 Circuit for Example 8.9

Example 8.9 Refer to the circuit in Figure 8.29. Solve for the total resistance as seen by the source R_T, and then solve for I_1, I_2, I_3, V_1, V_2, V_3, P_1, P_2, and P_3.

Solution

Total resistance: $R_T = R_1 + R_2 \| R_3 = 1.2 \text{ k}\Omega + \dfrac{12 \text{ k}\Omega (8 \text{ k}\Omega)}{12 \text{ k}\Omega + 8 \text{ k}\Omega}$

$$= 1.2 \text{ k}\Omega + \frac{96}{20} \text{ k}\Omega = 1.2 \text{ k}\Omega + 4.8 \text{ k}\Omega$$

$$= 6 \text{ k}\Omega$$

Total current, using Ohm's law:

$$I_1 = \frac{E}{R_T} = \frac{12 \text{ V}}{6 \text{ k}\Omega} = 2 \text{ mA}$$

Current in resistor R_3, using the current-divider principle:

$$I_3 = \frac{R_2}{R_2 + R_3} I_1 = \left(\frac{12 \text{ k}\Omega}{20 \text{ k}\Omega} \right) 2 \text{ mA} = 1.2 \text{ mA}$$

Current in resistor R_2, using KCL:

$$I_2 = I_1 - I_3 = 2 \text{ mA} - 1.2 \text{ mA} = 0.8 \text{ mA}$$

There is another way to solve for I_2 and I_3, once you know the total current I_1.

The voltage drop across resistor R_1, from Ohm's law,

$$V_1 = I_1 R_1 = 2 \text{ mA} (1.2 \text{ k}\Omega) = 2.4 \text{ V}$$

The voltage across resistors R_2 and R_3, from KVL,

$$V_2 = V_3 = E - V_1 = 12 \text{ V} - 2.4 \text{ V} = 9.6 \text{ V}$$

Currents I_2 and I_3 may now be found by Ohm's law:

$$I_2 = \frac{V_2}{R_2} = \frac{9.6 \text{ V}}{12 \text{ k}\Omega} = 0.8 \text{ mA}$$

$$I_3 = \frac{V_3}{R_3} = \frac{9.6 \text{ V}}{8 \text{ k}\Omega} = 1.2 \text{ mA}$$

Check the values for these currents by using KCL.

Does $I_2 + I_3$ equal I_1?

$$0.8 \text{ mA} + 1.2 \text{ mA} = 2.0 \text{ mA}$$

With current, voltage, and resistance all known, we may choose any form of the power equation to determine the power in each of the resistors.

$$P_1 = I_1 V_1 = 2 \text{ mA} (2.4 \text{ V}) = 4.8 \text{ mW}$$

or

$$P_1 = I_1^2 R_1 = (2 \times 10^{-3})^2 (1.2 \times 10^3) = 4.8 \text{ mW}$$

$$P_2 = I_2 V_2 = (0.8 \text{ mA})(9.6 \text{ V}) = 7.68 \text{ mW}$$

$$P_3 = I_3 V_3 = (1.2 \text{ mA})(9.6 \text{ V}) = 11.52 \text{ mW}$$

Example 8.10 Refer to the circuit in Figure 8.30. Solve for the total resistance "seen" from the power supply, and then solve for the current through each resistor.

Solution Studying Figure 8.30(a), we see that

$$I_9 = I_1$$

The value of the current returning to the source must equal the value of the current leaving the source.

$$I_5 = I_6 = I_7$$

This section of the circuit is a series circuit.

To reduce the circuit shown in Figure 8.30(a), study carefully Figure 8.30(b), (c), (d), and (e). We must start the reduction of the circuit on the right, with resistors R_5, R_6, and R_7. Because these are the only resistors in the circuit in Figure 8.30(a) that can be combined. The other resistors are connected in series-parallel combinations.

Resistors R_5, R_6, and R_7 are connected in series; therefore we may combine them into one resistor called R_D; see Figure 8.30(b).

$$R_D = R_5 + R_6 + R_7$$
$$= 15 \text{ k}\Omega + 10 \text{ k}\Omega + 15 \text{ k}\Omega = 40 \text{ k}\Omega$$

In Figure 8.30(b), resistor R_D is in parallel with resistor R_4. So these two resistors may be reduced to one resistor called R_E; see Figure 8.30(c).

$$R_E = R_4 \| R_D = \frac{(60 \text{ k}\Omega)(40 \text{ k}\Omega)}{60 \text{ k}\Omega + 40 \text{ k}\Omega} = 24 \text{ k}\Omega$$

In Figure 8.30(c), resistor R_E is in series with resistors R_3 and R_8. So these resistors may be reduced to one resistor called R_F; see Figure 8.30(d).

$$R_F = R_3 + R_E + R_8$$
$$= 12 \text{ k}\Omega + 24 \text{ k}\Omega + 12 \text{ k}\Omega = 48 \text{ k}\Omega$$

In Figure 8.30(d), resistor R_F is in parallel with resistor R_2. Therefore these two resistors may be reduced to the resistor called R_G; see Figure 8.30(e).

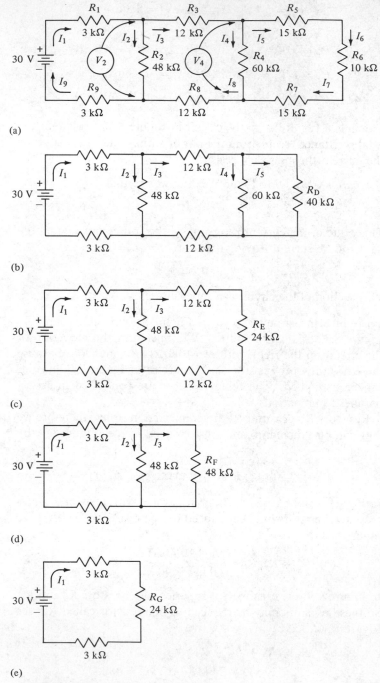

Figure 8.30 Circuit and circuit reductions for Example 8.10

$$R_G = R_2 \| R_F = \frac{48 \text{ k}\Omega}{2} = 24 \text{ k}\Omega$$

Figure 8.30(e) can be reduced even further, so that there is the power supply and only one resistor, R_T. To find R_T, the total resistance for the circuit, refer to Figure 8.30(e). Resistors R_1, R_G, and R_9 are all connected in series. Therefore

$$R_T = R_1 + R_G + R_9 = 3 \text{ k}\Omega + 24 \text{ k}\Omega + 3 \text{ k}\Omega = 30 \text{ k}\Omega$$

To find the current drawn from the source, I_1, we use Figure 8.30(e) and Ohm's law.

$$I_1 = \frac{E}{R_T} = \frac{30 \text{ V}}{30 \text{ k}\Omega} = 1 \text{ mA}$$

To find the remaining currents in the circuit, we just go backward, using the current-divider principle and KCL. Using the current-divider principle in Figure 8.30(d), we get

$$I_3 = \frac{R_2}{R_2 + R_F} I_1 = \left(\frac{48 \text{ k}\Omega}{96 \text{ k}\Omega} \right) 1 \text{ mA} = 0.5 \text{ mA}$$

Using KCL in Figure 8.30(d), we have

$$I_2 = I_1 - I_3 = 1.0 \text{ mA} - 0.5 \text{ mA} = 0.5 \text{ mA}$$

Using the current-divider principle in Figure 8.30(b), we obtain

$$I_5 = \frac{R_4}{R_4 + R_D} I_3 = \left(\frac{60 \text{ k}\Omega}{100 \text{ k}\Omega} \right) 0.5 \text{ mA} = 0.3 \text{ mA}$$

Using KCL in Figure 8.30(b), we get

$$I_4 = I_3 - I_5 = 0.5 \text{ mA} - 0.3 \text{ mA} = 0.2 \text{ mA}$$

Using KCL in Figure 8.30(a), we find that

$$I_8 = I_4 + I_7 \qquad \text{but } I_7 = I_5$$
$$= 0.2 \text{ mA} + 0.3 \text{ mA} = 0.5 \text{ mA}$$

We could find the voltage drop across each resistor by using Ohm's law.

Problem 8.17 Refer to the circuit in Figure 8.31. Given that $E = 16 \text{ V}$, $R_1 = 4 \text{ k}\Omega$, $R_2 = 12 \text{ k}\Omega$, and $R_3 = 6 \text{ k}\Omega$, solve for I_1, V_1, V_2, I_2, I_3, P_1, P_2, and P_3. (Figure 8.31 is on page 160.)

Problem 8.18 Refer to the circuit in Figure 8.31. Given that $E = 30 \text{ V}$, $V_3 = 15 \text{ V}$, $I_2 = 1 \text{ mA}$, and $R_1 = 5 \text{ k}\Omega$, solve for R_2, V_1, I_1, I_3, R_3, P_1, P_2, and P_3.

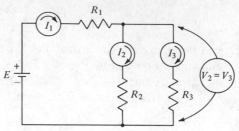

Figure 8.31   Circuit for Problems 8.17 and 8.18

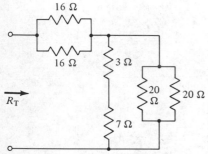

Figure 8.32   Circuit for Problem 8.19

Problem 8.19   Refer to the circuit in Figure 8.32. Find the total resistance R_T.

Problem 8.20   Refer to the circuit in Figure 8.33. Solve for the total resistance seen from the power supply, and then solve for the current in and the voltage drop across each resistor. Check your results by using KVL.

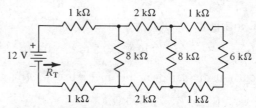

Figure 8.33   Circuit for Problem 8.20

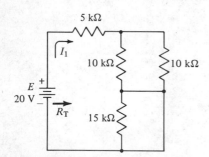

Figure 8.34   Circuit for Problem 8.21

Problem 8.21   Refer to the circuit in Figure 8.34. Solve for R_T and I_1. Note that a *short*, or a wire with zero resistance, is connected across the 15-kΩ resistor. This is not meant to trick you, because occasionally a resistor in a circuit is shorted, either accidentally or on purpose.

Problem 8.22   A two-socket convenience outlet has a 15-A fuse for protection against overloads; see Figure 8.35. An air conditioner with a rating of 1360 W at 120 V is plugged into one of

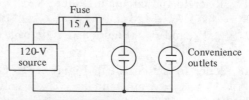

Figure 8.35   Circuit for Problem 8.22

the sockets. A hair dryer with a rating of 700 W at 120 V is then plugged into the other socket. Determine the amount of current that is required when both devices are operating at one time. If the air conditioner is running, will the hair dryer blow the fuse when it is turned on?

Open or Shorted Resistor

Objective 7   Determine what happens to resistance and voltage measurements in parallel and series-parallel resistive circuits when a resistor opens or is shorted.

In Objective 5 of Unit 7, we discussed what happens to resistance and voltage measurements in *series* resistive circuits when a resistor opens or is shorted. In this objective, you will apply the principles you learned to parallel and series-parallel resistive circuits. Before you try to solve the problems involving parallel and series-parallel circuits, study Examples 8.11, 8.12, and 8.13. You may want to review Objective 5 of Unit 7 before you study the examples.

Example 8.11   Using the circuit in Example 8.4, Figure 8.17, solve for the resistance and voltage measurements and the source current when the 7-kΩ resistor opens or is shorted.

Solution   Resistor R_2 opens.

$$R_2 = \infty \ \Omega \qquad \text{(open circuit)}$$

The source now "sees" a total or equivalent resistance of

$$R_T = R_{eq} = R_1 = 3 \ \text{k}\Omega$$

Since R_1 and R_2 were connected in parallel across the 21-V source, the voltage across R_1 is still equal to the source voltage.

$$V_1 = 21 \ \text{V}$$

Using Ohm's law, the source current is equal to

$$I_T = \frac{E}{R_T} = \frac{21 \ \text{V}}{3 \ \text{k}\Omega} = 7 \ \text{mA}$$

Resistor R_2 is shorted.

$$R_2 = 0 \ \Omega \qquad \text{(short circuit)}$$

The source now "sees" zero resistance.

$$R_{eq} = \frac{R_1 R_2}{R_1 + R_2} = \frac{3 \ \text{k}\Omega(0)}{3 \ \text{k}\Omega} = 0 \ \Omega$$

Since resistor R_2 is shorted, the power supply is also shorted. This causes the supply to try to furnish infinite current.

$$I_T = \frac{E}{R_T} = \frac{21}{0} = \infty \text{ A}$$

No power supply can supply infinite current. Therefore, unless the power supply is protected by a fuse or circuit breaker, it will be destroyed.

Before you go on to Examples 8.12 and 8.13, you should compare the results of Example 8.11 with Example 8.4. You can then see the effect an open or shorted parallel resistor has on a circuit. Also you should compare the answers obtained in Examples 8.12 and 8.13 with those obtained when there are no faults in the circuit. Figure 8.36(a) lists the voltages that should exist across the various resistors when the switch is closed.

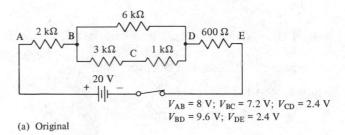

$V_{AB} = 8 \text{ V}; V_{BC} = 7.2 \text{ V}; V_{CD} = 2.4 \text{ V}$
$V_{BD} = 9.6 \text{ V}; V_{DE} = 2.4 \text{ V}$

(a) Original

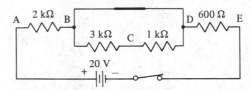

(b) The 6 kΩ resistor is shorted

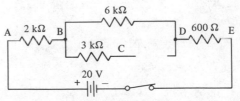

(c) The 1 kΩ resistor is open

Figure 8.36   Series-parallel resistive circuit

Example 8.12 The 6-kΩ resistor in the circuit in Figure 8.36(a) has shorted; see Figure 8.36(b). Solve for R_{BD}, R_T, V_{AB}, V_{BD}, V_{BC}, V_{CD}, and V_{DE}.

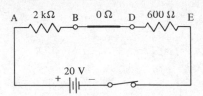

Figure 8.37 Reduction of circuit of Figure 8.36(b) when the 6-kΩ resistor is shorted

Solution Since the 6-kΩ resistor is shorted, the resistance from points B to D is equal to zero ohms. Therefore the circuit shown in Figure 8.36(b) appears as the simple two-resistor series circuit of Figure 8.37.

$R_{BD} = 0 \ \Omega$

$R_T = R_{AB} + R_{BD} + R_{DE} = 2000 \ \Omega + 0 \ \Omega + 600 \ \Omega = 2600 \ \Omega$

Since $R_{BD} = 0 \ \Omega$, $V_{BD} = V_{BC} = V_{CD} = 0$ V, and the 20 V must be dropped across the 2-kΩ and 600-Ω resistors.

Using the voltage-divider principle, we obtain

$$V_{DE} = \left(\frac{R_{DE}}{R_T}\right)E = \left(\frac{600}{2600}\right)20 \ V = 4.62 \ V$$

V_{AB} by KVL: $V_{AB} = E - V_{DE} = 20 - 4.62 = 15.38$ V

Example 8.13 The 1-kΩ resistor in the circuit in Figure 8.36(a) has opened; see Figure 8.36(c). Solve for R_{BD}, R_T, V_{AB}, V_{BD}, V_{BC}, V_{CD}, and V_{DE}.

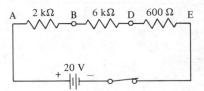

Figure 8.38 Reduction of circuit of Figure 8.36(c) when the 1-kΩ resistor is open.

Solution Since the 1-kΩ resistor has opened, no current can flow in the path B–C–D in the circuit in Figure 8.36(c). Therefore the circuit shown in Figure 8.36(c) appears as the three-resistor series circuit shown in Figure 8.38.

$R_{BD} = 6 \ k\Omega$

$R_T = R_{AB} + R_{BD} + R_{DE} = 2000 \ \Omega + 6000 \ \Omega + 600 \ \Omega = 8600 \ \Omega$

With no current in the 3-kΩ resistor, no voltage is dropped across it. Therefore $V_{BC} = 0$ V, and then $V_{CD} = V_{BD}$.

Using the voltage-divider principle, we find that

$$V_{AB} = \left(\frac{R_{AB}}{R_T}\right)E = \left(\frac{2000}{8600}\right)20 \ V = 4.65 \ V$$

$$V_{BD} = \left(\frac{R_{BD}}{R_T}\right)E = \left(\frac{6000}{8600}\right)20 \ V = 13.95 \ V$$

V_{DE} by KVL: $V_{DE} = E - V_{AB} - V_{BD}$
$$= 20 - 4.65 - 13.95 = 1.40 \ V$$

Problem 8.23 Given the parallel circuit shown in Figure 8.39, solve for the resistance and voltage measurements and the source current when the 1-kΩ resistor opens or is shorted.

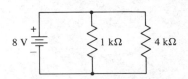

Figure 8.39 Circuit for Problem 8.23

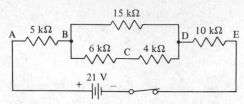

Figure 8.40   Circuit for Problems 8.24, 8.25, and 8.26

Problem 8.24   Given the series-parallel circuit in Figure 8.40, solve for R_{BD}, R_T, V_{AB}, V_{BD}, V_{BC}, V_{CD}, and V_{DE}. Use the voltage-divider principle to solve for V_{BC} or V_{CD}.

Problem 8.25   The 15-kΩ resistor in the circuit of Figure 8.40 has shorted. Solve for R_{BD}, R_T, V_{AB}, V_{BD}, V_{BC}, V_{CD}, and V_{DE}.

Problem 8.26   The 6-kΩ resistor in the circuit of Figure 8.40 has opened. Solve for R_{BD}, R_T, V_{AB}, V_{BD}, V_{BC}, V_{CD}, and V_{DE}.

In this objective we have asked you to determine the voltage measurements once you know which resistor is open or shorted. However, in trouble-shooting a circuit, you usually know the voltage readings and determine the fault from them. We have included a couple of problems on determining the fault. Since this is not part of this objective, we have specified them as *advanced*.

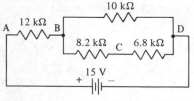

Figure 8.41   Circuit for Problems 8.27 and 8.28

Problem 8.27 (*Advanced*)   One of the resistors in the circuit shown in Figure 8.41 is either open or shorted. Using the following voltage measurements, determine which resistor it is and whether it is open or shorted: $V_{AB} = 10.91$ V, $V_{BC} = 4.09$ V, $V_{CD} = 0$ V, and $V_{BD} = 4.09$ V. [*Hint*: Calculate the theoretical voltage that would exist in the circuit if all resistors were good. Then, by comparing the actual readings with the theoretical readings and remembering what happens when a resistor opens or is shorted, you should be able to answer the question.]

Problem 8.28 (*Advanced*)   One of the resistors in the circuit shown in Figure 8.41 is either open or shorted. Using the following voltage measurements, determine which resistor it is and whether it is open or shorted: $V_{AB} = 8.18$ V, $V_{BC} = 0$ V, $V_{CD} = 6.82$ V, and $V_{BD} = 6.82$ V.

Test

1. Two resistors, 2 kΩ and 8 kΩ, are connected in parallel across a 16-V power supply.
    (a) Draw the electric circuit diagram.

(b) What is the voltage across the 8-kΩ resistor in the circuit?
(c) Determine the equivalent resistance.
(d) Determine the amount of supply current in the circuit.
(e) Determine the current in the 2-kΩ resistor.
2. Refer to the circuit in Figure 8.42. Find R_{eq}.

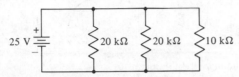

Figure 8.42 Circuit for Test Problem 2

3. Refer to the circuit shown in Figure 8.43. What is the voltage across R_2?

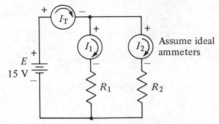

Figure 8.43 Circuit for Test Problems 3 through 7.

4. Refer to the circuit shown in Figure 8.43. Given that total current I_T is equal to 1.0 A and current I_2 is equal to 300 mA, find current I_1.
5. Refer to the circuit shown in Figure 8.43. Given that R_1 and R_2 are both equal to 15 kΩ, find R_{eq}.
6. Refer to the circuit shown in Figure 8.43. Given that I_1 is equal to 1.5 mA, what is R_1?
7. Refer to the circuit shown in Figure 8.43. Given that R_1 is equal to 5 kΩ, R_2 is equal to 10 kΩ, and I_T is equal to 4.5 mA, solve for I_2 using the current-divider principle.
8. Three resistors are connected in parallel across a dc power supply. They draw a total current of 17 mA from the power supply. The first resistor is 1 kΩ, the current in the second resistor is 5 mA, and the voltage drop across the third resistor is 10 V. Draw the circuit diagram and calculate the resistance of the second and third resistors.
9. Refer to the circuit in Figure 8.44. Determine the total resistance across the 20-V source, R_T, and also determine V_2.

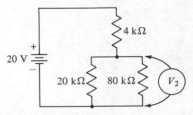

Figure 8.44 Circuit for Test Problem 9

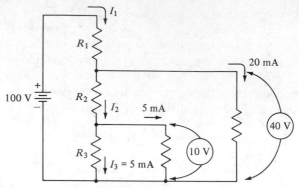

Figure 8.45   Circuit for Test Problem 10

10. Refer to the circuit in Figure 8.45. Find I_2, I_1, R_1, R_2, and R_3.

11. Refer to the circuit in Figure 8.46. Solve for R_T, I_1, I_2, I_3, I_4, I_5, P_1, P_6, and V_4.

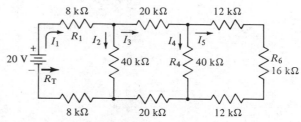

Figure 8.46   Circuit for Test Problem 11

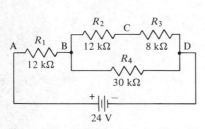

Figure 8.47   Circuit for Test Problem 12

12. Refer to the circuit shown in Figure 8.47. Find the voltages V_{AB}, V_{BC}, V_{CD}, and V_{BD} for the following conditions: (a) R_1 open (b) R_1 shorted (c) R_2 shorted (d) R_3 open   (e) R_4 shorted

Thévenin's Unit 9
and Norton's
Equivalent
Circuits

The equivalent circuit may be one of your most valuable tools for analyzing complicated circuits. An experienced technician can often look at a complicated schematic diagram and tell you what to expect when the circuit load conditions vary. It may take you years of practice to obtain this ability. However, you may shortcut much of this time if you learn to think of complicated circuits in terms of their equivalent circuits.

For example, by using Thévenin's theorem, you may replace the complicated electronic power-supply circuit shown enclosed in the box in Figure 9.1(a) by the equivalent circuit shown in the box of Figure 9.1(b). Then, as the load varies in Figure 9.1(b), you will have no trouble predicting the terminal voltage or load current in the circuit.

In this unit you will be working with *T*hévenin's *E*quivalent *C*ircuit (TEC) and *N*orton's *E*quivalent *C*ircuit (NEC). Using TEC and NEC, you will be able to find the equivalent circuit for many complicated circuits.

Objectives

After completing all the work associated with this unit, you should be able to:

1. Determine the TEC for a series-parallel resistive circuit. Then solve for the voltage across the load or the current through the load.

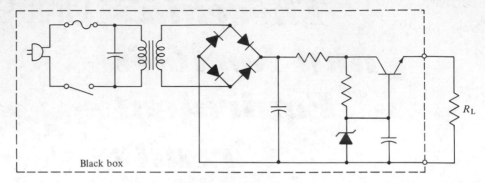

(a) Electronic power supply

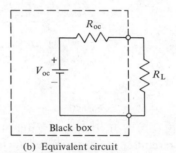

(b) Equivalent circuit

Figure 9.1　Electronic power supply and its equivalent circuit.

2. Determine the NEC for a series-parallel resistive circuit. Then solve for the voltage across the load or the current through the load.

3. Convert the TEC to the NEC.

4. (*Advanced*) Determine the TEC for such circuits as a binary ladder digital-to-analog converter, common-emitter transistor-amplifier circuit, or Wheatstone bridge circuit. Then solve for the voltage across the load or the current through the load.

　To show you how you may determine an equivalent circuit, we shall work mostly with series-parallel resistive circuits. However, the use of an equivalent circuit is not limited to series-parallel resistive circuits, as you see in Objective 4. Actually, to solve a single-supply series-parallel resistive circuit, you may not need to use an equivalent circuit. You may find that Ohm's law and KVL and KCL are sufficient. However, even with the series-parallel resistive circuit, using an equivalent circuit may save you calculation time. That is, you will save time if you only want the voltage across or the current through one resistor (usually designated as the *load* resistor).

Before we discuss Objective 1, solve the circuit shown in Figure 9.2 (Problem 9.1) using conventional methods. You can use this for comparison when you solve the same circuit using an equivalent circuit.

Problem 9.1 Find the current in the 40-Ω resistor in the circuit shown in Figure 9.2 using conventional methods (that is, Ohm's law and KVL and KCL).

Thévenin's Equivalent Circuit

Objective 1 Determine the TEC for a series-parallel resistive circuit. Then solve for the voltage across the load or the current through the load.

Thevenin's theorem allows us to simplify a circuit by replacing all the components and supplies *except the load* by a voltage source and a series resistor. The voltage source and series resistor are called Thévenin's equivalent circuit (TEC). The *load* may be a single component such as a resistor, or it may be several components including transistors and sources. This is discussed in more detail in Objective 4.

Figure 9.3(a) shows a series-parallel resistive circuit that may be simplified using Thévenin's theorem. The TEC with the load attached is shown in Figure 9.3(b). We call the circuit shown in Figure 9.3(b) *equivalent* to the circuit shown in Figure 9.3(a). Because, if the circuits to the left of the load in Figure 9.3 were placed in identical *black boxes*, the load would not be able to tell the difference. Electrically, the load would "see" the same terminal voltage and would draw the same current from each of the black boxes.

Briefly, to determine the TEC, we remove the load from the circuit [terminals A–B in Figure 9.3(a)] and determine the open-circuit voltage, and then determine the resistance across the open terminals A–B with the source replaced by its internal resistance. As you examine the equivalent circuit shown in Figure 9.3(b), you see a series circuit. Therefore, by replacing part of the original circuit by the TEC, we have made a circuit that is much simpler to analyze. However, we must know how to determine Thévenin's voltage source V_{oc} and Thévenin's series resistor R_{oc}. (The subscript oc stands for "open circuit.")

We shall now give you in detail the steps that you need to obtain the TEC.

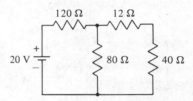

Figure 9.2 Circuit for Problem 9.1

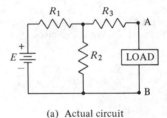

(a) Actual circuit

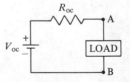

(b) Thévenin's equivalent circuit with the load attached

Figure 9.3 Series-parallel resistive circuit

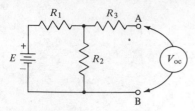

(a) Circuit for Step 1 and Step 2

(b) Circuit for Step 3

$R_{oc} = R_3 + R_1 \parallel R_2$

(c) TEC with load attached (Step 4)

$$V_{oc} = \left(\frac{R_2}{R_1 + R_2}\right) E$$

Figure 9.4 Steps to obtain the TEC for a series-parallel resistive circuit.

Step 1. Remove the element or elements in the circuit that were designated as the *load*; see Figure 9.4(a). It may help to label the terminals at which the *load* was removed (A and B).

Step 2. Determine the open-circuit voltage across terminals A–B; call it V_{oc}. Since the circuit is open at terminals A–B, no current will flow through resistor R_3. Therefore all we need to do is solve for the voltage across resistor R_2. We may use the voltage-divider principle:

$$V_{oc} = \frac{R_2}{R_1 + R_2} E \qquad (9.1)$$

Step 3. Replace the voltage supply E by its internal resistance; see Figure 9.4(b). For a dc voltage supply, either a battery or an electronic supply, the internal resistance is considered to be zero. This does not mean that the power supply has no internal resistance. But, when the power-supply voltage is measured under loaded conditions (current flowing), the internal resistance is considered to be zero. Therefore voltage sources are replaced by a *short* circuit. Current sources are replaced by an *open* circuit.

Find the series resistance, called R_{oc}, of Figure 9.4(b). R_{oc} is the resistance we see when we look back into terminals A–B after we have replaced the source or power supply by its internal resistance.

$$R_{oc} = R_3 + R_1 \parallel R_2 \qquad (9.2)$$

Step 4. Replace all the circuit components except those designated as the load by the TEC. In Figure 9.4(a) the part of the circuit to the left of terminals A–B is replaced by the TEC. Then attach the load to terminals A–B; see Figure 9.4(c). We can determine the voltage across the load by using the voltage-divider principle. We can determine the current in the load by using Ohm's law.

To see how you can use these steps to determine the TEC of the circuit shown in Figure 9.5(a), study Example 9.1. Then study Examples 9.2 and 9.3.

Example 9.1 Simplify the circuit shown in Figure 9.5(a) by using the TEC. Then solve for the voltage across and the current in the 60-Ω resistor connected between terminals A–B.

Solution Call the 60-Ω resistor connected between terminals A–B the load resistor.

Step 1. Remove the load; see Figure 9.5(b). We don't have to label the terminals at which the load was removed, because they were already labeled.

Step 2. Determine the open-circuit voltage across terminals A–B of Figure 9.5(b). From Equation (9.1), we obtain

$$V_{oc} = \frac{R_2}{R_1 + R_2} E = \frac{60}{60 + 60} (240 \text{ V}) = 120 \text{ V}$$

Step 3. Replace the 240-V supply by its internal resistance ($R_{int} = 0 \ \Omega$); see Figure 9.5(c). Use Equation (9.2) to find R_{oc}.

$$R_{oc} = R_3 + R_1 \| R_2 = 10 \ \Omega + 60 \ \Omega \| 60 \ \Omega = 10 \ \Omega + 30 \ \Omega = 40 \ \Omega$$

Step 4. Replace all the circuit components except those designated as the load by the TEC. Then attach the load. [See Figure 9.5(d).] The voltage across the 60-Ω load resistor R_L, using the voltage-divider principle, is

$$V_L = \frac{R_L}{R_{oc} + R_L} V_{oc} = \frac{60}{40 + 60} (120 \text{ V}) = 72 \text{ V}$$

We can find the current through the 60-Ω load resistor by using Ohm's law with the voltage across the load resistor known:

$$I = \frac{V_L}{R_L} = \frac{72 \text{ V}}{60 \ \Omega} = 1.2 \text{ A}$$

We can also find this current by using the total voltage in the circuit shown in Figure 9.5(d), V_{oc}, and the total circuit resistance:

$$I = \frac{V_{oc}}{R_{oc} + R_L} = \frac{120 \text{ V}}{40 \ \Omega + 60 \ \Omega} = 1.2 \text{ A}$$

Example 9.2 Find the current through and the voltage across the 10-Ω resistor in the circuit shown in Figure 9.6(a). Use the TEC.

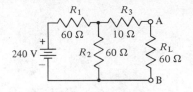

(a) Original circuit for Example 9.1

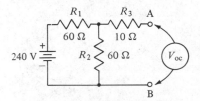

(b) Circuit for Steps 1 and 2

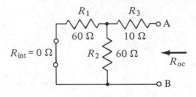

(c) Circuit for Step 3

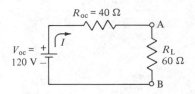

(d) TEC with 60 Ω load attached

Figure 9.5 Circuits for Example 9.1

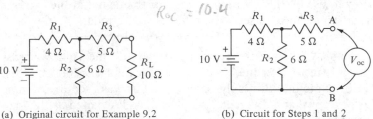

(a) Original circuit for Example 9.2

(b) Circuit for Steps 1 and 2

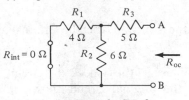

(c) Circuit for Step 3

(d) TEC with 10 Ω load attached

Figure 9.6 Circuits for Example 9.2

Solution   Call the 10-Ω resistor the load, and label the terminals of the load A and B.

Step 1.   Remove the load; see Figure 9.6(b). We now have an open circuit at terminals A–B, and no current flows through the 5-Ω resistor.

Step 2.   Determine the open-circuit voltage across terminals A–B of Figure 9.6(b).

$$V_{oc} = \frac{R_2}{R_1 + R_2} E = \frac{6}{4 + 6} (10 \text{ V}) = 6 \text{ V}$$

Step 3.   Replace the 10-V battery by its internal resistance ($R_{int} = 0 \ \Omega$); see Figure 9.6(c). Then determine the resistance across the open terminals A–B.

$$R_{oc} = R_3 + R_1 \| R_2 = 5 \ \Omega + 6 \ \Omega \| 4 \ \Omega = 5 + \frac{6(4)}{6 + 4} = 7.4 \ \Omega$$

Step 4.   Replace all the circuit components except those designated as the load by the TEC. Then attach the load. [See Figure 9.6(d).] The circuit is now a simple series circuit. Therefore we can now solve for the current in the 10-Ω resistor and the voltage across the 10-Ω resistor.

$$I = \frac{V_{oc}}{R_{oc} + R_L} = \frac{6 \text{ V}}{7.4 \ \Omega + 10 \ \Omega} = 0.345 \text{ A} = 345 \text{ mA}$$

$$V_L = IR_L = 0.345(10) = 3.45 \text{ V}$$

Any component or part of an electric or electronic circuit may be called the load in Thévenin's theorem. Example 9.3 shows how to use the TEC for a load resistor that is not at the extreme right-hand side of a series-parallel resistive circuit.

Example 9.3   Find the current through and the voltage across the 6-Ω resistor in the circuit shown in Figure 9.7(a). Use the TEC. Note that this is the same circuit as was used in Example 9.2, except that a different resistor is now designated as the load resistor.

Solution   Call the 6-Ω resistor the load, and label the terminals of the load A and B.

Step 1.   Remove the load; see Figure 9.7(b). Check the circuit shown in Figure 9.7(b) against the original circuit shown in Figure 9.7(a) with the 6-Ω load removed. The circuit may appear to be different, but it is the same.

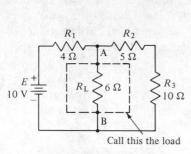

(a) Original circuit for Example 9.3

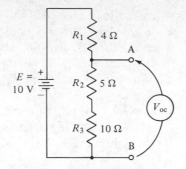

(b) Circuit for Steps 1 and 2

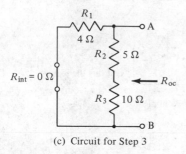

(c) Circuit for Step 3

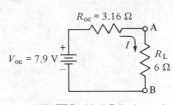

(d) TEC with 6 Ω load attached

Figure 9.7 Circuits for Example 9.3

Step 2. Determine the open-circuit voltage across terminals A–B of Figure 9.7(b).

$$V_{oc} = \frac{R_2 + R_3}{R_1 + R_2 + R_3} E = \frac{5 + 10}{4 + 5 + 10}(10 \text{ V}) = 7.9 \text{ V}$$

Step 3. Replace the 10-V battery by its internal resistance ($R_{int} = 0 \ \Omega$); see Figure 9.7(c). Then determine the resistance across the open terminals A–B.

$$R_{oc} = (R_2 + R_3)\|R_1 = (5 + 10)\|4 = \frac{15(4)}{15 + 4} = 3.16 \ \Omega$$

Step 4. Replace all the circuit components except those designated as the load by the TEC. Then attach the load. [See Figure 9.7(d).] We can now find the current in the 6-Ω resistor and the voltage across the 6-Ω resistor.

$$I = \frac{V_{oc}}{R_{oc} + R_L} = \frac{7.9 \text{ V}}{3.16 \ \Omega + 6 \ \Omega} = 0.862 \text{ A}$$

$$V_L = IR_L = 0.862(6) = 5.17 \text{ V}$$

The next example, Example 9.4, includes a *current source* in the circuit. A current source is an electronic power supply that has been designed to deliver a constant *current* to the load, even when the load changes in value. Remember, a voltage source is designed to deliver a constant *voltage* across the load, even when the load changes in value. As we stated earlier under the description of Step 3, we replace current sources by an open circuit when we determine R_{oc}.

Example 9.4 Find the current through and the voltage across the 10-kΩ resistor in the circuit shown in Figure 9.8(a). Also find the current in the 3-kΩ resistor. Use the TEC.

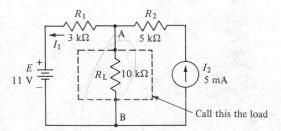

(a) Original circuit for Example 9.4

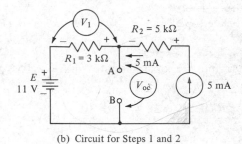

(b) Circuit for Steps 1 and 2

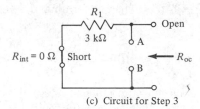

(c) Circuit for Step 3

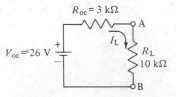

(d) TEC with the 10 kΩ load attached

Figure 9.8 Circuits for Example 9.4

Solution   Call the 10-kΩ resistor the load, and label the terminals of the load A and B.

Step 1.   Remove the load; see Figure 9.8(b). The 5 mA from the current source flows through resistors R_2 and R_1 because that is the only closed path.

Step 2.   Determine the open-circuit voltage across terminals A–B of Figure 9.8(b). Note the polarity of the voltage drop across R_1. Therefore

$$V_{oc} = E + V_1$$
$$= 11 + (5 \text{ mA})(3 \text{ k}\Omega) = 11 + 15 = 26 \text{ V}$$

Step 3.   Replace the voltage and current sources by their internal resistances; see Figure 9.8(c). The voltage source is replaced by a short circuit, and the current source is replaced by an open circuit. Then determine the resistance across the open terminals A–B.

$$R_{oc} = R_1 = 3 \text{ k}\Omega$$

Step 4.   Replace all the circuit components except those designated as the load by the TEC. Then attach the load. [See Figure 9.8(d).] The voltage across the 10-kΩ load R_L, using the voltage-divider principle, is

$$V_L = \frac{R_L}{R_{oc} + R_L} V_{oc} = \frac{10 \text{ k}\Omega}{13 \text{ k}\Omega} (26 \text{ V}) = 20 \text{ V}$$

We can find the current through the 10-kΩ load resistor by using Ohm's law.

$$I_L = \frac{V_L}{R_L} = \frac{20 \text{ V}}{10 \text{ k}\Omega} = 2 \text{ mA}$$

We know there is 5 mA in R_2 because of the constant current source. We just found that there is 2 mA flowing down through the 10-kΩ resistor. Therefore we can find the current in the 3-kΩ resistor R_1 by KCL.

$$I_1 = I_2 - I_L = 5 \text{ mA} - 2 \text{ mA} = 3 \text{ mA}$$

Problem 9.2   Find the current through and the voltage across the 40-Ω resistor in Figure 9.2 using the TEC.

Problem 9.3   In the circuit shown in Figure 9.9, call the 22-kΩ resistor the load. Find the current through and the voltage across the 22-kΩ resistor using the TEC.

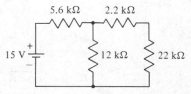

Figure 9.9   Circuit for Problem 9.3

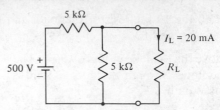

Figure 9.10   Circuit for Problem 9.4

Problem 9.4   For the circuit shown in Figure 9.10, find R_L so that the load current I_L is equal to 20 mA.

Problem 9.5   Find the current through and the voltage across the 80-Ω resistor in Figure 9.2 using the TEC.

Problem 9.6   Find the current through and the voltage across the 12-Ω resistor in Figure 9.2 using the TEC.

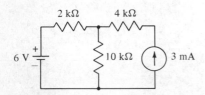

Figure 9.11   Circuit for Problem 9.7

Problem 9.7   Find the current through and the voltage across the 10-kΩ resistor in the circuit shown in Figure 9.11. Also find the current in the 2-kΩ resistor. Use the TEC.

The TEC of a series-parallel resistive circuit, Figure 9.3, may also be used to find the value required for a load resistor if maximum power is to be transferred into the load. From the *maximum power transfer theorem*, the maximum power in the load occurs when R_L is equal to R_{oc}. Study Example 9.5 for an application of this.

Example 9.5   Given that V_{oc} is equal to 6 V and R_{oc} is equal to 4 Ω, find R_L for maximum power into R_L. Also find the maximum power $P_{L,\,max}$ in the load.

   Solution

$$R_L = R_{oc} = 4\ \Omega$$

$$V_L = \frac{R_L}{R_{oc} + R_L}\, V_{oc} = \frac{4}{8}\,(6\ \text{V}) = 3\ \text{V}$$

Therefore

$$P_{L,\,max} = \frac{V_L^2}{R_L} = \frac{3^2}{4} = 2.25\ \text{W}$$

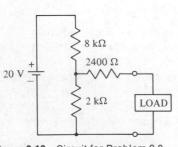

Figure 9.12   Circuit for Problem 9.8

Problem 9.8   In Figure 9.12, what should the load resistor equal so that maximum power may be delivered to it? What is the maximum power?

Norton's Equivalent Circuit

Objective 2 Determine the NEC for a series-parallel resistive circuit. Then solve for the voltage across the load or the current through the load.

We may also solve the series-parallel resistive circuit shown in Figure 9.13(a) by using Norton's equivalent circuit (NEC); see Figure 9.13(b). Norton's theorem allows us to simplify a circuit by replacing all the other components and supplies *except the load* by a current source and a parallel resistor.

Briefly, to determine the NEC, we remove the circuit component or components that we choose to call the load. Then we determine the resistance "seen" across the open circuit [terminals A–B in Figure 9.13(a)] and the short-circuit current I_{sc} that would flow if terminals A–B were shorted. This then gives the equivalent circuit shown in Figure 9.13(b). As you examine this equivalent circuit, you see a simple parallel circuit.

We shall now give you in detail the steps that are required to obtain the NEC.

Steps 1 and 3 are exactly the same as for the TEC; see Figure 9.14(a) and (c).

Step 2. Determine the short-circuit current that would flow if terminals A–B were shorted; see Figure 9.14(b). First we find the total current flowing from the source with terminals A–B shorted.

$$I_T = \frac{E}{R_T} \quad \text{where} \quad R_T = R_1 + R_2 \| R_3$$

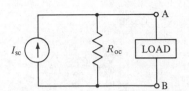

(a) Actual circuit

(b) Norton's equivalent circuit with the load attached

Figure 9.13 Series-parallel resistive circuit

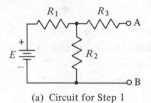

(a) Circuit for Step 1

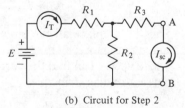

(b) Circuit for Step 2

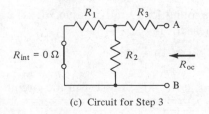

(c) Circuit for Step 3

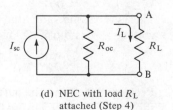

(d) NEC with load R_L attached (Step 4)

Figure 9.14 Steps to obtain the NEC for a series-parallel resistive circuit

Therefore, using the current-divider principle, we obtain

$$I_{sc} = \frac{R_2}{R_2 + R_3} I_T \tag{9.3}$$

Step 4. Replace all the circuit components except those designated as the load by the NEC. In Figure 9.14(a) the part of the circuit to the left of terminals A–B is replaced by the NEC. Then attach the load to terminals A–B; see Figure 9.14(d). The current in the load can be determined by the current-divider principle, and the voltage across the load can be determined by Ohm's law.

To see how we can use these steps to determine the NEC of the circuit shown in Figure 9.15, study Example 9.6.

Example 9.6   Solve the circuit given in Example 9.1, Figure 9.5(a), using the NEC.

Solution   Steps 1 and 3 are the same as for the TEC; see Example 9.1. Also see Figure 9.15(a) and (c). From Step 3 of Example 9.1, $R_{oc} = 40\ \Omega$.

Step 2.   Determine the short-circuit current that flows when terminals A–B are shorted; see Figure 9.15(b).

$$R_T = R_1 + R_2 \| R_3 = 60 + 10 \| 60 = 60 + \frac{10(60)}{70} = 68.57\ \Omega$$

$$I_T = \frac{E}{R_T} = \frac{240\ \text{V}}{68.57\ \Omega} = 3.5\ \text{A}$$

From Equation (9.3), we have

$$I_{sc} = \frac{R_2}{R_2 + R_3} I_T = \frac{60}{60 + 10} (3.5) = 3\ \text{A}$$

Step 4.   Replace all the circuit components except those designated as the load by the NEC. Then attach the load. [See Figure 9.15(d).] We can find the current in the 60-Ω load resistor by the current-divider principle:

$$I_L = \frac{R_{oc}}{R_{oc} + R_L} I_{sc} = \frac{40}{40 + 60} (3) = 1.2\ \text{A}$$

This is the same value that we found for I_L in Example 9.1. The voltage across the 60-Ω load resistor is

$$V_L = I_L R_L = 1.2(60) = 72\ \text{V}$$

Problem 9.9   Find the current in the 40-Ω resistor in the circuit shown in Figure 9.2 using the NEC.

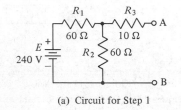

(a) Circuit for Step 1

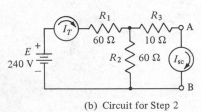

(b) Circuit for Step 2

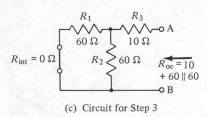

(c) Circuit for Step 3

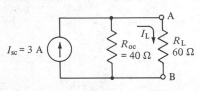

(d) NEC with 60 Ω load attached

Figure 9.15   Circuits for Example 9.6

Problem 9.10 Find the current in the 22-kΩ resistor of Figure 9.9 using the NEC.

Conversion of TEC to NEC

Objective 3 Convert the TEC to the NEC.

There is a relation between the TEC and the NEC. You may find that this is an easier way to find the NEC for a given circuit than the one given in Objective 2.

To determine the relation between the TEC and the NEC, study the TEC and NEC circuits shown in Figure 9.16. The load in both the TEC and the NEC must be the same. It must have the same voltage across it and the same current through it. Using the voltage-divider principle, we find that the voltage across the load in Figure 9.16(a) is

$$V_L = \frac{R_L}{R_{oc} + R_L} V_{oc}$$

The current through the load in Figure 9.16(b), by the current-divider principle, is

$$I_L = \frac{R_{oc}}{R_{oc} + R_L} I_{sc}$$

This then makes the voltage across the load in Figure 9.16(b) equal to

$$V_L = I_L R_L$$

$$V_L = \frac{R_{oc}}{R_{oc} + R_L} I_{sc} R_L$$

Since the voltage across the load must be the same, then

$$V_L \left[\text{Figure 9.16(a)} \right] = V_L \left[\text{Figure 9.16(b)} \right]$$

$$\frac{R_L}{R_{oc} + R_L} V_{oc} = \frac{R_{oc}}{R_{oc} + R_L} I_{sc} R_L$$

Therefore

$$V_{oc} = I_{sc} R_{oc}$$

or

$$I_{sc} = \frac{V_{oc}}{R_{oc}} \qquad (9.4)$$

You may remember Equation (9.4) by remembering Ohm's law.

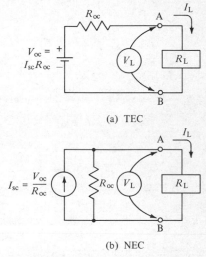

(a) TEC

(b) NEC

Figure 9.16 Relation between the TEC and the NEC

Example 9.7   For the circuit given in Example 9.1, solve for I_{sc} using the relation between the TEC and the NEC.

Solution   From Step 2 of Example 9.1, $V_{oc} = 120$ V. From Step 3 of Example 9.1, $R_{oc} = 40\ \Omega$. Therefore, using Equation (9.4),

$$I_{sc} = \frac{V_{oc}}{R_{oc}} = \frac{120\ \text{V}}{40\ \Omega} = 3\ \text{A}$$

Problem 9.11   From the TEC of the circuit in Figure 9.2, considering the 40-Ω resistor as the load, convert to the NEC. Then, using the NEC, find the current in the 40-Ω resistor.

Problem 9.12   From the TEC of the circuit in Figure 9.9, considering the 22-kΩ resistor as the load, convert to the NEC. This should be the same NEC found in Problem 9.10.

Applications of TEC

Objective 4 (*Advanced*)   Determine the TEC for such circuits as a binary ladder digital-to-analog converter, common-emitter transistor-amplifier circuit, or Wheatstone bridge circuit. Then solve for the voltage across the load or the current through the load.

Study Examples 9.8, 9.9, and 9.10. When you understand them, do the problems that follow the examples. When you have done this, you have completed this advanced objective.

Figure 9.17 shows a digital-to-analog converter circuit called a *binary ladder*. There are two values of voltages that may be applied to terminals X, Y, and Z for digital input. The digital input may be a certain dc voltage, shown as V in Figure 9.17, or zero volts. The output of the circuit is a single value of voltage known as the analog equivalent of the digital input.

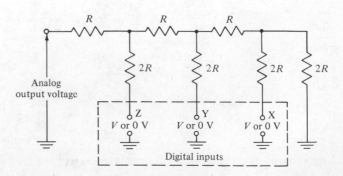

Figure 9.17   Binary ladder digital-to-analog converter

Example 9.8 Solve the binary-ladder circuit shown in Figure 9.18(a), which has a dc input of 4 V applied to terminals X and Y. That is, determine V_{out} of Figure 9.18(a) by repeated application of Thévenin's theorem.

Solution There is no complete circuit between point D and ground, and therefore no current flows through the resistor between points C and D. V_{out} is equal to the voltage at point C with respect to ground. To find the voltage at point C with respect to ground, we have to apply Thévenin's theorem twice.

For the first application of Thévenin's theorem, we replace everything to the right of terminals A–B by the TEC; see Figure 9.18(b). We break the circuit here because the circuit to the right

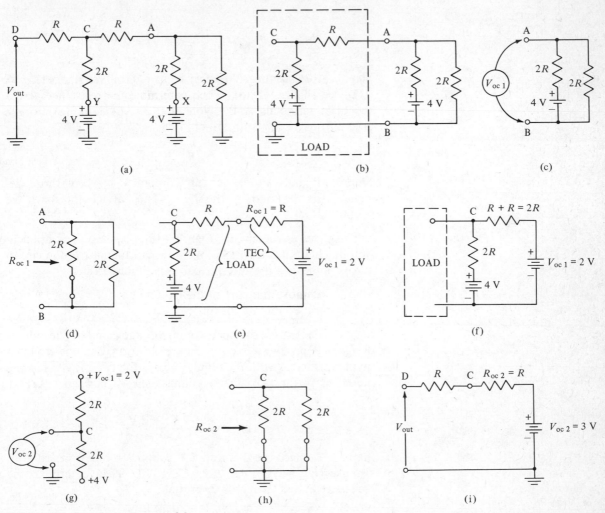

Figure 9.18 Circuits for Example 9.8

of A–B is simple enough to determine an open-circuit voltage. Whereas, if we try to consider the circuit to the right of C now, it is too complicated. We don't know how to work with two supplies with a resistor of $2R$ from point A to ground.

As you study the circuit reductions shown in Figure 9.18, note that the way the circuit is drawn is the reverse of most of the circuit diagrams you have used. The load is shown to the left on the diagram.

Step 1.   Remove the part of the circuit called the load; see Figure 9.18(b) and (c).

Step 2.   Determine the open-circuit voltage across the terminals designated as A–B in Figure 9.18(c). The voltage-divider principle is used.

$$V_{oc_1} = \frac{2R}{2R + 2R}(4\text{ V}) = 2\text{ V}$$

Step 3.   Replace the 4-V battery by its internal resistance ($R_{int} = 0\ \Omega$); see Figure 9.18(d). Then determine the resistance across the open terminals A–B. Note that we are looking into the circuit from the left to right.

$$R_{oc_1} = 2R\|2R = R$$

Step 4.   Replace all the circuit components except those designated as the load by the TEC. Then attach the load; see Figure 9.18(e).

This gives us the circuit, Figure 9.18(e), for a second application of Thévenin's theorem. We now call the part of the circuit across point C to ground the load; see Figure 9.18(f).

Step 1.   Remove the part of the circuit shown in Figure 9.18(f) as the load; see Figure 9.18(g).

Step 2.   Determine the open-circuit voltage across the terminals designated as C–ground in Figure 9.18(g). Use the voltage-divider principle, with the voltage across the two resistors as the difference in the two supply voltages shown in Figure 9.18(g).

$$V_{oc_2} = \frac{2R}{4R}(4\text{ V} - 2\text{ V}) + 2\text{ V} = 3\text{ V}$$

Step 3.   Replace the 2-V and 4-V supplies by their internal resistance ($R_{int} = 0\ \Omega$); see Figure 9.18(h). Then determine the resistance across terminals C–ground.

$$R_{oc_2} = 2R\|2R = R$$

Step 4.   Replace all the circuit components except those designated as the load by the TEC. Then attach the load; see Figure 9.18(i).

For this example, we do not actually have a load at point C because we don't have a complete circuit. The voltage at point D or at point C is equal to V_{oc_2} because there is no voltage drop in the resistors. Therefore

$$V_{out} = V_{oc_2} = 3 \text{ V}$$

For this example we did not need the value for R_{oc_2}. We performed the calculation only to show you how to determine R_{oc_2}. If we had a load at point C, R_{oc_2} would be required to calculate the current and voltage in the circuit.

The circuit shown in Figure 9.19(a) is known as a common-emitter transistor amplifier circuit. In Example 9.9, we are only working with a dc voltage applied to the circuit (this is known as the *bias voltage*).

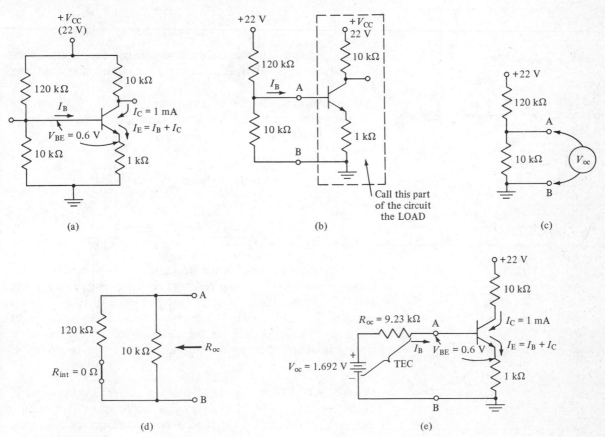

Figure 9.19   Circuits for Example 9.9

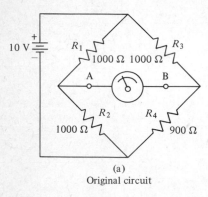

(a)
Original circuit

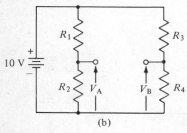

(b)

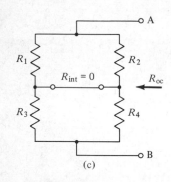

(c)

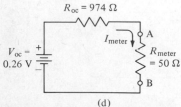

(d)

Figure 9.20 Circuits for Example 9.10

Example 9.9 Solve for the base current I_B in the transistor circuit of Figure 9.19(a).

Solution Since we don't know how to work with the transistor attached to the 120-kΩ and 10-kΩ voltage divider, we shall remove it from the circuit. Therefore we shall call this part of the circuit the load; see Figure 9.19(b).

Step 1. Remove the part of the circuit called the load; see Figure 9.19(c).

Step 2. Determine the open-circuit voltage across terminals A–B of Figure 9.19(c). The voltage-divider principle is used.

$$V_{oc} = \frac{10 \text{ k}\Omega}{10 \text{ k}\Omega + 120 \text{ k}\Omega} (22 \text{ V}) = 1.692 \text{ V}$$

Step 3. Replace the 22-V power supply by its internal resistance ($R_{int} = 0 \text{ }\Omega$); see Figure 9.19(d). Then determine the resistance across terminals A–B.

$$R_{oc} = 10 \text{ k}\Omega \| 120 \text{ k}\Omega = \frac{10 \text{ k}\Omega(120 \text{ k}\Omega)}{10 \text{ k}\Omega + 120 \text{ k}\Omega} = 9.23 \text{ k}\Omega$$

Step 4. Replace all the circuit components except those designated as the load by the TEC. Then attach the load; see Figure 9.19(e).

If we examine the part of the circuit (V_{oc}, R_{oc}, V_{BE}, and the 1-kΩ resistor), we see that it is a closed loop. We may apply KVL to solve for I_B.

$$V_{oc} = I_B R_{oc} + V_{BE} + I_E(1 \text{ k}\Omega)$$
$$= I_B R_{oc} + V_{BE} + (I_B + I_C)(1 \text{ k}\Omega)$$

Solving for I_B, we obtain

$$I_B = \frac{V_{oc} - V_{BE} - I_C(1 \text{ k}\Omega)}{R_{oc} + 1 \text{ k}\Omega} = \frac{1.692 \text{ V} - 0.6 \text{ V} - 1.0 \text{ V}}{9.23 \text{ k}\Omega + 1 \text{ k}\Omega} = 9 \text{ }\mu\text{A}$$

The circuit shown in Figure 9.20(a) is known as a *Wheatstone bridge*. Usually one of the resistors in the circuit is unknown, and the circuit is used to determine the value of the unknown resistor. The ammeter in the circuit deflects either to the right or to the left, depending on the direction of the current in it. If the bridge is balanced, the pointer of the meter is exactly in the middle.

Example 9.10 Using the TEC, find the current in the meter in the circuit of Figure 9.20(a). Also give the direction of deflection of the meter. Assume that the internal resistance of the meter is 50 Ω.

Solution   Call the meter in the circuit of Figure 9.20(a) the load.

Step 1.   Remove the load from the circuit; see Figure 9.20(b).

Step 2.   Determine the open-circuit voltage across the terminals A–B in Figure 9.20(b).   Use the voltage-divider principle.

$$V_A = \frac{R_2}{R_1 + R_2}(10 \text{ V}) \qquad V_B = \frac{R_4}{R_3 + R_4}(10 \text{ V})$$

$$V_{oc} = V_A - V_B$$

$$V_A = \frac{1000}{1000 + 1000}(10 \text{ V}) = 5 \text{ V}$$

$$V_B = \frac{900}{1000 + 900}(10 \text{ V}) = 4.74 \text{ V}$$

Therefore

$$V_{oc} = V_A - V_B = 5 - 4.74 = 0.26 \text{ V}$$

The meter deflects to the right because V_A is larger than V_B.

Step 3.   Replace the 10-V power supply by its internal resistance ($R_{int} = 0 \ \Omega$); see Figure 9.20(c).   Then determine the resistance across terminals A–B.

$$R_{oc} = R_1 \| R_2 + R_3 \| R_4 = \frac{1000(1000)}{1000 + 1000} + \frac{900(1000)}{900 + 1000}$$

$$= 500 + 474 = 974 \ \Omega$$

Step 4.   Replace all the circuit components except those designated as the load by the TEC.   Then attach the load; see Figure 9.20(d).

$$I_{meter} = \frac{V_{oc}}{R_{oc} + R_{meter}} = \frac{0.26 \text{ V}}{974 \ \Omega + 50 \ \Omega} = 0.254 \text{ mA}$$

Problem 9.13   Find the V_{out} of the circuit shown in Figure 9.21 by repeated application of Thévenin's theorem.

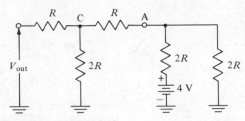

Figure 9.21   Circuit for Problem 9.13

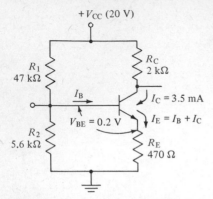

Figure 9.22   Circuit for Problem 9.14

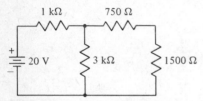

Figure 9.24   Circuit for Test Problems
1, 3, 5, and 7

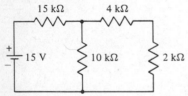

Figure 9.25   Circuit for Test Problems 2
and 6

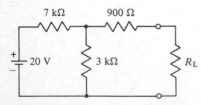

Figure 9.26   Circuit for Test Problem 4

Problem 9.14   Solve for the base current I_B in the transistor circuit of Figure 9.22.

Problem 9.15   Using the TEC, find the current in the meter in the circuit of Figure 9.23. Also give the direction of deflection of the meter. Assume that the internal resistance of the meter is 100 Ω.

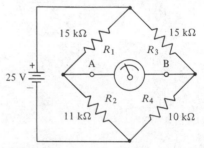

Figure 9.23   Circuit for Problem 9.15

Test

1. Find the current through and the voltage across the 1500-Ω resistor in Figure 9.24, using the TEC.
2. Find the current through and the voltage across the 2-kΩ resistor in the circuit shown in Figure 9.25, using the TEC.
3. Find the current through and the voltage across the 3-kΩ resistor in Figure 9.24, using the TEC.
4. In Figure 9.26, determine the value required for R_L to obtain maximum power into it. Find the power in R_L.
5. Find the current in the 1500-Ω resistor of Figure 9.24 using the NEC.
6. Find the current in the 2-kΩ resistor in the circuit shown in Figure 9.25, using the NEC.
7. From the TEC of Figure 9.24, considering the 1500-Ω resistor as the load, convert to the NEC. Then, using the NEC, find the current in the 1500-Ω resistor.
8. (*Advanced*) Determine V_{out} of the circuit shown in Figure 9.27 by repeated application of Thévenin's theorem.

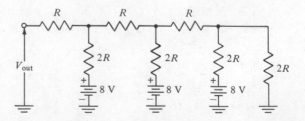

Figure 9.27   Circuit for Test Problem 8

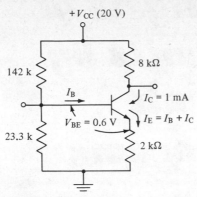

Figure 9.28   Circuit for Test Problem 9

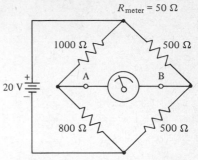

Figure 9.29   Circuit for Test Problem 10

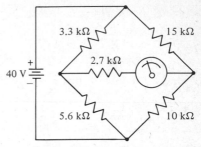

9. (*Advanced*) Solve for I_B in Figure 9.28.
10. (*Advanced*) Solve for the current in the meter in Figure 9.29.
11. (*Advanced*) Assume that the resistance of the ammeter in the circuit shown in Figure 9.30 is negligible.
    (a) Determine the current in the 2.7-kΩ resistor, using the TEC.
    (b) Replace the 2.7-kΩ resistor by a 5-kΩ resistor, and determine the current in the 5-kΩ resistor.

Figure 9.30   Circuit for Test Problem 11

Unit 10 *Analysis of Circuits by Circuit Theorems*

In this unit you will be introduced to two fundamental techniques for solving special types of circuits. You already know how to solve a circuit by applying Ohm's law and Kirchhoff's voltage and current laws. Also you can solve the circuit by converting it to a Thévenin or Norton equivalent and then solving a simple series or simple parallel circuit.

In this unit you will learn how to write loop equations and how to apply the superposition theorem in analyzing any circuit.

Objectives

After completing all the work associated with this unit, you should be able to:

1. Apply Kirchhoff's voltage law to a simple, single- (closed) loop electric circuit that may contain a number of voltage sources.
2. Write and solve several loop equations of a series-parallel circuit. This circuit may contain more than one power supply.
3. Apply the superposition theorem to solve for the current through or voltage across any particular element in a circuit containing more than one source of emf.
4. (*Advanced*) Write loop equations for a complicated circuit (three or more loops) containing one or more power supplies. Solve the loop equations for the unknown values.

Loop Equations Derived from Kirchhoff's Voltage Law

Objective 1   Apply Kirchhoff's voltage law to a simple, single-(closed) loop electric circuit that may contain a number of voltage sources.

Single Loop

When Kirchhoff's voltage law is applied to a closed-loop electric circuit, as shown in Figure 10.1, the algebraic sum of the emf's (voltage sources) must equal the algebraic sum of all the voltage drops. In Figure 10.1, the source voltages E_1 and E_2 are connected in series and are pointing in the same direction. This voltage-source direction results in a current I flowing in a clockwise direction. Furthermore, this current I flows through all the resistors in this series circuit, producing a net voltage drop of V_1, V_2, and V_3. Kirchhoff's voltage law states that the algebraic sum of V_1, V_2, and V_3 must equal the sum of all the voltage sources. Mathematically this can be expressed as

$$E_1 + E_2 = V_1 + V_2 + V_3 \tag{10.1}$$

where E_1, E_2 = voltage sources in closed loop

V_1 = voltage drop across resistor R_1

V_2 = voltage drop across resistor R_2

V_3 = voltage drop across resistor R_3

Applying Ohm's law, we can rewrite Equation (10.1) as

$$E_1 + E_2 = I[R_1 + R_2 + R_3] \tag{10.2}$$

and solving for current I, we can obtain

$$I = \frac{E_1 + E_2}{R_1 + R_2 + R_3} \tag{10.3}$$

The following example illustrates the application of Kirchhoff's voltage law (KVL) to a single closed-loop electric circuit. In it, you will learn not only how to calculate the current I flowing in the circuit, but also how to determine the voltage at any point with respect to some reference point (ground).

Example 10.1   Calculate the current I flowing in the circuit shown in Figure 10.2. Also find the voltage at $V_{a\text{-ground}}$ and $V_{b\text{-ground}}$.

Solution   (a) Since the power supply E_2 is larger than E_1, the current direction I is counterclockwise, as shown. As a result, the net applied voltage is $E_2 - E_1$ or 10 V.

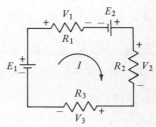

Figure 10.1   A closed-loop electric circuit

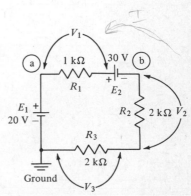

Figure 10.2   Circuit for Example 10.1

$$\therefore \quad (E_2 - E_1) = I(R_1 + R_2 + R_3)$$
$$30 - 20 = I(1\ k\Omega + 2\ k\Omega + 2\ k\Omega) = I(5\ k\Omega)$$
$$I = \frac{10\ V}{5\ k\Omega} = 2\ mA$$

(b) Applying Ohm's law, we obtain

$$V_1 = IR_1 = (2\ mA)(1\ k\Omega) = 2\ V$$
$$V_2 = IR_2 = (2\ mA)(2\ k\Omega) = 4\ V$$
$$V_3 = IR_3 = (2\ mA)(2\ k\Omega) = 4\ V$$

Thus the voltage at ⓐ to ground is $V_{a-ground} = E_1 = +20\ V$. The voltage at ⓑ to ground is $V_{b-ground} = -V_3 - V_2 = -8\ V$ or $V_{b-ground} = +E_1 + V_1 - E_2 = 20\ V + 2\ V - 30\ V = -8\ V$. Checking the proof of KVL:

$$E_2 - E_1 = V_1 + V_2 + V_3$$
$$30\ V - 20\ V = 2\ V + 4\ V + 4\ V$$
$$10\ V = 10\ V \qquad (\text{Checks okay})$$

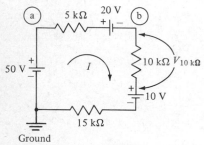

Ground

Figure 10.3   Circuit for Problem 10.1

Problem 10.1   A single-looped electric circuit is shown in Figure 10.3.   Find the following:

(a) What is current I?

(b) What is the voltage at point ⓐ with respect to ground?

$$V_{a-ground} = \underline{\hspace{3cm}}$$

(c) What is the voltage across the 10-kΩ resistor?

$$V_{10\ k\Omega} = \underline{\hspace{3cm}}$$

(d) What is the voltage at point ⓑ with respect to ground?

$$V_{b-ground} = \underline{\hspace{3cm}}$$

Several Loops

Objective 2   Write and solve several loop equations of a series-parallel circuit. This circuit may contain more than one power supply.

In Unit 8, the method you used to solve circuits such as the one shown in Figure 10.4 was to reduce all the resistors across the power supply to one equivalent resistor. In Unit 9 you converted the circuit to a Thévenin or Norton equivalent. This left the circuit with a battery and one resistor—a simple series circuit—in which to find the current. If you then wanted to know the currents flowing in resistors R_2, R_3, and R_4, you used the current-divider principle.

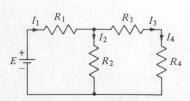

Figure 10.4   Sketch of a series-parallel circuit

There is yet another method that you may use to solve circuits of this type. This method is called *mesh* or *loop* equations. In this objective you will learn that loop equations are also helpful in solving circuits containing more than one power supply. A few examples illustrate this point.

Before we proceed to the examples, let us develop the loop equations using KVL for the circuit shown in Figure 10.4. Once we have developed these equations, let's see if we can learn to write these same equations in an easier way. The circuit of Figure 10.4, which will be used in our development of loop equations, is redrawn in Figure 10.5. You may recall that we have already analyzed this circuit in an earlier unit (Figure 8.29), and the answers to all the currents flowing through each resistor are listed on the right side of the figure. However, in Figure 8.29, the 2-kΩ and 6-kΩ resistors are combined into one 8-kΩ resistor. You may recall that in Objective 1 of this unit we needed a closed path or loop to apply KVL. Let us assume that I_1 is the current that is flowing in closed path 1, in a clockwise direction. This closed path is made up of the +12-V power supply, resistor R_1, and resistor R_2.

However, this circuit has another closed path. This second closed path is made up of resistors R_2, R_3, and R_4. We shall call the loop current in this second closed path I_2. Furthermore, we assume that the direction of I_2 is also clockwise. If an ammeter were inserted next to resistor R_1, the actual measured current would be loop current I_1. Furthermore, the current flowing through resistors R_3 and R_4 would be I_2. However, if an ammeter were inserted in series with R_2, the actual measured current there would be the difference between the loop currents I_1 and I_2. The reason for this is that loop current I_1 goes down through R_2, whereas loop current I_2 goes up through R_2. As a result, the actual current that is flowing through resistor R_2 is $I_1 - I_2$. We can see this another way. By applying KCL to node 1 of Figure 10.6, we have

$$I_1 = I + I_2 \qquad (10.4)$$

$$\therefore \quad I = I_1 - I_2 \qquad (10.5)$$

Writing the KVL equation for the first loop, we obtain

$$E = I_1 R_1 + (I_1 - I_2) R_2 \qquad \text{(first loop)} \qquad (10.6)$$

or

$$E = (R_1 + R_2) I_1 - R_2 I_2 \qquad (10.7)$$

We can obtain Equation (10.7) in an easier manner. As you write the equation around the loop, picture yourself standing

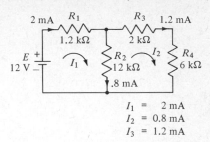

$$\begin{aligned} I_1 &= 2 \text{ mA} \\ I_2 &= 0.8 \text{ mA} \\ I_3 &= 1.2 \text{ mA} \end{aligned}$$

Figure 10.5 Practical series-parallel circuit

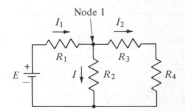

Figure 10.6 Relationship of loop currents

inside the loop. If you are looking around the loop in the direction of the arrow, the loop current I_1 is then flowing through all the resistors in the loop ($R_1 + R_2$). If you are inside loop 1, then loop current I_2 flows through resistor R_2 in the opposite direction. As a result, we can write the equation around loop 1 as

$$E = (R_1 + R_2)I_1 - R_2I_2 \qquad (10.8)$$

where E = total applied voltage in loop 1

     $+(R_1 + R_2)$ = total resistance in loop 1

         $-R_2$ = resistance common to loops 1 and 2

Comparing Equations (10.8) and (10.7), we see that they are both the same. Thus let us apply this easier technique to loop 2. Writing the KVL equation for loop 2, we obtain

$$0 = -R_2I_1 + (R_2 + R_3 + R_4)I_2 \qquad (10.9)$$

where 0 = total applied voltage in loop 2

       $-R_2$ = resistance common to loops 1 and 2

$R_2 + R_3 + R_4$ = total resistance in loop 2

Repeating Equations (10.7) and (10.9), we have

$$E = (R_1 + R_2)I_1 - R_2I_2 \qquad (10.10)$$

$$0 = -R_2I_1 + (R_2 + R_3 + R_4)I_2 \qquad (10.11)$$

Note that Equations (10.10) and (10.11) are two loop equations that were developed from KVL. Are there any more loops in Figures 10.4 and 10.5? The answer is yes. We can construct another loop from R_1, R_3, R_4, and E. But if you look at Equations (10.10) and (10.11), there are only two unknown variables, I_1 and I_2. Therefore we need only two equations to solve for any two unknowns. The two easiest loops to be used in writing loop equations should be made up of smallest loops. These loops are shown in Figure 10.5 by loop currents I_1 and I_2.

Let us now apply the technique of loop equations to the circuit shown in Figure 10.5.

Example 10.2 Solve for all the currents in the circuit shown in Figure 10.5.

Solution   KVL around the first loop involving current I_1:

$$12 \text{ V} = (1.2 \text{ k}\Omega + 12 \text{ k}\Omega)I_1 - (12 \text{ k}\Omega)I_2$$

and around the second loop involving current I_2:

$$0 \text{ V} = -(12 \text{ k}\Omega)I_1 + (12 \text{ k}\Omega + 2 \text{ k}\Omega + 6 \text{ k}\Omega)I_2$$

Simplifying the two loop equations, we have

$$12 \text{ V} = (13.2 \text{ k}\Omega)I_1 - (12 \text{ k}\Omega)I_2 \qquad \text{(loop 1)}$$

$$0 \text{ V} = -(12 \text{ k}\Omega)I_1 + (20 \text{ k}\Omega)I_2 \qquad \text{(loop 2)}$$

Thus we have two equations with two unknowns, I_1 and I_2. We can solve these two equations in a number of ways. In this example we shall describe two methods of solution (substitution and addition or subtraction) and compare the results with the answers given in Figure 10.5.

Method 1. Substitution method Rewriting the loop 2 equation, we get

$$(12 \text{ k}\Omega)I_1 = (20 \text{ k}\Omega)I_2 \qquad \text{or} \qquad I_2 = \frac{12 \text{ k}\Omega}{20 \text{ k}\Omega} I_1 = 0.6I_1$$

Substituting for I_2 in the loop 1 equation, we have

$$12 \text{ V} = (13.2 \text{ k}\Omega)I_1 - (12 \text{ k}\Omega)(0.6I_1)$$

or

$$12 \text{ V} = (13.2 \text{ k}\Omega)I_1 - (7.2 \text{ k}\Omega)I_1$$

$$12 \text{ V} = (6 \text{ k}\Omega)I_1 \qquad \therefore \quad I_1 = \frac{12 \text{ V}}{6 \text{ k}\Omega} = 2 \text{ mA}$$

Since

$$I_2 = 0.6I_1 = 0.6(2 \text{ mA}) \qquad \therefore \quad I_2 = 1.2 \text{ mA}$$

The current through resistor R_2 is $I_1 - I_2$, or

$$I = I_1 - I_2 = 2 \text{ mA} - 1.2 \text{ mA} = 0.8 \text{ mA}$$

These three currents agree with the answers given in Figure 10.5.

Method 2. Addition (or subtraction) method By this method, the loop 1 equation is multiplied by 5, while the loop 2 equation is multiplied by 3.

$$[12 = (13.2 \text{ k}\Omega)I_1 - (12 \text{ k}\Omega)I_2] \times (5)$$

$$[0 = -(12 \text{ k}\Omega)I_1 + (20 \text{ k}\Omega)I_2] \times (3)$$

Rewriting the above equations, we obtain

$$60 = (66 \text{ k}\Omega)I_1 - (60 \text{ k}\Omega)I_2 \qquad 0 = -(36 \text{ k}\Omega)I_1 + (60 \text{ k}\Omega)I_2$$

Adding both equations, we obtain

$$60 + 0 = (66 \text{ k}\Omega)I_1 - (36 \text{ k}\Omega)I_1 + 0$$

$$60 = (30 \text{ k}\Omega)I_1 \qquad \therefore \quad I_1 = \frac{60}{30 \text{ k}\Omega} = 2 \text{ mA}$$

Substituting I_1 in the loop 2 equation, we have

$$0 \text{ V} = -(12 \text{ k}\Omega)(2 \text{ mA}) + (20 \text{ k}\Omega)I_2$$

$$24 \text{ V} = (20 \text{ k}\Omega)I_2 \qquad \therefore \quad I_2 = 1.2 \text{ mA}$$

These currents agree with the given answers.

The following two examples illustrate the application of loop equations to circuits containing several supplies.

Example 10.3   Find the loop currents I_1 and I_2 in Figure 10.7, and also find the voltage drop across the 2-kΩ voltmeter.

Solution   Applying KVL to loop 1, we have

$$48 \text{ V} = (15 \text{ k}\Omega + 13 \text{ k}\Omega + 2 \text{ k}\Omega)I_1 - (2 \text{ k}\Omega)I_2$$

and KVL around loop 2 is

$$56 \text{ V} = -(2 \text{ k}\Omega)I_1 + (2 \text{ k}\Omega + 5 \text{ k}\Omega + 3 \text{ k}\Omega)I_2$$

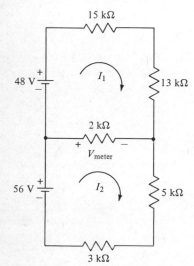

Figure 10.7   Circuit for Example 10.3

Rewriting the loop 1 and 2 equations and simplifying,

$$48 = (30 \text{ k}\Omega)I_1 - (2 \text{ k}\Omega)I_2 \qquad \text{(loop 1)}$$

$$56 = -(2 \text{ k}\Omega)I_1 + (10 \text{ k}\Omega)I_2 \qquad \text{(loop 2)}$$

Multiplying the loop 1 equation by 5 and adding it to the loop 2 equation,

$$48 \times 5 = (30 \text{ k}\Omega \times 5)I_1 - (2 \text{ k}\Omega \times 5)I_2$$

or

$$240 = (150 \text{ k}\Omega)I_1 - (10 \text{ k}\Omega)I_2 \qquad \text{(loop 1)}$$

$$\underline{56 = -(2 \text{ k}\Omega)I_1 + (10 \text{ k}\Omega)I_2} \qquad \text{(loop 2)}$$

$$296 = (148 \text{ k}\Omega)I_1$$

$$\therefore \quad I_1 = \frac{296 \text{ V}}{148 \text{ k}\Omega} = 2 \text{ mA}$$

Substituting I_1 back into the loop 1 equation,

$$48 = (30 \text{ k}\Omega)(2 \text{ mA}) - (2 \text{ k}\Omega)I_2 = 60 - (2 \text{ k}\Omega)I_2$$

$$-12 = -(2 \text{ k}\Omega)I_2 \qquad \therefore \quad I_2 = +6 \text{ mA}$$

The net current flow through 2 kΩ is the difference between the I_1 and I_2 loop currents. Since I_2 is larger than the current is,

$$I_2 - I_1 = (6 \text{ mA} - 2 \text{ mA}) = 4 \text{ mA}$$

$$V_{\text{meter}} = (I_2 - I_1)2 \text{ k}\Omega = (4 \text{ mA})(2 \text{ k}\Omega) = 8 \text{ V}$$

Checking KVL for loop 2,

$$56 \text{ V} = +8 \text{ V} + (6 \text{ mA})(5 \text{ k}\Omega) + (6 \text{ mA})(3 \text{ k}\Omega)$$

$$56 \text{ V} = 8 \text{ V} + 30 \text{ V} + 18 \text{ V}$$

$$56 \text{ V} = 56 \text{ V} \qquad \text{(Checks okay)}$$

Example 10.4 Find the current I flowing through the 4-kΩ resistor shown in Figure 10.8 and the voltage drop across the 4-kΩ resistor.

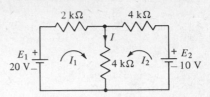

Figure 10.8 Circuit for Example 10.4

Solution Since both supplies E_1 and E_2 are positive, we shall assume that the loop currents I_1 and I_2 are flowing in the directions shown. Note that now current I, by KCL, is

$$I = I_1 + I_2$$

Writing KVL around loop 1, we obtain

$$20 \text{ V} = (2 \text{ k}\Omega)I_1 + (4 \text{ k}\Omega)(I_1 + I_2)$$

or simplifying,

$$20 \text{ V} = (6 \text{ k}\Omega)I_1 + (4 \text{ k}\Omega)I_2 \qquad \text{(loop 1)}$$

The loop 1 equation can be written in an easier way. If you are in loop 1, then loop current I_2 flows through the 4-kΩ resistor in the same direction as I_1. Thus we can write loop 1 as

$$20 \text{ V} = (2 \text{ k}\Omega + 4 \text{ k}\Omega)I_1 + (4 \text{ k}\Omega)I_2$$

Note that this equation is similar to the loop 1 equation we wrote earlier. Furthermore, note that the sign of the term $(+4 \text{ k}\Omega)I_2$ is positive, since I_2 and I_1 are both flowing in the same direction.

Likewise, writing KVL for loop 2, we obtain

$$10 \text{ V} = (4 \text{ k}\Omega)I_1 + (8 \text{ k}\Omega)I_2 \qquad \text{(loop 2)}$$

Collecting terms and rewriting, we get

$$20 \text{ V} = (6 \text{ k}\Omega)I_1 + (4 \text{ k}\Omega)I_2 \qquad \text{(loop 1)}$$
$$10 \text{ V} = (4 \text{ k}\Omega)I_1 + (8 \text{ k}\Omega)I_2 \qquad \text{(loop 2)}$$

Multiplying loop 1 by 2 and subtracting, we find

$$40 \text{ V} = (12 \text{ k}\Omega)I_1 + (8 \text{ k}\Omega)I_2$$
$$\underline{10 \text{ V} = (4 \text{ k}\Omega)I_1 + (8 \text{ k}\Omega)I_2} \qquad \text{(loop 2)}$$
$$40 \text{ V} - 10 \text{ V} = (12 \text{ k}\Omega - 4 \text{ k}\Omega)I_1 + 0$$
$$30 \text{ V} = (8 \text{ k}\Omega)I_1$$

$$\therefore \quad I_1 = \frac{30 \text{ V}}{8 \text{ k}\Omega} = 3.75 \text{ mA}$$

Substituting I_1 in the loop 2 equation,

$$10 \text{ V} = (4 \text{ k}\Omega)(3.75 \text{ mA}) + (8 \text{ k}\Omega)I_2$$
$$10 \text{ V} = 15 \text{ V} + (8 \text{ k}\Omega)I_2$$
$$-5 \text{ V} = (8 \text{ k}\Omega)I_2$$
$$I_2 = -0.625 \text{ mA}$$
$$I = I_1 + I_2 = 3.75 - 0.625 = 3.125 \text{ mA}$$
$$V = I(4 \text{ k}\Omega) = (3.125 \text{ mA})(4 \text{ k}\Omega)$$
$$V = 12.5 \text{ V}$$

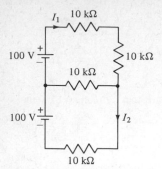

Figure 10.9 Circuit for Problem 10.2

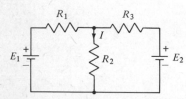

Figure 10.10 Typical circuit applying superposition theorem

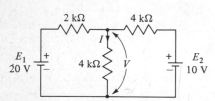

Figure 10.11 Circuit for Example 10.5

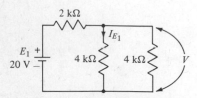

Figure 10.12 Circuit of Example 10.5 with $E_2 = 0$

Problem 10.2   From Figure 10.9, find:

$$I_1 = \underline{\hspace{3cm}}$$

$$I_2 = \underline{\hspace{3cm}}$$

Superposition Theorem

Objective 3   Apply the superposition theorem to solve for the current through or voltage across any particular element in a circuit containing more than one source of emf.

In this objective you will learn another way to solve complicated circuits; that is, by the superposition theorem. The superposition theorem allows you to solve for the voltage across or current through an element in a series-parallel circuit with two or more power supplies.

An example of the type of circuit you may encounter is shown in Figure 10.10.

The superposition theorem states that if we have a resistor or a *linear, bilateral* network, we can find the current through or the voltage across any element by algebraically adding the currents or voltages that each source would produce independently. The expression *linear* means that the characteristics of the circuit or circuit elements are *independent* of the voltage across or the current through them. Most resistors are linear; many electronic circuits or circuit elements may be considered to be linear in a certain range of application. Also, if a certain element does not change its characteristics when a current through or the voltage across it *changes direction*, then it is considered to be bilateral.

Applying the superposition theorem to Figure 10.10, we can find the current I flowing through resistor R_2 or the voltage across it by first replacing source E_2 by its internal resistance and using only one source E_1. In most cases, an ideal voltage source like E_2 becomes equal to 0 V, or a short, and an ideal current source becomes equal to 0 A, or an open circuit. The following two examples illustrate the application of the superposition theorem to a circuit containing two voltage sources and to a circuit containing a voltage and a current source.

Example 10.5   Find the current through and voltage across the 4-kΩ resistor shown in Figure 10.11.

Solution   (a) Consider only source E_1. Then $E_2 = 0$. The circuit of Figure 10.11 then looks like Figure 10.12.

We find the voltage across the 4 kΩ resistor by using the voltage-divider principle.

$$V_{4k\Omega} = \left(\frac{4\ k\Omega \| 4\ k\Omega}{4\ k\Omega \| 4\ k\Omega + 2\ k\Omega}\right) 20\ V = \left(\frac{2\ k\Omega}{2\ k\Omega + 2\ k\Omega}\right) 20\ V = 10\ V$$

or

$$I_{E_1} = \frac{V}{4\ k\Omega} = \frac{10\ V}{4\ k\Omega} = 2.5\ mA$$

This current I_{E_1} is the current flowing through the 4-kΩ resistor with only source E_1. (*Note*: $E_2 = 0$)

(b) Now, considering source $E_1 = 0$ and finding the current and voltage contribution of source E_2, we have the situation shown in Figure 10.13.

Applying the voltage-divider principle to the circuit of Figure 10.13, we find that the voltage across the 4 kΩ due to the source E_2 is

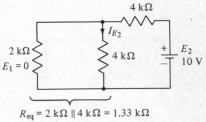

$R_{eq} = 2\ k\Omega \| 4\ k\Omega = 1.33\ k\Omega$

Figure 10.13   Circuit of Example 10.5 with $E_1 = 0$

$$V_{4\ k\Omega} = \left(\frac{1.33\ k\Omega}{1.33\ k\Omega + 4\ k\Omega}\right) 10\ V = \left(\frac{1.33\ k\Omega}{5.33\ k\Omega}\right) 10\ V = 2.495\ V$$

$$I_{E_2} = \frac{V_{4\ k\Omega}}{4\ k\Omega} \simeq \frac{2.5\ V}{4\ k\Omega} = 0.625\ mA$$

Thus, applying the superposition theorem, we obtain the total current through the 4-kΩ resistor contributed by both supplies E_1 and E_2 as

$$I = I_{E_1} + I_{E_2} = 2.5\ mA + 0.625\ mA$$
$$\therefore\quad I = 3.125\ mA$$

The actual voltage across the 4-kΩ resistor is

$$V = (4\ k\Omega)(I) = (4\ k\Omega)(3.125\ mA) = 12.5\ V$$

Note that these results agree with the answers of Example 10.4. In that example, we analyzed the circuit by loop equations.

Example 10.6   Find the current through and the voltage across the 2-kΩ resistor shown in Figure 10.14.

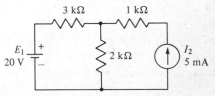

Figure 10.14   Circuit for Example 10.6

Solution   In this example, we have one voltage source E_1 and one current source I_2. In applying the superposition theorem, we shall first use supply E_1 and let the current source I_2 be open. The circuit looks like Figure 10.15. The voltage drop across 2 kΩ is

$$V_{2\ k\Omega} = \left(\frac{2\ k\Omega}{2\ k\Omega + 3\ k\Omega}\right) 20\ V = \left(\frac{2\ k\Omega}{5\ k\Omega}\right) 20\ V = 8\ V$$

$$\therefore\quad I_{2\ k\Omega} = \frac{V_{2\ k\Omega}}{2\ k\Omega} = 4\ mA$$

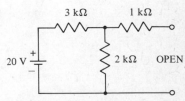

Figure 10.15   Circuit of Example 10.6 with current source I_2 removed

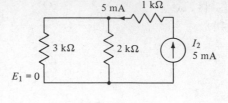

Figure 10.16 Circuit of Example 10.6
with voltage source $E_1 = 0$

To find the current and voltage contribution of current source I_2 (with $E_1 = 0$), we have the circuit shown in Figure 10.16. Applying the current-divider principle, we find that the current $I_{2\,k\Omega}$ is

$$I_{2\,k\Omega} = \left(\frac{3\ k\Omega}{2\ k\Omega + 3\ k\Omega}\right) 5\ mA = 3\ mA$$

$$\therefore\quad V_{2\,k\Omega} = I_{2\,k\Omega} \times 2\ k\Omega = 6\ V$$

Applying the superposition theorem, we know that the total current flowing through the 2-kΩ resistor is the algebraic sum of the currents $I_{2\,k\Omega}$ supplied by *both* supplies. It is

$$I_{2\,k\Omega} = 4\ mA + 3\ mA = 7\ mA$$

The voltage across the 2-kΩ resistor can be found by Ohm's law as

$$V_{2\,k\Omega} = I_{2\,k\Omega} \times 2\ k\Omega = 7\ mA \times 2\ k\Omega = 14\ V$$

or by the superposition theorem as

$$V_{2\,k\Omega} = 8\ V + 6\ V = 14\ V$$

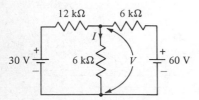

Figure 10.17 Circuit for Problem 10.3

Problem 10.3 Apply the superposition theorem to the circuit shown in Figure 10.17. Find the current and voltage indicated.

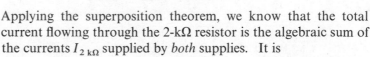

Problem 10.4 Apply the superposition theorem to the circuit shown in Figure 10.18. Find V.

Figure 10.18 Circuit for Problem 10.4

Problem 10.5 Apply the superposition theorem to the circuit shown in Figure 10.19. Find the current and voltage indicated.

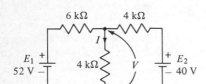

Figure 10.19 Circuit for Problem 10.5

Multiloop Equations of Complicated Circuits

Objective 4 (*Advanced*) Write loop equations for a complicated circuit (three or more loops) containing one or more power supplies. Solve the loop equations for the unknown values.

In Objective 2 you learned how to write loop equations for a simple circuit containing only two loops. In this objective you

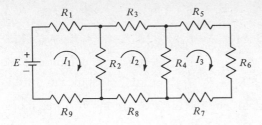

Figure 10.20 Generation of loop equations in a given circuit

will learn how to write loop equations for a complicated circuit (three or more loops). Referring to a typical circuit like the one shown in Figure 10.20, let us again write the loop equations by using KVL. After studying them, let's see if we can again write them in an easier manner. In Figure 10.20 you see three loop currents I_1, I_2, and I_3 drawn in the assumed direction. As you are writing the loop equations around the loop, picture yourself standing inside the loop. If you are inside loop 1, then the loop I_2 current through resistor R_2 is flowing in the direction opposite to that of the loop I_1 current. Applying KVL to loop 1, we obtain

First loop

$$E = I_1 R_1 + (I_1 - I_2)R_2 + I_1 R_9 \qquad (10.12)$$

or recombining terms, we can rewrite Equation (10.12) as

$$E = (R_1 + R_2 + R_9)I_1 - R_2 I_2 - (0)I_3 \qquad (10.13)$$

where E = total applied voltage in first loop

$R_1 + R_2 + R_9$ = total resistance in first loop

$-R_2$ = resistance common to loop 2

Writing the loop equation (no power supply) around loop 2:

Second loop

$$0 = (I_2 - I_1)R_2 + I_2 R_3 + (I_2 - I_3)R_4 + I_2 R_8 \quad (10.14)$$

or recombining terms, we can rewrite Equation (10.14) as

$$0 = -R_2 I_1 + (R_2 + R_3 + R_4 + R_8)I_2 - R_4 I_3 \quad (10.15)$$

where $-R_2$ = resistance common to loop 1

$R_2 + R_3 + R_4 + R_8$ = total resistance in second loop

$-R_4$ = resistance common to loop 3

Using the easier technique to write the loop equation around loop 3, we have the third loop.

Third loop

$$0 = (0)I_1 - R_4I_2 + (R_4 + R_5 + R_6 + R_7)I_3 \qquad (10.16)$$

where $-R_4$ = resistance common to loop 2

$R_4 + R_5 + R_6 + R_7$ = total resistance in third loop

Thus, summarizing the three equations for three loops, we have

$$E = (R_1 + R_2 + R_9)I_1 - R_2I_2 + 0I_3 \qquad \text{(loop 1)}$$
$$0 = -R_2I_1 + (R_2 + R_3 + R_4 + R_8)I_2 - R_4I_3 \quad \text{(loop 2)}$$
$$0 = 0I_1 - R_4I_2 + (R_4 + R_5 + R_6 + R_7)I_3 \qquad \text{(loop 3)}$$

You might be wondering if there is not another loop. Yes, there is, around the outside of the circuit. It starts with the power supply E and includes resistors $R_1, R_3, R_5, R_6, R_7, R_8,$ and R_9. But this loop is *not* required, since the three loop currents $I_1, I_2,$ and I_3 are sufficient to represent all the currents flowing through any given resistor. For instance, the current through resistor R_3 is just loop current I_2. The current through R_2 is the difference between the loop currents I_1 and I_2 $(I_1 - I_2)$.

In summary, the three closed loops result in three independent equations that one can solve for three independent loop currents $I_1, I_2,$ and I_3.

The following two examples demonstrate the writing of loop equations for two complicated circuits. The first circuit, in Example 10.7, was solved earlier by other conventional techniques. Here it is solved by loop equations. The answers given in the figure are compared with the example solutions.

The second example (Example 10.8) is the analysis of a Wheatstone bridge.

For Example 10.7, we present two methods of solution: substitution and determinants. (The use of determinants is considered to be an advanced topic and somewhat difficult for the average student.)

Example 10.7 Solve the circuit given in Example 8.10 using loop equations. Find all the currents.

Solution Referring to the circuit shown in Figure 10.21, we see that we have three loops. The three loop currents $I_A, I_B,$ and I_C

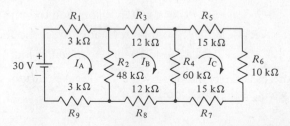

Figure 10.21 Circuit for Example 10.7

are all assumed to be flowing in a clockwise direction. Writing loop equations, we have

$$30 = (R_1 + R_2 + R_9)I_A - (R_2)I_B - (0)I_C \qquad \text{(loop A)}$$
$$0 = -(R_2)I_A + (R_2 + R_3 + R_4 + R_8)I_B - (R_4)I_C \quad \text{(loop B)}$$
$$0 = -(0)I_A - (R_4)I_B + (R_4 + R_5 + R_6 + R_7)I_C \quad \text{(loop C)}$$

Therefore

$$30 = (54 \text{ k}\Omega)I_A - (48 \text{ k}\Omega)I_B \qquad \text{(A)}$$
$$0 = -(48 \text{ k}\Omega)I_A + (132 \text{ k}\Omega)I_B - (60 \text{ k}\Omega)I_C \qquad \text{(B)}$$
$$0 = -(60 \text{ k}\Omega)I_B + (100 \text{ k}\Omega)I_C \qquad \text{(C)}$$

We can solve these three simultaneous equations in many ways. The methods of substitution and determinants are presented here.

Solution by substitution

Rewriting Equation (A), we have

$$(54 \text{ k}\Omega)I_A = 30 + (48 \text{ k}\Omega)I_B$$

Solving for I_A, we get

$$I_A = \frac{30 + (48 \text{ k}\Omega)I_B}{54 \text{ k}\Omega}$$

Rewriting Equation (C), we obtain

$$(100 \text{ k}\Omega)I_C = (60 \text{ k}\Omega)I_B$$

Solving for I_C, we have

$$I_C = \frac{60 \text{ k}\Omega}{100 \text{ k}\Omega} I_B$$

Substituting I_A and I_C into Equation (B), we obtain

$$0 = -(48 \text{ k}\Omega)\left[\frac{30 + (48 \text{ k}\Omega)I_B}{54 \text{ k}\Omega}\right] + (132 \text{ k}\Omega)I_B - (60 \text{ k}\Omega)\left(\frac{60 \text{ k}\Omega}{100 \text{ k}\Omega}\right)I_B$$
$$= -26.667 - (42.667 \text{ k}\Omega)I_B + (132 \text{ k}\Omega)I_B - (36 \text{ k}\Omega)I_B$$
$$26.667 = (53.333 \text{ k}\Omega)I_B$$
$$I_B = 0.5 \text{ mA} \qquad \text{(loop B)}$$
$$\therefore \quad (54 \text{ k}\Omega)I_A = 30 + (48 \text{ k}\Omega)(0.5 \text{ mA})$$
$$(54 \text{ k}\Omega)I_A = 54$$
$$I_A = 1 \text{ mA} \qquad \text{(loop A)}$$
$$\therefore \quad (100 \text{ k}\Omega)I_C = (60 \text{ k}\Omega)(0.5 \text{ mA})$$
$$I_C = 0.3 \text{ mA} \qquad \text{(loop C)}$$

Therefore, summarizing currents flowing through each resistor, we have:

$$I_1 = I_A = 1 \text{ mA}$$
$$I_2 = I_A - I_B = 0.5 \text{ mA}$$
$$I_3 = I_B = 0.5 \text{ mA}$$
$$I_4 = I_B - I_C = 0.2 \text{ mA}$$
$$I_5 = I_C = 0.3 \text{ mA}$$
$$I_6 = I_C = 0.3 \text{ mA}$$
$$I_7 = I_C = 0.3 \text{ mA}$$
$$I_8 = I_B = 0.5 \text{ mA}$$
$$I_9 = I_A = 1 \text{ mA}$$

Solution by determinants

$$I_A = \frac{\begin{vmatrix} 30 & -48 \text{ k}\Omega & 0 \\ 0 & 132 \text{ k}\Omega & -60 \text{ k}\Omega \\ 0 & -60 \text{ k}\Omega & +100 \text{ k}\Omega \end{vmatrix}}{\begin{vmatrix} 54 \text{ k}\Omega & -48 \text{ k}\Omega & 0 \\ -48 \text{ k}\Omega & +132 \text{ k}\Omega & -60 \text{ k}\Omega \\ 0 & -60 \text{ k}\Omega & +100 \text{ k}\Omega \end{vmatrix}}$$

$$= \frac{\begin{matrix}(30 \cdot 132 \text{ k}\Omega \cdot 100 \text{ k}\Omega) + (-48 \text{ k}\Omega \cdot -60 \text{ k}\Omega \cdot 0) + (0 \cdot 0 \cdot -60 \text{ k}\Omega) - (0 \cdot 132 \text{ k}\Omega \cdot 0) \\ - (0 \cdot -48 \text{ k}\Omega \cdot 100 \text{ k}\Omega) - (30 \cdot -60 \text{ k}\Omega \cdot -60 \text{ k}\Omega)\end{matrix}}{\begin{matrix}+(54 \text{ k}\Omega \cdot 132 \text{ k}\Omega \cdot 100 \text{ k}\Omega) + (-48 \text{ k}\Omega \cdot -60 \text{ k}\Omega \cdot 0) + (0 \cdot -48 \text{ k}\Omega \cdot -60 \text{ k}\Omega) \\ - (0 \cdot 132 \text{ k}\Omega \cdot 0) - (-48 \text{ k}\Omega \cdot -48 \text{ k}\Omega \cdot 100 \text{ k}\Omega) - (54 \text{ k}\Omega \cdot -60 \text{ k}\Omega \cdot -60 \text{ k}\Omega)\end{matrix}}$$

$$= \frac{(30 \cdot 132 \text{ k}\Omega \cdot 100 \text{ k}\Omega) - (30 \cdot -60 \text{ k}\Omega \cdot -60 \text{ k}\Omega)}{(54 \text{ k}\Omega \cdot 132 \text{ k}\Omega \cdot 100 \text{ k}\Omega) - (-48 \text{ k}\Omega \cdot -48 \text{ k}\Omega \cdot 100 \text{ k}\Omega) - (54 \text{ k}\Omega \cdot -60 \text{ k}\Omega \cdot -60 \text{ k}\Omega)}$$

$$= \frac{30(13,200 - 3600) \times 10^6}{712,800 \times 10^9 - 230,400 \times 10^9 - 194,400 \times 10^9}$$

$$= \frac{30(9600 \times 10^6)}{288,000 \times 10^9}$$

$$I_A = 1 \text{ mA}$$

Having found I_A, we can find the two remaining unknown currents I_B and I_C by substitution.

Example 10.8 The circuit shown in Figure 10.22 is a Wheatstone bridge. Using loop equations, find the current flow in the meter. The internal resistance of the meter is 50 Ω.

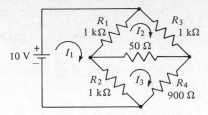

Figure 10.22 Circuit for Example 10.8

Solution

$$10 = (2 \text{ k}\Omega)I_1 - (1 \text{ k}\Omega)I_2 - (1 \text{ k}\Omega)I_3 \qquad \text{(loop 1)}$$

$$0 = -(1 \text{ k}\Omega)I_1 + (2.05 \text{ k}\Omega)I_2 - (0.05 \text{ k}\Omega)I_3 \quad \text{(loop 2)}$$

$$0 = -(1 \text{ k}\Omega)I_1 - (0.05 \text{ k}\Omega)I_2 + (1.95 \text{ k}\Omega)I_3 \quad \text{(loop 3)}$$

Adding loop 1 to two times loop 2, we obtain:

$$
\begin{array}{ll}
10 = \quad (2 \text{ k}\Omega)I_1 - \quad (1 \text{ k}\Omega)I_2 - (1.0 \text{ k}\Omega)I_3 & \text{(loop 1)} \\
\underline{0 = -(2 \text{ k}\Omega)I_1 + (4.1 \text{ k}\Omega)I_2 - (0.1 \text{ k}\Omega)I_3} & 2 \times \text{(loop 2)} \\
10 = \qquad\qquad\quad + (3.1 \text{ k}\Omega)I_2 - (1.1 \text{ k}\Omega)I_3 & \text{(A)}
\end{array}
$$

Subtracting loop 3 from loop 2, we obtain

$$0 = +(2.1 \text{ k}\Omega)I_2 - (2.0 \text{ k}\Omega)I_3 \qquad \text{(B)}$$

Multiplying Equation (A) by 2 and Equation (B) by 1.1, and then subtracting, gives us

$$
\begin{array}{ll}
20 = +(6.2 \text{ k}\Omega)I_2 \quad - (2.2 \text{ k}\Omega)I_3 & 2 \times \text{(A)} \\
\underline{0 = +(2.31 \text{ k}\Omega)I_2 - (2.2 \text{ k}\Omega)I_3} & 1.1 \times \text{(B)} \\
20 = +(3.89 \text{ k}\Omega)I_2 &
\end{array}
$$

$$\therefore \quad I_2 = 5.141 \text{ mA}$$

Substituting the value of I_2 into Equation (B), we find I_3:

$$0 = (2.1 \text{ k}\Omega)(5.141 \text{ mA}) - (2 \text{ k}\Omega)I_3$$

$$I_3 = \frac{(2.1 \text{ k}\Omega)(5.141 \text{ mA})}{2 \text{ k}\Omega} = 5.4 \text{ mA}$$

We don't need the solution for I_1 in order to solve this problem, but we can find it by substituting I_2 and I_3 into the loop 1 equation.

$$10 = (2 \text{ k}\Omega)I_1 - (1 \text{ k}\Omega)(5.141 \text{ mA}) - (1 \text{ k}\Omega)(5.4 \text{ mA})$$

$$10 = (2 \text{ k}\Omega)I_1 - 5.141 \text{ V} - 5.4 \text{ V}$$

$$20.541 = (2 \text{ k}\Omega)I_1$$

$$I_1 = \frac{20.541}{2 \text{ k}\Omega} = 10.27 \text{ mA}$$

The meter current is $(I_3 - I_2)$:

$$I_{meter} = 5.4 \text{ mA} - 5.141 \text{ mA} = 0.259 \text{ mA}$$

Problem 10.6   For the circuit shown in Figure 10.23, solve for the currents I_1, I_2, and I_3, using loop equations.

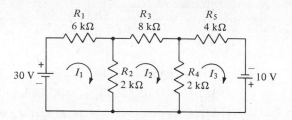

Figure 10.23   Circuit for Problem 10.6

Problem 10.7   Using loop equations, find the current flow I_m in the meter shown in Figure 10.24.   Assume that the meter resistance R_m is 100 Ω.

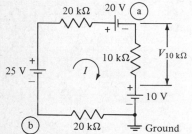

Figure 10.25   Circuit for Test Problem 1

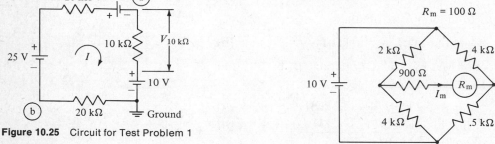

Figure 10.24   Circuit for Problem 10.7

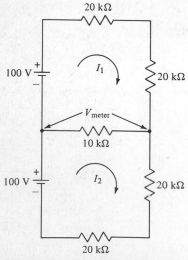

Figure 10.26   Circuit for Test Problem 2

Test

1. Find the current I flowing through the 10-kΩ resistor in the circuit shown in Figure 10.25.

   $I =$ _____

   $V_{10 \text{ k}\Omega} =$ _____

   $V_{a-b} =$ _____

2. Refer to Figure 10.26.   Find the values of the following.

   $I_1 =$ _____

   $I_2 =$ _____

   $V_{meter} =$ _____

3. Apply the superposition theorem to the circuit shown in Figure 10.27. Find current I and voltage V.

4. Solve the circuit shown in Figure 10.27 by means of loop equations. Find current I and voltage V and compare these values with your answers to Problem 3.

5. Solve for the current and voltage across a 1-kΩ resistor. See Figure 10.28.

6. Find the current flowing through the 2-kΩ resistor in Figure 10.29.

7. Apply the superposition theorem to the circuit shown in Figure 10.30. Find I and V.

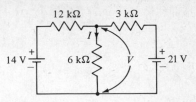

Figure 10.27   Circuit for Test Problems 3 and 4

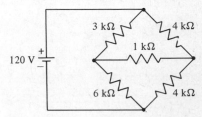

Figure 10.28   Circuit for Test Problem 5

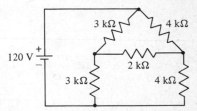

Figure 10.29   Circuit for Test Problem 6

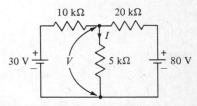

Figure 10.30   Circuit for Test Problem 7

Unit 11 *Capacitance*

Up to this point, you have been studying units containing circuits that have steady voltages and currents and that contain only resistive components. These *resistive* components have the property of *opposing* the flow of electric current. Thus they are all dissipative, that is, all the energy that is supplied to them is converted to heat and cannot be recovered.

In this unit and in a later unit you will be introduced to two new elements, the capacitor and the inductor. These capacitors and inductors are not important in the study of dc electric circuits, but play an important role in circuits in which currents and voltages are changing (ac circuits). Capacitors and inductors are different from resistors in that they can store electrical energy. As a result, at some later time this stored energy can be recovered and given back to some other circuit element.

In this unit you will learn about the capacitor. Here you will learn that *capacitance* is the property of an electric circuit that *opposes any change in voltage* or potential difference across the circuit.

Objectives

After completing all the work associated with this unit, you should be able to:

1. Recall the definition, electrical and letter symbols, and unit of measurement of a capacitor.

2. State when an electric circuit has a capacitance of one farad. Use the equation $C = Q/V$ to solve problems. Recall the property of capacitance by referring to the equation

$$i = C \frac{dv}{dt}$$

3. List the three physical factors that determine the capacitance of a capacitor. Relate these factors in a mathematical equation, and use this equation to solve problems.

4. Calculate the total capacitance and charge stored by each capacitor when two, three, or more capacitors are connected in parallel.

5. Calculate the total or equivalent capacitance, charge stored, and voltage across each capacitor when two, three, or more capacitors are connected in series.

6. (*Advanced*) Calculate the total or equivalent capacitance and voltage across each capacitor when several capacitors are connected in series-parallel combinations.

7. Calculate the energy stored by a capacitor.

Capacitors

Objective 1 Recall the definition, electrical and letter symbols, and unit of measurement of a capacitor.

A *capacitor*, sometimes called a *condenser*, consists basically of two conductors separated by an insulator. In an actual capacitor, the two conductors may be called *plates* and the insulating material between the plates is a *dielectric*. Figure 11.1 shows the electrical symbols used for capacitors. The letter symbol used for capacitance in equations is C. The unit of measurement of capacitance is the *farad*, abbreviated F. Since the farad is a very large unit, most capacitors are rated in microfarads or picofarads:

μF (microfarad, or 10^{-6} F) Sometimes symbolized as MF on capacitors

or

pF (picofarad, or 10^{-12} F) In the past the picofarad was symbolized $\mu\mu$F

The fixed capacitor in Figure 11.1, whose value cannot be varied, is available in many shapes and sizes. The most common types of fixed capacitors are mica, ceramic, paper tubular, and

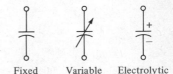

Fixed Variable Electrolytic

Figure 11.1 Electrical symbols used for capacitors

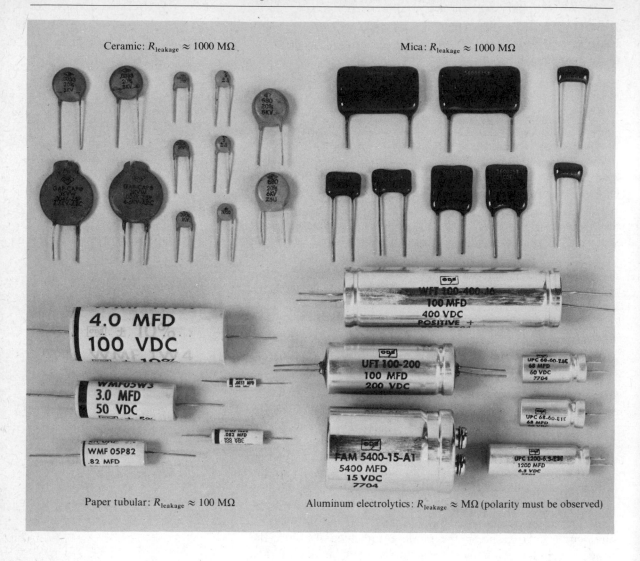

Ceramic: $R_{leakage} \approx 1000$ MΩ

Mica: $R_{leakage} \approx 1000$ MΩ

Paper tubular: $R_{leakage} \approx 100$ MΩ

Aluminum electrolytics: $R_{leakage} \approx$ MΩ (polarity must be observed)

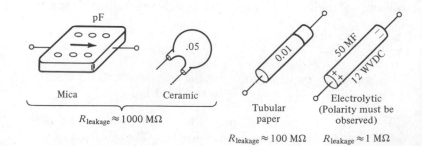

Figure 11.2 Examples of different types of capacitors

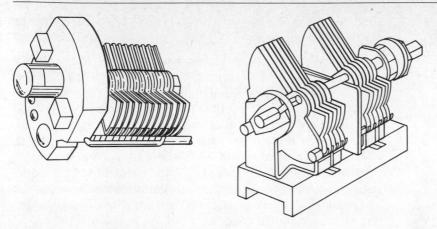

Figure 11.3 Variable capacitors

electrolytic. These are shown in Figure 11.2. Of these capacitors, the mica and ceramic ones have the smallest leakage currents; their resistances are of the order of $R_{\text{leakage}} \approx 1000$ MΩ.

The variable capacitor uses air as a dielectric. Its plates consist of several half-moon-shaped flat metal discs that are connected and intermeshed. This is shown in Figure 11.3. These plates can be rotated so as to have maximum or minimum area between the plates. Thus, by varying the area, one can vary the value of capacitance. Variable capacitors are used mostly in radio receivers for the purpose of tuning.

Very large capacitors, from 5 μF to ~200 μF, are usually electrolytic capacitors. This large capacitance value compared with the small physical size is due to the special type of construction used. They are constructed by electrolytic chemical action, and therefore proper polarity must be observed. A plus (+) mark is always placed on the capacitor. (See Figures 11.1 and 11.2.) Failure to connect an electrolytic capacitor properly in a circuit will damage the capacitor.

Problem 11.1 Draw the electrical symbols for a fixed capacitor and an electrolytic capacitor. Write the letter symbol used for capacitors, and include the basic unit of measurement for a capacitor.

Capacitor Color Codes

Capacitors are coded in the same way as resistors. There are two color-coding systems for capacitors:

(a) EIA system (Electronic Industries Association)

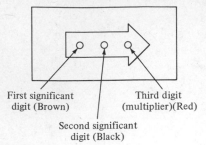

First significant digit (Brown)

Third digit (multiplier)(Red)

Second significant digit (Black)

Figure 11.4 A three-dot code for a mica capacitor (EIA system)

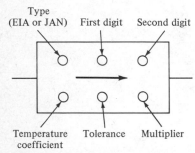

Type (EIA or JAN)   First digit   Second digit

Temperature coefficient   Tolerance   Multiplier

Figure 11.5 Six-dot EIA and JAN color-coding system for capacitors

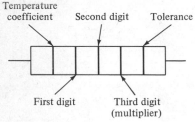

Temperature coefficient   Second digit   Tolerance

First digit   Third digit (multiplier)

Figure 11.6 Tubular ceramic capacitor

(b) JAN system (Joint Army Navy) (The JAN system is used by the military.)

A typical three-dot EIA mica capacitor with a 20% tolerance is shown in Figure 11.4. For instance, if the dots were brown, black, and red in the direction of the arrow, then the coded value of the capacitor would be 1000 pF (0.001 μF). It is understood that the coded value on mica capacitors is given in picofarads. Note that the same color code for resistors results in 1000 Ω.

An example of a six-dot JAN and EIA coding system for capacitors is shown in Figure 11.5. This system is used for coding mica and molded-paper capacitors. Starting from the upper left and reading clockwise, black (JAN) or white (EIA), orange, white, red, and silver are the codes for a 3900-pF capacitor with a 10% tolerance. The last dot refers to the temperature coefficient.

A tubular ceramic capacitor is usually coded in the form shown in Figure 11.6. The color code black, yellow, violet, brown, and silver would be interpreted as 470 pF, 10% tolerance, and zero temperature coefficient.

Definition of Capacitance

Objective 2 State when an electric circuit has a capacitance of one farad. Use the equation $C = Q/V$ to solve problems. Recall the property of capacitance by referring to the equation

$$i = C \frac{dv}{dt}$$

The basic function of a capacitor is to store electrical energy by placing *charge* (Q) on its plates. This energy is stored in the electrostatic field (dielectric) between the two plates. The more energy a capacitor can store, the larger the capacitance. Figure 11.7 shows a sketch of an air capacitor. It consists of two parallel metal plates of area A. These plates are separated by air of distance d. With the switch SW open, there is *no* charge Q on the plates, and thus no stored energy. Since

$$Q = i \times t \tag{11.1}$$

where Q = charge in coulombs (number of electrons)

i = instantaneous current for time t

t = time in seconds

and current i starts to flow when switch SW is closed, positive charge ($+Q$) is distributed on the top plate a. Likewise, an *equal* amount of negative charge ($-Q$) is distributed on plate b. When

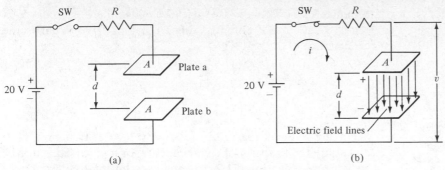

Figure 11.7 Sketch of air capacitor with parallel plates

positive charge is deposited on one plate and negative charge on the other plate, these charges create an electric field E between the plates. This electric field is shown in Figure 11.7(b) as electric field lines that *start* from the positive plate and *end* on the negative plate. The direction of these field lines is in the direction of conventional current flow.

For instance, if a positively charged particle were placed between the plates, it would be attracted to the negatively charged bottom plate and move in the direction of the electric field lines. Likewise, if a negative charge were placed in the electric field, it would move toward the positively charged top plate. If an electric field E is present, then there is a voltage v across the capacitor. As voltage v increases due to an increased field E created by more charge deposited on the plates, current i decreases and the capacitor starts to charge up slowly. We discuss this charging of the capacitor in more detail in a later unit. When the electrons stop flowing ($i = 0$), then the capacitor is said to be *fully charged*.

Let us now define capacitance. A capacitor has a *capacitance* of *one farad* if a *quantity of charge* of *one coulomb* is deposited on its plates when a potential difference of *one volt* is applied across its plates. Mathematically this can be expressed as

$$C = \frac{Q}{V} \tag{11.2}$$

where C = capacitance in farads

Q = quantity of charge in coulombs

V = potential difference in volts

Thus, keeping voltage constant, if the quantity of deposited charge Q is increased, the capacitance C increases. Let us now apply Equation (11.2) to the following example.

Example 11.1   Between two parallel plates there is a charge of 10×10^{-6} C and a potential difference of 100 V. What is the capacitance of the plates?

$$C = \frac{Q}{V} = \frac{10 \times 10^{-6} \text{ C}}{100 \text{ V}}$$

$$= 0.1 \times 10^{-6} \text{ F} \qquad \text{or} \qquad 0.1 \ \mu\text{F}$$

Having defined capacitance, let us now look at Equation (11.2) and study the property of capacitance. For a given capacitor C, we can use Equation (11.2) to calculate the total charge Q deposited on the plates. For instance, at any time t after switch SW is closed, we can find the charge Q on the plates by rewriting Equation (11.2) as

$$Q = CV \tag{11.3}$$

where C = capacitance in farads

V = the voltage across the capacitor in volts at time t

Q = the total charge deposited on the plates at time t

Using the same capacitor C, if at some later time $t + \Delta t$ ($\Delta t = $ a small increment of time after time t), the voltage across the capacitor is now $V + \Delta V$ ($\Delta V =$ additional voltage), then we can write the new charge as

$$Q + \Delta Q = C(V + \Delta V) \tag{11.4}$$

where $\Delta Q =$ additional charge added on plates during additional time Δt.

Subtracting Equation (11.3) from (11.4), we obtain

$$\Delta Q = C \, \Delta V \tag{11.5}$$

where $\Delta V =$ change in the voltage from time t to time $t + \Delta t$

$\Delta Q =$ change in the charge from time t to time $t + \Delta t$

Likewise, recalling that charge Q is related to current i by

$$Q = i \times t \tag{11.6}$$

and keeping current i constant, we can rewrite Equation (11.6) as

$$\Delta Q = i \times \Delta t \tag{11.7}$$

where $\Delta t =$ small increment of time later

$\Delta Q =$ additional charge deposited for time Δt

Substituting Equation (11.7) into (11.5), we get

$$i \, \Delta t = C \, \Delta V \tag{11.8}$$

or

$$i = C \frac{\Delta V}{\Delta t} \qquad (11.9)$$

Equation (11.5) states that a small incremental change (ΔQ) in charge Q on the plates of a capacitor produces a small incremental change (ΔV) in voltage across its plates. As the incremental changes of voltage and time (ΔV and Δt) are taken to be smaller and smaller, the ratio ($\Delta V/\Delta t$) approaches a *constant* value. When the increments ΔV and Δt are extremely small, most calculus books write the ratio $\Delta V/\Delta t$ as dv/dt. The expression dv/dt is interpreted to mean the *average rate of change* of voltage with respect to time t or the *derivative* of the voltage with respect to time t. We shall use this notation,

$$\frac{dv}{dt} \quad \text{or} \quad \frac{di}{dt}$$

extensively in the remaining units. A more detailed description of the *average rate of change* of any expression will be given in Unit 16, on Reactance. Let us now rewrite the expression in Equation (11.9) in calculus or derivative form.

Equation (11.9) can be rewritten in calculus form as

$$i = C \frac{dv}{dt} \qquad (11.10)$$

where dv/dt = rate of change of voltage with respect to time t

This equation states that *capacitance* is the *property* of an electric circuit that opposes any *change* in *voltage* or potential difference across the circuit.

Problem 11.2 One plate of a capacitor has 60 μC of charge. The applied voltage across the capacitor is 20 V. Find the amount of charge on the other plate; also find the capacitance C.

Problem 11.3 Recall the property of capacitance. Capacitance is the property of an electric circuit that *opposes* changes in
_____.

Physical Factors Governing Capacitance

Objective 3 List the three physical factors that determine the capacitance of a capacitor. Relate these factors in a mathematical equation, and use this equation to solve problems.

In Objective 2, you studied an air capacitor with a plate area A and distance between the plates d. Now let's see what would

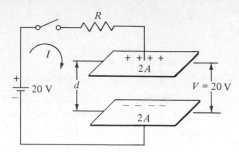

Figure 11.8 Capacitor with plate area doubled

happen if we doubled the area of the capacitor plate. With the plate area doubled, we have twice as much space available on each plate. (This is shown in Figure 11.8.) Thus *twice* as many electrons (Q) are deposited on the plates. Since the voltage V is the same as before and Q is doubled, we can show that capacitance is also doubled. Using Equation (11.2), we have

$$C = \frac{Q}{V} \qquad \text{or} \qquad 2C = \frac{2Q}{V} \qquad (11.11)$$

Now let's take this new capacitor of Figure 11.8 and place a dielectric material like glass between the plates, as shown in Figure 11.9. Glass in free space has an equal number of positive and negative charges. When we place this glass in the electric field between the plates, the positive charges inside the glass are forced by the electric field toward the negatively charged plate. Likewise, the negative charges are forced toward the positively charged top plate of the capacitor. This charge displacement in glass polarizes it, as shown in Figure 11.9. As a result of this polarization by the electric field, more positive charge is attracted (deposited) on the top plate, and more negative charge is attracted on the negative plate. The placement of a *dielectric* material between the plates of a capacitor *increases* the charge on each plate and *increases* the capacitance.

Now, if the distance d between the plates of the capacitor is *reduced*, what happens? Well, the glass dielectric is still polarized,

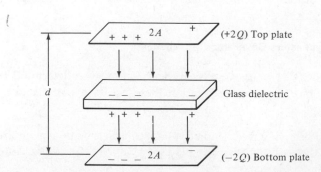

Figure 11.9 Dielectric placed between parallel plates of capacitor

as shown in Figure 11.9. But with the distance between the dielectric and plate reduced, more positive charge is attracted on the top plate and more negative charge is attracted on the bottom plate of the capacitor. Thus, if we *reduce* the *distance d* between plates of the capacitor, the charge on each plate *increases*, and as a result, capacitance *increases*.

We can summarize these three physical factors, plate area *A*, distance *d*, and dielectric, in the following statements.

The amount of capacitance is:

1. Directly proportional to the area of the plates, *A*.
2. Directly proportional to the dielectric constant of the material between the plates, *K*.
3. Inversely proportional to the distance or spacing between the plates, *d*.

We can summarize these three physical factors governing capacitance into one mathematical expression,

$$C \propto \frac{AK}{d} \qquad (11.12)$$

where C = capacitance in farads

$\propto$ = means "is proportional to"

A = cross-sectional area of parallel plate

K = dielectric constant

Table 11.1 gives the average dielectric constant *K* and dielectric strength for materials used in capacitors. This table compares

Table 11.1 Table of Average Dielectric Constants and Dielectric Strength

Dielectric	K	Dielectric strength (volts/mil)
Vacuum	1.0	
Air	1.0006	75
Teflon (plastic)	2	1500
Paper (paraffined)	2.5	500
Rubber	3	700
Transformer oil	4	400
Mica	5	5000
Porcelain (ceramic)	6	200
Bakelite (plastic)	7	400
Glass	7.5	3000
Barium-strontium titanate (ceramic)	7500	75

different dielectrics placed between the plates. All are related to vacuum, whose dielectric constant is 1.

Dielectric strength is the maximum voltage (usually expressed in volts/mil) that can be applied across each dielectric. Any voltage above this limit breaks down the dielectric. Because of this maximum voltage, all manufactured capacitors have a *voltage rating* specified on them.

We can calculate the maximum voltage rating of any capacitor if we know the dielectric used and the distance between the plates. Mathematically it is

$$V_{max} = \text{(dielectric strength)} \cdot (d) \qquad (11.13)$$

where V_{max} = maximum voltage rating of the capacitor

d = distance between plates, in mils

Therefore there are two important quantities to know about a capacitor, its capacitance and its voltage rating. The following example illustrates how to calculate the maximum voltage rating of a given capacitor.

Example 11.2 The capacitance of two metal plates separated by 0.004 in. (4 mils) of air is 0.5 μF. A potential of 200 V is applied across the plates. (a) What is the maximum allowable voltage across the capacitor? (b) If the dielectric is changed to mica, what are the new capacitance and maximum allowable voltage?

Solution (a) Referring to Table 11.1, the dielectric strength of air is 75 V/mil. Since the air gap between the plates is 4 mils, we can use Equation (11.13).

$$V_{max} = \text{(dielectric strength)} \cdot (d)$$
$$= 75 \text{ V/mil} \times 4 \text{ mils} = 300 \text{ V}$$

This is the maximum allowable voltage across the capacitor.

(b) If the dielectric is changed from air to mica, then the dielectric constant K changes from 1.0006 to 5. Since capacitance is directly proportional to the dielectric constant K [Equation (11.12)], we can write a proportion:

$$\frac{C_2}{C_1} = \frac{K_2}{K_1}, \qquad \frac{C_2}{0.5 \ \mu F} = \frac{5}{1}$$

$$\therefore \quad C = 5 \times 0.5 = 2.5 \ \mu F \qquad \text{with mica dielectric}$$

Using Equation (11.13), the maximum allowable voltage is

$$V_{max} = (5000) \cdot (4) = 20,000 \text{ V}$$

In summary, changing the dielectric from air to mica increases the capacitance by a factor of 5 and increases the breakdown voltage from 300 V to 20,000 V.

Example 11.3 A capacitor with parallel plates has a capacitance of 0.1 μF. When the spacing between the plates is doubled, what is the capacitance?

Solution Since capacitance is inversely proportional to the spacing between the plates, as shown by Equation (11.12),

$$C \propto \frac{AK}{d}$$

then as d increases, C decreases.

$$\therefore \quad C = \frac{0.1 \ \mu\text{F}}{2} = 0.05 \ \mu\text{F}$$

Problem 11.4 A capacitor with parallel plates has a capacitance of 0.05 μF when its dielectric is Teflon. When the dielectric is changed to porcelain (ceramic), what is the capacitance?

Problem 11.5 Refer to Figure 11.10. The capacitor C_1 has a capacitance of 50 pF. The dimensions and dielectric of capacitor C_2 are compared with those of C_1. What is the capacitance of C_2?

Problem 11.6 A potential difference of 6 V is applied across the plates of a 0.1 μF capacitor. How much charge is deposited on the plates of this capacitor?

We can use the following formulas to determine the capacitance of parallel-plate capacitors.

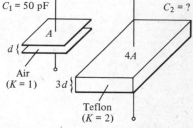

Figure 11.10 Capacitors of Problem 11.5

MKS units

$$C = \frac{8.85AK}{10^6 d} \tag{11.14}$$

where C = capacitance in microfarads

$\quad A$ = area of one plate in square meters

$\quad K$ = dielectric constant

$\quad d$ = distance between plates in meters

English practical units

$$C = \frac{0.225AK}{10^6 d} \tag{11.15}$$

where C = capacitance in microfarads

A = area of one plate in square inches

K = dielectric constant

d = distance between plates in inches

The following example illustrates how to apply the above two equations to a practical problem.

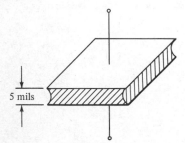

Figure 11.11 Capacitor for Example 11.4

Example 11.4 For the capacitor shown in Figure 11.11, mica is used as the dielectric. The thickness of the mica is 5 mils, and the dimensions of the plate are $\frac{1}{4}$ in. by $\frac{1}{4}$ in. (a) What is the capacitance? (b) What is the maximum voltage that may be applied to the capacitor?

Solution (a) Since the dimensions are in inches, we use the English practical units and Equation (11.15).

$$C = \frac{0.225AK}{10^6 d}$$

$$C = \frac{0.225(0.25 \times 0.25)5}{10^6(5 \times 10^{-3})} = 14.06 \times 10^{-6} \ \mu\text{F} \qquad \text{or} \qquad 14.06 \text{ pF}$$

(b) From the table on dielectric strengths, we know that mica has a dielectric strength of 5000 V/mil.

$$\therefore \quad V_{max} = 5000 \text{ V/mil} \times 5 \text{ mils} = 25{,}000 \text{ V}$$

Problem 11.7 What is the capacitance of the multiplate capacitor shown in Figure 11.12? [*Hint*: Total plate area is the sum of the individual plate areas. The total area is most easily found by counting the dielectric spaces between plates.]

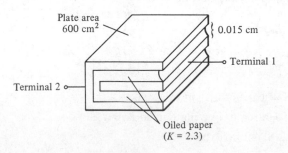

Figure 11.12 Capacitor for Problem 11.7

Plate area 600 cm^2

0.015 cm

Terminal 1

Terminal 2

Oiled paper ($K = 2.3$)

Problem 11.8 A tubular capacitor is made from two sheets of aluminum foil 1 in. wide by 4 ft long and two sheets of paraffin-impregnated paper $1\frac{1}{4}$ in. wide by 4 ft long by 2 mils thick, rolled

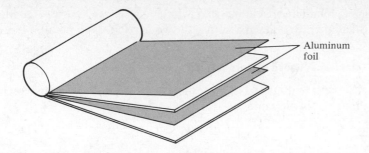

Figure 11.13 Capacitor for Problem 11.8

as shown in Figure 11.13. The paraffin-impregnated paper has a dielectric strength of 500 V/mil and a dielectric constant of 2.5. What are the capacitance and voltage rating of the capacitor?

Capacitors in Parallel

Objective 4   Calculate the total capacitance and charge stored by each capacitor when two, three, or more capacitors are connected in parallel.

In this objective you will learn how to find the equivalent capacitance of several capacitors connected in parallel. Figure 11.14(a) shows a sketch of two capacitors connected in parallel. When you connect capacitors in parallel, you effectively increase the area of the plates; see Figure 11.14(b). Recalling Equation (11.12), we have

$$C \propto \frac{KA}{d}$$

which means that the capacitance C is directly proportional to the cross-sectional area A. If the area of the plates is increased, so is the capacitance. Thus two capacitors in parallel result in

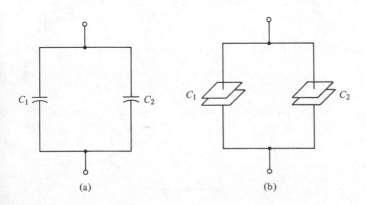

(a)   (b)

Figure 11.14 Two capacitors connected in parallel

an equivalent capacitance that is larger. Let's see how we actually calculate the total capacitance of two capacitors in parallel.

Total Capacitance of Parallel Capacitors

Referring to Figure 11.15, we note that the voltage across each capacitor is the same,

$$E = V_1 = V_2 \tag{11.16}$$

Furthermore, KCL is still valid for a parallel circuit. Therefore

$$I_T = I_1 + I_2 \tag{11.17}$$

Or, multiplying by time t, we have

$$I_T t = I_1 t + I_2 t \tag{11.18}$$

But charge Q is defined as

$$Q = It \tag{11.19}$$

Substituting Equation (11.19) into (11.18), we get

$$Q_T = Q_1 + Q_2 \tag{11.20}$$

where Q_T = total charge on both capacitors
$\quad Q_1$ = charge on capacitor C_1
$\quad Q_2$ = charge on capacitor C_2

But

$$C = \frac{Q}{V} \tag{11.21}$$

or

$$Q = CV \tag{11.22}$$

Substituting the relation of Equation (11.22) into Equation (11.20),

$$C_T E = C_1 V_1 + C_2 V_2 \tag{11.23}$$

But $E = V_1 = V_2$. Dividing Equation (11.23) by E, we get

$$\therefore \quad C_T = C_1 + C_2 \tag{11.24}$$

In general, when we connect *capacitors* in *parallel*, the total

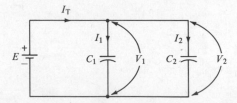

Figure 11.15 Equivalent capacitance of two parallel-connected capacitors

capacitance is the sum of all the individual capacitors, that is,

$$C_T = C_1 + C_2 + C_3 + \cdots. \qquad (11.25)$$

This is the same as resistors in series.

Example 11.5 Two capacitors are connected in parallel, as shown in Figure 11.16. (a) What is the total capacitance of this circuit? (b) What is the quantity of charge stored by each capacitor?

Solution

(a) $C_T = C_1 + C_2 = 5\ \mu F + 10\ \mu F = 15\ \mu F$

(b) Quantity of charge:

$$Q_{5\ \mu F} = CV = 5\ \mu F(10\ V) = 50\ \mu C$$
$$Q_{10\ \mu F} = CV = 10\ \mu F(10\ V) = 100\ \mu C$$

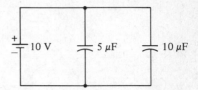

Figure 11.16   Circuit for Example 11.5

Problem 11.9 Three capacitors are connected in parallel as shown in Figure 11.17. (a) What is the total capacitance of this circuit? (b) What is the quantity of charge stored by each capacitor?

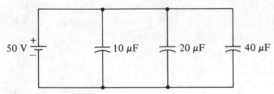

Figure 11.17   Circuit for Problem 11.9

Capacitors in Series

Objective 5   Calculate the total or equivalent capacitance, charge stored, and voltage across each capacitor when two, three, or more capacitors are connected in series.

Figure 11.18(a) shows a sketch of two capacitors connected in series. When you connect capacitors in series, you effectively increase the distance d between the plates. Recalling Equation (11.12), we have

$$C \propto \frac{KA}{d}$$

where d = distance between plates

C = capacitance

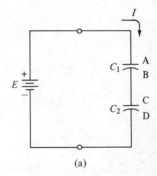

(a)

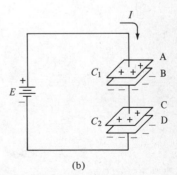

(b)

Figure 11.18   Two capacitors connected in series

Since distance *d* is *inversely* proportional to capacitance, increasing the distance between the plates decreases the capacitance. Thus two capacitors connected in series result in an equivalent capacitance that is smaller. Let's see how to calculate the equivalent capacitance of two capacitors connected in series.

Equivalent Capacitance of Capacitors in Series

Referring to Figure 11.19, we note that in a series circuit the current *I* is the same, that is,

$$I_T = I_1 = I_2 \tag{11.26}$$

Multiplying Equation (11.26) by time *t*, we obtain

$$I_T t = I_1 t = I_2 t \tag{11.27}$$

or since

$$Q = It \tag{11.28}$$

then

$$Q_T = Q_1 = Q_2 \tag{11.29}$$

Equation (11.29) shows that in a series-connected circuit, the charge *Q* on each capacitor is the *same*. Applying KVL to the series circuit of Figure 11.19, we get

$$E = V_1 + V_2 \tag{11.30}$$

Since

$$C = \frac{Q}{V} \quad \text{or} \quad V = \frac{Q}{C} \tag{11.31}$$

then we can rewrite Equation (11.30) as

$$\frac{Q_T}{C_T} = \frac{Q_1}{C_1} + \frac{Q_2}{C_2} \tag{11.32}$$

Since all the charges *Q* are equal [Equation (11.29)], then Equation (11.32) can be divided by *Q*, resulting in

$$\frac{1}{C_T} = \frac{1}{C_1} + \frac{1}{C_2} \tag{11.33}$$

or

$$C_T = C_{eq} = \frac{C_1 C_2}{C_1 + C_2} \tag{11.34}$$

Equation (11.34) can be remembered as the product over the sum, just as when *two* resistors are connected in parallel. Note that this equation applies *only* to *two* capacitors connected in series.

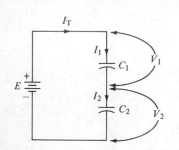

Figure 11.19 Equivalent capacitance of two series-connected capacitors.

The total or equivalent capacitance for more than two capacitors connected in series is:

$$\frac{1}{C_T} = \frac{1}{C_1} + \frac{1}{C_2} + \frac{1}{C_3} + \cdots \qquad (11.35)$$

This equation is the same as that for resistors connected in parallel.

Voltage Drops across Capacitors in Series

In this section, let us learn how to calculate the voltage V_1 and V_2 across each capacitor if we know the supply voltage. See Figure 11.19.

Since the charge Q on each capacitor is the same in a series circuit, Equation (11.29) is

$$Q_T = Q_1 = Q_2 \qquad (11.36)$$

Recalling that $Q = CV$, we can rewrite Equation (11.36) as

$$C_T E = C_1 V_1 = C_2 V_2 \qquad (11.37)$$

then

$$V_1 = \frac{C_T}{C_1} E \qquad (11.38)$$

or

$$V_2 = \frac{C_T}{C_2} E \qquad (11.39)$$

We can simplify Equations (11.38) and (11.39) if only *two* capacitors are present. Recall Equation (11.34) for two capacitors in series

$$C_T = \frac{C_1 C_2}{C_1 + C_2} \qquad (11.40)$$

Substituting Equation (11.40) into (11.38), we obtain

$$V_1 = \frac{\dfrac{C_1 C_2}{C_1 + C_2}}{C_1} E \qquad (11.41)$$

$$\therefore \quad V_1 = \frac{C_2}{C_1 + C_2} E \qquad (11.42)$$

Likewise, simplifying Equation (11.39), we obtain

$$V_2 = \frac{C_1}{C_1 + C_2} E \qquad (11.43)$$

These two equations are similar to the voltage-divider principle of resistors, but the opposite capacitor is used in the calculation.

If more than two capacitors are connected in series, we can find the voltage across any capacitor C_N in the following way: Since

$$Q_T = Q_1 = Q_2 = \cdots = Q_N \tag{11.44}$$

and

$$C_T E = C_1 V_1 = C_2 V_2 = \cdots = C_N V_N \tag{11.45}$$

then

$$C_T E = C_N V_N \tag{11.46}$$

or

$$\boxed{V_N = \frac{C_T}{C_N} E} \tag{11.47}$$

where C_N = capacitance value of Nth capacitor

C_T = total equivalent capacitance of all series-connected capacitors

V_N = voltage across Nth capacitor

E = supply voltage or total voltage

The following example illustrates the application of the theory learned in this objective.

Example 11.6   Three capacitors are connected in series, as shown in Figure 11.20. (a) What is the total or equivalent capacitance of this circuit? (b) What is the quantity of charge stored by each capacitor? (c) What is the voltage across each capacitor?

Solution   (a) We can calculate the total capacitance C_T using Equation (11.35),

$$\frac{1}{C_T} = \frac{1}{C_1} + \frac{1}{C_2} + \frac{1}{C_3} = \frac{1}{5\ \mu F} + \frac{1}{10\ \mu F} + \frac{1}{20\ \mu F}$$

$$= 0.2 \times 10^6 + 0.1 \times 10^6 + 0.05 \times 10^6 = 0.35 \times 10^6$$

$$\therefore \quad C_T = \frac{1}{0.35 \times 10^6} = 2.857\ \mu F$$

(b) Since the charge Q is the same in a series circuit,

$$Q_T = Q_1 = Q_2 = Q_3 = C_T E = (2.857\ \mu F)(140\ V)$$

$$Q_T = 399.98\ \mu C \approx 400\ \mu C \qquad \therefore \quad Q_T = Q_1 = Q_2 = Q_3 = 400\ \mu C$$

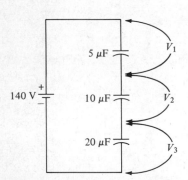

Figure 11.20   Circuit for Example 11.6

(c) The voltage across each capacitor is

$$V_1 = \frac{C_T}{C_1} E = \left(\frac{2.857 \ \mu F}{5 \ \mu F}\right) 140 \ V = 79.996 \ V \approx 80 \ V$$

$$V_2 = \frac{C_T}{C_2} E = \left(\frac{2.857 \ \mu F}{10 \ \mu F}\right) 140 \ V = 39.998 \ V \approx 40 \ V$$

$$V_3 = \frac{C_T}{C_3} E = \left(\frac{2.857 \ \mu F}{20 \ \mu F}\right) 140 \ V = 19.999 \ V \approx 20 \ V$$

Checking charge across one capacitor, we find that

$$Q_2 = C_2 V_2 = (10 \ \mu F)(40 \ V) = 400 \ \mu C$$

This checks with the answers in part (b).

Problem 11.10 Two capacitors are connected in series, as shown in Figure 11.21. (a) What is the total or equivalent capacitance of this circuit? (b) What is the quantity of charge stored on each capacitor? (c) What is the voltage across each capacitor?

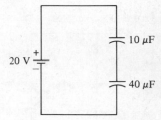

Figure 11.21 Circuit for Problem 11.10

Capacitors in Series-Parallel Combination

Objective 6 (*Advanced*) Calculate the total or equivalent capacitance and voltage across each capacitor when several capacitors are connected in series-parallel combinations.

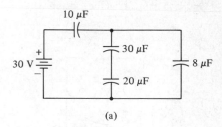

(a)

In this objective you will learn how to apply the material learned in Objectives 4 and 5. The following example illustrates the procedure to use in analyzing a series-parallel combination of capacitors.

Example 11.7 The series-parallel combination to be analyzed is shown in Figure 11.22. (a) What is the total capacitance of this circuit? (b) What is the voltage across each capacitor?

Solution (a) The 30 μF and 20 μF are in series,

$$\frac{(30)(20)}{50} = 12 \ \mu F$$

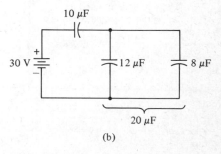

(b)

The parallel combination of 12 μF and 8 μF gives $12 + 8 = 20 \ \mu F$. This is shown in Figure 11.22(b). We can redraw Figure 11.22(b) to give Figure 11.22(c).

Now we have 10 μF in series with 20 μF, or

$$\frac{10(20)}{30} = 6.67 \ \mu F = C_T$$

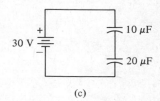

(c)

Figure 11.22 Circuits for Example 11.7

(b) The voltage across the 10-μF capacitor is $\frac{20}{30}$(30 V) = 20 V. By KVL, the voltage across the 20 μF (or the 8 μF in parallel with the series combination of 30 μF and 20 μF) is equal to 30 V − 20 V = 10 V. So 10 V is across the 8-μF capacitor. Since 10 V is also across the series combination of 30 μF and 20 μF, the voltage across the 30-μF capacitor is

$$\frac{20\ \mu F}{(30 + 20)\ \mu F}\ (10\ V) = 4\ V$$

The voltage across the 20-μF capacitor is 10 V − 4 V = 6 V.

Problem 11.11 Analyze the series-parallel combination shown in Figure 11.23. Find the total or equivalent capacitance of the circuit. Also find the voltage across each capacitor and the charge on the 0.015-μF capacitor.

Energy Stored by a Capacitor

Objective 7 Calculate the energy stored by a capacitor.

In Objective 2, you learned that capacitors store their energy in the electric field created between the plates. In this objective, let's see how we can calculate how much energy is stored in the field. When we close switch SW in Figure 11.24, current i_C will start to charge the capacitor. As the capacitor is charging, the voltage v_C across the capacitor increases up to the supply voltage E. A detailed study of the charging of a capacitor will be given in a later unit on *time constants*. A sketch of the resulting current i_C and voltage v_C versus time t is shown in Figure 11.25. Since energy W_C is related to power as

$$W_C = Pt \tag{11.48}$$

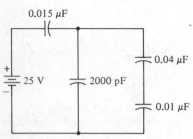

0.015 μF

25 V

2000 pF

0.04 μF

0.01 μF

Figure 11.23 Circuit for Problem 11.11

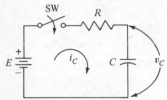

SW   R

E i_C C v_C

Figure 11.24 Circuit used in calculating the energy stored by a capacitor

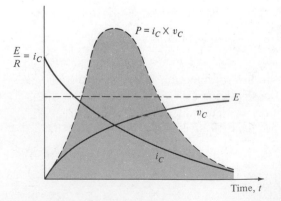

$P = i_C \times v_C$

$\frac{E}{R} = i_C$

E

v_C

i_C

Time, t

Figure 11.25 Calculation of energy stored by a capacitor

where P = power in watts or vars

t = time in seconds

W_C = energy stored by capacitor (joules)

and

$$P = v_C i_C \qquad (11.49)$$

Then we can write Equation (11.48) as

$$W_C = (v_C i_C)t \qquad (11.50)$$

where v_C = instantaneous voltage at time t

i_C = instantaneous current at time t

Knowing the relationship of current i_C and voltage v_C as a function of time, we can calculate power P as a function of time ($P = v_C \cdot i_C$). This product P is shown dashed on the same graph. The shaded area under the power curve represents the energy W_C stored by the capacitor. By using calculus, we can compute the area underneath the curve to be

$$\boxed{W_C = \tfrac{1}{2}CE^2} \quad \text{joules} \qquad (11.51)$$

where C = capacitance in farads

E = supply voltage in volts

W_C = energy stored by capacitor, in joules

Example 11.8 Calculate the energy stored by the capacitor after the switch has been closed. See Figure 11.26.

Solution After the switch has been closed, the capacitor charges up to the supply voltage E. The energy stored by the capacitor is

$$W_C = \tfrac{1}{2}CE^2 = \tfrac{1}{2}(8 \times 10^{-6})(50)(50) = 10{,}000 \times 10^{-6} \text{ J} = 0.01 \text{ J}$$

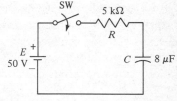

Figure 11.26 Circuit for Example 11.8

Problem 11.12 Calculate the energy stored by the capacitive circuit shown in Figure 11.26. Assume that the power supply $E = 60$ V, resistance $R = 20$ kΩ and capacitor $C = 15$ μF.

Test

1. Complete the sentence:
   Capacitance is the property of an electric circuit that

   opposes _____.
2. A capacitor with parallel plates has a capacitance of 0.1 μF. We decrease the cross-sectional area of the plates by $\tfrac{1}{2}$. What is the capacitance?

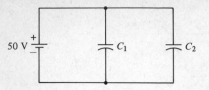

Figure 11.27 Circuit for Test Problems 3–5

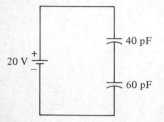

Figure 11.28 Circuit for Test Problems 6–8

3. What is the total capacitance of the circuit shown in Figure 11.27, given that $C_1 = 0.01$ μF and $C_2 = 0.05$ μF?
4. What is the quantity of charge on the 0.01-μF capacitor in Test Problem 3?
5. Given that, in Figure 11.27, $C_1 = 0.01$ μF and $C_2 = 4000$ pF, what is the total capacitance?
6. What is the total or equivalent capacitance of the circuit shown in Figure 11.28?
7. What is the quantity of charge stored by each capacitor in Figure 11.28?
8. What is the voltage across the 60-pF capacitor of Test Problem 6?
9. What is the total capacitance of the circuit shown in Figure 11.29? What is the voltage across the 30-pF capacitor?
10. Calculate the energy (in joules) stored by capacitor C_1 in Figure 11.30.

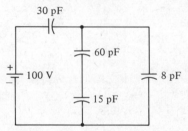

Figure 11.29 Circuit for Test Problem 9

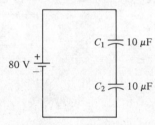

Figure 11.30 Circuit for Test Problem 10

Magnetic Unit 12
Circuits

In our age of technology, the property of magnetism is used in many ways. Looking around us, we can see computer memories, computer magnetic tapes, car ignition systems, television, radio, telephone, tape recorders, transformers, power generators, and motors—all employing the property of magnetism. In this unit we cover only enough material on magnetism to help you to better understand electric circuits. For a more in-depth study of magnetism, you should refer to the list of references at the end of this unit.

Objectives

After completing all the work associated with this unit, you should be able to:

1. Describe magnetism and the magnetic effects associated with a permanent magnet.
2. Determine the direction of a magnetic field around a current-carrying conductor (straight conductor and solenoid) by using the right-hand rule; or, if given the direction of the magnetic field, determine the direction of conventional current flow.
3. Identify the electric circuit analogy, the symbol, and the unit used for the following magnetic-circuit quantities: (a) magnetic flux (flux), (b) magnetomotive force (mmf), (c) permeability, (d) reluctance, (e) flux density, (f) magnetic field intensity.
4. Determine the permeability of magnetic materials at a specified flux density from a set of magnetization curves.

5. Given a specified number of turns of wire wound around a toroid-shaped core or cylindrical core (magnetic or nonmagnetic), solve for the current I when the total flux ϕ in the core is specified.
6. (*Advanced*) Given a specified number of turns of wire wound around a magnetic toroid-shaped core with an air gap, solve for the current I when the total flux ϕ in the core is specified.

Magnetism is a very old phenomenon. Around 600 B.C., the ancient Greeks knew about the attractive power of magnets. However, it wasn't until the sixteenth century that any experimental work was done on magnets. At this time, the English physician Gilbert realized that a magnetic field existed around the earth. Further studies by Oersted in 1819 led to the realization that current-carrying conductors could cause magnetic effects. Ampère's studies of the magnetic field around current-carrying loops led to the theory of magnetism on an atomic scale.

Since there is such a vast amount of material available on the topic of magnetism, we shall limit our discussion to two main areas: the permanent magnet and the magnetic field set up by a current-carrying conductor. In Objective 1, we discuss the permanent magnet, while in Objective 2, we discuss the magnetic field set up by a current-carrying conductor.

A Permanent Magnet

Objective 1   Describe magnetism and the magnetic effects associated with a permanent magnet.

Figure 12.1 shows a sketch of a permanent magnet. It has a north and a south pole, and magnetic lines of force are present around it. These lines of force are drawn dashed because you cannot see the actual lines. But you *can* see the effect of these lines. Furthermore, these lines of force are continuous, and are coming from the north pole and going toward the south pole. They are also continuous between the south and north poles inside the magnet.

The source of magnetism in magnetic materials lies in their atomic structure. It is theorized that the electrons in all elements spin on their own axes and also revolve about the nucleus of the atom. The presence of this electron spin has led physicists to think that this spin of electrons has a magnetic moment and creates a magnetic field. Certain magnetized materials like iron, nickel, and cobalt (also called *ferromagnetic materials*) very often have a magnetic field around them. It is speculated that their field is created by this electron spin.

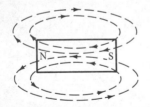

Figure 12.1   Sketch of a permanent magnet and its magnetic field

Referring back to the permanent magnet, let us now study what happens when we place two identical magnets close together. If we place the north and south magnetic poles of the two magnets opposite each other, as shown in Figure 12.2, then the two magnets are attracted to each other. This phenomenon takes place because magnetic lines of force are *continuous* and tend to go through paths of least resistance. Since magnetic materials have less resistance than air, these lines of force take the *easiest* path—through the magnetic material. Furthermore, since the direction of the magnetic lines of force of both magnets is the same, the magnets are attracted toward each other.

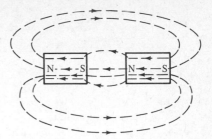

Figure 12.2 Sketch of interaction of two permanent magnets

The same phenomenon takes place when we place a permanent magnet near a piece of metal. Since the magnetic lines tend to go through the metal, these lines magnetize the metal. The point of the metal closest to the north pole of the permanent magnet becomes the south pole, while the part closest to the south pole of the permanent magnet becomes the north pole. This is shown in Figure 12.3. Since this configuration becomes similar to Figure 12.2, the magnetic pieces are attracted to each other.

Let us now consider a permanent magnet placed near a pile of metal nails. The closest nail becomes a magnet and is attracted to the permanent magnet. This nail, in turn, can magnetize and attract other nails. The number of nails that can attracted depends on the strength of the original permanent magnet. The stronger the magnet, the larger the number of nails; the weaker the magnet, the smaller the number of nails.

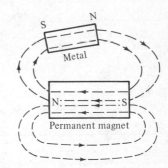

Figure 12.3 Metal in vicinity of permanent magnet

Figure 12.4 shows two opposing magnets. Here the north pole of the first magnet is placed near the north pole of the second. Since the magnetic lines of force go from north to south and are opposing, the two magnets are repelled, and are forced apart.

Problem 12.1   Recall the phenomenon of magnetism and explain why metal objects are attracted to permanent magnets.

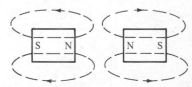

Figure 12.4 Sketch of two permanent magnets with like poles

Magnetic Field of Current-Carrying Conductor

Objective 2   Determine the direction of a magnetic field around a current-carrying conductor (straight conductor and solenoid) by using the right-hand rule; or, if given the direction of the magnetic field, determine the direction of conventional current flow.

In this objective, we discuss the second area of magnetism, the creation of a magnetic field by a current-carrying conductor.

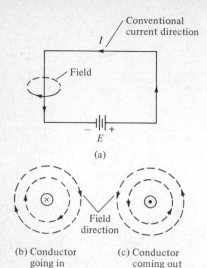

(a)

(b) Conductor
going in

(c) Conductor
coming out

⊗ Denotes current *I* flowing into page

⊙ Denotes current *I* flowing out of page

Figure 12.5 Sketch of magnetic lines of
force around current-carrying
conductor

The entire theory of magnetism and static magnetic fields is
based on the works of Biot, Savart, and Ampère. Their work led
to the observation that any current-carrying conductor creates a
magnetic field. If you refer to the electric circuit shown in Figure
12.5(a), you can see that a dc power supply *E* is connected across a
straight conductor.

The current *I* flowing in a straight conductor is pointing in the
conventional direction, and a magnetic field is created around the
conductor. To establish the direction of the magnetic lines of
force (field), you can use the right-hand rule. We can state the
right-hand rule as follows.

Right-Hand Rule

If you place your right hand over the conductor and place your
thumb in the direction of conventional current flow, the other
fingers of your hand point in the direction of the magnetic lines
of force (field direction).

The following example illustrates the right-hand rule.

Example 12.1 Draw the direction of magnetic field lines around
a conductor whose current is (a) flowing into the paper (⊗ denotes
current flowing in). (b) flowing out of the paper (⊙ denotes
current flowing out).

Solution (a) If you place your right hand over Figure 12.5(b)
and place your thumb on the mark ⊗, the remaining fingers of
your hand are pointing in a *clockwise* direction, as shown in the
figure.

(b) With current flowing out of the paper, place your right hand
over Figure 12.5(c) and point your thumb in an upward direction
(away from the page), as shown by ⊙. Your remaining fingers
are pointing in a *counterclockwise* direction, as shown in the
figure.

Note: If the current *I* flowing in the wire is stopped (*I* = 0), then
the magnetic field around the conductor collapses and is *zero*.
If the current *I* is increased, then the magnetic field increases
around the conductor and becomes stronger.

Now let's see what would happen to the magnetic field if the
conductor of wire were wrapped around a cylinder. If we first
study the field generated by *one* loop of wire, we can see what will
happen to the field when *more than one* turn of wire is wrapped
around the cylinder.

Figure 12.6 shows the field developed by one turn of current-
carrying wire. The current *I* is assumed to be flowing in a

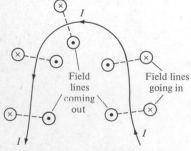

Figure 12.6 Field developed by one
turn of wire

counterclockwise direction. When you apply the right-hand rule, you note that the field lines inside the loop are all coming out, while the field lines outside the loop of wire are all going into the page. Thus the field inside the loop is in one direction and is concentrated.

Now, when more than one turn of wire is wrapped around a cylinder, the configuration looks like Figure 12.7.

The magnetic lines of force are concentrated within the cylinder. Furthermore, each turn of wire creates a magnetic field in the same direction. Since all the turns of wire have the same direction of current flow, the resulting magnetic field generated inside the cylinder is all in the same direction.

To establish the direction of the magnetic lines of force in Figure 12.7, we place the right hand over the cylinder. The fingers are now pointing in the direction of current flow. As a result, the thumb is pointing in the direction of the magnetic lines of force. Figure 12.7 shows this.

In the first objective, you learned how a permanent magnet attracted a metal material toward it. Let us now discuss the same phenomenon, but this time create a magnet by wrapping current-carrying conductors around a metal cylinder. This device is called an *electromagnet*.

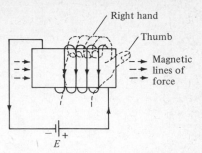

Figure 12.7 Sketch of magnetic lines of force developed by multiple turns wrapped around a cylinder

Electromagnet

Figure 12.8 shows an electromagnet. A current-carrying conductor is wrapped around a cylinder of magnetic material. Applying the right-hand rule, you note that the created magnetic lines of force are in a clockwise direction. Thus the cylinder becomes a magnet with a north and south pole. This type of magnet is called an *electromagnet*, since the current-carrying conductor around the cylinder creates the magnetic lines of force. When a metal material is brought close to the electromagnet, the

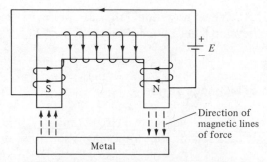

Figure 12.8 Sketch of an electromagnet

resulting magnetic lines of force tend to flow through the metal and attract the metal toward the electromagnet, just as if the electromagnet were a permanent magnet. Recall that this attraction of a magnet was described in Objective 1.

There are many applications of electromagnets. Scrap metal dealers often mount an electromagnet on a crane and use it to pick up and move scrap metal from one location to another. Electromagnets are also used in electric relays and solenoids. Another application of an electromagnet is the audio speaker. This is described briefly in the following paragraph.

Audio Speakers

The operation of the speaker involves the principles discussed in both Objective 1 and Objective 2. The audio speaker consists of a permanent magnet and a cone mounted on a metal material. (This is shown in Figure 12.9.) The metal material is wrapped with a current-carrying conductor whose terminals are designated A and B. If a constant voltage signal is applied to terminals A–B, then current flows through the conductor in a certain direction. This flow of current makes the metal an electromagnet with a north and south pole. If the created lines of force are in the same direction as those of the permanent magnet, then the two magnets (permanent magnet and electromagnet) are *attracted*, and the cone moves to the left.

If the polarity of the voltage signal at terminals A–B is reversed, then the electromagnet has magnetic lines of force that oppose those of the permanent magnet; then the two magnets *repel* each other. This results in the cone moving to the right.

In summary, if an alternating audio voltage signal is applied to terminals A–B, the cone moves back and forth, moving a volume of air and thus creating sound.

Problem 12.2 Determine the direction of the magnetic field for the circuit shown in Figure 12.10.

Problem 12.3 Which way should the current flow (conventional current) to develop a magnetic field direction as shown in Figure 12.11?

Definition of Magnetic Terms

Objective 3 Identify the electric-circuit analogy, the symbol, and the unit used for the following magnetic circuit quantities:

Figure 12.9 Audio speaker

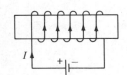

Figure 12.10 Sketch for Problem 12.2

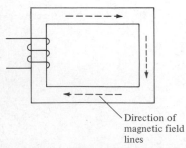

Figure 12.11 Sketch for Problem 12.3

(a) magnetic flux (flux), (b) magnetomotive force (mmf), (c) permeability, (d) reluctance, (e) flux density, (f) magnetic field intensity.

Since the theory of magnetism was developed in the eighteenth century and was based on ideas parallel to the theory of electric circuits, we shall try to develop an analogy between an electric circuit and a magnetic circuit. Before we present the equivalent parameters for the magnetic circuit, let us define some of the terms used in magnetic circuits.

Magnetic flux (flux)   The magnetic lines of force that are created by a current-carrying conductor or permanent magnet are defined as *magnetic flux lines*, or *flux*. The symbol for flux is ϕ. The units for flux are webers (abbreviated Wb) in MKS units, and lines in English units. [Note: 1 Wb $= 10^8$ lines.]

Magnetomotive force (mmf)   The driving force that creates the magnetic lines of force (flux) is defined as the magnetomotive force. It depends on the current flow I through the conductor and on the number of turns N of wire used. The symbol for magnetomotive force (mmf) is $\mathscr{F}$.

$$\mathscr{F} = NI \qquad\qquad (12.1)$$

where $N =$ number of turns

$I =$ current flow in amperes

The unit for $\mathscr{F}$ is ampere-turns, or At.

Permeability   Since the number of flux lines established depends on the type of material used, permeability is a measure of how easily a particular material allows the creation of flux lines. Some materials are better "conductors" of magnetic flux (high permeability), while other materials are poor conductors of magnetic flux (low permeability). The symbol for permeability is μ (Greek letter mu)

The unit for permeability is webers per ampere-turn-meter.

Now that we have defined some of the terms for a magnetic circuit, we are ready to make a comparison between an electric circuit and a magnetic circuit. Figure 12.12 (page 237) shows an electric circuit and a magnetic circuit. This figure also includes a comparison of electric circuit and magnetic circuit parameters. You are already familiar with the electric circuit, which consists of an electromotive force E, current I, and resistance R. These parameters are related by Ohm's law as

$$R = \frac{E}{I} \qquad (12.2)$$

where E = electromotive force in volts

I = current in amperes

The equivalent parameters for the magnetic circuit are the magnetomotive force $\mathscr{F}$, flux ϕ, and reluctance $\mathscr{R}$. The equivalent Ohm's law for a magnetic circuit is:

$$\mathscr{R} = \frac{\mathscr{F}}{\phi} \qquad (12.3)$$

where $\mathscr{F}$ = magnetomotive force in ampere-turns

ϕ = flux in webers

At this point we shall define the remaining terms in this objective.

Reluctance   Just as resistance R in electric circuits is the opposition to the flow of electrons, reluctance $\mathscr{R}$ is the opposition of a material to the setting up of the flux ϕ. The symbol for reluctance is $\mathscr{R}$. The units for reluctance are ampere-turns per weber.

In electric circuits the value of resistance R depends on the physical characteristics length, area, and resistivity. These are related by

$$R = \rho \frac{l}{A} \qquad (12.4)$$

where ρ = resistivity of material (Greek letter rho)

l = length of material

A = cross-sectional area of material

In magnetic circuits the value of reluctance $\mathscr{R}$ depends on the physical characteristics length, arca, and permeability. These parameters are related by

$$\mathscr{R} = \frac{l}{\mu A} \qquad (12.5)$$

where μ = permeability of the material

l = length of magnetic circuit

A = cross-sectional area of magnetic circuit

Note that the reluctance is inversely proportional to permeability; that is, if the permeability is very high, then the reluctance $\mathscr{R}$ is low.

Flux density   The number of flux lines per unit area is defined as flux density. The symbol for flux density is B. The units for flux density B are webers per square meter. The mathematical relationship defining flux density is

$$B = \frac{\phi}{A} \tag{12.6}$$

where ϕ = flux in webers

   A = unit area in square meters (m^2)

Magnetic field intensity   The magnetic field intensity is also called magnetomotive force per unit length. Since the magnetomotive force $\mathscr{F}$ is related to the number of turns N and the current I ($\mathscr{F} = NI$), the field intensity created depends on how many turns are wound per unit length. With a loosely wound coil, the field strength is small, but in a tightly wound coil, with many turns per unit length, the field strength is large.

The symbol for magnetic field intensity is H. The units for H are ampere-turns per meter. The mathematical relationship for H is

$$H = \frac{\mathscr{F}}{l} = \frac{NI}{l} \tag{12.7}$$

where $\mathscr{F}$ = magnetomotive force

   l = length in meters

   N = number of turns of wire

   I = current in wire in amperes

Figure 12.12 and Table 12.1 give a summary of these terms.

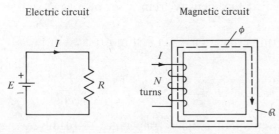

Figure 12.12   Sketch of electric circuit and equivalent magnetic circuit

Table 12.1

E = electromotive force (volts)	$\mathscr{F} = NI$ = magnetomotive force (ampere-turns)
I = electric current (amperes)	ϕ = total flux lines (webers)
R = resistance = $\rho \dfrac{l}{A}$ (ohms)	$\mathscr{R}$ = reluctance = $\dfrac{l}{\mu A}$

where ρ = resistivity (ohm-meters)
  l = length (meters)
  A = cross-sectional area
   (square meters)

where μ = permeability
  l = length (meters)
  A = cross-sectional area
   (square meters)

Ohm's law:

$$R = \frac{E}{I}$$

Ohm's law:

$$\mathscr{R} = \frac{\mathscr{F}}{\phi} = \frac{NI}{\phi} = \frac{\text{ampere-turns}}{\text{weber}}$$

Flux density: $B = \dfrac{\phi}{A}$

Magnetic field intensity:

$$H = \frac{\mathscr{F}}{l} = \frac{NI}{l}$$

Permeability: $\mu = \dfrac{B}{H}$

Problem 12.4 List the electric-circuit analogy to the following magnetic-circuit parameters: (a) magnetic flux, (b) magneto-motive force ($\mathscr{F}$), (c) permeability (μ), (d) reluctance ($\mathscr{R}$).

Determination of Permeability μ from $B-H$ Curves

Objective 4 Determine the permeability of magnetic materials at a specified flux density from a set of magnetization curves.

The term permeability was first introduced in Equation (12.5) as

$$\mu = \frac{l}{\mathscr{R} A} \qquad \text{or} \qquad \mathscr{R} = \frac{l}{\mu A} \qquad (12.8)$$

Since we cannot measure reluctance $\mathscr{R}$ readily, we must try to relate permeability to some other measurable parameters. Let us refer to Table 12.1 and relate permeability to other parameters like B and H. Ohm's law for magnetic circuits is

$$\mathscr{F} = \phi \mathscr{R} \tag{12.9}$$

and

$$\mathscr{F} = Hl \tag{12.10}$$

$$\phi = BA \tag{12.11}$$

$$\mathscr{R} = \frac{l}{\mu A} \tag{12.12}$$

Substituting Equations (12.10), (12.11), and (12.12) into Equation (12.9), we have

$$Hl = (BA)\left(\frac{l}{\mu A}\right) \tag{12.13}$$

Simplifying Equation (12.13), we have

$$H = \frac{B}{\mu} \tag{12.14}$$

or

$$\boxed{\mu = \frac{B}{H}} \tag{12.15}$$

Therefore, if you know the flux density B and the magnetic field intensity H, you can calculate the permeability μ of the material. Figure 12.13 shows an illustration of how you can measure B and H in a practical setup.

In this setup, a current I is applied to N turns of a toroid. If we know the number of turns N per unit length, we can calculate the magnetization field intensity H by using

$$H = \frac{\mathscr{F}}{l} = \frac{NI}{l} \tag{12.16}$$

Note that flux ϕ is flowing in the clockwise direction, because of the right-hand rule discussed in Objective 3. To measure B,

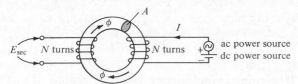

Figure 12.13 Practical setup to measure B and H

we must first relate the flux ϕ and area A. Since

$$B = \frac{\phi}{A} \qquad (12.17)$$

and the secondary voltage E_{sec} is proportional to ϕ, we can measure the voltage E_{sec} and relate this to the measured B. This generation of a voltage E_{sec} is based on Faraday's law, which we shall discuss in more detail in Unit 13 on inductance.

Thus, in general, the magnetic field intensity H is an independent variable over which we have control. The resulting flux ϕ and permeability μ depend on the type of material used in the toroidal core. Furthermore the permeability μ depends on the flux density B, which in turn depends on the flux ϕ and the cross-sectional area A.

A plot of the independent variable H versus B is called a *magnetization curve*. These curves are provided by manufacturers for different types of materials and are used to calculate the permeability of the material at any particular point. In this objective you will learn how to determine the permeability of any particular material at any specified flux density B.

Before we actually determine the permeability of any particular material from its magnetization curve, let us first understand the magnetization curve better. A typical B–H magnetization curve for a ferromagnetic material like iron is shown in Figure 12.14. Many authors use ferromagnetic domain (magnetic region) theory to describe this B–H magnetization curve. Using this theory, let us depict the magnetic material at point A as four magnetic blocks. With no external current or magnetic field, each block has a random magnetic field pointing in a net direction.

The magnetizing field H of Figure 12.14 is plotted on the horizontal axis, while the flux density B is plotted on the vertical axis. Since H is proportional to current I [Equation (12.16)], and H at point A is zero, current I is also zero. With current I being zero, *no* internal magnetic field is created in the iron. As a result, the four magnetic domains are pointing in all directions, resulting in an internal net magnetic field of zero ($B = 0$).

As the current I is increased (H is increased), one of the magnetic domains grows larger than the other magnetic domains. This results in a dominant magnetic field at point B of Figure 12.14 pointing in a specific direction. At point C, as current I is increased, the magnetization curve tends to flatten out toward a horizontal line. At this point all the magnetic domains are pointing in the same direction, the direction of the magnetizing field. This point C indicates that magnetization of the sample

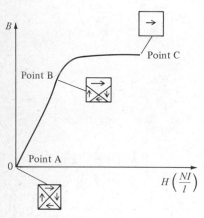

Figure 12.14 Description of B–H magnetization curve

has reached the point of *saturation.* Any additional increase in current I beyond this point will not increase B.

Some typical magnetic curves for sheet steel, cast steel, and cast iron are shown in Figures 12.15 and 12.16. Note that the horizontal axes represent the magnetic field intensity H in ampere-turns per inch in English practical units or ampere-turns per meter in MKS units. The vertical axes represent the flux density B in webers per square inch or webers per square meter.

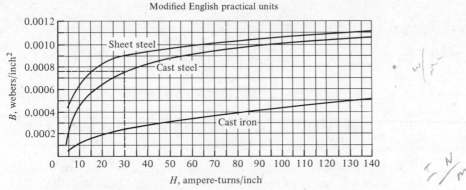

Figure 12.15 *B–H* magnetization curves of sheet steel, cast steel, and cast iron in modified English practical units

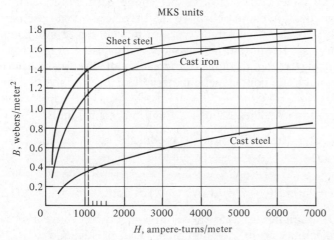

Figure 12.16 *B–H* magnetization curves for sheet steel, cast steel, and cast iron in MKS units.

The following two examples present the method of determining the permeability μ of any partical material from the *B–H* magnetization curve. You can see that the permeability μ is not constant, but depends on which point we are on the magnetization curve.

Example 12.2 Find the permeability of cast steel when run at a flux density B of 7.6×10^{-4} Wb/in^2.

Solution (a) Since the flux density B is in Wb/in^2, we must use English practical units. Referring to Figure 12.15, we find that for $B = 7.6 \times 10^{-4}$ Wb/in^2, we have $H = 30$ At/in. for cast steel.

(b) Since the permeability μ is

$$\mu = \frac{B}{H} = \frac{7.6 \times 10^{-4} \text{ Wb/in}^2}{30 \text{ At/in.}} = 2.53 \times 10^{-5}$$

$$\therefore \quad \mu = 2.53 \times 10^{-5} \text{ Wb/At-in.}$$

Example 12.3 Find the permeability of sheet steel when run at a flux density of 1.4 Wb/m^2.

Solution (a) Since flux density B is in Wb/m^2, we must use MKS units. Referring to Figure 12.16, we find that H is 1100 At/m.

(b) $\mu = \dfrac{B}{H} = \dfrac{1.4}{1100} = 1.27 \times 10^{-3}$ Wb/At-m

You have learned how to determine permeability μ from a magnetization curve. Now we shall see how we can use permeability to classify different types of magnetic and nonmagnetic materials. Recall that permeability is a measure of how easily a particular material allows the creation of flux lines. The permeability of a nonmagnetic material like air (free space) is given as

$$\mu = 3.2 \text{ lines/At-in.} \qquad \text{English practical units}$$
$$\mu = 3.2 \times 10^{-8} \text{ Wb/At-in.} \qquad \text{modified English practical units} \tag{12.18}$$

or

$$\mu = 4\pi \times 10^{-7} \text{ Wb/At-m} \qquad \text{MKS units} \tag{12.19}$$

Other materials that are nonmagnetic and have the same permeability as air are aluminum, glass, plastic, and wood.

Materials that have permeabilities slightly less than that of air are called *diamagnetic*. In these materials, the net magnetic flux density B in the interior material body is less than the external field. Some materials in this class are copper, bismuth, zinc, silver, lead, and mercury.

Materials that are slightly magnetic and have permeabilities larger than that of air are called *paramagnetic*. In these materials, the flux density B inside the material is slightly larger than the external field. Examples of paramagnetic materials are aluminum, oxygen, and platinum.

Materials that have very large paramagnetic effects are considered to be *ferromagnetic*. Examples of ferromagnetic materials are iron, nickel, and cobalt. The permeability of ferromagnets is 100, or 1000 times that of air.

Problem 12.5 Find the permeability of sheet steel at a flux density B of 0.0006 Wb/in^2. Refer to Figures 12.15 and 12.16.

Determination of Current in a Toroid-Shaped Core

Objective 5 Given a specified number of turns of wire wound around a toroid-shaped core or cylindrical core (magnetic or nonmagnetic), solve for the current I when the total flux ϕ in the core is specified.

To solve for the current required in a coil with a specified core for a specified flux in the core, two methods may be used.

1. If the core is nonmagnetic, or μ is specified, you may find I by going from μ to solve for $\mathscr{R}$, then to $\mathscr{F}$, and finally to I.
2. If you are given a set of magnetization or B–H curves, you can find H from the curve because you know B. Then you obtain $\mathscr{F}$ from $(H \cdot l)$, and from $\mathscr{F}$, you can find I.

The two methods are illustrated in the following examples.

Without Air Gap

Example 12.4 Given that 1000 turns are wound around the plastic ring core shown, what current must flow in the turns to develop a total flux in the ring of 1.9×10^{-4} Wb? (See Figure 12.17.)

Solution 1   (a) The cross-sectional area of the core is

$$A = l \times w = 0.5 \times 0.5 = 0.25 \text{ in}^2$$

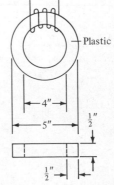

Figure 12.17   Sketch for Example 12.4

The average length of the magnetic path is

$$l = \text{circumference of} \sim 4.5\text{-in. diameter} = \pi d$$
$$= \pi(4.5 \text{ in.}) = 14.137 \text{ in.}$$

(b) Since plastic is nonmagnetic and the units are given in inches, we use English practical units for permeability μ. The permeability of plastic is the same as that for free air:

$$\mu = 3.2 \times 10^{-8} \text{ Wb/At-in.} \qquad \text{modified English practical units}$$

(c) The reluctance $\mathscr{R}$ can be found by

$$\mathscr{R} = \frac{l}{\mu A} = \frac{14.137}{(3.2 \times 10^{-8})(0.25)} = 1.767 \times 10^9 \text{ At/Wb}$$

(d) Since the magnetomotive force $\mathscr{F}$ for a magnetic circuit is

$$\mathscr{F} = \phi \mathscr{R} = (1.9 \times 10^{-4})(1.767 \times 10^9) = 3.3573 \times 10^5 \text{ At}$$

then using

$$\mathscr{F} = NI$$

we have

$$I = \frac{\mathscr{F}}{N} = \frac{3.3573 \times 10^5}{1000} = 335.7 \text{ A}$$

Solution 2 We can also use a second method of solving for the current. It is as follows:

(a) The flux density B is

$$B = \frac{\phi}{A} = \frac{1.9 \times 10^{-4}}{0.25} = 7.6 \times 10^{-4} \text{ Wb/in}^2$$

(b) Using the relationship $B = \mu H$, we have

$$H = \frac{B}{\mu} = \frac{7.6 \times 10^{-4}}{3.2 \times 10^{-8}} = 2.375 \times 10^4 \text{ At/in.}$$

(c) Since

$$\mathscr{F} = Hl = NI$$

we first find

$$\mathscr{F} = Hl = (2.375 \times 10^4)(14.137) = 3.3575 \times 10^5 \text{ At}$$

and

$$I = \frac{\mathscr{F}}{N} = \frac{3.3575 \times 10^5}{1000} \qquad \therefore \quad I = 335.7 \text{ A}$$

The following example illustrates the application of the second method of finding current in a given magnetic circuit using the *B–H* curve.

Example 12.5 Given that 1000 turns are wound around this cast-steel core, what current must flow in the turns to develop a total flux in the core of 1.9×10^{-4} Wb? (See Figure 12.18.)

Solution (a) Since the same size core is used as in Example 12.4, the length of the magnetic path and the cross-sectional area are the same as in Example 12.4.

$$\therefore \quad l = 14.137 \text{ in.} \qquad A = 0.25 \text{ in}^2$$

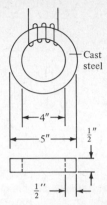

Figure 12.18 Sketch for Example 12.5

(b) Flux density $B = \dfrac{\phi}{A} = \dfrac{1.9 \times 10^{-4} \text{ Wb}}{0.25 \text{ in}^2} = 7.6 \times 10^{-4} \text{ Wb/in}^2$

(c) Using the magnetization or *B–H* curves (modified English practical units) for cast steel shown in Figure 12.15, we find that

$$H = 30 \text{ At/in.}$$

(d) We know that the magnetomotive force $\mathscr{F}$ is equal to

$$\mathscr{F} = H \cdot l = 30 \text{ At/in.} \times (14.137 \text{ in.}) = 424 \text{ At}$$

But

$$\mathscr{F} = NI \qquad \therefore \quad I = \frac{\mathscr{F}}{N} = \frac{424 \text{ At}}{1000 \text{ t}} = 424 \text{ mA}$$

Problem 12.6 Given that 800 turns are wound around a toroid made of sheet steel. What current must flow in the turns to develop a flux density *B* of 1.4 Wb/m²? From Example 12.3, the permeability of sheet steel is 1.27×10^{-3} Wb/At-m. (See Figure 12.19.)

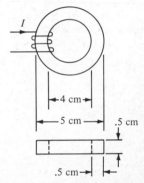

Figure 12.19 Sketch for Problem 12.6

With Air Gap

Objective 6 (*Advanced*) Given a specified number of turns of wire wound around a magnetic toroid-shaped core with an air gap, solve for the current *I* when the total flux ϕ in the core is specified.

Before we start to discuss the magnetic circuit with an air gap, let us first study the equivalent effect in a dc electric circuit. The given electric circuit is shown in Figure 12.20, with resistor R_1 of 10 Ω and a 40-V dc source. The current *I* flowing in the

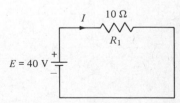

Figure 12.20 Electric circuit example used for study of equivalent magnetic circuit.

circuit can be calculated by Ohm's law as

$$I = \frac{E}{R_1} = \frac{40 \text{ V}}{10 \, \Omega} = 4 \text{ A} \qquad (12.20)$$

Now if we modify the circuit in Figure 12.20 by adding another resistor R_2 in series, it will look like Figure 12.21.

If we want the same current I of 4 A to flow through R_1, what should be the applied voltage E?

With the new resistor $R_2 = 20 \, \Omega$ and $I = 4$ A, we have the voltage V_2, by Ohm's law, as

$$V_2 = I \times R_2 = 4 \times 20 \, \Omega = 80 \text{ V}$$

and

$$V_1 = I \times R_1 = 4 \times 10 \, \Omega = 40 \text{ V}$$

Applying KVL to the circuit, we have

$$E = V_1 + V_2 = 40 \text{ V} + 80 \text{ V}$$
$$\therefore \quad E = 120 \text{ V} \qquad (12.21)$$

Therefore, you see that if we add more resistance to a series circuit when we want to maintain the same current, we must increase the electromotive force E.

The same holds true for a magnetic circuit with an air gap. The reluctance $\mathscr{R}_m$ of the magnetic material in Example 12.4 was calculated from the equation

$$\mathscr{R}_m = \frac{l}{\mu A} \qquad (12.22)$$

where l = length of the magnetic material

μ = permeability of the magnetic material

A = cross-sectional area of material

$\mathscr{R}_m$ = reluctance of magnetic material

If an *air gap* is *added* to a magnetic circuit, the total reluctance of the complete magnetic circuit is *increased*. The reluctance of the air gap can be calculated by the equation

$$\mathscr{R}_A = \frac{l_A}{\mu_A A_A} \qquad (12.23)$$

where l_A = length of air gap

μ_A = permeability of air

A_A = cross-sectional area of air gap

$\mathscr{R}_A$ = reluctance of air gap

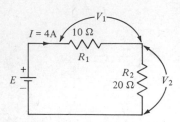

Figure 12.21   Modified circuit of Figure 12.20

The new total reluctance of the magnetic circuit with an air gap is then the sum of both reluctances, or

$$\mathscr{R}_{\text{total}} = \mathscr{R}_m + \mathscr{R}_A \qquad (12.24)$$

If an air gap is added to a magnetic circuit and the same total flux ϕ is to be maintained, then the magnetomotive force $\mathscr{F}$ must be *increased*. The same was true for an electric circuit. There, the electromotive force E had to be *increased* to maintain a constant current I.

In the magnetic circuit with an air gap, the total magnetomotive force $\mathscr{F}$ is

$$\mathscr{F}_{\text{total}} = \mathscr{F}_m + \mathscr{F}_A \qquad (12.25)$$

where $\mathscr{F}_m = H_m l_m$ = magnetomotive force for magnetic material
$\quad\;\; \mathscr{F}_A = H_A l_A$ = magnetomotive force for air gap

In the next example, since flux spreads out as it passes through the air gaps, we shall use a *rule of thumb* in calculating the dimensions of the air gap.

Example 12.6 The cast-steel core of Example 12.5 has an air gap of 0.05 in. What current is required to develop a total flux in the core of 1.9×10^{-4} Wb? (See Figure 12.22.)

Solution The magnetomotive force $\mathscr{F}$ required to set up this flux in the cast steel, from Example 12.5, is 424 At. Now we have to find out how much magnetomotive force is required to set up the same flux in the air gap.

Since flux spreads out as it passes through the air in the air gap (*fringing flux*), the cross section of the magnetic circuit in the air gap is larger than in the cast steel. If the air gap is small, we can assume that the dimensions increase by an amount equal to the air gap. (This assumption is a rule of thumb used by many authors.)

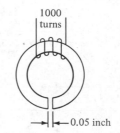

Figure 12.22 Sketch for Example 12.6

(a) The effective cross-sectional area of the air gap is

$$A = (0.5 + 0.05)(0.5 + 0.05) = 0.3025 \text{ in}^2$$

The length of the air gap is

$$l_A = 0.05 \text{ in.}$$

(b) The flux density B_A in the air gap is

$$B_A = \frac{\phi}{A} = \frac{1.9 \times 10^{-4}}{0.3025} = 6.281 \times 10^{-4} \text{ Wb/in}^2$$

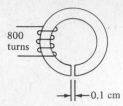

Figure 12.23 Sketch for Problem 12.7

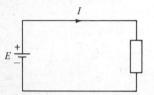

Figure 12.24 Sketch for Test Problem 1

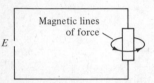

Figure 12.25 Sketch for Test Problem 2

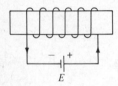

Figure 12.26 Sketch for Test Problem 3

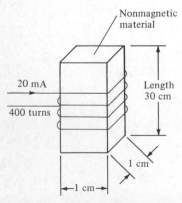

Figure 12.27 Sketch for Test Problem 6

(c) The magnetic field intensity H_A for air is

$$H_A = \frac{B_A}{\mu_A} = \frac{6.281 \times 10^{-4}}{3.2 \times 10^{-8}} = 1.963 \times 10^4 \text{ At/in.}$$

(d) Since $\mathscr{F} = Hl$, we have for air

$$\mathscr{F}_A = H_A l_A = (1.963 \times 10^4)(0.05) = 981.5 \text{ At}$$

(e) The total magnetomotive force required for this circuit is then

$$\mathscr{F}_{\text{total}} = \mathscr{F}_{\text{cast steel}} + \mathscr{F}_A$$

Since $\mathscr{F}_{\text{cast steel}} = 424$ from Example 12.5,

$$\therefore \quad \mathscr{F}_{\text{total}} = 424 \text{ At} + 981.5 \text{ At} = 1405 \text{ At}$$

and

$$I = \frac{\mathscr{F}_{\text{total}}}{N} = \frac{1405 \text{ At}}{1000 \text{ t}} = 1.405 \text{ A}$$

Problem 12.7 If the sheet-steel core of Problem 12.6 has an air gap of 0.1 cm, what current is now required to maintain a flux density B in the core of 1.4 Wb/m^2? (See Figure 12.23.)

Test

1. Draw the direction of the magnetic lines of force around the straight conductor shown. (Figure 12.24)
2. Draw the direction of conventional current in the circuit shown. (Figure 12.25)
3. Indicate the direction of the magnetic lines of force in the solenoid. Draw the north and south poles. (Figure 12.26)
4. Calculate the reluctance of a solenoid made up of 300 turns of wire carrying 4 A. The total flux flowing in the magnetic circuit is 3×10^{-3} Wb.
5. The reluctance of a magnetic circuit is 4.8×10^5 At/Wb. What current must be passed through a 1000-turn solenoid to produce a total flux of 0.2 mWb?
6. The solenoid shown in Figure 12.27 has 400 turns. Calculate the reluctance of the magnetic circuit.
7. Calculate the permeability μ of cast steel at a flux density of $B = 1$ Wb/m^2. Refer to Figure 12.16.
8. Given that 800 turns are wound around the plastic ring core shown, what current must flow in the turns to

develop a total flux in the ring of 1.5×10^{-4} Wb? (Figure 12.28)

9. Given that 800 turns are wound around a toroid made of cast steel, what current must flow in the turns to develop a total flux of 1.5×10^{-4} Wb? (Figure 12.29)

10. Given that the cast steel core of Test Problem 9 has an air gap of 0.1 in., what current is now required to develop a total flux in the core of 1.5×10^{-4} Wb? (Figure 12.30)

List of References

Robert L. Boylestad, *Introductory Circuit Analysis*, third edition, Columbus, Ohio: Merrill Publishing Co., 1977

Herbert W. Jackson, *Introduction to Electric Circuits*, fourth edition, Englewood Cliffs, N.J.: Prentice-Hall, 1976

E. A. Nesbitt, "Ferromagnetic Domains," prepared by Bell Telephone Laboratories for educational use, Murray Hill, N.J.: Bell Telephone Laboratories, 1962

J. Rosenblatt and M. H. Friedman, *Direct and Alternating Current Machinery*, New York: McGraw-Hill, 1963

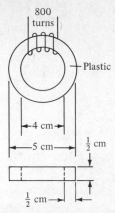

Figure 12.28 Sketch for Test Problem 8

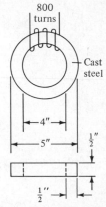

Figure 12.29 Sketch for Test Problem 9

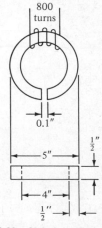

Figure 12.30 Sketch for Test Problem 10

Unit 13 *Inductance*

In this unit you will learn about the inductor. Here you will learn that *inductance* is the property of an electric circuit that *opposes* any change in the current through a circuit. Furthermore, you will learn that many properties of inductors are similar to those of capacitors.

Objectives

After completing all the work associated with this unit, you should be able to:

1. Recall the electrical and letter symbols and the unit of measurement for an inductor.

2. State when an electric circuit has an inductance of one henry. This would include an explanation of the relation

$$L = \frac{v_L}{di/dt}$$

3. List the four physical factors that determine the inductance of an inductor. Relate these factors in a mathematical equation, and use this equation to solve problems.

4. Calculate the total inductance when two, three, or more inductors are connected in series.

5. Calculate the total or equivalent inductance when two, three, or more inductors are connected in parallel.

6. Calculate the energy stored by an inductor.

Inductors

Objective 1   Recall the electrical and letter symbols and the unit of measurement for an inductor.

An inductor basically consists of a coil of wire, with or without a magnetic core. Although this coil of wire may be wrapped around a nonmagnetic form such as a plastic rod or cylinder, it is often wrapped around ferromagnetic material in the form of a rod, toroid, or some other shape. This ferromagnetic material increases the inductance without changing the number of turns or the physical dimensions of the coil.

Figure 13.1 shows the electrical symbols used for inductors. The letter symbol used for inductance is *L*. The basic unit of inductance of an inductor is the *henry*. The abbreviation of henry is H. The three common ratings used for most inductors (also called coils and chokes) are

$$\mu\text{H (microhenry or } 10^{-6} \text{ H)}$$

$$\text{mH (millihenry or } 10^{-3} \text{ H)}$$

$$\text{H (henry)}$$

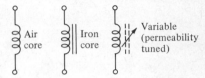

Figure 13.1   Electrical symbol for inductors

Most inductors that are rated in henries have a large number of turns of wire surrounding an iron core.

Problem 13.1   Draw the electrical symbol for an iron-core inductor, and write the letter symbol and basic unit of measurement for an inductor.

Definition of Inductance

Objective 2   State when an electric circuit has an inductance of one henry. This would include an explanation of the relation

$$L = \frac{v_{\text{L}}}{di/dt}$$

In Unit 12 on magnetism, you learned that a magnetic field exists around a permanent magnet or around an electric conductor (straight, solenoid, or any other shape) when it is carrying current. The larger the current, the stronger the magnetic field or the larger the number of lines of flux (ϕ). In this objective you will learn about the properties of a *varying* magnetic field.

In 1831 Joseph Henry, an American physicist, and Michael Faraday, a British scientist, found that electric currents could be made to flow in a conductor placed in a magnetic field. Current flows if a conductor is moved through a magnetic field so that

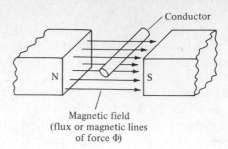

Magnetic field
(flux or magnetic lines
of force Φ)

Figure 13.2 Example demonstrating
principle of induction

it cuts *through* the flux or magnetic lines of force. Likewise, if
a conductor is held stationary and the magnetic field is moved
so that the lines of flux cut through the conductor, a voltage is
induced across the conductor. This property of an electric circuit is known as *induction*. The principle of induction is used
in the operation of transformers, magnetic sources, and other
electrical devices.

We can better understand this principle if we refer to Figure
13.2. In this illustration, a magnetic field is created by a magnet
possessing a North and South pole. The flux lines are coming
from the North pole and going toward the South pole. If a
conductor moves parallel to the lines of flux, *no* voltage is induced across the conductor. Likewise, for a stationary conductor,
if the magnetic field doesn't move or change, *no* voltage is induced.

On the other hand, if the conductor in Figure 13.2 is moved
in a direction perpendicular to the flux lines ($d\phi/dt$ = max), then
a voltage is induced in the conductor and current flows. Likewise, if the lines of flux are moving or changing with respect to
time ($d\phi/dt$), as shown in Figure 13.3, then a stationary conductor
in the vicinity has an induced voltage. Since Faraday was the
first to describe these phenomena, the law of electric induction
is known as *Faraday's law*.

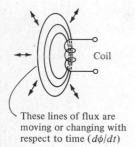

These lines of flux are
moving or changing with
respect to time ($d\phi/dt$)

Figure 13.3 Lines of flux moving near
a stationary conductor

Faraday's Law

Basically this law states that if a coil of N turns is placed in
the vicinity of a changing magnetic field, the voltage that is induced across the coil is equal to

$$v_L = N \frac{d\phi}{dt} \qquad (13.1)$$

where v_L = induced voltage across coil

   N = number of turns in coil

$\dfrac{d\phi}{dt}$ = rate of change of flux lines

The expression $d\phi/dt$ really means the rate of change of the
flux lines for a certain time period, or time rate of change. A

small change in the flux lines divided by the time required to cause this change can be expressed mathematically as

$$\frac{d\phi}{dt} = \frac{\text{a change (small) in the flux lines}}{\text{the time } (t) \text{ required to cause this change}} \quad (13.2)$$

Since all coils have inductance, let's see what happens when we have a solenoid or a coil of wire. The following paragraph tries to describe inductance as the property of *opposing any change in current.*

First, for anything to happen, since we have a stationary coil, the magnetic field or lines of flux must move or change. See Figure 13.4. How do we get the lines of flux to move? Since current sets up the magnetic lines of flux, for the flux to move, the current must change (either increase or decrease).

But what happens when the current changes? When it does change, the lines of flux change, and these changing lines of flux induce a voltage across the coil. Well—we must have had a voltage across the coil in the first place or no current would have flowed!

Now the question is: What happens when this induced voltage is added to the original circuit voltage? This induced voltage can't really be added to the original circuit voltage, or more current would flow, which would mean more induced voltage, which would mean more current, and so forth, and the whole thing would take off in perpetual motion, something for nothing . . . which is impossible!

Therefore the polarity of this induced voltage (also called *counter* emf) must be such that the current that flows as a result of it develops a flux that opposes any change in the original flux. This law is called *Lenz's law.* So this coil (inductor) has the property that it opposes any change in current in the circuit.

Therefore we can now define inductance. We say that a circuit has an inductance of *one henry* when a current changing at the rate of one ampere per second (*di/dt*) induces a voltage (counter emf) of one volt in the circuit. Mathematically, it is

$$L = \frac{v_{\mathrm{L}}}{\dfrac{di}{dt}} \quad (13.3)$$

where L = inductance in henries.

v_{L} = induced voltage (volts)

$\dfrac{di}{dt}$ = rate of change in current (one ampere per second)

Problem 13.2 Recall the property of inductance. Inductance is the property of an electric circuit that *opposes* change in

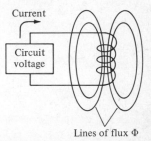

Current

Circuit voltage

Lines of flux Φ

Figure 13.4 Illustration of Lenz's law

Physical Factors Governing Inductance

Objective 3 List the four physical factors that determine the inductance of an inductor. Relate these factors in a mathematical equation, and use this equation to solve problems.

In Objective 2 you learned about the properties and characteristics of inductance. In this objective you will learn about the physical factors that affect inductance. Since the property of inductance is so closely tied to magnetics, you should have a firm grasp on some magnetic principles like permeability, magnetization curves, and flux. If you feel that you are weak in these areas, refer back to Unit 12 on magnetism and review these concepts.

There are four factors that affect inductance. The amount of inductance L is:

1. Directly proportional to the *square* of the number of turns in the coil N^2.
2. Directly proportional to the permeability of the magnetic circuit μ.
3. Directly proportional to the cross-sectional area of the magnetic circuit (inside cross-sectional area of the coil for air-core coils) A.
4. Inversely proportional to the length of the magnetic circuit (length of the coil for air-core coils) l.

We can summarize these four physical factors governing inductance in one mathematical expression:

$$L = \frac{N^2 \mu A}{l} \tag{13.4}$$

where $L =$ inductance in henries

$\mu =$ permeability of core material

$A =$ cross-sectional area of core material

$l =$ length of the inductor or coil

$N =$ number of turns

The value of permeability μ for nonmagnetic material or free space is given as

$\mu = 4\pi \times 10^{-7}$ MKS units (webers per ampere-turn for 1 m^3 of material or space)

$\mu = 3.2 \times 10^{-8}$ English practical units (webers per ampere-turn for 1 in^3 of material or space)

The following two statements should clarify whether the MKS units or the English practical units for permeability should be used.

MKS units

If the dimensions of the cross-sectional area and length are given in square meters and meters, respectively, then use MKS units for permeability. If the dimensions are given in centimeters, convert them to meters. (100 cm = 1 m or 1 cm = 0.01 m)

English practical units

If the dimensions of the cross-sectional area and length are given in square inches and inches, respectively, then use English practical units for permeability.

The following two examples illustrate the application of Equation (13.4) to two practical problems.

Example 13.1 A choke is made of 2500 turns of wire wound on a sheet steel core. The cross section of the core is 0.5 in. by 1 in., and the average path length is 5 in. The permeability of the sheet-steel core is 1.66×10^{-5} Wb/At-in. What is the inductance of the choke? (Figure 13.5)

Solution Using Equation (13.4) for inductance, we obtain

$$L = \frac{N^2 \mu A}{l} = \frac{(2500)^2(1.66 \times 10^{-5})(0.5 \times 1.0)}{5} = 10.375 \text{ H}$$

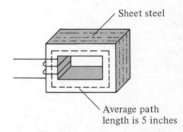

Figure 13.5 Sketch for Example 13.1

Example 13.2 A plastic cylinder of 1 cm outside diameter by 5 cm long has 1000 turns of wire wrapped around its entire length. What is its inductance? (Figure 13.6)

Solution We assume that there is no leakage flux between the turns. Since the units for area and length are in terms of square meters and meters, then the permeability μ for plastic is

$$\mu = 4\pi \times 10^{-7} \text{ Wb/At-m} \qquad \text{MKS units}$$

Figure 13.6 Sketch for Example 13.2

Furthermore, converting all the dimensions to meters by the relationship 1 cm = 0.01 m, we obtain

$$L = \frac{N^2 \mu A}{l} = \frac{(1000)^2(4\pi \times 10^{-7})\left[\frac{\pi}{4}(0.01)^2\right]}{0.05} = 1.974 \text{ mH}$$

Problem 13.3   Find the inductance of 300 turns of wire wound on a nonmagnetic cylinder (Figure 13.6) whose diameter is 2 cm and whose coil length is 10 cm.

It is sometimes desirable to calculate the number of turns N needed for a given specified inductance L. Rewriting Equation (13.4) and solving for N^2, we get

$$N^2 = \frac{L \times l}{\mu A} \tag{13.5}$$

or

$$N = \sqrt{\frac{L \times l}{\mu A}} \tag{13.6}$$

where N = number of turns

L = inductance in henries

l = length of coil

μ = permeability

A = cross-sectional area of core material

Try to apply the above equation to the following problem.

Problem 13.4   How many turns must be wound on an air core 1 in. long and $\frac{1}{2}$ in. in diameter to obtain an inductance of 30 mH?

Example 13.3   An inductor of 10 mH has a cross-sectional area of $\frac{1}{2}$ in^2. The physical parameters of permeability μ, length l, and turns N are kept constant. What is the new inductance, given that the cross-sectional area is 1 in^2?

Solution   Referring to Equation (13.4), we know that inductance L is directly proportional to cross-sectional area A.

$$\therefore \quad L \propto A$$

Setting up a proportion, we obtain

$$\frac{L_2}{L_1} = \frac{A_2}{A_1} \qquad \text{or} \qquad \frac{L_2}{10 \text{ mH}} = \frac{1 \text{ in}^2}{0.5 \text{ in}^2}$$

$$L_2 = \frac{1 \text{ in}^2}{0.5 \text{ in}^2} \times 10 \text{ mH} \qquad \therefore \quad L_2 = 20 \text{ mH}$$

Thus doubling the cross-sectional area doubles the inductance.

Example 13.4 We have two physically identical coils, and the first coil has turns $N_1 = 10$ and inductance $L_1 = 5$ mH. What is the inductance of the second coil if it has turns $N_2 = 20$?

Solution Since inductance L is proportional to N^2, writing a proportion

$$\frac{L_2}{L_1} = \frac{N_2^2}{N_1^2} = \left(\frac{N_2}{N_1}\right)^2$$

$$\frac{L_2}{5 \text{ mH}} = \left(\frac{20}{10}\right)^2 = (2)^2 = 4 \qquad \therefore \quad L_2 = 4(5 \text{ mH}) = 20 \text{ mH}$$

Thus doubling the turns on an inductor increases the inductance by a *factor* of 4.

Problem 13.5 A 180-mH inductor has 9000 turns. What is the inductance when the number of turns is reduced to 3000?

Problem 13.6 A 3-H inductor has 2000 turns. Given that we want to increase the inductance to 5 H, how many turns must we add?

Inductors in Series

Objective 4 Calculate the total inductance when two, three, or more inductors are connected in series.

In this objective you will learn how to replace a number of series-connected inductors with *one* equivalent inductor. In the development we shall assume that there is no mutual coupling between the inductors; that is, the flux developed by one inductor does not reach the other inductor. Figure 13.7 shows a sketch of two inductors in series.

To find the total equivalent inductance of two inductors in series, we recall that the current I is common in a series circuit and KVL still holds. That is,

$$E = V_1 + V_2 \qquad (13.7)$$

where V_1 = voltage across inductor L_1 and V_2 = voltage across inductor L_2

We can express the total inductance L_T shown in Figure 13.7(b) as

$$L_T = \frac{E}{di/dt} \qquad (13.8)$$

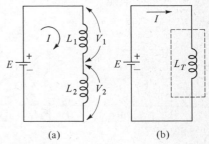

Figure 13.7 Two inductors connected in series.

since $E = V_1 + V_2$. Substituting Equation (13.8) into (13.7), we obtain

$$L_T = \frac{V_1 + V_2}{di/dt} = \frac{V_1}{di/dt} + \frac{V_2}{di/dt} \qquad (13.9)$$

Since the current I is common, then di/dt is also common. Applying the definition of inductance,

$$L = \frac{V}{di/dt} \qquad (13.10)$$

to Equation (13.9), we get

$$L_T = L_1 + L_2 \qquad (13.11)$$

Note that the equation for inductors in series is similar to that used for resistors in series. Therefore we can conclude that for n inductors in series, the equivalent total inductance is

$$L_T = L_1 + L_2 + L_3 + \cdots \qquad (13.12)$$

Example 13.5   Find the total inductance L_T of the circuit shown in Figure 13.8. Assume that there is no mutual coupling between inductors and that the current in the circuit has reached its final value.

Solution   Applying Equation (13.11), we have

$$L_T = L_1 + L_2$$
$$L_T = 2\,H + 3\,H = 5\,H$$

Problem 13.7   Find the total inductance L_T of three inductors in series. The inductor values are 4 mH, 7 mH, and 9 mH. Assume that there is no mutual coupling between inductors.

Inductors in Parallel

Objective 5   Calculate the total or equivalent inductance when two, three, or more inductors are connected in parallel.

In this objective you learn that the equivalent inductance L_T for parallel inductors is similar to that used for resistors in parallel. Again we assume in the development of these relations that there is no mutual coupling between inductors. Figure 13.9 shows a sketch of two inductors connected in parallel.

To find the total equivalent inductance of two inductors in parallel, we recall that the voltage E or V is common in a parallel

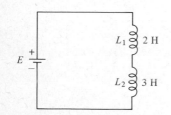

Figure 13.8   Circuit for Example 13.5

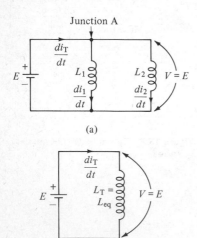

Figure 13.9   Two inductors connected in parallel

circuit and that KCL must still hold at any junction. That is,

$$\frac{di_T}{dt} = \frac{di_1}{dt} + \frac{di_2}{dt} \quad \text{at junction A} \qquad (13.13)$$

where $\dfrac{di_1}{dt}$ = rate of change of current i_1 through L_1

$\dfrac{di_2}{dt}$ = rate of change of current i_2 through L_2

We can calculate the total equivalent inductance $L_T = L_{eq}$ in Figure 13.9 as

$$L_T = L_{eq} = \frac{E}{di_T/dt} \qquad (13.14)$$

Substituting Equation (13.14) into (13.13), we get

$$L_T = L_{eq} = \frac{E}{(di_1/dt) + (di_2/dt)} \qquad (13.15)$$

Inverting both sides of Equation (13.15), we get

$$\frac{1}{L_T} = \frac{(di_1/dt) + (di_2/dt)}{E} = \frac{(di_1/dt)}{E} + \frac{(di_2/dt)}{E} \qquad (13.16)$$

Again applying the definition of inductance, we obtain

$$\frac{1}{L_T} = \frac{1}{L_1} + \frac{1}{L_2} \qquad (13.17)$$

or

$$L_T = L_{eq} = \frac{L_1 L_2}{L_1 + L_2} \qquad (13.18)$$

The general equation for more than two inductors in parallel can be expressed as:

$$\boxed{\frac{1}{L_T} = \frac{1}{L_1} + \frac{1}{L_2} + \frac{1}{L_3} + \frac{1}{L_4} + \cdots} \qquad (13.19)$$

Example 13.6 Calculate the equivalent inductance L_{eq} for the circuit shown in Figure 13.10. Assume no mutual coupling.

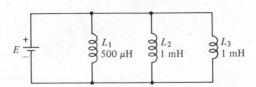

Figure 13.10 Circuit for Example 13.6

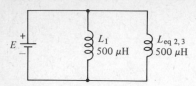

Figure 13.11 Sketch for solution of Example 13.6

Solution Simplifying L_2 and L_3, we get

$$L_{eq2,3} = \frac{L_2 L_3}{L_2 + L_3} = \frac{(1 \text{ mH})(1 \text{ mH})}{2 \text{ mH}} = 0.5 \text{ mH} \quad \text{or} \quad 500 \text{ } \mu\text{H}$$

Note that the procedure is similar to that for replacing two parallel resistors. The new equivalent circuit is shown in Figure 13.11. Again applying Equation (13.18), we get

$$L_{eq} = \frac{L_1 L_{eq2,3}}{L_1 + L_{eq2,3}} = \frac{(500 \text{ } \mu\text{H})(500 \text{ } \mu\text{H})}{500 \text{ } \mu\text{H} + 500 \text{ } \mu\text{H}}$$

$$= 250 \text{ } \mu\text{H}$$

Problem 13.8 A 300-mH inductor is connected in parallel with a 0.7-H inductor. Assume that there is no mutual coupling. What is the equivalent inductance?

Energy Stored by an Inductor

Objective 6 Calculate the energy stored by an inductor.

In an earlier objective you learned that an inductor stores energy in a magnetic field. In this objective you learn how to calculate the amount of energy stored in an inductor. When we close switch SW in Figure 13.12, current i_L starts to flow through inductor L. The voltage v_L across inductor L starts to decay, as shown in Figure 13.13. A detailed study of the current and voltage across inductors is discussed in the next unit on time constants.

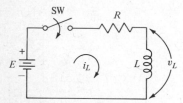

Figure 13.12 Circuit used for calculating energy stored by an inductor

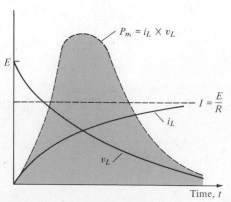

Figure 13.13 Calculation of energy stored by an inductor

Since energy W_m is related to power as

$$W_m = P_m t \qquad (13.20)$$

where W_m = magnetic energy stored by inductor

P_m = magnetic power in watts or vars

t = time in seconds

and

$$P_m = i_L \cdot v_L$$

then we can rewrite Equation (13.20) as

$$W_m = (i_L v_L)t \qquad (13.21)$$

where v_L = instantaneous voltage at time t

i_L = instantaneous current through inductor

Knowing the relationship of current i_L and v_L as a function of time, we can calculate P_m as a function of time. This product P_m is drawn dashed on the graph of Figure 13.13. The shaded area under the power curve represents the energy W_m stored by the inductor. Using calculus, we can compute the area underneath the curve to be

$$\boxed{W_m = \tfrac{1}{2}LI^2} \qquad \text{joules} \qquad (13.22)$$

where L = inductance in henries

I = steady-state current in circuit

W_m = magnetic energy stored by inductor

The following example illustrates how to calculate the amount of energy stored by an inductor.

Example 13.7 Calculate the energy stored by the inductor after the switch in Figure 13.14 has been closed.

Solution After switch SW is closed, we can calculate the steady-state current I to be

$$I = \frac{E}{R} = \frac{30 \text{ V}}{5 \text{ k}\Omega} = 6 \text{ mA}$$

Applying Equation (13.22), we obtain

$$W_m = \tfrac{1}{2}LI^2 = \tfrac{1}{2}(3)(6 \text{ mA})^2 = 54 \ \mu\text{J}$$

Problem 13.9 Calculate the energy stored by inductor L after the switch in Figure 13.15 has been closed.

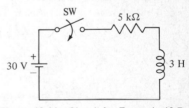

Figure 13.14   Circuit for Example 13.7

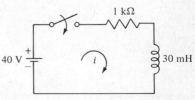

Figure 13.15   Circuit for Problem 13.9

Test

1. Complete the sentence:
   Inductance is the property of an electric circuit that

   opposes _____

   _____ .

2. An inductor has an inductance of 100 mH. Suppose that we double the number of turns on the coil without changing any dimensions or permeability of the magnetic circuit. What is the new inductance?

3. A 20-mH inductor is connected in series with a 30-mH inductor. Assuming that there is no mutual coupling, what is the total inductance?

4. A 2-H inductor is connected in series with a 500-mH inductor. Assuming that there is no mutual coupling, what is the total inductance?

5. A 20-mH inductor is connected in parallel with a 20-mH inductor. Assuming that there is no mutual coupling, what is the total inductance?

6. A 2-H inductor is connected in parallel with a 500-mH inductor. Assuming that there is no mutual coupling, what is the total inductance?

7. Find the equivalent inductance L_T shown in Figure 13.16

8. (*Bonus*) What is the approximate inductance of an air-core coil of 100 turns with an inside diameter of 1 cm and a length of 3 cm? Assume no leakage flux.

9. Calculate the energy stored by the inductor L (Figure 13.17) after the switch has been closed.

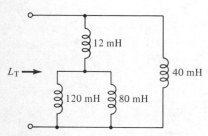

Figure 13.16   Circuit for Test Problem 7

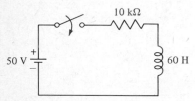

Figure 13.17   Circuit for Test Problem 9

Time Constants Unit 14
in Series RC
and RL Circuits

In the previous units you studied the characteristics of capacitors and inductors. You learned that these elements store energy—the capacitor in its electrical field, and the inductor in its magnetic field. Since energy is related to voltage and current, this unit on time constants describes the manner in which the voltage and current reach a specific value. Studying *RC* and *RL* circuits, you will learn that it takes time for the voltage and current to reach this specific value. Furthermore, this specific time depends on the values of the resistance and capacitance or inductance. This unit presents many examples that illustrate the technique used in predicting the voltage and current level at any specified time.

Objectives

After completing all the work associated with this unit, you should be able to:

1. Recall from memory the exponential curve and its values at $x = 0, 1, 2,$ and 5 or greater.
2. (*Capacitor discharge*) Given a simple series *RC* circuit with an initial charge on the capacitor, calculate the voltage across the capacitor at a specified time as the capacitor is discharging. (The universal exponential curve may be used.)
3. (*Capacitor charge*) Given a series *RC* circuit with the switch open and no initial voltage v_C across the capacitor, calculate the

voltage across the capacitor at a specified time after the switch has been closed and the capacitor is charging.

4. (*Advanced*) Given a complicated series *RC* circuit with a switch to charge and discharge the capacitor, with an initial charge present across the capacitor, calculate the voltage across the capacitor at specified times.

5. Given a series *RL* circuit with the switch open, calculate the *current* through the inductor and the voltage across the inductor at a specified time after the switch has been closed.

6. In a series *RL* circuit, calculate the voltage across a switch at the instant the switch is opened.

7. (*Advanced*) Given a series *RL* circuit with a pulsed input, calculate the current, the voltage drop across the resistor, and the voltage drop across the inductor at specified times.

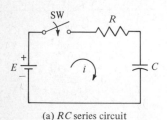

(a) *RC* series circuit

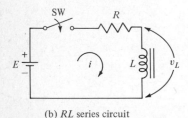

(b) *RL* series circuit

Figure 14.1   Series *RC* and *RL* circuits

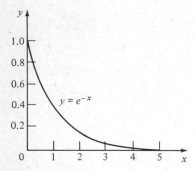

Figure 14.2   Sketch of an exponential curve

Exponential Curve

Objective 1   Recall from memory the exponential curve and values at $x = 0$, 1, 2, and 5 or greater.

Experiments have shown us that the current i in the circuit of Figure 14.1(a) and the voltage v_L across the inductor in Figure 14.1(b) follow an exponential curve.

What is an exponential curve?

Figure 14.2 shows the exponential curve. When $x = 0$, the vertical scale y has a value of 1. As x increases, y gets smaller and smaller. This continues until $x = 5$, where for all practical purposes y is zero. The mathematical equation for this exponential curve is

$$y = e^{-x} \tag{14.1}$$

Before we start studying *RC* or *RL* circuits, let's study the exponential curve in more detail. First, what is e? [Some books use ε (Greek letter epsilon) instead of e.] The symbol e is an irrational number whose value is

$$e = 2.71828 \ldots \tag{14.2}$$

An *irrational number* is a real number that cannot be expressed as a quotient of integers. Furthermore, the decimal numeral for an irrational number never ends and never repeats. Other examples of irrational numbers are π, $\sqrt{2}$, and $\sqrt{5}$.

Since the value of e is given in Equation (14.2), let us try to evaluate the exponential curve at different values. This will be accomplished in the first problem.

Problem 14.1 Use your calculator and figure out the values for the equation $y = e^{-x}$ for $x = 0, +1, +2, +3, +4$, and $+5$.

$$e^{-0} = 2.718^{-0} \qquad\qquad = 1$$

$$e^{-1} = (2.718)^{-1} = \frac{1}{2.718} \quad =$$

$$e^{-2} = (2.718)^{-2} = \frac{1}{(2.718)^2} =$$

$$e^{-3} = (2.718)^{-3} = \frac{1}{(2.718)^3} =$$

$$e^{-4} = (2.718)^{-4} = \frac{1}{(2.718)^4} =$$

$$e^{-5} = (2.718)^{-5} = \frac{1}{(2.718)^5} \cong 0 \qquad \text{(very small)}$$

Plot the equation $y = e^{-x}$ for $x = 0, +1, +2, +3, +4, +5$ on Figure 14.3. Then compare your graph with the exponential curve shown in Figure 14.2.

Problem 14.2 Subtract the values calculated in Problem 1 from 1.0000.

$$1 - e^{-0} = 1 - \underline{\quad 1 \quad} = 0$$

$$1 - e^{-1} = 1 - \underline{\qquad\qquad} =$$

$$1 - e^{-2} = 1 - \underline{\qquad\qquad} =$$

$$1 - e^{-3} = 1 - \underline{\qquad\qquad} =$$

$$1 - e^{-4} = 1 - \underline{\qquad\qquad} =$$

$$1 - e^{-5} = 1 - \underline{\qquad\qquad} \approx 1$$

Plot the equation $y = 1 - e^{-x}$ on Figure 14.3. Compare your graph with that shown in Figure 14.4 on the next page.

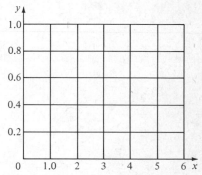

Figure 14.3 Sketch for Problems 14.1 and 14.2

Universal Exponential Curve

If you refer to Figure 14.3 of Problems 1 and 2, you find that you have drawn two curves for the equations

$$y = e^{-x} \qquad\qquad (14.3)$$

and

$$y = 1 - e^{-x} \qquad\qquad (14.4)$$

Comparing the curves of Figure 14.3 with the universal exponential curves of Figure 14.4, you can see that both curves agree.

Instead of labeling the vertical or y scale (Figure 14.4) 0 to 1, we could label it 0 to 100%. The equation $y = e^{-x}$ represents the current i in the circuit of Figure 14.1(a) or the voltage v_L across the inductor in Figure 14.1(b).

The equation $y = 1 - e^{-x}$ represents the *charging* voltage across the capacitor of Figure 14.1(a) or the current rise in Figure 14.1(b). We study this in more detail in later objectives. When $x = 5$, the capacitor is considered to be fully charged, or maximum current is flowing in the inductor. The universal exponential curve is used to tell you what percentage of the maximum value y you have at a certain value of x.

Before we go to the next objective, let us learn how to use the universal exponential curves of Figure 14.4. The following examples illustrate the application of these curves.

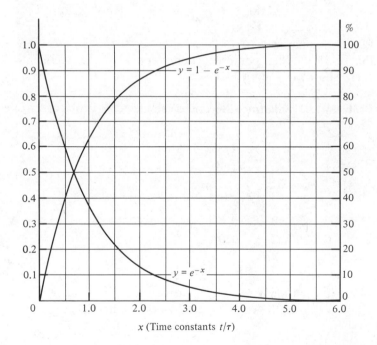

x (Time constants t/τ)

x	e^{-x}	x	e^{-x}	x	e^{-x}	x	e^{-x}
0.0	1.000000	1.0	0.367879	3.0	0.049787	5.0	0.006738
0.1	0.904837	1.2	0.301194	3.2	0.040762	5.5	0.004087
0.2	0.818731	1.4	0.246597	3.4	0.033373	6.0	0.002479
0.3	0.740818	1.6	0.201897	3.6	0.027324	6.5	0.001503
0.4	0.670320	1.8	0.165299	3.8	0.022371	7.0	0.000912
0.5	0.606531	2.0	0.135335	4.0	0.018316	7.5	0.000553
0.6	0.548812	2.2	0.110803	4.2	0.014996	8.0	0.000335
0.7	0.496585	2.4	0.090718	4.4	0.012277	8.5	0.000203
0.8	0.449329	2.6	0.074274	4.6	0.010052	9.0	0.000123
0.9	0.406570	2.8	0.060810	4.8	0.008320	10.0	0.000045

Figure 14.4   Universal exponential curves

Examples and Problems
Using the Universal Curve

Example 14.1 Find the value of y for $x = +3$, where $y = e^{-x}$.
Use the universal exponential curve.

Solution
(a) Referring to Figure 14.4 and looking at the curve $y = e^{-x}$ at
the point $x = 3$, we find that $y \approx 0.05$.
(b) A more accurate method is to use the table given below the
graph of Figure 14.4. Here, for $x = +3$, $e^{-x} = 0.049787$.

Example 14.2 Evaluate $y = 1 - e^{-x}$ at $x = +3$.

Solution
(a) Referring to the curve $y = 1 - e^{-x}$ in Figure 14.4 and looking
at the point $x = 3$, we find that

$$y \approx 0.95$$

(b) A more accurate way of evaluating the equation $y = 1 - e^{-x}$
is to first evaluate e^{-x} at $x = 3$ by using the table below Figure
14.4. Thus

$$\text{for } x = 3 \quad e^{-x} = 0.049787$$
$$\therefore \quad y = 1 - e^{-x} = 1 - 0.049787 = 0.950213$$

Problem 14.3 Evaluate y for $y = e^{-x}$ at $x = +2$.   *(Answer: $y = 0.135$)*

Problem 14.4 Evaluate y for $y = 1 - e^{-x}$ at $x = +4$.   *(Answer: $y = 0.9817$)*

It is sometimes important to know for what value of x we have
a certain given value of y. The following two examples illustrate
the use of the universal exponential curve to find a value of x
for a given y.

Example 14.3 Find the value of x where $y = e^{-x}$ and $y = 0.1653$.

Solution
(a) Referring to the curve $y = e^{-x}$ in Figure 14.4, we look at
$y = 0.1653$. If we draw a horizontal line at $y = 0.1653$, we find
that it intersects the curve $y = e^{-x}$ at a certain x value. Projecting
the intersection point down, we have $x \approx 1.8$.
(b) A more accurate method is to use the table given below
Figure 14.4. Looking at the table, for $e^{-x} = 0.1653$, we find that
$x = 1.8$. You may also use the tables in Appendix C.

Example 14.4 For $y = 1 - e^{-x}$, find the value for x such that $y = 0.632$.

Solution

(a) Referring to the curve for $y = 1 - e^{-x}$ in Figure 14.4, we locate $y = 0.632$. If we draw a horizontal line, we find that it intersects the curve $y = 1 - e^{-x}$ at a certain value of x. The x value graphically is $x \approx 1.0$

(b) Using the table and evaluating the equation, we get

$$y = 1 - e^{-x}, \qquad 0.632 = 1 - e^{-x}, \qquad -0.368 = -e^{-x}$$

$$\therefore \quad x = 1.0$$

Problem 14.5 Find the value for x where $y = 1 - e^{-x}$ and *(Answer: $x = 1.6$)* $y = 0.8$.

Series *RC* Circuit

Objective 2 (*Capacitor discharge*) Given a simple series *RC* circuit with an initial charge on the capacitor, calculate the voltage across the capacitor at a specified time as the capacitor is discharging. (The universal exponential curve may be used.)

Capacitor Discharge

The capacitor shown in Figure 14.5 has an initial charge on it. If the switch SW in this circuit has remained in position 1 for a long time, or at least 5 time constants, then the capacitor has the supply voltage E across it. This voltage, v_C or E, is considered to be the initial voltage. (Note the polarity of the initial charge on C.) When the switch is thrown to position 2, the voltage across the capacitor decreases with time. In this position, the resistor R provides a path for the electrons (charge Q) to leak off the capacitor plates. If the resistor R were infinite, then the capacitor would remain charged indefinitely. For a finite value of resistance R, let's see what kind of curve could represent the discharge voltage across the capacitor. The following paragraph explains how

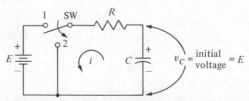

Figure 14.5 Series *RC* circuit with initial charge across the capacitor

the voltage across the capacitor decreases as the capacitor discharges.

Recall that the charge on a capacitor is related to the voltage across the plates by the relationship

$$Q_T = Cv_C \qquad (14.5)$$

where Q_T = total charge on capacitor plates

v_C = voltage across the capacitor (instantaneous)

Furthermore, since charge Q and current i are related by

$$Q = i \times t \qquad (14.6)$$

where i = rate of charge flow (instantaneous amperes)

t = time in seconds

then, as the capacitor is discharged, charge Q is removed from the plates of the capacitor at a charge rate of i. At the instant the switch SW closes to position 2, the current i (rate of removing charge Q) is found by Ohm's law to be

$$i = \frac{v_C}{R} \qquad \text{where initial voltage } v_C = E \qquad (14.7)$$

At this instant charge rate i is a maximum. As time goes by, charge Q is removed from the capacitor, and voltage v_C drops across the capacitor. Since i is related to voltage v_C [Equation (14.7)], as voltage v_C across the capacitor drops, so does the rate of charge flow i. With a smaller discharge rate i, less charge Q is removed, and the voltage v_C across the capacitor drops at a slower rate. This phenomenon continues until the capacitor is fully discharged.

This voltage across the capacitor follows the exponential curve e^{-x}, as shown in Figure 14.6. The vertical axis represents the discharge voltage v_C across the capacitor. At the instant of closing the switch, $x = 0$ and $v_C = E$. The horizontal axis could represent time, or—to be more general—it may represent the number of time constants x. The number of time constants x may be found by Equation (14.8). Here

$$x = \frac{t}{\tau} \qquad (14.8)$$

where x = number of time constants (no units)

t = actual time elapsed, in seconds

τ = (Greek letter tau) time constant of the circuits, in seconds

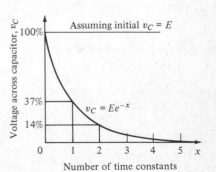

Figure 14.6 Exponential curve of discharge voltage across capacitor

For a series RC circuit, the time constant

$$\tau = R \cdot C \qquad (14.9)$$

where R = resistance in ohms

$\quad\;\; C$ = capacitance in farads

$\quad\;\; \tau$ = time constant in seconds

Since x in Equation (14.8) equals (t/τ) and both t and τ are measured in units of seconds, x has no units.

Referring back to Figure 14.6, since the capacitor had an initial voltage v_C or E at $x = 0$, then the time t is also equal to 0. At the end of one time constant ($\tau = RC$, $x = 1$), the capacitor discharges to 36.8% ($\approx 37\%$) of its initial value. At the end of two time constants ($x = 2$), the capacitor discharges to 13.5% ($\approx 14\%$) of its initial value. The capacitor is considered fully discharged at the end of five time constants ($x = 5$). At this time, the capacitor has no voltage across it, and no current is flowing in the circuit.

The following example illustrates the calculation of the discharge voltage across a capacitor at any specified time.

Example 14.5 The switch in Figure 14.7 has remained in position 1 long enough to charge the capacitor C up to 20 V. Determine the voltage across the capacitor 1.5 ms after the switch has been thrown to position 2.

Solution (See Figure 14.8.)

(a) The time constant τ of this circuit is

$$\tau = RC = (100 \times 10^3)(0.01 \times 10^{-6}) = 1 \times 10^{-3} \text{ s}$$
$$\tau = 1 \text{ ms}$$

(b) For time $t = 1.5$ ms, the number of time constants is

$$x = \frac{t}{\tau} = \frac{1.5 \text{ ms}}{1 \text{ ms}} = 1.5$$

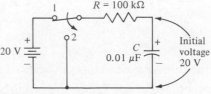

Figure 14.7   Circuit for Example 14.5

(c) Since the capacitor has an initial voltage of 20 V and is *discharging*, we use the e^{-x} curve in Figure 14.4, or the table, for $x = 1.5$ and obtain $e^{-x} = 0.22$.

$$\therefore \quad v_C = E_{init}(e^{-x}) = 20(0.22) = 4.4 \text{ V}.$$

Problem 14.6 In the circuit shown in Figure 14.7, change the values to $R = 1 \text{ M}\Omega$ and $C = 0.5 \text{ }\mu\text{F}$. Assume that the capacitor has an initial charge on it of 20 V. What is the voltage across the capacitor 0.35 s after the switch has been thrown to position 2?

Capacitor Charge

Objective 3 (*Capacitor charge*) Given a series *RC* circuit with the switch open and no initial voltage v_C across the capacitor, calculate the voltage across the capacitor at a specified time after the switch has been closed and the capacitor is charging.

The switch in the circuit shown in Figure 14.9 is open. In this objective, we assume that the capacitor has no voltage v_C across it when the switch SW is open. Since charge q_T on a capacitor is related to its voltage by the equation

$$q_T = C v_C \tag{14.10}$$

where q_T = instantaneous charge on capacitor plates

C = capacitance in farads

v_C = voltage across capacitor

then with no voltage v_C on the plates, there is no charge q_T on its plates. In other words, the capacitor has an initial charge of 0. When the switch SW is closed, charge Q is deposited on the plates of the capacitor. Since charge is related to voltage [Equation (14.10)], the voltage v_C across the capacitor increases, but it does not increase linearly. Let us see why.

Applying KVL to the circuit shown in Figure 14.9, we obtain

$$E = v_R + v_C \tag{14.11}$$

where E = total applied input voltage

v_R = voltage across resistor

v_C = voltage across capacitor

This means that at any instant of time the sum of the voltage drops must always equal the applied voltage. For example, at the instant switch SW is closed, since there is no charge ($v_C = 0$) on the capacitor, all the voltage is across the resistor R. Therefore

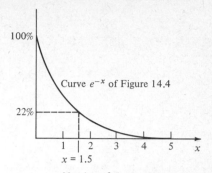

Curve e^{-x} of Figure 14.4

Number of time constants

Figure 14.8 Sketch of solution of Example 14.5

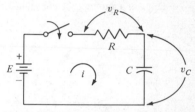

Figure 14.9 Series *RC* circuit with zero initial voltage across capacitor

the initial value of current i can be calculated as

$$I_{\max} = \frac{v_R}{R} \qquad \text{where initial voltage } v_R = E$$

where $I_{\max}$ = maximum instantaneous current flow when switch is closed

Since charge Q is related to current i by

$$Q = i \times t \qquad (14.12)$$

where i = current flow in amperes

$\qquad t$ = time in seconds

then after switch SW has been closed for some time t, so much charge Q is deposited on the plates at a rate i. Since charge is related to voltage [Equation (14.10)], the voltage across the capacitor is increasing. However, as the voltage across C increases, this means less voltage across R, and therefore less current i in the circuit. This is shown in Figure 14.10.

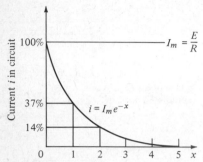

Figure 14.10 Current flowing in series *RC* circuit while capacitor is charging

After five time constants ($x = 5$), the capacitor is considered to be fully charged and the capacitor voltage is equal to the supply voltage ($v_C = E$). Since the voltage across the resistor at this time is zero, current i must be zero. Applying Ohm's law, we can calculate the voltage across the resistor to be

$$v_R = iR = \left[I_{\max}e^{-x}\right]R = \left[\frac{E}{R}e^{-x}\right]R \qquad (14.13)$$

$$\therefore \quad v_R = Ee^{-x} \qquad (14.14)$$

Equation (14.14) is shown plotted in Figure 14.11. Thus, while the capacitor is charging, the voltage across the resistor R is dropping and follows an exponential curve. After 5 time constants ($x = 5$), the capacitor is fully charged ($i = 0$) and the voltage v_R across the resistor is zero.

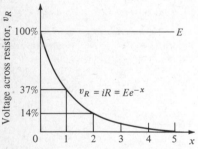

Figure 14.11 Voltage across resistor while capacitor *C* is charging

We can compute the voltage across the capacitor by applying KVL to the circuit shown in Figure 14.9. Here

$$E = v_R + v_C \qquad (14.15)$$

or the voltage across the capacitor is

$$v_C = E - v_R \qquad (14.16)$$

But referring to Figure 14.11 and Equation (14.14), we find

$$v_R = Ee^{-x} \qquad (14.17)$$

Therefore, substituting Equation (14.17) into (14.16), we obtain

$$v_C = E - Ee^{-x} = E(1 - e^{-x}) \qquad (14.18)$$

Equation (14.18) represents the voltage across a charging capacitor. This equation is plotted in Figure 14.12. Note that the

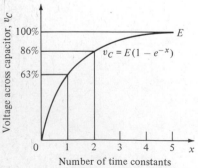

Figure 14.12 Voltage across capacitor as capacitor is charging

voltage across the capacitor follows an exponential curve of the form $(1 - e^{-x})$.

As in the previous objective, we can find the number of time constants x by

$$x = t/\tau \qquad (14.19)$$

where x = number of time constants

t = actual time elapsed in seconds

τ = time constant in seconds $(\tau = RC)$

R = resistance in ohms

C = capacitance in farads

The vertical axis in Figure 14.12 represents the voltage v_C across a capacitor. At the instant the switch is closed (time $t = 0$, $x = 0$), the voltage v_C across the capacitor is zero. At the end of one time constant $(x = 1)$, the capacitor is charged up to 63.2% $(\approx 63\%)$ of its final value. After two time constants $(x = 2)$, the capacitor is charged up to 86.5% of its final value. Finally, after five time constants $(x = 5)$, the capacitor is considered to be fully charged and has 100% of its value $(v_C = E)$.

But what is the voltage across the resistor as the capacitor is charging? We known that KVL must hold at all times, so referring to Figure 14.9, we have

$$E = v_R + v_C \qquad (14.20)$$

By referring to Figures 14.11 and 14.12 and using Equation (14.20), let's determine what the voltages across the resistance and capacitance are at different times.

At $x = 0$,

$$v_C = 0\% \qquad \text{and} \qquad v_R = 100\%$$
$$\therefore \quad v_C(0\%) + v_R(100\%) = E(100\%)$$

At one time constant,

$$x = 1 \qquad v_C = 63\% \qquad \text{and} \qquad v_R = 37\%$$
$$\therefore \quad v_C(63\%) + v_R(37\%) = E(100\%)$$

At five time constants,

$$x = 5 \qquad v_C = 100\% \qquad \text{and} \qquad v_R = 0\%$$
$$v_C(100\%) + v_R(0\%) = E(100\%)$$

The above three illustrations verified that KVL must hold at all times, i.e., that the sum of all the voltages must add up to the supply voltage E.

The following example illustrates how to calculate the voltage across any capacitor at a specified time.

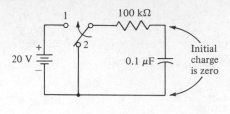

Figure 14.13 Circuit for Example 14.6

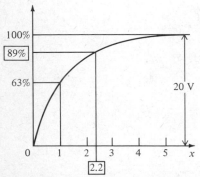

Figure 14.14 Sketch for solution c of Example 14.6

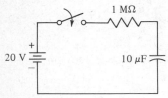

Figure 14.15 Circuit for Problem 14.7

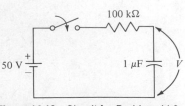

Figure 14.16 Circuit for Problem 14.8

Example 14.6 The capacitor in Figure 14.13 has no initial charge on it. What is the voltage across the capacitor 22 ms after the switch has been thrown to position 1?

Solution

(a) The time constant for this circuit is

$$\tau = RC = 100 \times 10^3 \times 0.1 \times 10^{-6} = 10 \times 10^{-3} \text{ s}$$

$$\tau = 10 \text{ ms}$$

(b) The number of time constants x is found by

$$x = \frac{t}{\tau} = \frac{22 \text{ ms}}{10 \text{ ms}} = 2.2$$

(c) Referring to the universal curve $(1 - e^{-x})$ of Figure 14.14, we find that the capacitor has charged up to 89% of its final value of the supply voltage 20 V.

$$v_C = 20(1 - e^{-2.2}) = 20(0.89) = 17.8 \text{ V}$$

Problem 14.7 The capacitor in this circuit (Figure 14.15) has no initial charge across it. (a) What is the voltage or potential difference across the capacitor at the instant the switch is closed? (b) What is the voltage across the capacitor 15 s after the switch is closed? (c) How long does the capacitor take to charge to the supply voltage of the circuit?

Sometimes we want to find the time it takes a capacitor to charge up to a specified voltage. Before trying to do the following problem, go back and review Example 14.4.

Problem 14.8 The capacitor in this circuit (Figure 14.16) has no initial voltage across it. (a) What is the time constant τ for this circuit? (b) How long does it take for the voltage across the capacitor to reach 45 V (90% of supply voltage)?

Before proceeding to the next objective, let us try to relate the equation for capacitance

$$i = C\frac{dv}{dt} \qquad (14.21)$$

to this unit on time constants. Recall that this equation was presented in the earlier unit on capacitance. In studying this unit, we have found that charge Q is deposited on the plates of a charging capacitor at a rate i. This was given by Equation (14.12) as

$$Q = i \times t \qquad (14.22)$$

Moreover, recalling the definition of capacitance, with

$$Q = CV \qquad (14.23)$$

we can state that for a given capacitance C, a change in the charge (ΔQ) produces a change in the voltage (ΔV). (The symbol Δ is the Greek capital letter delta, and it means "change in" or "increment.") Or mathematically,

$$\Delta Q = C \Delta V \qquad (14.24)$$

where ΔQ = small change in charge on the plates of the capacitor

ΔV = small change in voltage across the capacitor

Since this change in charge (ΔQ) takes place in a certain increment of time (Δt), then, rewriting Equation (14.22), we have

$$\Delta Q = i(\Delta t) \qquad (14.25)$$

where Δt = small increment of time

i = constant charge rate for small increment of time (current)

Substituting Equation (14.25) into Equation (14.24), we obtain

$$i \Delta t = C \Delta V \qquad (14.26)$$

We can rewrite Equation (14.26) as

$$i = C \frac{\Delta V}{\Delta t} \qquad (14.27)$$

or, in calculus notation,

$$i = C \frac{dv}{dt} \qquad (14.28)$$

Let us see what Equations (14.27) and (14.28) actually mean. At the instant the switch is closed in Figure 14.9, we have the largest possible change of voltage with respect to time (dv/dt is largest). This means that i must be the largest possible at this time. This is shown in Figure 14.10. After five time constants, the capacitor is fully charged, and voltage across it does not change ($dv/dt = 0$). At this time, since dv/dt equals zero, current i is also equal to zero. As a result, Equations (14.27) and (14.28) mathematically relate the actual current and voltage of a charging capacitor.

Capacitor Charge and Discharge

Objective 4 (*Advanced*)   Given a complicated series *RC* circuit with a switch to charge and discharge the capacitor, with an initial charge present across the capacitor, calculate the voltage across the capacitor at specified times.

Figure 14.17 shows a series *RC* circuit with an initial voltage across the capacitor. We are now assuming that capacitor *C* has some initial charge (or voltage) on it before the swtich is thrown to position 1. Notice that the direction of the initial charge is positive. This initial charge could also have a negative value. The initial charge assumed in Figure 14.17 could result if the switch weren't allowed to remain in position 2 long enough for the capacitor to discharge completely.

If we throw the switch to position 1, with an initial charge on the capacitor, the capacitor can still only charge up to the battery voltage *E*. Recall that *E* is the only supply in the circuit. However, with an initial voltage the capacitor has a head start! This is shown in Figure 14.18.

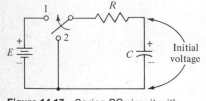

Figure 14.17   Series *RC* circuit with initial voltage across capacitor *C*.

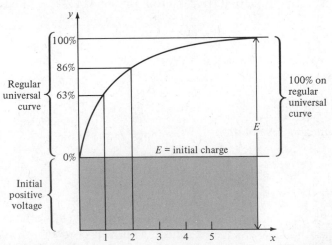

Figure 14.18   Application of the universal exponential curve to a series *RC* circuit with the capacitor having an initial voltage

Referring to the figure, notice that 100% on the universal exponential curve is no longer E but $(E - \text{initial voltage})$. The reason for this is explained as follows.

Referring to Figure 14.17, we know that KVL holds for all time, that is,

$$E = v_R + v_C \tag{14.29}$$

Since $v_C = v_{\text{initial}}$ at time $t = 0$, the voltage across the resistor R is

$$v_R = E - v_C \tag{14.30}$$

or

$$v_R = E - v_{\text{initial}} \tag{14.31}$$

Applying Ohm's law, we find that the initial charging current i is

$$i = \frac{v_R}{R} = \frac{E - v_{\text{initial}}}{R} \tag{14.32}$$

Since current i is a rate of charge flow, then

$$i = \frac{Q}{t}$$

where Q = charge (coulombs on capacitors)

t = time in seconds

If the capacitor had *no initial charge* on it ($v_{\text{initial}} = 0$), the current i would then be

$$i = \frac{E}{R} \qquad \text{where } v_{\text{initial}} = 0 \tag{14.33}$$

Comparing the current in Equation (14.32) to that in Equation (14.33), we find that the currents are different; thus we obtain different charging curves.

The following example illustrates how to calculate the voltage across a capacitor at any specified time. An initial voltage across a capacitor is given.

Example 14.7 The capacitor in Figure 14.19 has an initial voltage of $+5\,\text{V}$. What is the voltage across the capacitor 18 ms after the switch has been thrown to position 1?

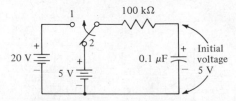

Figure 14.19 Circuit for Example 14.7

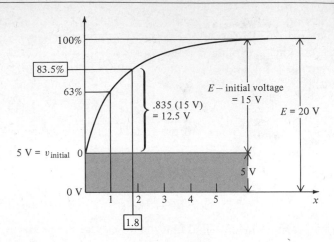

Figure 14.20 Sketch for solution for
Example 14.7

Solution

(a) The time constant τ for this circuit is

$$\tau = RC = 100 \times 10^3 \times 0.1 \times 10^{-6} = 10 \text{ ms}$$

(b) Since the actual time t is 18 ms, the number of time constants is

$$x = t/\tau = \frac{18 \text{ ms}}{10 \text{ ms}} = 1.8$$

(c) The value of $x = 1.8$ on the universal chart $(1 - e^{-x})$ is about 83.5% of the 100% value. However, the 20 V is not 100% on the universal chart.

$$100\% = E - v_{\text{initial}} = 20 - 5 = 15 \text{ V}$$

Therefore, after a time of 18 ms ($x = 1.8$), the capacitor charges up 0.835 (15) = 12.5 V above the initial charge.

$$\therefore \quad v_C = v_{\text{initial}} + v = 5 \text{ V} + 12.5 \text{ V} = 17.5 \text{ V}$$

This analysis is illustrated in Figure 14.20.

Problem 14.9   The capacitor in Figure 14.21 has an initial charge of *negative* 10 V across it. Make a sketch of the voltage across a number of time constants. Start the time equal to zero when the switch is thrown to position 1. What is the voltage across the capacitor 1 s after the switch has been thrown to position 1?

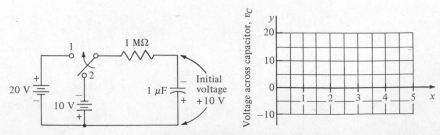

Figure 14.21   Sketch for Problem 14.9

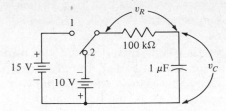

Figure 14.22 Circuit for Example 14.8

Example 14.8 The capacitor in Figure 14.22 has *no* initial charge across it. The switch is thrown to position 1 for 100 ms. It is then thrown to position 2 for 100 ms, and finally back to position 1 for 1 min. Make a sketch of the voltage across the capacitor and resistor plotted against the number of time constants. What are the voltages across the capacitor at the end of 100, 200, and 300 ms and at the end of 10 min?

Solution

(a) The time constant of the circuit is

$$\tau = RC = (100 \times 10^3)(1 \times 10^{-6}) = 100 \text{ ms}$$

(b) The number of time constants in 100 ms is

$$x = \frac{t}{\tau} = \frac{100 \text{ ms}}{100 \text{ ms}} = 1$$

In position 1, after 100 ms ($x = 1$), the capacitor charges up to 63.2% of maximum voltage. Thus

$$v_C = 0.632(15 \text{ V}) = 9.48 \text{ V}$$

When the switch is thrown to position 2, if it were left in position 2 long enough, the capacitor would first discharge from the 9.48 V down to zero (see Figure 14.23). Then it would charge in

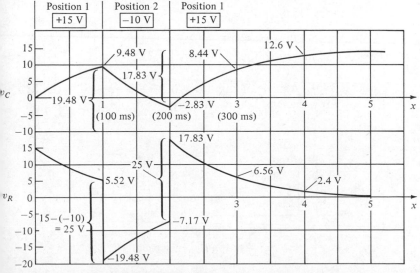

Figure 14.23 Sketch for solution of Example 14.8

the opposite direction down to -10 V, which is the voltage across the circuit when the switch is in position 2. However, do we have enough time for this to happen?

How long (in time constants) does the switch remain in position 2? As we figured out earlier, 100 ms is *one* time constant. From the discharge curve, we know that the voltage across the capacitor drops only 63.2% of the total charge in one time constant. The voltage across the capacitor drops 63.2% of the distance from $+9.48$ V to -10 V (a distance of 19.48 V). Or the voltage starts from $+9.48$ V and drops 63.2% of 19.48 V in *one* time constant. Mathematically

$$9.48 \text{ V} - 0.632(19.48) = 9.48 \text{ V} - 12.31 \text{ V} = -2.83 \text{ V}$$

Therefore, at the end of 100 ms in position 2, the voltage across the capacitor is -2.83 V. See the v_C graph in Figure 14.23.

When the switch is thrown back to position 1, the capacitor charges to the supply voltage ($+15$ V) if given enough time. For a 10-min time, the number of time constants x is

$$x = \frac{10 \text{ min}}{100 \text{ ms}} = \frac{10(60) \text{ s}}{0.1 \text{ s}} = 6000$$

This is many times greater than five, so the capacitor has 15 V across it at the end of 10 minutes in position 1.

Let's see what the voltage is across the capacitor after it has remained in this position (position 1) for only 100 ms (one time constant). Start:

$$-2.83 \text{ V} + 0.632[15 - (-2.83)] = -2.83 \text{ V} + 11.269 \text{ V}$$
$$= 8.44 \text{ V} \qquad \text{(see Figure 14.23)}$$

To plot the v_R curve, see Figure 14.23, and remember KVL:

$$\text{Supply voltage} = v_C + v_R$$

Position 1 [supply voltage $= +15$ V] $\therefore$ $v_R = 15 - v_C$

Position 2 [supply voltage $= -10$ V] $\therefore$ $v_R = -10 - v_C$

Example 14.9   A pulse of amplitude of 10 V is applied to the *RC* network for 3 ms every 20 ms (Figure 14.24). Make a sketch of

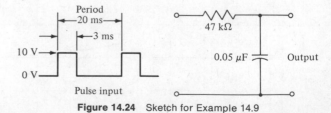

Figure 14.24   Sketch for Example 14.9

the voltage across the capacitor (output) and resistor plotted against time. Label voltage levels on the sketch.

Solution   The time constant of the circuit is

$$\tau = RC = (47 \times 10^3)(0.05 \times 10^{-6}) = 2.35 \text{ ms}$$

The number of time constants for 3 ms *charge* is

$$x_c = \frac{t}{\tau} = \frac{3 \text{ ms}}{2.35 \text{ ms}} = 1.277$$

and the number of time constants for 17 ms *discharge* is

$$x_d = \frac{t}{\tau} = \frac{17 \text{ ms}}{2.35 \text{ ms}} = 7.23$$

[Since $x_d > 5$ time constants, the capacitor discharges to zero.]
To determine the voltage across the resistor, we use KVL:

$$\text{Input voltage} = v_C + v_R$$

For a 3-ms pulse, the input voltage $= 10$ V at $t = 0$ ms.

$$v_R = 10 - v_C$$

When $v_C = 0$, then $v_R = 10$

At $t = 3$ ms, when $v_C = 7.2$ V, then $v_R = 10 - 7.2 = 2.8$ V.
For 17 ms, the input voltage $= 0$ V

$$v_R = 0 - v_C$$
$$v_R = -v_C \qquad \text{(see Figure 14.25)}$$

After 20 ms, a 10-V pulse is again applied, and the analysis is repeated.

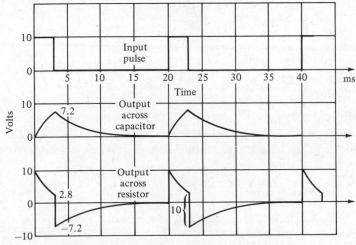

Figure 14.25   Sketch for solution of Example 14.9

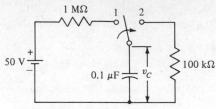

Figure 14.26   Circuit for Problem 14.10

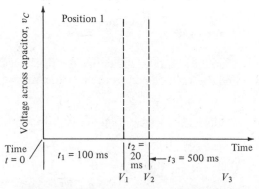

Figure 14.27   Output voltage v_C versus time t.

Problem 14.10   The capacitor in the given circuit (Figure 14.26) has an initial voltage of 0 V. Assuming that the switch is thrown to position 1, then to position 2, and then back to position 1, calculate the voltage across the capacitor at the following specified times and plot the waveforms on an attached graph (Figure 14.27).

Position 1, time $t_1 = 100$ ms
Position 2, time $t_2 =\ \ 20$ ms
Position 1, time $t_3 = 500$ ms

Series *RL* Circuit

Objective 5   Given a series *RL* circuit with the switch open, calculate the current through the inductor and the voltage across the inductor at a specified time after the switch has been closed

Current Rise in Inductor

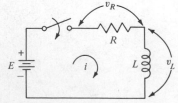

Figure 14.28   Series *RL* circuit with switch initially in open position

The switch in the circuit of Figure 14.28 is open and no current is flowing. With zero current there is no magnetic field around the inductor. When the switch is closed, the current i increases following the curve of $1 - e^{-x}$. We can express the current i mathematically as

$$i = I_{max}[1 - e^{-x}] \qquad (14.34)$$

where I_{max} = maximum current $\left(I_{max} = \dfrac{E}{R} \right)$

Figure 14.29 shows Equation (14.34). Notice that the current rise in a series *RL* circuit (Figure 14.29) is similar to the voltage rise across the capacitor in a series *RC* circuit (Figure 14.12). Similarly the current fall in a series *RL* circuit is similar to the voltage fall across the capacitor (capacitor discharge) in a series *RC* circuit (Figure 14.6).

The horizontal axis x in Figure 14.29 represents the number of time constants, and can be expressed as

$$x = t/\tau \qquad (14.35)$$

where x = number of time constants

t = actual time elapsed

τ = time constant of the circuit

Notice that this is similar to the *RC* series circuit discussed earlier. For a series *RL* circuit, the time constant

$$\tau = \frac{L}{R} \qquad (14.36)$$

where R = resistance in ohms

L = inductance in henries

τ = time constant in seconds

Recall that for a series *RC* circuit, the time constant is

$$\tau = RC \qquad (14.37)$$

At the instant the switch is closed, the current in this circuit is zero. Since the inductor opposes any change in the current through it, the current in the circuit cannot change instantaneously. At the end of one time constant ($x = 1$), the current in the inductor rises to 63.2% ($\approx 63\%$) of its final value. At the end of two time constants, the current is at 86.5% of its final value. The current through the inductor is considered to be at its maximum value at the end of five time constants.

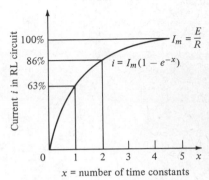

Figure 14.29 Exponential rise of current in a series *RL* circuit

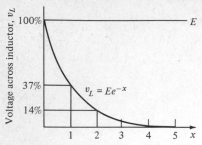

Figure 14.30 The voltage across an inductor in a series *RL* circuit.

Voltage Fall across Inductor

In Objective 1 we pointed out that the voltage v_L across an inductor [Figure 14.1(b)] follows an exponential curve e^{-x}. This is shown again in Figure 14.30.

Let us find out why the curve is decaying. In Unit 13, on inductance, you learned that the voltage v_L across an inductor can be expressed as

$$v_L = L \frac{di}{dt} \qquad (14.38)$$

where L = inductance in henries

$\dfrac{di}{dt}$ = rate of change of the current with respect to time

Since the rate of change of current i with respect to time is largest at $t = 0$ ($x = 0$), the maximum counter emf is induced in the inductor at this time. (See Figure 14.29.) Furthermore, since KVL states that

$$E = v_L + v_R \qquad (14.39)$$

and at $x = 0$, the current $i = 0$ ($v_R = 0$), then

$$v_L = E \qquad \text{at } t = 0 \qquad (14.40)$$

This is shown in Figure 14.30. As there is less change in the current in the coil, the voltage v_L developed across the coil falls to zero. This occurs at $x = 5$, where the current i has reached its maximum steady-state value.

The following example illustrates the current rise and voltage fall in an inductor.

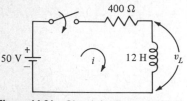

Figure 14.31 Circuit for Example 14.10

Example 14.10 Given a series *RL* circuit with $R = 400\ \Omega$ and $L = 12$ H, plot the current rise i and voltage fall v_L (Figure 14.31). Specify the current i and voltage v_L at $x = 1$, 3, and 5 ($x =$ number of time constants).

Solution

(a) The maximum steady-state current is

$$I_{max} = \frac{E}{R} = \frac{50\text{ V}}{400\ \Omega} = 125\text{ mA}$$

(b) The time constant τ of the circuit is

$$\tau = \frac{L}{R} = \frac{12\text{ H}}{400\ \Omega} = 0.03 = 30\text{ ms}$$

(c) Since the current i follows the equation

$$i = I_{max}(1 - e^{-x}) = 125 \text{ mA}(1 - e^{-x})$$

and the voltage v_L across the inductor is

$$v_L = Ee^{-x} = 50e^{-x}$$

where $x = \dfrac{t}{\tau} = \dfrac{t}{30 \text{ ms}}$

Let's tabulate some values for time 0, 30, 90, and 150 ms.

t (milliseconds)	x	$i = 125(1 - e^{-x})$	$v_L = 50e^{-x}$
0	0	0.0 mA	50 V
30	1	79.02 mA	18.39 V
90	3	118.78 mA	2.49 V
150	5	125 mA	0 V

(d) Figure 14.32 shows a plot of the current i and voltage v_L.

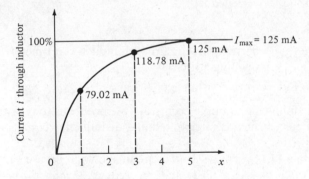

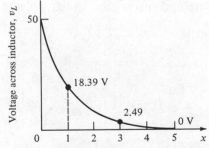

Figure 14.32   Sketch of solution d of Example 14.10

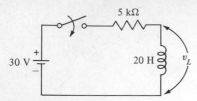

Figure 14.33 Circuit for Problem 14.11

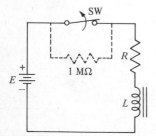

Figure 14.34 Series *RL* circuit with switch closed

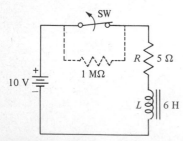

Figure 14.35 Circuit for Example 14.11

Problem 14.11 Given a series *RL* circuit with $R = 5\text{ k}\Omega$ and $L = 20$ H (Figure 14.33), calculate the current i and voltage v_L at time $t = 8$ ms, 12 ms, and 20 ms.

Voltage across Open Switch

Objective 6 In a series *RL* circuit, calculate the voltage across a switch at the instant the switch is opened.

A sketch of a series *RL* circuit with a switch is shown in Figure 14.34. Assume that the switch is closed and that a steady-state current I_{max} is flowing through the inductor. Since the ON resistance of the switch is 0 Ω, we can compute the steady-state current flowing through the inductor L as

$$I_{max} = \frac{E}{R} \tag{14.41}$$

At the instant the switch is opened, the current in the circuit must continue to flow because inductance has the property of opposing changes in current. Assuming that the resistance of the open switch is R_{SW}, then the voltage across the switch terminals when they are opened is

$$V_{switch} = I_{max} R_{SW} \tag{14.42}$$

where I_{max} = current flowing through inductor at time of opening switch

R_{SW} = open resistance of switch

The following example points out the problems to be aware of when current flow is interrupted in a dc inductive circuit.

Example 14.11 The series *RL* circuit shown in Figure 14.35 has the switch closed. Steady-state current is flowing through the 6-H inductor. The resistance of the open switch is 1 MΩ. Calculate the voltage developed across the switch when the switch is opened.

Solution
(a) While the switch is closed, the current in the circuit is given by Equation (14.41) as

$$I_{max} = \frac{E}{R} = \frac{10\text{ V}}{5\text{ }\Omega} = 2\text{ A}$$

(b) Since the resistance of the open switch is $R_{SW} = 1$ MΩ, then the voltage is

$$V_{switch} = I_{max} R_{SW} = (2\text{ A}) \times 10^6 = 2 \times 10^6\text{ V}$$

Thus this voltage is present across the switch when the current is interrupted. This voltage causes an arc when the switch is opened and could break down the insulation of the inductor.

This particular problem of arcing can be solved by placing a capacitor across the switch.

Problem 14.12   The switch in Figure 14.36 is closed and a steady-state current is flowing in the series dc inductive circuit. Assuming an open-switch resistance of 500 kΩ, what is the voltage across the switch when the terminals are opened?

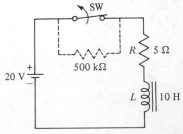

Figure 14.36   Circuit for Problem 14.12

Pulsed Input

Objective 7 (*Advanced*)   Given a series *RL* circuit with a pulsed input, calculate the current, the voltage drop across the resistor, and the voltage drop across the inductor at specified times.

In this objective you study the characteristics of a series *RL* circuit with a pulsed input. This is illustrated in the following example.

Example 14.12   A 12-V pulse with a pulse width of 10 ms is applied to a series *RL* network every 100 ms. Sketch the following waveforms: current *i*, voltage v_L, and voltage v_R.

Solution (See Figure 14.37.)
(a) The time constant of the circuit is

$$\tau = \frac{L}{R} = \frac{30 \text{ mH}}{6 \text{ }\Omega} = 5 \text{ ms}$$

(b) The maximum possible current I_{max} in this circuit is

$$I_{max} = \frac{E}{R} = \frac{12 \text{ V}}{6 \text{ }\Omega} = 2 \text{ A}$$

(c) With the 12-V applied pulse, the number of time constants x_a is

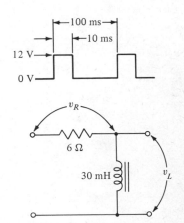

Figure 14.37   Sketch for Example 14.12

$$x_a = \frac{t_a}{\tau} = \frac{10 \text{ ms}}{5 \text{ ms}} = 2 \qquad \text{(applied pulse)}$$

With no pulse, the number of time constants x_b is

$$x_b = \frac{t_b}{\tau} = \frac{90 \text{ ms}}{5 \text{ ms}} = 18 \qquad \text{(no pulse)}$$

[*Note*: With *x* greater than 5, the steady-state condition has been reached.]

(d) The equation for current rise in a series *RL* circuit when a 12-V pulse is applied is

$$i = I_{\max}(1 - e^{-x_a})$$

where $I_{\max}$ = maximum current flowing through the circuit

x_a = number of time constants when pulse is applied

$$i = 2(1 - e^{-2}) = 2(1 - 0.1353) = 2(0.8647)$$
$$i = 1.729 \text{ A}$$

The voltage v_R across the resistor is

$$v_R = i \cdot R = (1.729 \text{ A}) \cdot (6 \text{ }\Omega)$$
$$v_R = 10.37 \text{ V}$$

while the voltage across the inductor is, by KVL,

$$E = v_L + v_R$$

or $\qquad v_L = E - v_R = 12 - 10.37 = 1.63 \text{ V}$

A sketch of the waveforms for i, v_R, and v_L is shown in Figure 14.38.

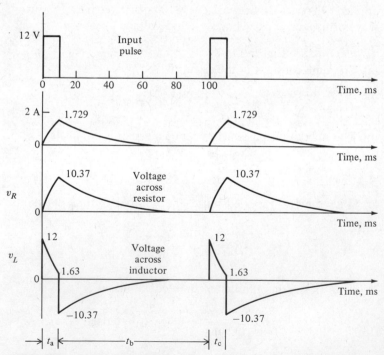

Figure 14.38 Sketch of waveforms for i, v_R, and v_L for series *RL* circuit of Example 14.12

(e) After the pulse is removed, the applied voltage is zero. The current now starts to decay to zero. The equation for current i is

$$i = I_p(e^{-x_b})$$

where I_p = current flowing at the instant the pulse is removed
(1.729 A)

x_b = number of time constants with no pulse

$\therefore \quad i = 1.729(e^{-18}) = 1.729(0) = 0 \text{ A}$

$v_R = i \cdot R = (0 \text{ A}) \cdot (6 \, \Omega) = 0 \text{ V} \quad \text{and} \quad v_L = E - v_R = 0 - 0 = 0 \text{ V}$

Referring to Figure 14.37, note that KVL must hold at all times. For instance, at the end of time t_a, the 12-V pulse is removed. Then

$$E = 0 \quad \text{and} \quad v_R = 10.37 \text{ V}$$

since

$$v_L = E - v_R$$

then

$$v_L = 0 - 10.37 = -10.37 \text{ V}$$

In summary, during time t_a,

$$E = 12 = v_R + v_L$$

and during time t_b,

$$E = 0 = v_R + v_L$$

Problem 14.13 A 100-V pulse with a pulse width of 6 ms is applied to a series RL network every 20 ms (Figure 14.39). Sketch the following waveforms: (a) current i, (b) voltage v_R, (c) voltage v_L. Label all critical points.

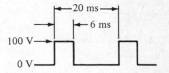

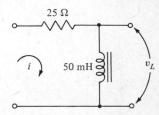

Figure 14.39 Sketch for Problem 14.13

Test

1. For the exponential curve $y = e^{-x}$, give the values for y when x equals the following:

x	y
1	
2	
3	
5	

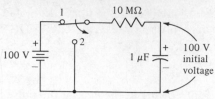

Figure 14.40   Circuit for Test Problem 2

2. The initial voltage across the capacitor in Figure 14.40 is 100 V. What is the voltage across the capacitor 9 s after the switch has been thrown to position 2?

3. There is *no* initial charge across the capacitor in Figure 14.41. What is the voltage across the capacitor 200 μs after the switch has been thrown to position 1?

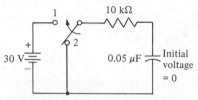

Figure 14.41   Circuit for Test Problem 3

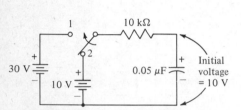

Figure 14.42   Circuit for Test Problem 4

4. In Figure 14.42, the initial voltage across the capacitor is +10 V. What is the voltage across the capacitor 200 μs after the switch has been thrown to position 1?

5. A choke with an inductance of 15 H and a resistance of 500 Ω is connected through a switch to a 20-V source of emf.
   (a) After the switch is closed, how long does it take the instantaneous current in the choke to reach a magnitude of 30 mA?
   (b) What is the instantaneous current 90 ms after the switch is closed?

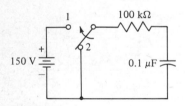

Figure 14.43   Circuit for Test Problem 6

6. The switch is now resting in position 2 in the circuit of Figure 14.43.
   (a) The switch is thrown to position 1 for 5 ms. What is the voltage across the capacitor at the end of this 5 ms?
   (b) The switch is then thrown to position 2 for 10 ms. What is the voltage across the capacitor at the end of this 10 ms?

7. In Figure 14.44, the switch is thrown to postion 1 for 0.6 s. Then it is thrown to position 2 for 20 s. What is the voltage across the capacitor at each of these times?

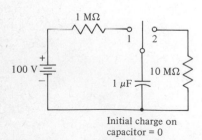

Figure 14.44   Circuit for Test Problem 7

<h1 style="text-align: center;">AC and the Unit 15
Sine Wave</h1>

In all the preceding units you have been learning about the direct current (dc) electric circuit. In this circuit the source of electromotive force or voltage is a battery, a constant-voltage power supply. Although this type of circuit is used to power flashlights, automobiles, portable radios, and electronic calculators, it is not suitable for general home or industrial use. For instance, what form of electric power is used in the light bulbs in your home? To power your electric washer?

The other form of electric power is *alternating current* (ac). In this circuit the source of electromotive force is a generator that produces *changing* voltage. This alternating voltage varies between two extremes. Some examples of dc and ac circuits are shown in Figure 15.1(a) and (b). Figure 15.1(a) shows an electrical representation of a flashlight powered by a dc source. The source of electric power E is a dry-cell battery. In Figure 15.1(b) the electric circuit is that of a washing machine. The source of electric power is the standard 120-V household ac power. This power drives the ac motor of the washing machine. In this unit you will learn more about the ac source of power.

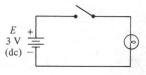

(a) Flashlight circuit

(b) Electric washer circuit

Figure 15.1 Example of dc and ac circuits

Objectives

After completing all the work associated with this unit, you should be able to:

1. Sketch the sine curve and show on the curve the values at angles of 0°, 30°, 60°, 90°, 180°, 270°, and 360° from memory.

2. Briefly describe how an ac voltage is generated.

3. Define the following terms: cycle, period, rotating vector or phasor, reference axis, phase angle, instantaneous value, peak value, and frequency.

4. Calculate the unknown quantity for a sine wave of the form $e = E_{max} \sin \phi$ when either E_{max} and ϕ or E_{max} and e are known.

5. Given the time a phasor takes to make one complete revolution (360°), calculate the frequency, and vice versa.

6. Define the terms *radian* and *angular velocity* (ω).

7. Calculate the angular velocity (ω) in radians per second for a given frequency in hertz, and vice versa.

8. Calculate the instantaneous voltage at various times for a sine wave whose equation is given in the form $e = E_{max} \sin \omega t$.

9. Define the terms *effective*, or *rms*, *value* and *average value* of a sine wave. Calculate the effective, or rms, value and the average value of a sine wave whose peak value or peak-to-peak value is given, and vice versa.

10. Calculate the average (or dc) voltage or current of a half-cycle sine wave.

Sine Wave

Objective 1   Sketch the sine curve and show on the curve the values at angles of 0°, 30°, 60°, 90°, 180°, 270°, and 360° from memory.

Alternating voltage and current (ac) are not constant or steady like direct voltage and current (dc), but vary with time. The time variations of voltage and current are usually sine or cosine functions. In this unit we limit our discussions to the sine wave. The cosine wave will be introduced in a later unit.

Figure 15.2 shows a sine wave which has a value of 0 at 0 degrees, 1 at 90 degrees, 0 at 180 degrees, −1 at 270 degrees, and 0 at 360 degrees. Note that the sine wave repeats itself every 360 degrees. Therefore you can say that 360° is identical to 0° and draw a second identical sine wave.

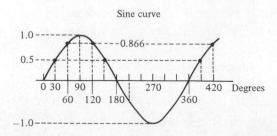

Figure 15.2   Diagram of the sine wave expressed in degrees

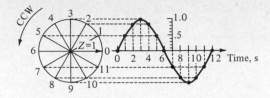

Figure 15.3 Method used to construct a sine wave

Figure 15.3 shows a simple method for constructing a sine curve. Take a pointer one unit long and rotate it around the pivot counterclockwise (CCW) through 360°, or until it is back where it started. The tip of the pointer completes a circle with a radius equal to the length of the pointer. Let's say it takes 12 s for the pointer to make one complete revolution.

You can see the times marked on the circle in Figure 15.3. Also in Figure 15.3 you can see the times marked off along the horizontal axis. If we project the various points on the circle horizontally to intersect the vertical lines drawn on the time scale, we can draw a sine curve by joining all the points.

Instead of marking the points in terms of time as we did in Figure 15.3, we could mark them in degrees. When the pointer completes a full circle, it has rotated through 360°. Therefore, instead of 0 s, 1 s, 2 s, 3 s, and so on, we mark $360°/12 = 30°$ per second. For example, 3 s corresponds to 90°, 6 s corresponds to 180°, and 9 s corresponds to 270°. This is shown in Figure 15.4.

We can now write an expression for the sine curve in degrees.

$$x = \sin \phi \qquad (15.1)$$

where ϕ = angle of the sine wave (horizontal axis)

x = instantaneous value of the sine wave at the specified angle (vertical axis)

Figure 15.4 now looks like the sine curve in Figure 15.2. From the curves shown in Figures 15.3 and 15.4, we can make a table giving the value of the sine curve at the various angles. To do this we measure the vertical distance to the point on the circle or on the sine curve at each angle. If you do this for different

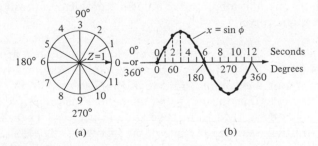

(a)   (b)

Figure 15.4 Sine curve in degrees

Table 15.1   Values of Sine Curve
Evaluated at Different Angles of ϕ

Angle ϕ (phi)	$\sin \phi$
0°	0
30°	0.5
60°	0.866
90°	1.0
120°	0.866
150°	0.5
180°	0
210°	−0.5
240°	−0.866
270°	−1.0
300°	−0.866
330°	−0.5
360° (0°)	0

angles of ϕ, you obtain the value of the sine. This is shown in Table 15.1.

We can write the values of the sine ($\sin \phi$) listed in Table 15.1 in the form

$$x = \sin \phi \qquad (15.2)$$

where x = vertical distance of the sine wave from the horizontal axis

The standard sine curve has a maximum amplitude of 1. Recall that it was generated by a pointer one unit long. However, the length of the pointer (radius of that circle) could be any value, say Z. Then the sine curve would have a maximum amplitude of Z. Then the equation for the sine curve would be

$$x = Z \sin \phi \qquad (15.3)$$

where Z = maximum amplitude of sine curve

In this section you learned how to generate a sine wave using a pointer one unit long. In the following problem you will construct a sine curve.

Problem 15.1   Take a pointer one unit long and rotate it around the pivot so that it makes a complete circle. Assume that it takes 24 s for the pointer to make a complete revolution. Plot with respect to time the curve that the pointer makes as it travels through its cycle. Mark the time scale in degrees. Then, from your drawing, estimate the sine of the following angles: sine 0°,

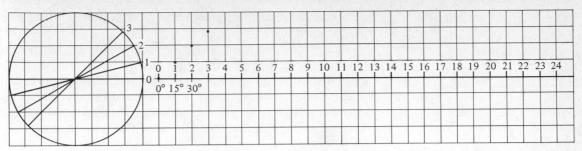

Figure 15.5 Figure for Problem 15.1

15°, 30°, 45°, 60°, 75°, 90°, 105°, 180°, 195°, 270°, 345°, 360°.
(See Figure 15.5)

We had you determine the sine of specific angles. To determine the sine of any angle, you do not have to use the method of Problem 15.1 each time. Values of the sine for angles between 0° and 90° have been tabulated. You can find them in various handbooks. We have included tables for the sine of the angle in Appendix B. Take the time to check your answers to Problem 15.1 for angles of 0°, 30°, 60°, and 90°. Note that this table only goes to 90°. What do you do for answers above 90°? This last question is discussed in more detail in Objective 4.

Generation of ac Voltage

Objective 2 Briefly describe how an ac voltage is generated.

In the first objective you learned how to generate the sine wave. Now let's see how a sine wave of voltage is actually generated. A magnetic field is set up in the air gap between the North and South poles of a magnet (see Figure 15.6). The North pole is

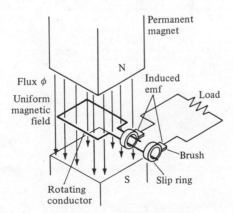

Figure 15.6 ac voltage generator

on top, and the magnetic flux lines (ϕ) go from the North pole to the South pole.

An electric conductor in the form of a loop is rotated continuously in the magnetic field. We learned in the unit on inductance that the amount of voltage induced per turn is proportional to the amount of flux cut per unit time ($d\phi/dt$). To apply this induced voltage (emf) to a fixed load, we must use slip rings. These slip rings are circular and keep the rotating conductor from twisting.

As the rotating conductor completes one complete revolution, the induced voltage across the fixed load is a sine wave. This sine wave is transferred from the circular slip rings through brushes. The actual generation of the ac voltage is shown in Figure 15.7, in which we denote the rotating conductor by a hollow and a filled circle.

When the rotating conductor is parallel to the flux lines [Figure 15.7(a) and (e)], no flux lines are being cut ($d\phi/dt = 0$); therefore no voltage is being induced into the conductor. In the positions shown in parts (c) and (g), the rotating conductor is moving perpendicular to the flux lines, and the maximum $d\phi/dt$ occurs. Notice in part (c) that the open-circle end of the conductor is moving from right to left through the field. We'll call this direction *positive*.

In Figure 15.7(g) the open-circle end is moving from left to right, so the direction is *negative*. In the positions shown in parts (b), (d), (f), and (h), the rotating conductor is cutting through

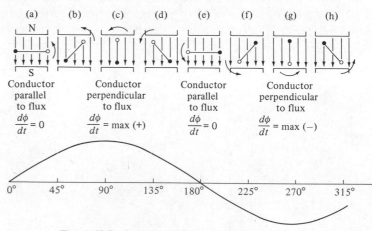

Figure 15.7   Actual ac voltage generation into the load

the flux; however, it is not doing so at the same rate of change as when the conductor is perpendicular, so the amount of induced voltage (emf) is proportionally less.

When a voltage is induced into the circuit shown in Figure 15.6, a current flows to the load. Recall the previous statement about $d\phi/dt$. The magnitude and direction of the current depend on the induced voltage. Did you notice in Figure 15.7 that the induced voltage is in the form of a sine wave? Therefore the equation for the induced voltage is

$$e = E_{max} \sin \phi \qquad (15.4)$$

where e = instantaneous value of voltage at any instant in time or angle

ϕ (Greek letter phi) = the angle we have moved from our reference or 0° point

E_{max} = maximum or peak value of sine wave

Note that Equation (15.4) is similar to Equation (15.1) and that the waveform of Figure 15.7 is similar to that of Figure 15.4. Before you go to the next objective, see if you can do Objective 2.

Definition of Sine-Wave Terms

Objective 3 Define the following terms: cycle, period, rotating vector or phasor, reference axis, phase angle, instantaneous value, peak value, and frequency.

We discussed how the rotating pointer could give us a sine wave. We said, for example, that the pointer completed one revolution or cycle in 12 s; see Figure 15.8.

If the pointer continues to rotate at this rate, it repeats the same circle five times each minute, or five times in 60 s. A good example of this rotating pointer is the second hand on your watch. However, the second hand rotates clockwise, whereas our pointer rotates counterclockwise. At this point we can define some of the terms listed in the objective.

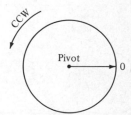

Figure 15.8 Rotating pointer

Cycle One complete rotation of the pointer in Figure 15.8. For example, one complete rotation of the loop in Figure 15.6 is defined as one cycle.

Period (*T*) The time it takes for the pointer to complete one circle (cycle). For example, if a pointer makes 10 revolutions in 100 s, the period T for 1 revolution is 10 s.

Vector or Phasor Since the pointer represents a magnitude (length) and direction (angle), it may be called a vector. The pointer rotates around a pivot point, so it is a rotating vector. A special name is given to the rotating vector: a *phasor*.

Frequency (f) The number of cycles completed in one second (cycles/second). The unit of frequency is called the *hertz* (Hz).

Using frequency, we can find the period:

$$T = \frac{1}{f} \tag{15.5}$$

where T = period in seconds (seconds/cycle)

f = cycles per second (hertz)

The definitions of reference axis, phase angle, instantaneous value and peak value can best be illustrated by a diagram, shown in Figure 15.9. Referring to Figure 15.9, the definitions are as follows.

Reference axis The axis on which the phasor is at 0 time or 0°.

Phase angle The angle the phasor makes with respect to the reference axis. For example, ϕ_2 is the phase angle given in Figure 15.9.

Instantaneous value The value of the sine wave at any instant of time. For example, in Figure 15.9, a phasor at point 3 has a negative instantaneous value.

Peak value The peak value of the sine wave is defined as E_{max}; in Figure 15.9 this occurs at an angle of 90° and 270°.

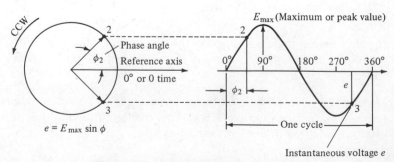

Figure 15.9 Diagram of phasor definitions

Example 15.1   The pointer (phasor) makes one complete revolution in 10 ms. What are the period and frequency of the phasor?

Solution   We are given the period from the definition:

$$T = 10 \text{ ms}$$

Therefore we find the frequency:

$$f = \frac{1}{T} = \frac{1}{10 \text{ ms/cycle}} = 0.1 \times 10^3 = 100 \text{ cycles/s} = 100 \text{ Hz}$$

Problem 15.2   The pointer (phasor) makes one complete revolution in $\frac{1}{60}$ s. What are the period and the frequency of the phasor?

Sine Wave of Form $e = E_{max}$ sin ϕ

Objective 4   Calculate the unknown quantity for a sine wave of the form $e = E_{max}$ sin ϕ when either E_{max} and ϕ or E_{max} and e are known.

To complete Objective 4, you must work with the material you learned in Objectives 1 and 3. But before you start making calculations, you have to study the sine curve in more detail. As we have discussed, the general equation for a sine wave is $e = E_{max}$ sin ϕ, or $v = V_{max}$ sin ϕ, if you prefer using v to e. ϕ is in degrees.

Let us try to evaluate the sine curve at angles of 30°, 150°, 240°, and 300°. Figure 15.10 shows the phasor diagram at these angles. The sine of 30° is denoted by the phasor in position 1, while the sine of 150° is denoted by the phasor in position 2. Likewise the sine of 240° is in position 3, and the sine of 300° is in position 4. The same points are specified on the sine curve.

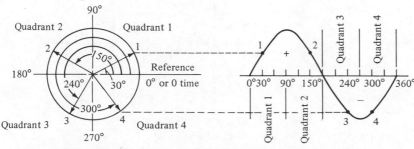

Figure 15.10   Diagram of sine wave at 30°, 150°, 240°, and 300°

To help us discuss the phasor, let's break up the phasor diagram into four quadrants. (Quadrant 1 is from 0° to 90°, Quadrant 2 is from 90° to 180°, Quadrant 3 is from 180° to 270°, and Quadrant 4 is from 270° to 360°.) Referring to the sine curve, we can see that it is positive in Quadrants 1 and 2 and negative in Quadrants 3 and 4. Now let's evaluate sin 30°. Looking at the sine curve at position 1 in Figure 15.10, or referring to the sine table in Appendix B, we find that

$$\sin 30° = +0.5$$

Notice that the value is positive because it is in Quadrant 1. Now let's look at the phasor at position 2. Position 2 is 150° from the reference angle of 0°. If we look at the sine table in Appendix B, we notice that it only goes up to 90°. What do we do? To simplify our discussion of 150°, let's redraw the phasor diagram into four quadrants and show them as in Figure 15.11.

The phasor magnitude Z is set equal to 1 to simplify the discussion. What was Z? Comparing Z of Figure 15.11 to E_{max} of Figure 15.10, we note that they are the same. For instance, if

$$e = E_{max} \sin \phi \qquad \text{and} \qquad x = Z \sin \phi \qquad (15.6)$$

then

$$\sin \phi = \frac{e}{E_{max}} \qquad \text{and} \qquad \sin \phi = \frac{x}{Z} \qquad (15.7)$$

where e or x = instantaneous value of sine curve at angle ϕ
(vertical distance from horizontal axis)

E_{max} or Z = maximum or peak value of sine wave

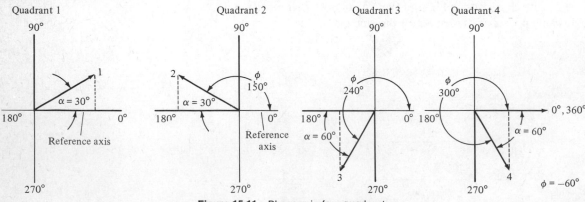

Figure 15.11   Phasors in four quadrants

Recall that phasor 2 is 150° away from the reference axis of 0°. Its vertical distance x is the same as that for phasor 1. Let's look at this more closely.

If we define the horizontal axis as the line going through 0° and 180°, what angle does phasor 2 make with the horizontal axis? The angle is

$$\alpha = 180° - \phi = 180° - 150° = 30°$$

We can also see this on the sine curve shown in Figure 15.10. Here again point 1 on the sine curve is the same positive vertical distance from the axis as point 2. Summarizing,

$$\sin \alpha = \sin \phi \quad \text{or} \quad \sin 30° = \sin 150°$$

means that the value of $\sin 30°$ has the same value as $\sin 150°$.

Now let's look at phasor 3. Phasor 3 is 240° from the reference axis of 0°, and is therefore in Quadrant 3. The sign of x in Quadrant 3 is negative, but what is its value? Let's look at the angle phasor 3 makes with the horizontal axis. The angle is

$$\alpha = \phi - 180° = 240° - 180° = 60° \qquad \therefore \sin 60° = -\sin 240°$$

That is, the value of $\sin 240°$ is equal to the negative value of $\sin 60°$.

Phasor 4 is 300° from the reference axis of 0°. It is in Quadrant 4, negative, and the angle phasor 4 makes with the horizontal axis is

$$\alpha = 360° - \phi = 360° - 300° = 60° \qquad \therefore \sin 60° = -\sin 300°$$

That is, the value of $\sin 300°$ is equal to the negative value of $\sin 60°$.

It should be noted here that we can write phasor 4 in another form. We have been looking at the phasor rotating in the counterclockwise direction. In this direction the phasor is $+300°$ from the reference axis of 0°. What if we looked at it in the clockwise direction? What would be the angle? Rotating the phasor clockwise from 0°, we would move 60° in the negative direction. Therefore, in the clockwise direction, phasor 4 moves $-60°$ from 0°.

Let us now apply what we have learned to a specific example. Here we evaluate the value of the sine at 30°, 150°, 240°, and 300°.

Example 15.2 For the sine wave $e = E_{max} \sin \phi$, where $E_{max} = 10$ V (Figure 15.12), find the instantaneous voltage for the following angles:

(a) $\phi = 30°$ (b) $\phi = 150°$ (c) $\phi = 240°$ (d) $\phi = 300°$

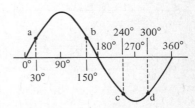

Figure 15.12 Sine wave for Example 15.2

Solution

(a) $e = E_{max} \sin \phi = 10 \sin 30° = 10(0.5) = 5$ V

(b) $e = E_{max} \sin \phi$ (90° to 180°)   since e is in quadrant 2

$e = 10 \sin 150°$ $\sin \phi = \sin \alpha$

$= 10(\sin 30°)$ $\sin \phi = \sin (180 - \phi)$

$= \sin (180 - 150°)$

$= 10(+0.5)$ $\sin \phi = \sin 30°$

$= 5$ V

(c) $e = E_{max} \sin \phi$ (180° to 270°)   since e is in quadrant 3

$= 10 \sin 240° = 10(-\sin 60°)$ $\sin \phi = -\sin(240 - 180°)$

$= 10(-0.866)$ $= -\sin 60°$

$= -8.66$ V   [*Note*: Sin ϕ is negative
                                                   in quadrants 3 and 4]

(d) $e = E_{max} \sin \phi$ (270° to 360°)   since e is in quadrant 4

$= 10 \sin 300°$ $\sin \phi = -\sin (360° - 300°)$

$= 10(-\sin 60°)$ $= -\sin 60°$

$= 10(-0.866)$

$= -8.66$ V

Problem 15.3   For the sine wave $e = E_{max} \sin \phi$, where $E_{max} =$ 10 V, find the instantaneous voltage for the following angles: (a) $\phi = 65°$ (b) $\phi = 170°$ (c) $\phi = 215°$ (d) $\phi = 306°$ Also locate the angles on a sine curve as in Example 15.2.

Example 15.3   For the sine wave $e = E_{max} \sin \phi$, where $E_{max} =$ 20 V, find the angles (between 0° and 360°) for the following instantaneous voltages: (a) 2 V   (b) -18 V

Solution   As you see from the sketch of the sine wave (Figure 15.13), there are two angles between 0° and 360° for each of the instantaneous voltages.

$$e = E_{max} \sin \phi \qquad \sin \phi = \frac{e}{E_{max}}$$

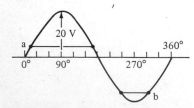

Figure 15.13   Sine wave for Example 15.3

The angle ϕ is the angle whose sine is e/E_{max}, which is written in short notation as:

$$\phi = \arcsin \left(\frac{e}{E_{max}}\right) \qquad \text{or} \qquad \phi = \sin^{-1} \left(\frac{e}{E_{max}}\right)$$

(a) For 2 V,

$$\sin \phi = \frac{2}{20}$$

$$\therefore \quad \phi = \arcsin \frac{2}{20} = \sin^{-1} \frac{2}{20} = \sin^{-1}(0.1)$$

$$= 5.74° \text{ for the smaller angle}$$

For (90° to 180°), using $\alpha = 5.74°$,

$$\phi = 180° - \alpha = 180° - 5.74° = 174.26° \qquad \text{for the larger angle}$$

(b) For −18 V,

$$\alpha = \sin^{-1}\left(\frac{18}{20}\right) = \sin^{-1}(0.9) = 64.2°$$

For (180° to 270°),

$$\phi = \alpha + 180° = 64.2 + 180 = 244.2°$$

For (270° to 360°),

$$\phi = 360° - \alpha = 360 - 64.2 = 295.8°$$

Problem 15.4   For the sine wave $e = E_{max} \sin \phi$, where $E_{max} = 20$ V, find the angles for the following instantaneous voltages:
(a) 10 V   (b) −15 V   (c) −17.32 V
Also sketch the sine wave and locate the instantaneous voltages and angles on it.

Sine Wave of Form $e = E_{max} \sin (360\,ft)°$

Objective 5   Given the time a phasor takes to make one complete revolution (360°), calculate the frequency, and vice versa.

Sometimes we want to solve for the instantaneous voltage, but we know only the frequency and the time elapsed. How can we turn these into an angle in degrees? Let us look at what we know so far. The instantaneous voltage e was written in the form

$$e = E_{max} \sin \phi \qquad (15.8)$$

where E_{max} = maximum or peak value of the sine wave
ϕ = angle of the sine wave

Note that the angle ϕ has a value between $0°$ and $360°$. In other words,

$$0° \leq \phi \leq 360°$$

Now can we write this angle ϕ in another form incorporating frequency and time? Recall that one complete cycle is equal to $360°$. The period T is the time it takes to complete one cycle. Using this, we can say:

$$\phi \text{ (degrees)} = 360° \cdot \left(\frac{t}{T} \right) \tag{15.9}$$

where t = time elapsed (seconds)

T = period of one cycle (seconds)

$\left(\dfrac{t}{T} \right)$ = fraction of one full cycle completed

Therefore, combining Equations (15.8) and (15.9), we have

$$e = E_{max} \sin (\phi)° = E_{max} \sin \left(360 \cdot \frac{t}{T} \right)° \tag{15.10}$$

Note that ϕ is in degrees and that $(360 \cdot t/T)$ is in degrees. Moreover, from Equation (15.5), we can solve for f:

$$f = \frac{1}{T} \tag{15.11}$$

where f = frequency in cycles per second

T = period in seconds

Substituting Equation (15.11) into (15.10), we then have

$$\boxed{e = E_{max} \sin \phi = E_{max} \sin (360 \cdot t/T)° = E_{max} \sin (360 \cdot f \cdot t)°}$$

$$\tag{15.12}$$

Therefore now, if we know the frequency and the time elapsed, we can convert this information into an angle in degrees. We do this by using Equation (15.12).

Example 15.4   A 60-Hz sine wave has a peak or maximum value of 100 V. (a) What is the instantaneous voltage 15 ms after the instantaneous voltage passes through 0 V? (b) What is the period of this sine wave?

Solution

(a) $e = E_{max} \sin (360 \cdot f \cdot t)°$

$= 100 \sin [360 \cdot (60)(15 \times 10^{-3})]°$

$= 100 \sin 324°$

From Figure 15.11, $\phi = 324°, \alpha = 360° - \phi = 36°$

$\sin \phi = -\sin \alpha = -\sin 36°$

$e = 100(-\sin 36°)$

$= -100(0.5878) = -58.78$ V

(b) $T = \dfrac{1}{f} = \dfrac{1}{60} = 16.67$ ms

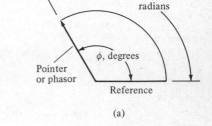

(a)

Problem 15.5   A 60-Hz sine wave has a peak or maximum value of 50 V. What is the instantaneous voltage 9 ms after the instantaneous voltage passes through 0 V?

Sine Wave of Form $e = E_{max} \sin \omega t$

Objective 6   Define the terms *radian* and *angular velocity* (ω).

Definition of Radian and Angular Velocity ω

In rotating machinery like generators or motors, it is sometimes extremely difficult to specify a specific angle. So instead of using the angle of rotation in degrees as a measurement, it is more common to use angular distance and angular velocity. The angular distance, which is the distance the conductor or free end of the rotating vector has moved from the reference point, is measured in *radians*. (See Figure 15.14.)

Let us define a radian. We know that the circumference C of a circle is:

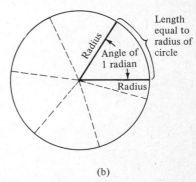

(b)

Figure 15.14   Definition of radian

$$C = \pi d = 2\pi r$$

where $\pi = 3.1416$

d = diameter

r = radius of circle ($d = 2r$)

If we lay off the radius of the circle along the circumference, the angle formed is defined as one radian; see Figure 15.14(b). Now let's find out how many radii we can place around the circumference of a circle. Since circumference

$$C = (2\pi)r = 2(3.14)r = 6.28r$$

we have 6.28 radii to complete one circle, or 2π radii. Since a radian is the angle formed by one radius of the circle laid off along the circumference, we have 2π radians in one complete revolution. Therefore

$$2\pi \text{ radians} = \text{one revolution (one cycle)}$$

but

$$1 \text{ cycle} = 1 \text{ circle} = 1 \text{ revolution} = 360°$$

$$\therefore \quad \boxed{2\pi \text{ radians} = 360°} \quad \boxed{\pi \text{ radians} = 180°} \quad (15.13)$$

and

$$1 \text{ radian} = \frac{360°}{2\pi} = \frac{360°}{6.28} = 57.3°$$

Now we can mark off our sine wave in degrees and radians; see Figure 15.15.

Above the axis we have marked off the sine wave in degrees. One complete cycle (revolution of phasor) is 360°. Below the axis we have marked off the sine wave in radians. Note that 360° equals 2π radians (6.28 radians).

Referring back to Figure 15.14(a), instead of expressing angular distance in degrees, we can now express it in radians. Now to find velocity, we divide distance by time.

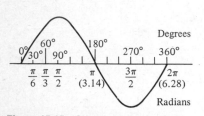

Figure 15.15 Sine wave marked off in radians

Angular velocity ω (lower-case Greek letter omega), which is angular distance (radians) divided by time (seconds), can be expressed in radians per second. Our equation for instantaneous voltage using frequency f and actual time t was Equation (15.12),

$$e = E_{max} \sin (360 \cdot f \cdot t)°$$

This equation gave us an angle measured in degrees. The $360f$ part gives us the degrees per second:

$$\left(\frac{360°}{\text{cycle}}\right)\left(f \frac{\text{cycle}}{\text{s}}\right)$$

Since there are 2π radians in 360°, Equation (15.12) in radian measurement is

$$e = E_{max} \sin (360ft)° \qquad \text{degree measurement}$$
$$e = E_{max} \sin (2\pi ft) \qquad \text{radian measurement}$$

Now let's define

$$\omega = 2\pi f$$

Then

$$e = E_{max} \sin \omega t \qquad\qquad (15.14)$$

where ω is radians/second

Before you go to the next objective, see if you can define a radian and angular velocity ω.

Application

Objective 7 Calculate the angular velocity ω in radians per second for a given frequency in hertz, and vice versa.

In this objective we try to apply the definitions of radians and angular velocity ω that we learned in Objective 6.

Example 15.5 Write the equation of a 60-Hz sine wave using radian measurement.

Solution

$$e = E_{max} \sin \omega t = E_{max} \sin 2\pi f t$$
$$= E_{max} \sin 2\pi(60)t = E_{max} \sin 377t$$

Example 15.6 Find the frequency and period of the following sine waves: (a) $v = V_{max} \sin 2000t$ (b) $e = 10 \sin 3140t$.

Solution   When an equation is written in this form, it is understood (unless you are told otherwise) that radian measurement is used. (See Figure 15.16.)

(a) General form $v = V_{max} \sin \omega t$

$$\therefore \quad \omega = 2\pi f = 2000$$

$$f = \frac{2000}{2\pi} = 318 \text{ cycles/s} = 318 \text{ Hz}$$

$$T = \frac{1}{f} = \frac{1}{318} = 3.14 \text{ ms}$$

(b) $\omega = 2\pi f = 3140$

$$f = \frac{3140}{2\pi} = 500 \text{ Hz}$$

$$T = \frac{1}{500 \text{ cycles/s}} = \frac{1000 \text{ ms}}{500 \text{ cycles}}$$

$$T = 2 \text{ ms}$$

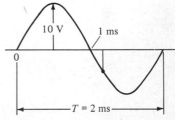

Figure 15.16   Sine wave for Example 15.6

Problem 15.6   Find the frequency and period of the following sine waves:

(a) $v = V_{max} \sin 100t$

(b) $e = 100 \sin 6280t$

(c) $e = 5 \sin (10^6\pi)t$

(d) $i = (100 \text{ mA}) \sin 377t$

Objective 8   Calculate the instantaneous voltage at various times for a sine wave whose equation is given in the form $e = E_{max} \sin \omega t$.

Objective 8 is very similar to Objective 4. In Objective 4, the sine wave had the form

$$e = E_{max} \sin \phi$$

where E_{max} = maximum or peak value of the sine wave

ϕ = angle in degrees

and you learned to evaluate the sine of an angle in degrees by looking up its value in the sine table in Appendix B. In Objective 8 we have the sine wave expressed in radian measurement. Here we must first change the radian measurement to degrees by recalling Equation (15.13),

$$2\pi \text{ rad} = 360°$$

Using proportions, we can say

$$\frac{\text{Angle (degrees)}}{360°} = \frac{\text{number of radians}}{2\pi \text{ rad}}$$

or, rewriting, we have

$$\boxed{\text{Angle (degrees)} = (\text{number of radians}) \times \frac{360°}{2\pi \text{ rad}}} \quad (15.15)$$

Now let's apply Equation (15.15) to Example 15.7.

Example 15.7   For the sine wave $e = (10 \text{ V}) \sin 3140t$, find the instantaneous value at the following times:

(a) 1 ms   (b) 1.25 ms   (c) 7.5 ms

Solution

(a) $e = 10 \sin 3140t$ Radian measurement is used

General equation: $e = E_{max} \sin \omega t$

$= E_{max} \sin 2\pi ft$

$$\text{Angular distance (in radians)} = \omega t = 3140 \frac{\text{rad}}{\text{s}} \times 1 \text{ ms}$$

$$= 3.14 \text{ rad}$$

To be able to use sine tables, we must change radians to degrees. Referring to Equation (15.15),

$$\text{Angle (degrees)} = (\text{number of radians}) \times \frac{360°}{2\pi \text{ rad}}$$

$$\phi = 3.14 \text{ rad} \times \frac{360°}{2\pi \text{ rad}} = 180°$$

$$\therefore \quad e = 10 \sin 180° = 10 \sin 0° = 10(0) = 0 \text{ V}$$

(b) 1.25 ms

$$\text{Angular distance (radians)} = 3140 \times 1.25 \times 10^{-3} = 3.925 \text{ rad}$$

$$\text{Angle (degrees)} = 3.925 \text{ rad} \times \frac{360°}{2\pi \text{ rad}} = 225°$$

$$\alpha = \phi - 180° = 225° - 180° = 45°$$

$$\therefore \quad e = 10 \sin 225° = 10(-\sin 45°)$$

$$= -10(0.707) = -7.07 \text{ V}$$

(c) 7.5 ms

$$\text{Angular distance (radians)} = \omega t = 3140 \times 7.5 \times 10^{-3}$$

$$= 23.55 \text{ rad}$$

$$\text{Angle (degrees)} = 23.55 \text{ rad} \times \frac{360°}{2\pi \text{ rad}} = 1350°$$

This means that we have gone through several cycles or revolutions. (We go through 360° on each revolution.)

$$\frac{1350°}{360°/\text{cycle}} = 3.75 \text{ cycles}$$

Or

$$360° + 360° + 360° + 0.75(360°) = 1350°$$

So we have gone 0.75 (360°) = 270° of the fourth cycle; therefore

$$e = 10 \sin 270° = 10(-\sin 90°) = 10(-1) = -10 \text{ V}$$

Problem 15.7 For the sine wave $e = (10 \text{ V}) \sin 6280t$, find the instantaneous value at the following times:
(a) 0.1 ms   (b) 0.75 ms   (c) 1.5 ms   (d) 3.3 ms

Effective and Average Value of Sine Wave

Objective 9 Define the terms *effective*, or *rms*, *value* and *average value* of a sine wave. Calculate the effective, or rms, value and the average value of a sine wave whose peak value or peak-to-peak value is given, and vice versa.

Since the terms effective, or rms, and average value of a sine wave are often used in describing ac circuits, let us define these two terms now.

Definition of Effective and Average Value

Average (*one cycle*)   The average value of the sine wave shown in Figure 15.17(b) is zero. When you examine this figure, you notice that over one complete cycle there is just as much negative as positive current. If a dc ammeter were connected to measure this current, it would read zero because a dc instrument reads average values.

Effective value (*rms*)   How do we compare an ac ampere to a dc ampere? Just by looking at Figure 15.17(a), we can see that 1.0 ($I_{max} = I_p = 1.0$) ac ampere would not be the same as 1.0 dc ampere. A dc ampere is a constant, steady, and continuous current. For the ac ampere of Figure 15.17(a), there are only two times during the complete cycle that 1.0 A actually flows. At all other times, the current is less than 1.0 A. At first, to come up with a comparison, we might think of averaging the ac current over one complete cycle. However, we know that the average of a sine wave over one complete cycle is zero! What we do is compare the dc and ac current of Figure 15.17(a) on the basis of heating effect.

The effective ac ampere is defined as that current that produces the same heat per time (power P) in a pure resistance as a dc ampere. Power is equal to I^2R, where I means dc or effective

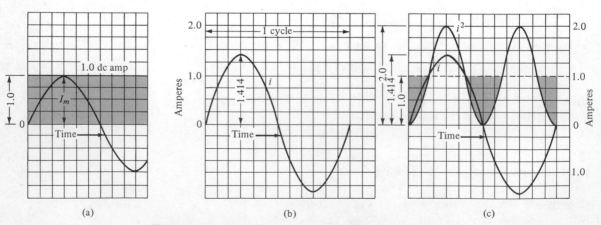

Figure 15.17   Determination of effective value of a sine wave

ac ampere. So that the mathematics is easy, let us assume that the resistor is equal to 1 Ω, so that $P = I^2R = I^2 (1 \, \Omega)$; power in this case is equal to I^2.

In Figure 15.17(b) you see an ac current of $i = 1.414 \sin \omega t$ plotted against time. The peak or maximum current is 1.414 A. Since power is equal to I^2R, let us square the instantaneous i curve; this is plotted in Figure 15.17(c). The maximum value of this i^2 curve is 2.0 A $[1.414^2 = (\sqrt{2})^2 = 2.0]$.

Examine the i^2 curve in Figure 15.17(c). It lies entirely above the zero axis because the squares of negative values are positive. The frequency of the i^2 curve is twice that of the original wave. Notice that there is a horizontal axis of symmetry at 1.0 A. Therefore the average value of the i^2 wave is 1.0 A, as shown by the dashed line. The areas above the dashed line just fit into the shaded valley below it. In general, since the average heating value varies as the average square of the peak current $I_p^2/2$, we can find the average of the i^2 curve. Therefore the average of i^2 is

$$I^2 = \frac{I_p^2}{2} \qquad (15.16)$$

or

$$I = \sqrt{\frac{I_p^2}{2}} = \frac{I_p}{\sqrt{2}} = \frac{I_p}{1.414} = 0.707 I_p \qquad (15.17)$$

Since we are taking the square root of the average or mean value of the square of the instantaneous current curve, the effective value is also known as the *root-mean-square* or *rms* value. That is,

$$\boxed{I = I_{\text{rms}} = I_{\text{eff}} = 0.707 I_p} \qquad (15.18)$$

or

$$\boxed{I_p = \frac{I}{0.707} = 1.414 I} \qquad (15.19)$$

where I = effective or equivalent dc value of current

I_p = peak or maximum value of the sine wave

Similar relationships can be expressed for voltage V.

$$V_p = 1.414 V \qquad (15.20)$$

or

$$V = \frac{V_p}{1.414} = 0.707 V_p \qquad (15.21)$$

where V = effective (rms) or equivalent dc value of voltage

V_p = peak voltage of the sine wave

For peak-to-peak measurements such as those made with an oscilloscope,

$$V_{p-p} = 2 \cdot V_p \qquad (15.22)$$

where V_{p-p} = peak-to-peak voltage

V_p = peak voltage

Therefore, substituting Equation (15.22) into (15.21),

$$V = V_{rms} = \frac{V_p}{1.414} = \frac{V_{p-p}}{2}\left(\frac{1}{1.414}\right) \qquad (15.23)$$

$$\therefore \quad V_{rms} = \frac{V_{p-p}}{2.828} \qquad (15.24)$$

Problem 15.8 Given the current curve in Figure 15.18, with maximum value of 2 A. Square the i curve. Find the average of the squared curve. Average of i^2 curve = _____. Since this average is the effective value *squared*, find the effective value

$I =$ _____. Compare this calculation to

$$I = \frac{I_p}{\sqrt{2}} = \frac{2\,A}{\sqrt{2}} = \sqrt{2}\,A = 1.414\,A$$

Having learned the definitions of the terms effective, or rms, value and average value of a sine wave, let us apply them to a few problems. Keep in mind that for ac voltages and currents, if the subscripts p–p or p do not follow the value given, then it is understood that the value is the *effective* or *rms* value.

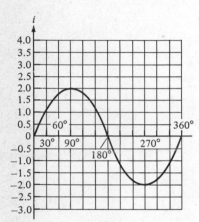

Figure 15.18 Sketch for Problem 15.8

Calculations Over One Cycle

Example 15.8 A wall socket has 120 V ac. What are the peak voltage and the peak-to-peak voltage?

Solution   A 120-V ac from the wall socket means 120 V rms. The 120 V rms can be converted to peak voltage V_p by using Equation (15.20).

$$V_p = 1.414(120) = (\sqrt{2})(120) = 169.68 \approx 170\text{ V}$$

The peak-to-peak voltage V_{p-p} is

$$V_{p-p} = 2V_p = 2(170) \approx 340\text{ V}_{p-p}$$

or

$$V_{p-p} = (2\sqrt{2})\,V = 2.828(120) = 340\text{ V}_{p-p}$$

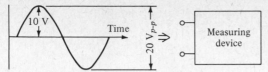

Figure 15.19 Waveform for Example 15.9

Example 15.9 Suppose that we have a 60-Hz voltage of the form

$$e = 10 \sin 2\pi ft = 10 \sin 377t \qquad \text{(radian measurement)}$$

Let us see what the measuring device would read if it were a(n)
(a) scope   (b) dc voltmeter   (c) ac voltmeter

Solution
(a) A scope would read a peak-to-peak value. Since the maximum voltage of the sine wave is 10, the scope would read $20V_{p-p}$.
(b) The dc voltmeter reads the average. The average value of a sine wave over a complete cycle is 0 V.
(c) The ac voltmeter reads effective (rms) voltage. Since

$$V_{\text{eff}} = 0.707V_{\text{peak}} = (0.707)(10)$$
$$V_{\text{eff}} = 7.07 \text{ V}$$

Example 15.10 The equation of an induced voltage is $e = 50 \sin 377t$. If the load is a 100-Ω resistor, what current will flow through it?

Solution When we have a question as to *what* current, it is understood that we are talking about effective current. Therefore, to find effective current, we must know the effective (rms) voltage. The value of voltage (50) given by the equation is the peak or maximum value of the sine wave.

$$E = 0.707E_{\text{max}} \text{ or } E = \frac{E_{\text{max}}}{1.414}$$

$$E = 0.707(50) = 35.35 \text{ V}$$

$$\therefore \quad I = \frac{E}{R} = \frac{35.35 \text{ V}}{100 \text{ }\Omega} = 0.3535 \text{ A} \qquad \text{or} \qquad 353.3 \text{ mA}$$

Remember that the capital letters I, E, and V mean dc or effective ac quantities.

Example 15.11 A 128-Ω soldering iron has a power rating of 100 W (average power). What is the peak voltage across the iron?

Solution The power equations for ac current

$$P = I^2R, \qquad P = \frac{E^2}{R}, \qquad \text{and} \qquad P = EI$$

give average or effective power when effective, or rms, values are used for E or I.

If we solve for E by using $P = E^2/R$, then E is an effective value, $E = \sqrt{PR}$.

$$E = \sqrt{PR} = \sqrt{100(128)} = 113.14 \text{ V}$$

[*Note*: We don't have to write rms after the answer because it is understood.]

$$E_{max} = (\sqrt{2})E = (1.414)(113.14) = 159.98 \approx 160 \text{ V}_p$$

Example 15.12   If an oscilloscope is used to view a 50-V, 60-Hz sine wave, what is the smallest vertical scale that may be used?

Solution   To view a sine wave on an oscilloscope, we need to see the whole picture, that is, from the top peak to the bottom peak. In other words, peak-to-peak (p–p) reading.

The 50 V is an effective rms value; therefore

$$V_{p-p} = 2(\sqrt{2}) \text{ V} = 2.828(50) = 141.4 \text{ V}_{p-p}$$

Therefore we must be able to see at least 141.4 V_{p-p} on the vertical scale.

Problem 15.9   An oscilloscope is used to view a 115-V, 60-Hz sine wave. What is the smallest vertical scale that may be used?

Example 15.13   An iron has a rating of 120 V, 1100 W. What is the peak current drawn by the iron?

Solution

$$P = VI$$

$$I = \frac{P}{V} = \frac{1100 \text{ W}}{120 \text{ V}} = 9.167 \text{ A}$$

$$I_{max} = \sqrt{2}I = 1.414(9.167) = 12.96 \text{ A}_p$$

Problem 15.10   A toaster has a rating of 120 V, 1375 W. What is the peak current drawn by the toaster?

Problem 15.11   An alternating voltage is described by the equation $e = 1414 \sin 3140t$

(a) What is the maximum instantaneous voltage value? _____

_____

(b) What is the effective (rms) value? _____

(c) What is the peak-to-peak value? _____

(d) What is the frequency in hertz (cycles per second)? _____

_____

(e) What is the period? _____

(f) What is the instantaneous emf when $t = 2$ ms? _____

_____

Calculations Over a Half-Cycle

Objective 10 Calculate the average (or dc) voltage or current of a half-cycle sine wave.

In Objective 9, you learned how to calculate the effective or rms value and the average value of a sine wave over one complete cycle. In this objective, you will learn how to calculate the average value of a sine wave over a half-cycle. Recall that the average value of any current or voltage is the value that a dc meter displays when reading the sine wave. This average (or dc) value is very important in later courses, where you will be studying power supplies and the conversion of ac signals into dc signals.

Figure 15.20 shows the type of waveforms that appear at the output of the power supplies. Figure 15.20(a) shows a sine wave

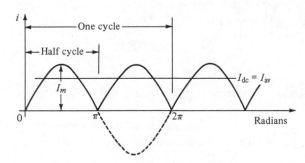

(a) Rectified sine wave (full wave)

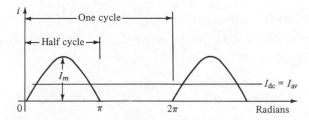

(b) Rectified sine wave (half wave)

Figure 15.20 Waveforms appearing at output of power supply

that has been rectified (the negative portion of the sine wave in the interval π to 2π has been brought over to the positive side). Figure 15.20(b) shows another type of waveform. Here the negative portion of the sine wave has been clipped off (removed), leaving only the positive portion in the time interval 0 to π rad. Let us first calculate the average (dc) value of the waveform shown in Figure 15.20(b) over a half-cycle of 0 to π radians.

The average value of any waveform can be calculated by the following relationship:

$$\text{Average value} = \frac{\text{area under the waveform}}{\text{unit length}} \qquad (15.25)$$

We can use Equation (15.25) to calculate the average value of the waveform shown in Figure 15.20(b). Using calculus, we find the area under the curve to be equal to

$$\text{Area of half-cycle} = 2I_{max} \qquad (15.26)$$

where I_{max} = maximum value of sine wave

Since the unit length of a half-cycle is π rad, the average current I_{av} for a half-cycle is

$$I_{av} = \frac{\text{area}}{\text{length}} = \frac{2I_{max}}{\pi} = 0.637I_{max} \qquad (15.27)$$

This average (or dc) current for a half-cycle is sometimes referred to as I_{HWA} (*half-wave average*). Thus

$$I_{dc} = I_{av} = I_{HWA} = \frac{2I_{max}}{\pi} = 0.637I_{max} \qquad (15.28)$$

Let us now see if we can apply Equation (15.28) to a few examples.

Example 15.14 What would a dc meter indicate if the maximum amplitude of the waveform shown in Figure 15.20(a) is 100 V?

Solution Since the dc value (or average) is defined by Equation (15.25) to be the total area divided by length, then

Area of one half-cycle $= 2V_{max} = 2(100) = 200$
Total area of two half-cycles $= 2(2V_{max}) = 2(2)(100) = 400$
Total length of cycle $= 2\pi$

$$\therefore \quad V_{av} = V_{dc} = \frac{2(2)V_{max}}{2\pi} = \frac{2}{\pi}V_{max} = 0.637V_{max}$$

$$V_{av} = V_{dc} = 0.637(100 \text{ V}) = 63.7 \text{ V}$$

Example 15.15 Calculate the average voltage V_{av} of the waveform shown in Figure 15.21 over one cycle.

Solution

Area of one half-cycle $= 2V_{max} = 2(100) = 200$

Total area of two half-cycles $= 200 + 0 = 200$

Total length of one cycle $= 2\pi$

$$V_{av} = V_{dc} = \frac{2V_{max}}{2\pi} = \frac{200}{2\pi} = \frac{100}{\pi} = 31.8 \text{ V}$$

or, in general,

$$V_{av} = V_{dc} = \frac{2V_{max}}{2\pi} = \frac{V_{max}}{\pi} = 0.318 V_{max}$$

Note that the average value of the waveform of Example 15.15 is half the value of the waveform of Example 15.14 (0.637 versus 0.318).

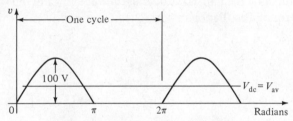

Figure 15.21 Waveform for Example 15.15

Problem 15.12 Calculate the half-wave average current I_{HWA} if the peak amplitude I_{max} of the waveform of Figure 15.20(a) is 40 A.

Test

1. An alternating voltage is described by the equation $e = 50 \sin 628t$.
   (a) Sketch the sine wave. Indicate maximum value of sine.
   (b) What is the maximum instantaneous voltage value?

   _____

   (c) What is the effective, or rms, value? _____
   (d) What is the peak-to-peak value? _____
   (e) What is the frequency in hertz (cycles/s)? _____

(f) What is the period? _____

(g) What is the instantaneous emf, when $t = 5.833$ ms?

_____

Find radians, degrees, and then emf.

2. A generator produces a voltage $V = (50 \text{ V}) \sin \phi$. Find the angles (between $0°$ and $360°$) for an instantaneous voltage of 30 V. Also sketch the sine wave and locate the instantaneous voltage and angles on it.

3. A 200-Hz sine wave has a peak (maximum) value of 50 V. What is the instantaneous voltage 13 ms after the instantaneous voltage passes through 0 V?

4. Suppose that we have a voltage wave of the form

$$e = 200 \sin 2512t$$

What would the following measuring devices read?
(a) scope   (b) dc voltmeter   (c) ac voltmeter

5. An air conditioner has a rating of 220 V and a power rating of 5 kW. What is the peak current drawn by the unit?

6. Calculate the half-wave average voltage V_{HWA} of the waveform of Test Problem 4.

Reactance Unit 16

In studying dc circuits, you learned to apply Ohm's law to resistors. In this unit you will learn to apply Ohm's law to ac circuits employing resistors, capacitors, and inductors.

Objectives

After completing all the work associated with this unit, you should be able to:

1. Recall what is meant by a phase relationship between two sine waves. Draw from memory two sine waves that are out of phase.
2. Recall what is meant by rate of change (derivative), and on a sine curve point out the following: maximum positive rate of change, maximum negative rate of change, and zero rate of change. Also be able to specify the value of the slope at the maximum positive rate of change.
3. Draw from memory the sine waves showing the phase relationship between voltage across and current in a resistor, capacitor, and inductor. Use a sine wave of current or voltage as reference, and draw the other sine wave.
4. Calculate the capacitive reactance (X_C) or inductive reactance (X_L) when the voltage across and the current through the reactive component are known.

319

5. Recall the operators that are attached to the inductive and capacitive reactive components.

6. Calculate the capacitive reactance (X_C) or inductive reactance (X_L) when the frequency and capacitance (C) or inductance (L) are known.

Phase Relationships of Two Sine Waves

Objective 1   Recall what is meant by a phase relationship between two sine waves. Draw from memory two sine waves that are out of phase.

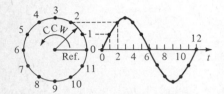

Figure 16.1   Rotation of phasor at zero time

So far we have only discussed the rotation of one pointer (phasor). Also we always started at zero time in the horizontal position, so that we generated a sine curve (Figure 16.1). If we start our zero time at point 3 on the circle in Figure 16.1, we generate a cosine curve. A cosine curve is just a sine curve shifted to the left by 90°. Later in this section we prove that

$$\cos \phi = \sin (\phi + 90°)$$

Let's see what kind of curve we would generate if we rotated two pointers (phasors) together. Assume that the hour and minute hand are set for 15 minutes after 3 o'clock.

When the two hands are locked and are moved together counterclockwise (CCW), both curves go through zero and the maximum values at the same time. These curves are said to be *in phase*. Let us denote the length of the short phasor by B_m and the length of the long phasor by A_m. Even though both curves are in phase, the amplitudes may be different. This is shown in the sine curves of Figure 16.2.

Let's set our clock hands at 3 o'clock. This is shown in Figure 16.3. The long pointer of an amplitude A_m at zero time is at point 3, and its waveform is shown dashed. The short pointer of amplitude B_m at zero time is at point 0. Note that the phasors A_m and B_m are 90° apart. If we lock the phasors in this position

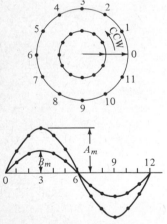

Figure 16.2   Two phasors rotating together

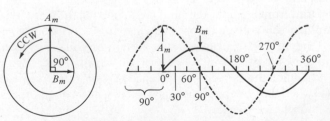

Figure 16.3   Two phasors out of phase by 90°

(90° apart) and rotate them counterclockwise (CCW), then the long phasor A_m leads the short phasor B_m by 90°. Referring to the sine waves generated at 0°, we have the long phasor A_m at the peak of the dashed sine wave, while the solid sine wave at the same time is at zero. When the dashed sine wave gets to zero at 90°, the solid sine wave has reached its maximum value of B_m. Therefore we can see that the dotted sine wave A_m is leading the solid sine wave B_m by 90°. We can write the expression for the long phasor as

$$A_m \sin(\omega t + 90°) \qquad \text{(long phasor)} \qquad (16.1)$$

and for the short phasor as

$$B_m \sin(\omega t) \qquad \text{(short phasor)} \qquad (16.2)$$

where A_m = peak value of sine wave

B_m = peak value of sine wave

Note that at the same instant of time t the sine wave in Equation (16.1) is 90° further than the sine wave in Equation (16.2). Let us examine both equations in more detail. For example, let $t = 0$. Then Equation (16.1) is

$$A_m \sin[\omega(0) + 90°] = A_m \sin 90°$$

Referring to the sine tables in Appendix B, we get

$$\sin 90° = 1 \quad \text{and} \quad A_m \sin(\omega(0) + 90°) = A_m \sin 90° = A_m(1)$$

Let us now examine $\cos \omega t$ at time $t = 0$. From the tables in Appendix B, we find that $\cos 0° = 1$. We can check Equations (16.1) and (16.2); we find that

$$A_m \sin(\omega t + 90°) = A_m \cos \omega t \qquad (16.3)$$

Apply what you have learned here by working Problem 16.1.

Problem 16.1 Sketch the curves generated by the phasors (Figure 16.4). What is the phase relationship between phasors? Between A_m and B_m? Which one is leading?

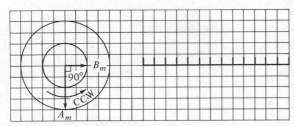

Figure 16.4 Diagram for Problem 16.1

Objective 2   Recall what is meant by rate of change (derivative), and on a sine curve point out the following: maximum positive rate of change, maximum negative rate of change, and zero rate of change. Also be able to specify the value of the slope at the maximum positive rate of change.

Before we start discussing rate of change, let us first ask, "Why do we want to look at rate of change?" You may recall that in the unit on inductance, $v_L = L \, di/dt$; we had found that a voltage would appear across an inductor if we had a change in the current. Since ac voltages are sine waves and sine waves are continuously changing, we must first understand the term di/dt.

Rate of Change (Derivative) of a Sine Wave

What do we mean when we say dv/dt, di/dt, or dx/dt? We are asking "What is the rate of change of v, i, or x with respect to time?" When we say dv/dt, or rate of change of voltage, we really mean *time* rate of change. A small change in voltage is divided by the time required to cause the change:

$$\frac{dv}{dt} = \frac{\text{a small change in voltage } v}{\text{the time } t \text{ required to cause the change}}$$

The changes we are discussing here are very small changes; actually the d means a *very* small change or a vanishing change. Therefore texts often refer to these as the instantaneous rate of change. The expression dv/dt in mathematics is called the *derivative*; in this case it is the derivative of voltage with respect to time. The d/dt part is just a shorthand notation for saying "What is the rate of change with respect to time of the term following the d/dt?" So dv_C/dt, where v_C is the voltage across a capacitor, is really saying "What is the rate of change of voltage across the capacitor with respect to time?"

As another example, if x represents distance, then dx/dt says "What is the rate of change of distance with respect to time?" You may know this as velocity or speed.

$$\text{Velocity} \quad v = \frac{dx}{dt}, \quad \text{acceleration} \quad a = \frac{dv}{dt} \quad \text{or} \quad a = \frac{d^2x}{dt^2}$$

And if we take the rate of change of velocity with respect to time, we have acceleration, or the second derivative of distance with respect to time.

Problem 16.2   What does di_L/dt mean? (i_L is the current in an inductor or coil.)

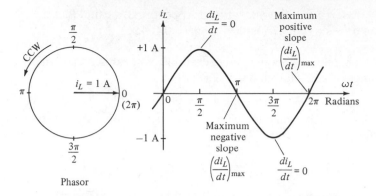

Figure 16.5 Phasor of current flowing through inductor

Let us examine the sketch of the sine wave with a peak value of one ampere of current flowing through an inductor; see Figure 16.5.

The current phasor of length 1 is rotating CCW. It starts from $0°$ and makes a complete revolution. Let us trace out the sine curve generated by this phasor. Think of it as a hill. What are the locations where the change in current with respect to time di_L/dt is the greatest? (At what points are we going up or down the steepest part of the hill?)

Referring to Figure 16.5, as we move from 0 to the right, the current is increasing, or we can say it has a *positive change* or *positive slope*. It continues to increase until we reach $\pi/2$. At $\pi/2$, the current levels off and the slope is equal to zero ($di_L/dt = 0$). From $\pi/2$ to $3\pi/2$, the current is decreasing, or we say that it has a negative slope. It decreases until we reach $3\pi/2$. At $3\pi/2$, the current again levels off and the slope is zero ($di_L/dt = 0$). Then from $3\pi/2$ it increases again. The maximum positive slope occurs at 0 and 2π rad, and, as we shall show in the next section, the slope is equal to $+\omega$ at 0 and 2π rad, and the maximum negative slope is $-\omega$ at π and 3π.

The equation for the sine curve in Figure 16.5 is

$$i_L = 1 \sin \omega t$$

Let's look at $\sin \omega t$ around the point of 0 rad. Can we evaluate the slope of $\sin \omega t$ around 0 rad? In the next example, for small values of ωt, we can show you that

$$\sin \omega t \approx \omega t \qquad \text{around 0 radians.}$$

Example 16.1 For $6°$ (or $\omega t = 0.1047$ radians) find the value of $\sin \omega t$.

Solution Recall that π rad $= 180°$

$$\omega t = \frac{\pi}{30} \text{ rad} = 6° \qquad \text{or} \qquad \omega t = \frac{\pi}{30} \text{ rad} = 0.10472 \text{ rad}$$

Referring to the sine table in Appendix B, we find that the sine of 6° is sin (0.10472 rad) = sin 6° = 0.1045.

Therefore, from the above example, you can see that when the angle is small (around 0 or 2π rad)

$$\sin \omega t \approx \omega t \qquad \text{for small } \omega t \qquad (16.4)$$

Then to determine the change in current (slope) with respect to time around 0 rad,

$$\frac{\sin \omega t}{t} \cong \frac{\omega t}{t} = +\omega$$

Thus the maximum positive slope of i_L at 0 or 2π rad is $+\omega$. We can use a similar proof for the sine wave at π and 3π to show that the maximum negative slope at those points is $-\omega$.

If we plot the values of di/dt on the sine curve for current shown in Figure 16.5, we obtain Figure 16.6. Let's examine the sketch of the rate of change of the current (di_L/dt) with respect to time and compare it with i_L. (i_L is dashed, and di_L/dt is a solid curve).

At 0 rad, $\qquad \dfrac{di_L}{dt} = +\omega \qquad i_L = 0$

At $\pi/2$ rad, $\quad \dfrac{di_L}{dt} = 0 \qquad i_L = 1$

At π rad, $\qquad \dfrac{di_L}{dt} = -\omega \qquad i_L = 0$

At $3\pi/2$ rad, $\quad \dfrac{di_L}{dt} = 0 \qquad i_L = -1$

At 2π rad, $\qquad \dfrac{di_L}{dt} = +\omega \qquad i_L = 0$

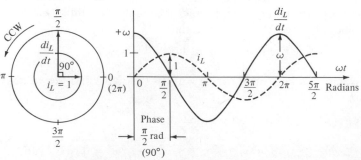

Figure 16.6   Plots of current i_L and di_L/dt

Summarizing, we notice that the curve of i_L is a sine wave. In like manner, the curve for its slope (di_L/dt) is also a sine wave, but out of phase with i_L by 90°.

Let us now examine the phasor diagram of i_L and di_L/dt in Figure 16.6. The length of phasor $i_L = 1$, and it starts at 0° or 0 rad. At the same time, the sine curve di_L/dt is at its maximum value of $+\omega$. This places the phasor di_L/dt 90° out of phase with phasor i_L, resulting in the phasor di_L/dt *leading* phasor i_L by 90° or $\pi/2$ radians.

The mathematical expressions for i_L and for di_L/dt can be written in the form

$$i_L = 1 \sin \omega t \tag{16.5}$$

and

$$\frac{di_L}{dt} = +\omega \sin \left(\omega t + \frac{\pi}{2} \right) \text{ or } +\omega \cos (\omega t) \tag{16.6}$$

where 1 = maximum amplitude of i_L

$+\omega$ = maximum amplitude of $\dfrac{di_L}{dt}$

$+\dfrac{\pi}{2}$ = phase angle, $\dfrac{di_L}{dt}$ leading i_L by 90°

Now suppose that the maximum amplitude of $i_L = 10$; what would be the maximum amplitude of di_L/dt? (Answer: 10 ω)

In the next problem, you will be asked to recall what you have learned in this objective.

Problem 16.3 List the points in Figure 16.7 where the sine curve has:

(a) Maximum positive rate of change = _____, _____

(b) Maximum negative rate of change = _____, _____

(c) Zero rate of change = _____, _____, _____

(d) Value of maximum positive rate of change _____

(e) Value of maximum negative rate of change _____

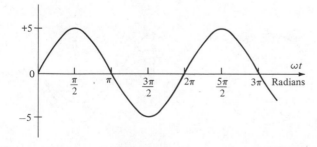

Figure 16.7 Sketch for Problem 16.3

Phase Relationship between Current and Voltage

Objective 3 Draw from memory the sine waves showing the phase relationship between voltage across and current in a resistor, capacitor, and inductor. Use a sine wave of current or voltage as reference and draw the other sine wave.

In this objective we study the phase relationship between voltage and current for a circuit containing only a resistor, only a capacitor, or only an inductor.

Circuit Containing Resistance *R*

Let us now refer to Figure 16.8 and study the phase relationship between current i_R and applied voltage e in a circuit that contains only a resistor. Here R = resistance in ohms (constant), e = applied ac voltage, and E_m = maximum value (amplitude) of ac sine voltage.

R is a constant in this circuit, because as long as temperature is constant, R is determined by physical factors only. The equation for the applied ac voltage source is

$$e = E_m \sin \omega t = E_m \sin (2\pi f)t \qquad (16.7)$$

where $\omega = 2\pi f = $ rad/s

E_m = maximum (peak) value of voltage

We can calculate the instantaneous current through the resistor R by using Ohm's law. It is

$$i_R = \frac{e}{R} = \frac{E_m}{R} \sin \omega t \qquad (16.8)$$

or

$$i_R = I_m \sin \omega t \qquad (16.9)$$

where $I_m = E_m/R$ and I_m is the maximum (peak) value of current

Figure 16.9 shows a sketch of the movement of the phasors E_m and I_m, and also includes a sketch of the sine voltage e and sine current i_R generated by these phasors.

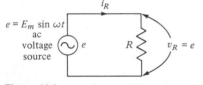

Figure 16.8 ac voltage applied across resistor *R*

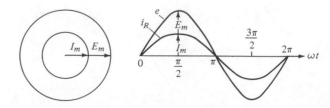

Figure 16.9 Phase relationship of voltage and current in a resistor

Notice that the phasors I_m and E_m are in phase and generate two sine waves that are in phase. The maximum amplitude of current is I_m, while the maximum amplitude of voltage is E_m.

Circuit Containing Inductance *L*

From our unit on inductance,

$$v_L = L \frac{di}{dt} \tag{16.10}$$

To set up a voltage across the coil (the counter emf), we must have a change in current. To illustrate this point, let's refer to Figure 16.10.

In Figure 16.10, L is assumed to be a constant. The applied voltage e is a varying sine-wave voltage of maximum amplitude E_m. The changing current i_L flows through the inductor L and sets up a voltage v_L across it.

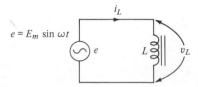

Figure 16.10 ac voltage applied across an inductor

First let's plot the sine wave of current i_L in the circuit. This sine wave is shown solid in Figure 16.11. It is drawn starting from 0° or 0 rad. A plot of the voltage v_L across the inductor is also drawn dashed on the same figure. Notice that the voltage v_L and current i_L are not in phase. Let's see why.

The equation for current through the inductor is

$$i_L = I_\mathrm{m} \sin \omega t \tag{16.11}$$

Using Equation (16.10), we can calculate the voltage v_L across the inductor. Referring to this equation, we notice that v_L is related to inductance L and to the change in the current di_L/dt. Recalling what we learned in Objective 2 (Figure 16.6), we notice that di_L/dt leads i_L by 90° or $\pi/2$ rad. Therefore the equation for di_L/dt is

$$\frac{di_L}{dt} = \omega I_\mathrm{m} \left[\sin \left(\omega t + \frac{\pi}{2} \right) \right] \tag{16.12}$$

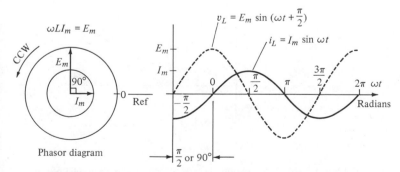

Figure 16.11 Phase relationship of voltage and current in an inductor

Notice that di_L/dt leads i_L by $\pi/2$ radians, or 90°. Moreover, the maximum amplitude of di_L/dt is ω times the maximum amplitude of i_L. Now to calculate v_L, combining Equations (16.10) and (16.12), we obtain

$$v_L = L\frac{di_L}{dt} = \omega L I_{\mathrm{m}}\left[\sin\left(\omega t + \frac{\pi}{2}\right)\right] \qquad (16.13)$$

We can rewrite Equation (16.13) in the form

$$v_L = E_{\mathrm{m}}\left[\sin\left(\omega t + \frac{\pi}{2}\right)\right] \qquad (16.14)$$

where

$$E_{\mathrm{m}} = \omega L I_{\mathrm{m}} \qquad (16.15)$$

Note that E_{m} is the maximum value of the voltage sine curve shown in Figure 16.11.

Let's now examine the phasor diagram of the sine curves in Figure 16.11. Since the voltage v_L across an inductor leads the current i_L by 90°, the phasor E_{m} must lead the phasor I_{m} by 90°. Notice that the length of E_{m} is equal to $\omega L I_{\mathrm{m}}$. If the phasor I_{m} is drawn on the reference axis, which is 0° or 0 rad, we can consider I_{m} a *reference phasor*. Then the phase of E_{m} with respect to I_{m}, the reference phasor, is +90° or $\pi/2$ rad.

It should be noted here that we could also consider the phasor E_{m} as the reference phasor. If phasor E_{m} is the reference phasor at 0°, and if it is leading the current I_{m} by 90°, then the phasor diagram is slightly changed; see Figure 16.12.

Since the phasor is rotating CCW, the voltage v_L is still leading the current i_L by 90°. The sine curves of Figures 16.11 and 16.12 are identical, and the voltage v_L leads i_L by 90° or $\pi/2$ radians. Since v_L in Figure 16.12 is the reference phasor, it crosses zero at 0 and π rad. Recall that in Figure 16.11 the current i_L was the reference phasor at 0° or 0 rad, and it crossed zero at 0 and π rad.

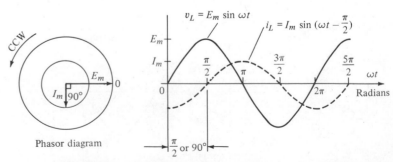

Figure 16.12   Voltage and current waveforms in an inductor

To help us remember that the applied voltage v_L across an inductor leads the current i_L, we can use the phrase

ELI

ELI means that the voltage E across an inductor L leads the current I by 90° or $\pi/2$ rad.

It is impossible to build an inductor without having some resistance. However, the resistance may be so small that we can neglect it. We shall discuss the effect of resistance in a later unit.

Circuit Containing Capacitance C

From our unit on capacitance, we know that

$$i_C = C \frac{dv_C}{dt} \tag{16.16}$$

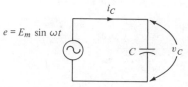

Figure 16.13 ac voltage applied across a capacitor

For current to flow in this circuit (Figure 16.13), we must have a change in the voltage.

In Figure 16.13, C is constant in this circuit. The changing voltage v_C across capacitor C causes a current i_C to flow through the capacitor. Let's plot the sine wave of voltage v_C across C. This sine wave is shown solid in Figure 16.14. A plot of the instantaneous current i_C flowing through the capacitor is shown dashed.

To obtain the current and voltage waveforms in Figure 16.14, we approach the circuit in a manner similar to our approach to inductance. The equation for the voltage across the capacitor is

$$v_C = E_m \sin \omega t$$

Using Equation (16.16), we can calculate the current i_C through the capacitor. Since

$$\frac{dv_C}{dt} = \omega \left[E_m \sin \left(\omega t + \frac{\pi}{2} \right) \right] \tag{16.17}$$

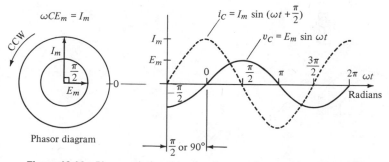

Figure 16.14 Phase relationship of voltage and current in a capacitor

Reactance

then again combining Equations (16.16) and (16.17), we obtain

$$i_C = C\frac{dv_C}{dt} = \omega C E_m \left[\overset{\bullet}{\sin} \left(\omega t + \frac{\pi}{2} \right) \right] \qquad (16.18)$$

We can rewrite Equation (16.16) as

$$i_C = I_m \left[\sin \left(\omega t + \frac{\pi}{2} \right) \right], \qquad (16.19)$$

where

$$I_m = \omega C E_m \qquad (16.20)$$

Note again that I_m is the maximum value of the current sine wave in Figure 16.14.

Let's examine the phasor diagram of the sine curves in Figure 16.14. Since the current i_C through a capacitor leads the applied voltage v_C by 90°, the phasors I_m and E_m are 90° out of phase. Furthermore, if E_m is the reference phasor at 0°, then I_m is leading it by 90°. The amplitude of the phasor I_m is equal to $\omega C E_m$.

To help us remember that the current i_C through a capacitor leads the applied voltage v_C, we can use the phrase

ICE

ICE means that the current I through capacitor C leads the voltage E by 90°. Therefore, to remember the phase relationship of voltage across and current in inductors or capacitors, remember the phrase

ELI the *ICE* man

Problem 16.4   Given the circuit in Figure 16.15, draw the phase relationship of current to applied voltage.

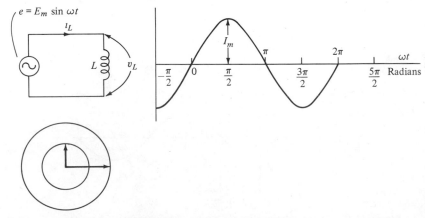

Figure 16.15   Sketch for Problem 16.4

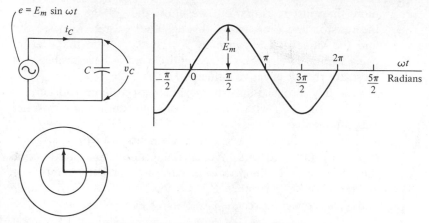

Figure 16.16 Sketch for Problem 16.5

Problem 16.5 Given the circuit in Figure 16.16, draw the phase relationship of current to applied voltage.

Computation of ac Reactance

Objective 4 Calculate the capacitive reactance (X_C) or inductive reactance (X_L) when the voltage across and the current through the reactive component are known.

Inductive or Capacitive Circuit

Before we discuss the term reactance, we must introduce you to the term *impedance*. Back in our section on dc circuits, we called resistance the property of an electric circuit that opposes the flow of current through the circuit. Now in ac circuits, we have resistors, capacitors, and inductors to worry about in our circuit. We are going to have to think up some new names! When an ac circuit has more components than just resistors, we say that the whole combination of components presents *impedance* to the flow of current.

Total opposition to the flow of current in an ac circuit is impedance,

$$Z = \frac{E \text{ (source)}}{I \text{ (source)}} \qquad (16.21)$$

where the units are E = volts, I = amperes, and Z = ohms.

This is Ohm's law for ac circuits.

We shall discuss impedance in much more detail in Unit 18.

As in dc circuits, if an ac circuit contains only resistance, we can find the resistance R by $R = E/I$. Or if the circuit contains more than just resistors, but we know the voltage drop across the resistor and the current through the resistor, then we can find the value of resistance by $R = V_R/I_R$. Note that we are using capital letters for voltage and current. This implies that these are rms or effective values.

You may also use maximum values for voltage and current and still come out with correct results, but do *not* mix rms and maximum values.

Now we still have two other elements to worry about—capacitors and inductors. We use the term *reactance* to describe their opposition to the flow of ac current. The letter symbol for reactance is X. Therefore X_L is known as *inductive reactance* and X_C as *capacitive reactance*.

$$X_L = \frac{V_L}{I_L} \tag{16.22}$$

where V_L = voltage across the inductor (rms)

    I_L = current through the inductor (rms)

$$X_C = \frac{V_C}{I_C} \tag{16.23}$$

where V_C = voltage across the capacitor

    I_C = current through the capacitor

The units for X_L and X_C are ohms.

Even though all these terms $(R, X_L, X_C, \text{and } Z)$ are measured in ohms, we cannot numerically add the *resistive* ohms to the *reactive* ohms to determine impedance. This is what makes ac circuits more difficult to study than dc circuits.

Let's see why we can't numerically add the resistive ohms to the reactive ohms. Recall from Figures 16.8 and 16.9 that, for a pure resistance, the ac current i_R in the circuit is in phase with the applied voltage e. Here the applied voltage

$$e = E_m \sin \omega t \tag{16.24}$$

$$i_R = \frac{e}{R} = \frac{E_m \sin \omega t}{R} = I_m \sin \omega t, \qquad \text{where } I_m = \frac{E_m}{R} \tag{16.25}$$

Now for an inductive circuit the current and applied voltage are out of phase by $\pi/2$ radians (90°); see Figures 16.10 and 16.11. From Equations (16.11) and (16.14), we have

$$i_L = I_m \sin \omega t \tag{16.26}$$

and

$$v_L = E_m \sin \left(\omega t + \frac{\pi}{2} \right) \tag{16.27}$$

Since the voltages across an inductor and a resistor are not in phase with their currents, they cannot be added algebraically. Likewise the resistance and reactance cannot be added algebraically. We shall discuss this complex problem in a later unit.

From Equations (16.22) and (16.23), it would appear that the magnitude or amplitude of the applied emf would have an effect on the reactance of the ac circuit. However, the magnitude of the applied emf has *no* effect on the reactance. For example, if we double the maximum value of the applied emf E_m (without changing frequency), the slope of the current waveform must be twice as great. This means that the maximum value I_m of current must also double. Since both E_m and I_m are double, the reactance does not change.

Summarizing, to calculate the reactance X of an individual component like an inductor L, we use equation (16.22).

$$X_L \text{ (ohms)} = \frac{V_L}{I_L}$$

while we find the reactance of a capacitor from Equation (16.23).

$$X_C \text{ (ohms)} = \frac{V_C}{I_C}$$

Now what about the phase relationship between current i_L and applied voltage v_L—does it change? The answer is no! Let's see why. Substituting Equation (16.27) into (16.22), we obtain

$$X_L = \frac{v_L}{i_L} = \frac{E_m \sin (\omega t + \pi/2)}{I_m \sin (\omega t)} \qquad (16.28)$$

Note: The amplitudes of v_L and i_L are E_m and I_m, respectively, while the phase relationship between voltage v_L and current i_L is 90° or $\pi/2$ radians out of phase. In the next objective, you will learn how to attach an operator to inductive and capacitive reactive components to indicate the current and voltage phase relations.

Example 16.2 An inductor allows 2 mA of current to flow when it is connected across a 12-V, 60-Hz source. What is the inductive reactance? Assume that the resistance of the inductor is negligible.

Solution $\quad X_L = \dfrac{V_L}{I_L} = \dfrac{12 \text{ V}}{2 \text{ mA}} = 6 \text{ k}\Omega$

Problem 16.6 An inductor allows 12 mA of current to flow when it is connected across a 76-V, 60-Hz source. What is the inductive reactance? Assume that the resistance of the inductor is negligible.

Example 16.3 A capacitor allows 0.5 mA of current to flow when it is connected across a 14-V, 60-Hz source. What is the capacitive reactance?

Solution $\quad X_C = \dfrac{V_C}{I_C} = \dfrac{14\text{ V}}{0.5\text{ mA}} = 28\text{ k}\Omega$

Problem 16.7 A capacitor allows 4.4 mA of current to flow when it is connected across a 11-V, 60-Hz source. What is the capacitive reactance?

Example 16.4 A current of $i = (50\text{ mA})\sin \omega t$ flows when a 50-V, 60-Hz source is connected across a capacitor. What is the capacitive reactance?

Remember: If the words peak or peak-to-peak are not included with the voltage specification, then it is understood to be an rms value. The general equation for a sine wave of current is $i = I_m \sin \omega t$; therefore 50 mA is equal to I_m.

Solution $\quad I_{\text{rms}} = \dfrac{50\text{ mA}}{1.414} = 35.36\text{ mA} \qquad X_C = \dfrac{V_C}{I_C} = \dfrac{50\text{ V}}{35.36\text{ mA}} = 1.414\text{ k}\Omega$

Use of *j* Operator *

Objective 5 Recall the operators that are attached to the inductive and capacitive reactive components.

To indicate the phase relationship of voltage and current, we use the symbol *j* (which we call an *operator*) to indicate the phase relationship of the applied voltage to the current. For inductors, the voltage leads current (recall *ELI*) by 90°. So we use

$$+j$$

in front of the numerical value in ohms for inductive reactance.

For capacitors, the voltage lags current (recall *ICE*) by 90°. So we use

$$-j$$

in front of the numerical value in ohms for capacitive reactance.

---

* We introduce the *j* operator here to remind you that capacitive and inductive reactance are different from pure resistance. We shall introduce the terms *impedance*, *vector*, *vector sum* $(5 + j4 \neq 9)$, and *phasor* in Unit 18, after you have mastered the mathematics of vectors (in Unit 17).

For resistors, the voltage is in phase with current, so we use *no* symbol in front of the numerical value of resistance.

In summary, the symbol *j* makes the distinction between reactance and resistance.

Example 16.5   Express the answers to Examples 16.2, 16.3, and 16.4 using the *j* operator.

Solution   Example 16.2 with *j* operator $X_L = +j6 \text{ k}\Omega$

Example 16.3 with *j* operator $X_C = -j28 \text{ k}\Omega$

Example 16.4 with *j* operator $X_C = -j1.414 \text{ k}\Omega$

Problem 16.8   Express the answers to Problems 16.6 and 16.7 using the *j* operator.

Use of Frequency *f*, Inductance *L*, or Capacitance *C*

Objective 6   Calculate the capacitive reactance (X_C) or the inductive reactance (X_L) when frequency and capacitance (C) or inductance (L) are known.

We found that we could find the amount of inductive reactance or capacitive reactance by measuring the voltage across and current through the inductor or capacitor, then dividing the voltage by the current. There is another way to determine the inductive or capacitive reactance (if *L* and *C* are known) without taking laboratory measurements.

***Inductance L* (*henries*)**   To calculate inductive reactance (X_L), let's refer to Equation (16.15), in which

$$E_m = \omega L I_m$$

$$\therefore \quad X_L = \frac{\omega L I_m}{I_m} = \omega L \quad \text{ since } X_L = \frac{E_m}{I_m}$$

$$X_L = 2\pi f L \quad \text{ since } \omega = 2\pi f \tag{16.29}$$

where *L* is measured in henries, *f* in hertz (cycles/s), and X_L in ohms.

Note: Equation (16.29) can be written with the *j* operator to emphasize the phase relationship between voltage and current (remember the phrase *ELI*). This form is

$$X_L = +j2\pi f L \quad \text{ with the } j \text{ operator} \tag{16.30}$$

Capacitance C (farads)   To calculate capacitive reactance (X_C), let's refer to Equation (16.20), in which

$$I_m = \omega C E_m$$

$$\therefore \quad X_C = \frac{E_m}{\omega C E_m} = \frac{1}{\omega C} \quad \text{since } X_C = \frac{E_m}{I_m}$$

$$X_C = \frac{1}{2\pi f C} \quad \text{since } \omega = 2\pi f \qquad (16\,31)$$

where C is measured in farads, f in hertz (cycles/s), and X_C in ohms.

Note: Equation (16.31) can be written with the j operator to emphasize the phase relationship between voltage and current (recall the phrase *ICE*). This form is

$$X_C = -j\,\frac{1}{2\pi f C} \qquad \text{with the } j \text{ operator} \qquad (16.32)$$

Example 16.6   What is the inductance of the inductor of Example 16.2?

Solution $X_L = 2\pi f L$ and $2\pi(60) \approx 377$

$$\therefore \quad L = \frac{X_L}{2\pi f} = \frac{6000}{377} = 15.9 \text{ H}$$

Example 16.7   What is the capacitance of the capacitor of Example 16.3?

Solution $X_C = \dfrac{1}{2\pi f C}$

$$C = \frac{1}{2\pi f X_C} = \frac{1}{377(28 \text{ k}\Omega)} = 0.095 \times 10^{-6} \text{ F}$$

$$= 0.095 \; \mu\text{F}$$

Example 16.8   What is the reactance of a 0.1-μF capacitor at 60 Hz?

Solution $X_C = \dfrac{1}{2\pi f C} = \dfrac{1}{377(0.1 \times 10^{-6})} = \dfrac{10^6}{37.7}$

$$= 26.5 \text{ k}\Omega \qquad \text{with } j \text{ operator } -j26.5 \text{ k}\Omega$$

Problem 16.9   What is the inductance of the inductor in Problem 16.6?

Problem 16.10 What is the capacitance of the capacitor in Problem 16.7?

Problem 16.11 What is the reactance of a 1.0-μF capacitor at 60 Hz?

Problem 16.12 What is the reactance of a 0.05-μF capacitor at 1000 Hz?

Problem 16.13 What is the reactance of a 14-H inductor at 60 Hz?

Problem 16.14 What is the reactance of a 30-mH inductor at 100 kHz?

Problem 16.15 Determine the frequency at which a 0.5-μF capacitor has a reactance of 132.6 kΩ.

Problem 16.16 At 100 kHz, what size capacitor is required for its reactance to equal in magnitude the reactance of a 30-mH inductor?

Problem 16.17 What are the X_L and X_C in a dc circuit ($f = 0$ Hz)?

Problem 16.18 What are the relative values of X_L and X_C at very high frequencies ($f \approx \infty$)?

Test

1. Draw sine waves showing the phase relationship between voltage across and current in a resistor, capacitor, and inductor.
2. A capacitor allows 2 mA current to flow when it is connected across a 20-V, 60-Hz source. What is the capacitive reactance? (Include the j operator.)
3. A current of $i = (28.28$ mA$)$ sin $377t$ flows when a 40-V, 60-Hz source is connected across an inductor. What is the inductive reactance?
4. What is the reactance of a 1.0-μF capacitor at 60 Hz?
5. What is X_L in a dc circuit?
6. Determine the frequency at which a 10-H choke has a reactance of 5 kΩ.
7. At 1 MHz, what size inductor is required for its reactance to equal that of a 200-pF capacitor?

Unit 17 *Vector Algebra: Phasors*

In later units you will be working with circuits like those shown in Figure 17.1(a) and (b). In our study of dc circuits, you learned that when you have three resistors connected in series across a dc supply, Kirchhoff's voltage law (KVL) applies. That is, the sum of the voltage drops across each resistor is equal to the supply voltage. KVL still applies in ac circuits like the one shown in Figure 17.1(a). However, now, as you remember from Unit 16, the voltage across the inductor V_L leads the current through it, and the voltage across the capacitor V_C lags the current into it. Since the voltages across each component shown in Figure 17.1(a) are not in phase with the current that is common in the circuit, you cannot numerically add the voltage drops and come up with the input voltage. What you have to do is *vectorially add* the voltages.

We can make a similar statement about the circuit shown in Figure 17.1(b). Kirchhoff's current law (KCL) still applies in

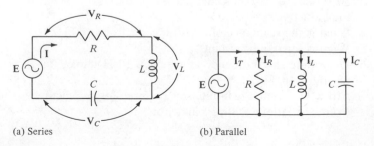

Figure 17.1 ac series and parallel *RLC* circuit

(a) Series   (b) Parallel

338

this parallel ac circuit. However, to find the total current I_T, you have to find the *vector sum* of I_R, I_L, and I_C.

We mentioned the words *vectorially add* and *vector sum* above. That is what this unit is about—learning vector algebra, so that you may use it as a tool for solving ac circuits.

Objectives

After completing all the work associated with this unit, you should be able to:

1. Define the terms scalar, vector, phasor, polar notation, and rectangular notation of a phasor.
2. Draw a graph showing the real and imaginary axes (label) and locate any phasor on it (in either polar or rectangular notation).
3. Given a phasor written in polar notation, convert it to rectangular notation, and vice versa.
4. Add and subtract phasors (complex numbers) written in rectangular notation.
5. Multiply and divide phasors written in polar and rectangular notation.

Definition of Terms

Objective 1   Define the terms scalar, vector, phasor, polar notation, and rectangular notation of a phasor.

A Little Background

We can divide physical quantities as follows.

A scalar is a quantity that is specified by its magnitude only. Examples of scalars are mass, length, temperature, energy, and electric charge. The volume of a tank is expressed in cubic feet. The temperature of a room is expressed in degrees. Scalars are ordinary numbers and obey all the rules of algebra. In dc circuits, when two resistors are connected in series, we can find the total resistance by adding the value of each resistor.

A vector is a quantity that must be specified by both a magnitude and a direction. Examples of vectors are force, acceleration, and electric and magnetic fields. A force from a push or a pull is not completely specified unless the direction as well as the magnitude of the force is given. When two or more forces are added, they are not necessarily added algebraically, but must be combined in a way that considers their direction as well as their magnitude.

Vectors are designated in print in a number of ways:

$$\text{Boldface type } \mathbf{E_T} = \mathbf{V}_R + \mathbf{V}_L \qquad (17.1)$$

$$\text{Regular italic type } E_T = V_R + V_L \qquad (17.2)$$

$$\text{Dot over the letter symbol } \dot{E}_T = \dot{V}_R + \dot{V}_L \qquad (17.3)$$

$$\text{Bar over the letter symbol } \bar{E}_T = \bar{V}_R + \bar{V}_L \qquad (17.4)$$

$$\text{Arrow over the letter symbol } \vec{E}_T = \vec{V}_R + \vec{V}_L \qquad (17.5)$$

Vectors obey the laws of vector analysis or complex numbers, which we shall discuss in this unit. We cannot add vector quantities by algebraic addition; we must add them by *vector addition*. If regular italic type is used, we must define the quantities as vectors, and for the addition operation, we must perform vector addition.

*A **phasor*** is a rotating vector with constant magnitude and constant angular velocity. Examples of phasors are ac voltage and ac current, as were used in the unit on ac current and the sine wave. Phasors obey the laws of vector analysis.

Polar Notation

Phasors may be designated by using polar notation:

$$\text{Magnitude (or length) } \underline{/\text{angle}°} \qquad (17.6)$$

You are already familiar with the polar notation of a phasor. When we discussed the voltage and current phasors in previous units, we always gave a length and showed them at a certain angle. For the phasor shown in Figure 17.2, the polar notation or polar coordinate form is the magnitude of the phasor followed by the angle from the horizontal reference, or $5\underline{/53.1°}$.

$$\text{Phasor } \mathbf{E_T} = E_T = |E_T| \ \underline{/\phi} \qquad (17.7)$$

where $|E_T|$ = magnitude of phasor and $\underline{/\phi}$ = angle of phasor.

If we use a letter to designate a phasor such as $\mathbf{E_T}$, we can indicate the magnitude of the phasor by using a vertical bar on each side of the letter, or simply by using the italic E with the appropriate subscript.

Phasors may also be designated using rectangular notation:

$$\text{Horizontal component} \pm j \text{ vertical component} \qquad (17.8)$$

The j in front of a number specifies it as the vertical component. For the phasor that is 5 units long at the angle shown in Figure 17.3, if we project it down to the horizontal scale, we get 3 units of length for the horizontal component. If we project the phasor over to the vertical scale, we get 4 units of length for the vertical component. The phasor may then be written in rectangular notation as $3 + j4$. This is a $+j$ because the vertical component goes

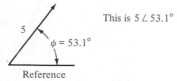

This is $5 \angle 53.1°$

$\phi = 53.1°$

Reference

Figure 17.2 Polar notation

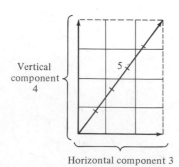

Vertical component 4

5

Horizontal component 3

Figure 17.3 Rectangular notation
$3 + j4$

up. If the vertical component goes down, it is specified by a $-j$ in front of the number. The sign in front of the horizontal component is $+$ if it falls to the right and $-$ if it falls to the left.

If you were given the horizontal and vertical components of a phasor, could you find the magnitude or length and angle of the phasor? Sure you could! Just go backward.

To get some practice in graphically changing from polar to rectangular and rectangular to polar notation, do Problems 17.1 and 17.2.

Problem 17.1   Given the phasors shown in polar form in Figure 17.4, project the phasor down to the horizontal axis and over to the vertical axis. When you have done this, express the phasor in rectangular notation.

Problem 17.2   Given the horizontal and vertical components of the phasors shown in Figure 17.5, complete the rectangle and draw the actual phasor. Then express the phasor in polar notation. To determine the magnitude, take a separate piece of paper, lay the edge along the scale drawn in Figure 17.5, and mark the scale off on your paper. Now you may use this piece of paper to determine the length or magnitude of the phasor. Angles are shown on the sketch.

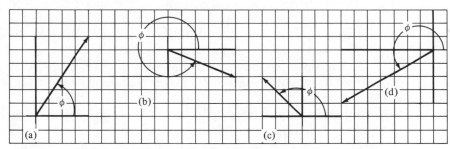

Figure 17.4   Figure for Problem 17.1

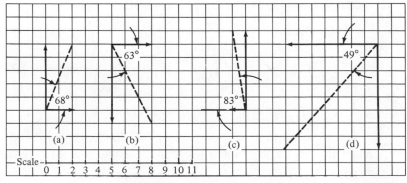

Figure 17.5   Figure for Problem 17.2

Objective 2 Draw a graph showing the real and imaginary axes (label) and locate any phasor on it (in either polar or rectangular notation).

The *j* Part of the Rectangular Form

Let's learn a little bit more about the *j* part of the rectangular notation. The rectangular notation for the phasor shown in Figure 17.3 is $3 + j4$. You may know the notation $3 + 4i$, which is known as a *complex* number. The $j = i = \sqrt{-1}$.

In electricity *i* is reserved for current, so we use *j* in the complex number. The *i* or *j* part of the expression is known as the *imaginary part*. It is not imaginary in the sense that it doesn't exist. It does exist! However, it got the name imaginary because it was thought at one time that square roots of negative numbers did not exist, or that they did not exist on the horizontal axis, which was known as the *real* axis. If it isn't real, it must be imaginary!

$$\sqrt{25} = 5 \qquad \text{but} \qquad \sqrt{-25} = 5\sqrt{-1} = 5i$$

Figure 17.6 shows a rectangular coordinate system. The horizontal line is known as the real axis. To the right of the vertical line we have positive values on the horizontal axis. To the left, we have negative values. The vertical line is known as the imaginary axis. Above the horizontal line the values are designated as *j*. Below, the values are designated by a $-j$. Therefore the horizontal part of a phasor is known as the *real* part, and the vertical part is known as the imaginary part.

So a complex number is a quantity consisting of a real part and an imaginary part, with the imaginary part designated by plus or minus *j*.

Sometimes you have to multiply two complex numbers together, or multiply a real number times a complex number. What happens then with the *j*'s? Read through the following calculations and see if you can follow what is happening on the graph shown in Figure 17.6.

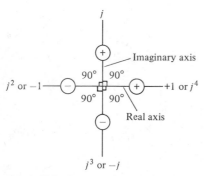

Figure 17.6 Rectangular coordinate system

$$1 \text{ at angle } 0°$$
$$j \cdot 1 = j \qquad \text{at angle } 90°$$
$$j \cdot j = j^2 = (\sqrt{-1})^2 = -1 \qquad \text{at angle } 180°$$
$$j \cdot j^2 = j^3 = j(-1) = -j \qquad \text{at angle } 270°$$
$$j \cdot j^3 = j^4 = 1 \qquad \text{at angle } 360° \text{ or back to } 0°$$

When you multiply a quantity by *j*, you shift it through a 90° angle in a counterclockwise direction. When you multiply a quantity by j^2, or -1, you shift it through a 180° angle. Figure

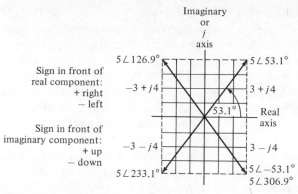

Figure 17.7 Summary of relations between polar and rectangular notation

17.7 shows a summary of the relations between polar and rectangular notation.

Resultant of Two Out-of-Phase Sine Waves

In this section you can see that finding the resultant of two out-of-phase sine waves is similar to changing rectangular notation to polar notation. Let's take two phasors 90° apart and let them generate the sine waves; see Figure 17.8(a). The short horizontal phasor could represent the real part of a complex number because

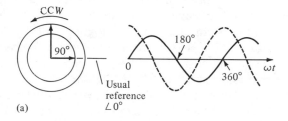

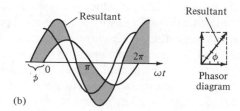

Figure 17.8 Resultant of two out-of-phase sine waves

it lies along the horizontal line which is usually used as a reference point. The long vertical phasor represents the imaginary part.

Now add the two sine waves together. That is, at several values of ωt or the angle in degrees, add the instantaneous value of one sine wave to the instantaneous value of the other; see Figure 17.8(b). Then project the phasor of the resultant sine wave over to the phasor diagram shown in Figure 17.8(b). Examine Figure 17.8(b) again. You can see that the resultant sum is a sine wave of the same frequency as the two sine waves that you added. Also you can see that you could have obtained the resultant phasor more easily than you did by adding the sine waves point by point. All you had to do was to complete the rectangle formed by the horizontal and vertical phasors, then draw the line from the center of the circle to the corner of the rectangle. Therefore you may use a phasor diagram to show the resultant of the addition of two sine waves.

For an application of why you would want to find the resultant of two sine waves, refer to the circuit shown in Figure 17.9(a). Remember from a previous unit that the voltage drop across a resistor in an ac circuit is in phase with the current through it, whereas the voltage drop across an inductor leads the current through it (*ELI*). Using current as a reference, you could probably draw the phasor diagram as shown in Figure 17.9(b). Therefore, for the circuit shown in Figure 17.9(a), the solid sine wave in Figure 17.8(a) could represent V_R, with V_L represented by the other sine wave. You find the input voltage applied to the circuit of Figure 17.9(a) by adding the two sine waves (KVL) or finding the resultant phasor.

When we use a phasor diagram to show the resultant of the addition of two sine waves, we show the phasors as fixed or stationary, as in the phasor diagram of Figure 17.8(b). The angle between the resultant phasor and the horizontal phasor is the same as the phase angle between the resultant sine wave and the sine wave that starts at 0°; see Figure 17.8(b).

When a phasor is used to generate a sine wave, the length of the phasor is the peak or maximum value of the sine wave. However, when a phasor is represented as *fixed*, so that a resultant phasor may be found, all the phasors are divided by the $\sqrt{2}$, so that the lengths are in rms values. Effective values (rms) are used almost exclusively on fixed or stationary phasors because most problems are stated using rms quantities. Therefore using rms values on the fixed phasors saves the trouble of convering from rms to peak and then back again to rms for the final answer. Therefore, when you use voltage or current phasors to determine resultant voltage or current in a circuit, they are given in rms values.

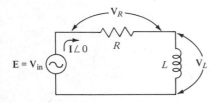

(a) Series *RL* circuit

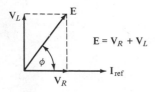

(b) Phasor diagram

Figure 17.9 Application of a phasor diagram to a series *RL* circuit

Conversion from Polar to Rectangular Notation

Objective 3 Given a phasor written in polar notation, convert it to rectangular notation, and vice versa.

To help you change from polar to rectangular notation, mathematically you will use some functions from trigonometry. In trig all types of triangles are used, but you need to know only the functions for a right triangle, that is, a triangle in which one of the angles equals 90°; see Figure 17.10. To identify the sides of the triangle, we'll use R for the horizontal side, X for the vertical side, and Z for the hypotenuse. Even though these letters may be different from the ones you learned in math, the choice of letters for the sides of the triangle is important. As we shall describe in more detail in future units, the triangle is an impedance triangle, where R is resistance, X is reactance, and Z is impedance. For the angle ϕ (Greek letter phi):

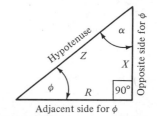

Figure 17.10 Impedance triangle

$$\text{sine } \phi = \frac{\text{opposite side}}{\text{hypotenuse}}, \qquad \sin \phi = \frac{X}{Z} \qquad (17.9)$$

$$\text{cosine } \phi = \frac{\text{adjacent side}}{\text{hypotenuse}}, \qquad \cos \phi = \frac{R}{Z} \qquad (17.10)$$

$$\text{tangent } \phi = \frac{\text{opposite side}}{\text{adjacent side}}, \qquad \tan \phi = \frac{X}{R} \qquad (17.11)$$

$$\text{cotangent } \phi = \frac{\text{adjacent side}}{\text{opposite side}}, \qquad \cot \phi = \frac{R}{X} \qquad (17.12)$$

Since the sum of all the angles in a triangle equals 180°, then:

$$\alpha = 90° - \phi$$

$$\sin \phi = \cos \alpha \qquad \text{or} \qquad \cos \phi = \sin \alpha$$

$$\tan \phi = \cot \alpha \qquad \text{or} \qquad \cot \phi = \tan \alpha$$

From the trig relationships, if you are given the Z and the angle ϕ, you may find R and X by:

$$\left. \begin{array}{l} R = Z \cos \phi \\ X = Z \sin \phi \end{array} \right\} \quad \begin{array}{l} \text{Conversion of polar notation} \quad (17.13) \\ Z \underline{/\phi} \text{ into rectangular notation} \quad (17.14) \end{array}$$

You can apply similar relations to other phasors when you want to break down a phasor into its horizontal and vertical components. Remember, phasors must originate from a pivot point. Therefore, when you work with phasors, the diagram looks like Figure 17.11. However, you are still finding the sides of a triangle.

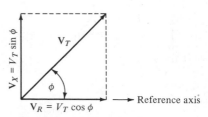

Figure 17.11 Phasors

The equation for the resultant voltage in a circuit may be

$$|V_T| \underline{/\phi} = |V_T| \cos \phi + j|V_T| \sin \phi \qquad (17.15)$$

or

$$V_T \underline{/\phi} = V_T \cos \phi + j V_T \sin \phi \qquad (17.16)$$

Remember that the vertical lines that mean magnitude only are often omitted, with it being understood that you are using the magnitude of the resultant.

For example, if the rms magnitude of a resultant voltage is 100 V and its phase angle is 45°, the other two phasors may be found by:

$$100 \underline{/+45°} = 100 \cos 45° + j100 \sin 45°$$
Polar notation

$$= 100(0.707) + j100(0.707)$$

$$= 70.7 + j70.7$$
Rectangular notation

Also see Examples 17.1, 17.2, and 17.3.

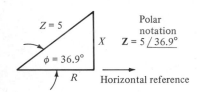

Figure 17.12   Diagram for Example 17.1

Example 17.1 $Z = 5$, $\phi = 36.9°$
Find R and X and write the **Z** in rectangular notation.

$$R = Z \cos \phi = 5 \cos 36.9° = 5(0.8) = 4$$
$$X = Z \sin \phi = 5 \sin 36.9° = 5(0.6) = 3$$
$$\mathbf{Z} = R + jX = 4 + j3$$

We have a $+$ in front of the j because the angle was $+$, which made **X** go up from the reference line.

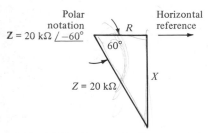

Figure 17.13   Diagram for Example 17.2

Example 17.2 $Z = 20$ kΩ, $\phi = -60°$
Find R and X and write the **Z** in rectangular notation.

$$R = Z \cos \phi = 20 \text{ kΩ} \cos 60° = 20 \text{ kΩ}(0.5) = 10 \text{ kΩ}$$
$$X = Z \sin \phi = 20 \text{ kΩ} \sin 60° = 20 \text{ kΩ}(0.866) = 17.32 \text{ kΩ}$$
$$\mathbf{Z} = R - jX = 10 \text{ kΩ} - j17.32 \text{ kΩ}$$

Example 17.3   Express the phasor $\mathbf{V}_T = 24$ V $\underline{/+70°}$ in rectangular notation. Draw a phasor diagram.

It is understood that 24 V is an rms quantity (because no p or p–p was attached).

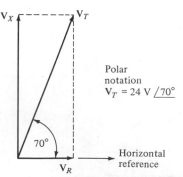

Figure 17.14   Diagram for Example 17.3

$$V_R = V_T \cos \phi = 24 \cos 70° = 24(0.3420) = 8.2 \text{ V}$$
$$V_X = V_T \sin \phi = 24 \sin 70° = 24(0.9397) = 22.6 \text{ V}$$
$$\mathbf{V}_T = V_R + jV_X = 8.2 + j22.6 = 24 \text{ V} \underline{/70°}$$

Conversion from Rectangular to Polar Notation

There are several ways to convert from rectangular to polar notation. You must decide which way is easiest or clearest to you after you read through the discussion and examples.

Solving for the Angle First, Then for the Magnitude

To solve for the angle first, you use

$$\tan \phi = \frac{X}{R} \quad \text{or} \quad \phi = \tan^{-1}\frac{X}{R}$$

Let's explain the meaning of the last notation.

$$\tan \phi = \frac{X}{R} \begin{cases} \text{this says that the tangent of the angle } \phi \\ \text{is equal to } X \text{ divided by } R \end{cases}$$

$$\phi = \arctan \frac{X}{R} \begin{cases} \text{this says that } \phi \text{ is equal to the angle whose} \\ \text{tangent is } X \text{ divided by } R. \text{ But instead of} \\ \text{saying "angle whose" each time, we say "arc."} \end{cases}$$

$$\phi = \tan^{-1}\frac{X}{R} \begin{cases} \text{this is even a shorter notation than arc.} \\ \text{This last statement still means that } \phi \text{ is} \\ \text{the angle whose tangent is } X/R. \end{cases}$$

When we know the phase angle, we can find the magnitude of Z by

$$Z = \frac{X}{\sin \phi}, \quad Z = \frac{R}{\cos \phi} \quad \text{or} \quad Z = \sqrt{R^2 + X^2}$$

Solving for the Magnitude First, Then for the Angle

To solve for Z (magnitude), you may use the Pythagorean theorem:

$$Z^2 = R^2 + X^2 \quad \text{or} \quad Z = \sqrt{R^2 + X^2}$$

Then use trig to find the angle:

$$\sin \phi = \frac{X}{Z} \quad \text{or} \quad \cos \phi = \frac{R}{Z}$$

$$\phi = \sin^{-1}\left(\frac{X}{Z}\right) \quad \text{or} \quad \phi = \cos^{-1}\left(\frac{R}{Z}\right)$$

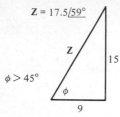

$\mathbf{Z} = 17.5\underline{/59°}$

$\phi > 45°$

Figure 17.15 Diagram for Example 17.4

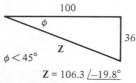

$\phi < 45°$

$\mathbf{Z} = 106.3 \underline{/-19.8°}$

Figure 17.16 Diagram for Example 17.5

Example 17.4 Change the rectangular notation of the phasor $9 + j15$ to polar notation.

$$\phi = \tan^{-1}\left(\frac{15}{9}\right) = \tan^{-1}(1.667) = 59°$$

$$Z = \frac{R}{\cos\phi} = \frac{9}{\cos 59°} = \frac{9}{0.515} = 17.5 \qquad \mathbf{Z} = 17.5\ \underline{/59°}$$

Example 17.5 Change the rectangular notation of the phasor $100 - j36$ to polar notation.

$$\phi = \tan^{-1}\left(\frac{36}{100}\right) = \tan^{-1}(0.36)$$

$$= -19.8° \qquad \text{because the sign in front of the } j \text{ is negative}$$

$$Z = \frac{X}{\sin\phi} = \frac{36}{\sin 19.8°} = \frac{36}{0.3387} = 106.3 \qquad \mathbf{Z} = 106.3\ \underline{/-19.8°}$$

To convert from polar to rectangular notation or from rectangular to polar notation, use a calculator with trig tables.

Look over these previously discussed methods and run through Examples 17.1 through 17.5, using an electronic calculator, if you have one. Then for additional practice, do the following problems.

Convert the following phasors from polar notation to rectangular notation.

Problem 17.3 $20\ \underline{/19°} =$

Problem 17.4 $20\ \underline{/73°} =$

Problem 17.5 $20\ \underline{/147°} =$

Problem 17.6 $20\ \underline{/236°} =$

Problem 17.7 $20\ \underline{/276°} =$

Problem 17.8 $20\ \underline{/4°} =$

Convert the following phasors from rectangular notation to polar notation.

Problem 17.9 $44.1 + j23.5 =$

Problem 17.10 $17.92 + j46.7 =$

Problem 17.11 $-8.68 + j49.2 =$

Problem 17.12 $-49.2 - j8.68 =$

Problem 17.13 $48.3 - j12.94 =$

Problem 17.14 $49.9 + j2.61 =$

Vector (Phasor) Arithmetic

Objective 4   Add and subtract phasors (complex numbers) written in rectangular notation.

Addition and Subtraction

First we must express the phasors in *rectangular* notation. Then we can add or subtract them algebraically. The real parts and the j parts are added or subtracted independently.

Addition

$$\mathbf{Z}_1 = R_1 + jX_1$$
$$\mathbf{Z}_2 = R_2 + jX_2$$
$$\mathbf{Z}_3 = \mathbf{Z}_1 + \mathbf{Z}_2 = (R_1 + R_2) + j(X_1 + X_2)$$

Example 17.6 $(4 + j3) + (6 + j2) = 10 + j5$

Example 17.7 $(5 + j10) + (3 - j4) = 8 + j6$

Example 17.8 $(-6 + j8) + (8 - j12) = 2 - j4$

Example 17.9 $5\ \underline{/36.9°} + 4\ \underline{/-60°} = 4 + j3 + 2 - j3.464$
$$= 6 - j0.464$$

Subtraction

$$\mathbf{Z}_3 = \mathbf{Z}_1 - \mathbf{Z}_2 = (R_1 - R_2) + j(X_1 - X_2)$$

Example 17.10 $(10 + j20) - (6 + j3) = 4 + j17$

Example 17.11 $(7 - j5) - (6 + j10) = 1 - j15$

Example 17.12 $(-3 - j4) - (9 - j6) = -12 + j2$

Example 17.13

$$15\ k\Omega\ \underline{/120°} - 10\ k\Omega\ \underline{/45°} = -7.5\ k\Omega + j13\ k\Omega$$
$$- (7.07\ k\Omega + j7.07\ k\Omega)$$
$$= -14.57\ k\Omega + j5.93\ k\Omega$$

Problem 17.15 $(30 + j20) + (-25 + j80) =$

Problem 17.16 $(-10 - j40) - (-40 + j30) =$

Problem 17.17 $20 \underline{/30°} + 10 \underline{/60°} =$

Problem 17.18 $5 \underline{/-60°} + 8 \underline{/45°} =$

Problem 17.19 $12 \underline{/125°} + 15 \underline{/15°} =$

Problem 17.20 $100 \underline{/10°} - 90 \underline{/20°} =$

Problem 17.21 $35 \underline{/160°} - 25 \underline{/220°} =$

Problem 17.22 $12 \underline{/25°} - 18 \underline{/-35°} =$

Objective 5   Multiply and divide phasors written in polar and rectangular notation.

Multiplication

When phasors or complex quantities are expressed in polar notation, multiply the magnitudes and add the angles.

$$V_1 \underline{/\phi_1} \times V_2 \underline{/\phi_2} = V_1 V_2 \underline{/\phi_1 + \phi_2}$$

Example 17.14 $(3 \underline{/30°})(10 \underline{/45°}) = 30 \underline{/75°}$

Example 17.15 $(10 \underline{/45°})(5 \underline{/-45°}) = 50 \underline{/0°}$

Example 17.16 $(5 \underline{/-30°})(4 \underline{/10°}) = 20 \underline{/-20°}$

Problem 17.23 $(9 \underline{/225°})(5 \underline{/-145°}) =$

Problem 17.24 $(20 \underline{/143°})(6 \underline{/83°}) =$

Problem 17.25 $(36 \underline{/78°})(123 \underline{/-43°}) =$

When phasors or complex quantities are expressed in rectangular notation, multiply the terms as you would the algebraic quantities $(a - b)(c - d)$. To simplify the results, remember that $-j^2 = 1$ and $j^2 = -1$.

$$
\begin{array}{l}
\quad R_1 \quad\; + jX_1 \\
\quad R_2 \quad\; + jX_2 \\
\hline
R_1 R_2 + jR_2 X_1 \\
\quad\quad\quad + jR_1 X_2 + j^2 X_1 X_2 \\
\hline
R_1 R_2 + j(R_1 X_2 + R_2 X_1) + j^2 X_1 X_2
\end{array}
$$

But $j^2 = -1$

$$\therefore \quad (R_1 R_2 - X_1 X_2) + j(R_1 X_2 + R_2 X_1)$$

Example 17.17

$$(3 + j4)(8 - j4) = 24 - j12 + j32 - j^2 16$$
$$= 24 + j20 - (-1)(16) = 40 + j20$$

Example 17.18

$$(5 - j4)(-10 + j20) = -50 + j100 + j40 - j^2 80$$
$$= 30 + j140$$

Problem 17.26 $(15 + j20)(7 + j6) =$

Problem 17.27 $(5 - j4)(3 + j9) =$

Problem 17.28 $(24 - j20)(8 - j15) =$

Problem 17.29 $(-2 - j3)(4 + j12) =$

Division

When phasors or complex quantities are expressed in polar notation, divide the magnitudes and subtract the angles.

$$V_1 \underline{/\phi_1} \div V_2 \underline{/\phi_2} = \frac{V_1}{V_2} \underline{/\phi_1 - \phi_2}$$

Example 17.19 $\dfrac{12 \underline{/90°}}{4 \underline{/30°}} = 3 \underline{/60°}$

Example 17.20 $\dfrac{86 \underline{/27°}}{12 \underline{/63°}} = 7.17 \underline{/-36°}$

Example 17.21 $\dfrac{48 \underline{/10°}}{5 \underline{/-15°}} = 9.6 \underline{/25°}$

Conjugate   To find the conjugate of a complex number, change the sign of the imaginary part. Do nothing with the sign of the real part.

$$\text{Conjugate of } a + jb = a - jb$$
$$\text{Conjugate of } -a - jb = -a + jb$$

The conjugate is used to divide complex numbers written in rectangular notation.

When complex quantities are expressed in rectangular notation, you first have to get rid of the j's in the denominator, so that the denominator contains only a real number. Then you can simplify the fraction by algebraic methods. To get rid of the j's in the denominator, you multiply the denominator by its conjugate. So that the value of the fraction is not changed, the numerator must also be multiplied by the same complex quantity as the denominator.

Example 17.22

$$\frac{3 + j4}{5 - j2}$$

Multiply both the numerator and denominator by the conjugate of the denominator.

$$\frac{(3 + j4)(5 + j2)}{(5 - j2)(5 + j2)}$$

Calculation for the denominator:
$(5 - j2)(5 + j2)$
$= 25 - j\cancel{10} + j\cancel{10} - j^2 4$
$= 25 - (-1)4$

$$= \frac{15 + j20 + j6 + j^2 8}{25 + 4}$$

$$= \frac{7 + j26}{29}$$

$$= 0.241 + j0.897$$

Problem 17.30 $\dfrac{48 \,\underline{/45°}}{12 \,\underline{/15°}} =$

Problem 17.31 $\dfrac{20 + j30}{7 - j6} =$

Problem 17.32 $\dfrac{10 \,\underline{/53.1°}}{4 - j3} =$

(*Hint:* You may prefer to change the rectangular notation to polar and then use the rule for dividing polar notation.)

Application of Vector Algebra to Electric Circuits

We'll be using a different configuration of circuits for the problems in this section. Although the circuits and mathematical relations given will probably be new to you, they are given so that you may see actual applications of vector algebra. As you use vector algebra as a tool to solve the circuits, you will see why you must master vector algebra. The circuits and the mathematical relations given are discussed in detail in later units: however, you should now be able to solve the mathematical relations even though you may not be sure how they are developed.

Problem 17.33 In Figure 17.17, $\mathbf{E} = \mathbf{V}_R + \mathbf{V}_C$ (by KVL). If $\mathbf{V}_R = 10 \text{ V } \underline{/0°}$ and $\mathbf{V}_C = 10 \text{ V } \underline{/-90°}$, solve for $\mathbf{E}$ and express it in polar notation.

$$\mathbf{E} = 10 \text{ V } \underline{/0°} + 10 \text{ V } \underline{/-90°} = 10 - j10$$

$$=$$

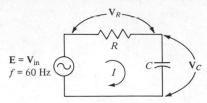

Figure 17.17 Circuit for Problem 17.33

Problem 17.34 In Figure 17.18, $\mathbf{E} = \mathbf{V}_R + \mathbf{V}_L + \mathbf{V}_C$ (by KVL). If $\mathbf{V}_R = 15 \text{ V } \underline{/0°}$, $\mathbf{V}_L = 20 \text{ V } \underline{/90°}$, and $\mathbf{V}_C = 10 \text{ V } \underline{/-90°}$, solve for $\mathbf{E}$, which will be in rectangular notation. Then express $\mathbf{E}$ in polar notation.

$$\mathbf{E} = 15 \text{ V } \underline{/0°} + 20 \text{ V } \underline{/90°} + 10 \text{ V } \underline{/-90°} = 15 + j20 - j10$$

$$=$$

Problem 17.35 In Figure 17.19, $\mathbf{I}_T = \mathbf{I}_R + \mathbf{I}_C$ (by KCL). If $\mathbf{I}_R = 2 \text{ mA } \underline{/0°}$ and $\mathbf{I}_C = 3 \text{ mA } \underline{/90°}$, solve for $\mathbf{I}_T$ and express it in polar notation.

$$\mathbf{I}_T =$$

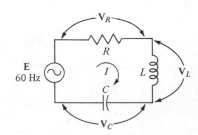

Figure 17.18 Circuit for Problem 17.34

Problem 17.36 In Figure 17.20, $\mathbf{I}_T = \mathbf{I}_R + \mathbf{I}_L$ (by KCL). If $\mathbf{I}_T = 10 \text{ mA}$ and $\mathbf{I}_L = 6 \text{ mA } \underline{/-90°}$, solve for the magnitude of $\mathbf{I}_R$.

$$\mathbf{I}_T = \mathbf{I}_R + \mathbf{I}_L$$
$$I_T^2 = I_R^2 + I_L^2$$
$$I_R = \sqrt{I_T^2 - I_L^2}$$
$$=$$

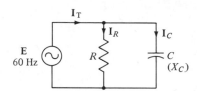

Figure 17.19 Circuit for Problem 17.35

Problem 17.37 In Figure 17.21, $\mathbf{V}_R = \mathbf{IR}$ and $\mathbf{V}_L = \mathbf{IX}_L$, where $X_L = 2\pi f L$ and is at an angle of $+90°$ with respect to $\mathbf{R}$. Given that $\mathbf{I} = 8 \text{ mA } \underline{/0°}$, $\mathbf{R} = 2 \text{ k}\Omega \underline{/0°}$, and $L = 1 \text{ H}$, solve for $\mathbf{V}_R$ and $\mathbf{V}_L$.

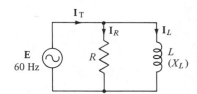

Figure 17.20 Circuit for Problem 17.36

Problem 17.38 In Figure 17.21,

$$\mathbf{I} = \frac{\mathbf{E}}{\mathbf{Z}} = \frac{\mathbf{E}}{R + jX_L}$$

Given that $E = 24 \text{ V}$, $R = 20 \text{ k}\Omega$, and $X_L = 12 \text{ k}\Omega$, find the magnitude of the denominator and then solve for the magnitude of $\mathbf{I}$.

Problem 17.39 In Figure 17.17, $\mathbf{I} = \mathbf{V}_R/\mathbf{R}$. Given that $\mathbf{V}_R = 10 \text{ V } \underline{/0°}$ and $\mathbf{R} = 2 \text{ k}\Omega \underline{/0°}$, solve for $\mathbf{I}$

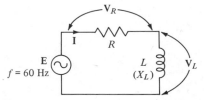

Figure 17.21 Circuit for Problem 17.37

Problem 17.40 In Figure 17.17, $\mathbf{X}_C = \mathbf{V}_C/\mathbf{I}$. Given that $\mathbf{V}_C = 30 \text{ V } \underline{/-90°}$ and $\mathbf{I} = 2 \text{ mA } \underline{/0°}$, solve for $\mathbf{X}_C$.

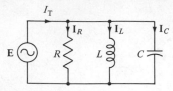

Figure 17.22 Circuit for Problem 17.41

Problem 17.41 In Figure 17.22,

$$\mathbf{I}_R = \frac{\mathbf{E}}{\mathbf{R}}, \qquad \mathbf{I}_L = \frac{\mathbf{E}}{jX_L}, \qquad \text{and} \qquad \mathbf{I}_C = \frac{\mathbf{E}}{-jX_C}.$$

Given $\mathbf{E} = 50$ V $\underline{/0°}$, $\mathbf{R} = 25$ kΩ $\underline{/0°}$, $X_L = 10$ kΩ, and $X_C = 5$ kΩ solve for each current.

Problem 17.42 In Figure 17.18, $\mathbf{Z} = R + jX_L - jX_C$. Given that $R = 4$ kΩ, $X_L = 3$ kΩ, and $X_C = 10$ kΩ, solve for $\mathbf{Z}$ and express in polar notation.

Problem 17.43 In Figure 17.19,

$$\mathbf{Z} = \frac{R\ \underline{/0°}\ X_C\ \underline{/-90°}}{R - jX_C} \text{ (product over sum)}$$

Given that $R = 10$ kΩ and $X_C = 5$ kΩ,

$$\mathbf{Z} = \frac{10\ \text{k}\Omega\ \underline{/0°}\ (5\ \text{k}\Omega\ \underline{/-90°})}{10\ \text{k}\Omega - j5\ \text{k}\Omega}$$

Change the denominator to polar notation, then solve for $\mathbf{Z}$ in polar notation.

Problem 17.44 In Figure 17.20,

$$\mathbf{Z} = \frac{R\ \underline{/0°}\ X_L\ \underline{/90°}}{R + jX_L}$$

Given that $R = 8$ kΩ and $X_L = 6$ kΩ, then

$$\mathbf{Z} = \frac{8\ \text{k}\Omega\ \underline{/0°}\ (6\ \text{k}\Omega\ \underline{/90°})}{8\ \text{k}\Omega + j6\ \text{k}\Omega}$$

Change the numerator into rectangular notation after taking the product, then use the conjugate of the denominator to solve for $\mathbf{Z}$. Express $\mathbf{Z}$ in rectangular notation.

Test

1. Carry out the instruction given in Objective 2. Plot the phasor $-5 + j4$.
2. Convert the following phasors from polar notation to rectangular notation.
   (a) $50\ \underline{/+150°} =$
   (b) $70\ \underline{/+300°} =$
   (c) $30\ \underline{/-135°} =$
   (d) $40\ \underline{/-210°} =$

3. Convert the following from rectangular notation to polar notation.
   (a) $+9 - j12 =$
   (b) $-4 - j4 =$
   (c) $+3 + j4 =$
   (d) $-2 + j4 =$

4. $(-5 + j6) + (12 - j4) =$

5. $(25 - j20) - (20 - j25) =$

6. $(4\ \underline{/+60°})(10\ \underline{/-15°}) =$

7. $\dfrac{36\ \underline{/+10°}}{12\ \underline{/-15°}} =$

8. $30\ \underline{/+35°} - 20\ \underline{/-25°} =$

9. $(8 + j7)(-6 - j6) =$

10. $\dfrac{3 - j5}{5 + j4} =$

Unit 18 **RC *and* RL Series Circuits**

In this unit you will learn how to apply Kirchhoff's voltage law and Ohm's law to simple *RC* and *RL* ac circuits. This involves the application of vector algebra to resistive and reactive components.

Objectives

After completing all the work associated with this unit, you should be able to:

1. Draw the voltage phasor diagram for an *RC* or *RL* series circuit. Label all phasors, including the resultant.
2. Draw the impedance triangles for an *RC* or *RL* series circuit. Label the sides.
3. Draw the power triangle for an *RC* or *RL* series circuit. Label the sides by name and magnitude.
4. Calculate the reactance, impedance, current, and voltage drop of an *RC* or *RL* series circuit when the resistor, reactive component, and applied input voltage are known.
5. Calculate the current, reactance, and impedance of an *RC* or *RL* series circuit when the voltage drops across the resistor and the reactive component are known.

Series *RC* and *RL* Circuits

Objective 1   Draw the voltage phasor diagram for an *RC* or *RL* series circuit. Label all phasors, including the resultant.

Series *RC* Circuit

Figure 18.1 shows a sketch of a typical *RC* series circuit. The applied ac voltage is defined as V_{in}, the voltage across resistor R is $\mathbf{V}_R$, and the voltage across capacitor C is $\mathbf{V}_C$. In your study of dc circuits, you learned that the current $\mathbf{I}$ was the same through all the series elements. Since this also holds true for series ac circuits, let's use the current as a reference and see what the voltage $\mathbf{V}_R$ across the resistor and the voltage $\mathbf{V}_C$ across the capacitor look like. Kirchhoff's voltage law (KVL) still applies around the closed loop in Figure 18.1.

Kirchhoff's voltage law states that the sum of all the voltage drops must be equal to the applied voltage. However, now we can't just add the maximum value of voltage across the resistor to the maximum value across the capacitor because, if we refer to Figure 18.2(a), we see that these maximums don't occur at the same time. On the other hand, we could add the instantaneous values of v_R and v_C and find the voltage across the series combination of R and C, but that is very tedious!

Recall that in the previous units we found that the voltage sine waves could be represented by stationary phasors (rms values). Then, from these voltage phasors, we could find the resultant phasor. Since current is common in all series circuits, we use it as a reference. From Figure 18.2(a) we see that the voltage v_R across the resistor reaches a maximum when current does. Therefore the phasor $\mathbf{V}_R$ lies along the same reference lines as our reference phasor $I \underline{/0°}$ or $\mathbf{I}$; see Fig. 18.2(b). The maximum value of v_R in Fig. 18.2(a) across R occurs at 90°, whereas voltage v_C across the capacitor becomes maximum 90° later in time, or at 180°. Therefore the voltage v_C across the capacitor lags the voltage v_R across the resistor by 90°, or i (the current through resistor and capacitor) leads v_C by 90°.

Recall the phrase *ICE*. The lengths of the phasors $\mathbf{V}_R$ and $\mathbf{V}_C$ are expressed as rms (effective) values of voltage. When we use stationary phasors, rms values are usually more convenient because most circuit values are given in terms of rms values. Figure 18.2(b) is known as the *voltage phasor* or *vector diagram* of the circuit. (Note that the resultant phasor is not shown.)

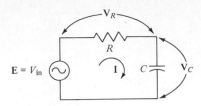

Figure 18.1 *RC* series circuit

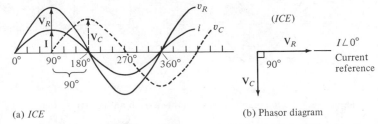

(a) *ICE* (b) Phasor diagram

Figure 18.2 Voltage and current sine waves in an *RC* series circuit

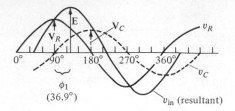

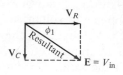

(a) Sum of the instantaneous values of v_R and v_C (b) Voltage phasor diagram including resultant phasor

Figure 18.3   Resultant voltage in an *RC* series circuit

We may find the voltage across the series *RC*, also known as v_{in}, by adding the instantaneous values of v_R and v_C [see Figure 18.3(a)], or we could complete the rectangle formed by the phasors $\mathbf{V}_C$ and $\mathbf{V}_R$ and find the diagonal [see Figure 18.3(b)]. This diagonal is known as the *resultant phasor* $\mathbf{E}$. Angle ϕ_1 represents the angle by which the resultant phasor $\mathbf{E}$ lags the phasor $\mathbf{V}_R$.

Series *RL* Circuit

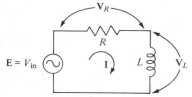

Figure 18.4   *RL* series circuit

Figure 18.4 shows a sketch of a typical *RL* series circuit. The applied voltage is $\mathbf{E}$, the voltage drop across resistor R is $\mathbf{V}_R$, and the voltage drop across L is $\mathbf{V}_L$. The resistance of the inductor coil L is considered to be small compared with $\mathbf{X}_L$ of the inductor, so that we may neglect the resistance of the inductor. The voltage and current sine waves in an *RL* series circuit are shown in Figure 18.5(a). Notice that the voltage v_L across the inductor leads the current i by 90°. (Recall the phrase *ELI*.) Since the current i is common to both L and R, we can construct a voltage phasor diagram, as shown in Figure 18.5(b). Again the reference phasor $I \underline{/0°}$ or $\mathbf{I}$ is in phase with the voltage phasor $\mathbf{V}_R$, while the phasor $\mathbf{V}_L$ leads the current $\mathbf{I}$ by 90°. The resultant phasor $\mathbf{E}$ makes an angle ϕ_2 with $\mathbf{V}_R$ and $\mathbf{I}$.

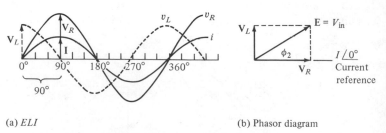

(a) *ELI*   (b) Phasor diagram

Figure 18.5   Voltage and current sine waves in an *RL* series circuit

Problem 18.1 Draw the voltage phasor diagrams for the circuits in Figure 18.6.

Impedance Triangles

Objective 2 Draw the impedance triangles for an *RC* or *RL* series circuit. Label the sides.

Let's take the phasor or voltage vector diagrams of Figures 18.3 and 18.5 and divide each side by current **I** (see Figure 18.7).

Notice in Figure 18.7 that we made triangles. We did not use the phasor $\mathbf{V}_C$ of Figure 18.3(b) or the phasor $\mathbf{V}_L$ of Figure 18.5(b) as one side of the triangle. Remember that current **I** is the same through each component in a series *RC* or *RL* circuit. (Notice that we used the vertical line that we call $\mathbf{V}_C/\mathbf{I} = X_C$ in Figure 18.7(a) and the vertical line that we call $\mathbf{V}_L/\mathbf{I} = X_L$ in Figure 18.7(b). We did this so that we could get a *triangle*, not a phasor diagram.)

These triangles in Figure 18.7 do not have arrows on the legs because they no longer represent phasors or vectors. These diagrams are called *impedance* triangles. Although impedance is not a phasor quantity, it is obtained mathematically by taking the vector sum of resistance and reactance. The angles ϕ_1 and ϕ_2 shown in Figure 18.7 are the same ϕ_1 and ϕ_2 shown in Figures 18.3 and 18.5. The impedance triangles of Figure 18.7 show you how to handle resistance and reactance to determine impedance in an *RC* or *RL* series ac circuit.

Two methods used in calculating impedance Z

Let's refer to a typical impedance triangle, drawn in Figure 18.8. This triangle is for an *RC* series circuit.

Remember how you can find the third side of a right triangle if you know two of them? One method is the use of Pythagoras' theorem:

$$Z^2 = R^2 + X^2 \tag{18.1}$$

$$|Z| = \sqrt{R^2 + X^2} \tag{18.2}$$

and

$$\phi = \arctan \frac{X}{R} \tag{18.3}$$

Another method is to find the angle ϕ first.

$$\phi = \arctan \frac{X}{R} \tag{18.4}$$

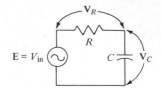

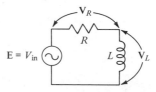

Figure 18.6 Circuit for Problem 18.1

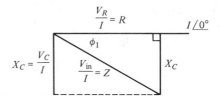

(a)

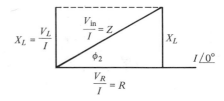

(b)

Figure 18.7 Impedance triangles for *RC* and *RL* series circuits

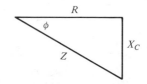

Figure 18.8 Impedance triangle for *RC* series circuit.

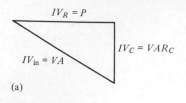

(a)

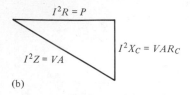

(b)

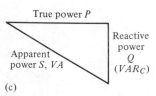

(c)

Figure 18.9 Power triangle with capacitive reactance

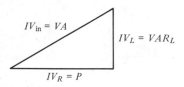

(a)

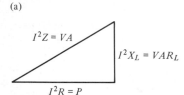

(b)

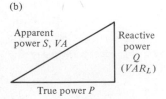

(c)

Figure 18.10 Power triangle with inductive reactance

and then

$$|Z| = \frac{X}{\sin \phi} \qquad (18.5)$$

or

$$|Z| = \frac{R}{\cos \phi} \qquad (18.6)$$

Problem 18.2 Draw the impedance triangle for each circuit in Problem 18.1 and label the sides.

Power Triangles

Objective 3 Draw the power triangle for an *RC* or *RL* series circuit. Label the sides with name and magnitude.

In Objective 2, we constructed an impedance triangle by taking the voltage phasor diagram and dividing each side by the current *I*. To construct a power triangle, we take the voltage phasor diagram and multiply each side by the current *I* [see Figures 18.9(a) and 18.10(a)].

Since power = $I \times V_R$ or $I^2 \times R$, we could have used the impedance triangles constructed in Objective 2 and multiplied each of their sides by I^2 [see Figures 18.9(b) and 18.10(b)]. The resulting power triangle is still the same, as shown in Figures 18.9(c) and 18.10(c). Note that on both the horizontal legs of Figures 18.9 and 18.10 we have power *P*. We learned that power in resistive circuits was true power that caused heat. Well, this is still true regarding the horizontal line in Figures 18.9(c) and 18.10(c), which is shown as *P* (true power).

We also know that energy is stored in an inductor or capacitor and that power is energy per unit time. Furthermore, this stored energy in an inductor or capacitor is not dissipated in the form of heat, so it is not true power. Since we say that a capacitor or inductor presents *reactance* to an ac circuit, let's call this *reactive power* or Volt Ampere Reactive (*VAR*). The subscript *C* or *L* shows that it is capacitive or inductive reactive power. This is shown in Figures 18.9 and 18.10. The unit of measurement of *VAR* is var.

Recall that to obtain the hypotenuse of the power triangle, we multiplied the actual input voltage V_{in} by the current *I* flowing in the circuit. From what we learned in dc circuits, this also means power. But not all this power generates heat, since a part of it is stored in the reactive element. Thus we give it the name *apparent power*. This is the power we are apparently supplying to the circuit. This is called *volt ampere* or *VA*.

Let us summarize the similarities between the voltage phasor diagram, impedance triangle, and power triangle, using as an example a series *RL* circuit. In Figure 18.11(a) we have constructed a voltage phasor diagram.

The voltage $\mathbf{V}_R$ is in phase with current $\mathbf{I}$, and $\mathbf{V}_L$ is leading the current by 90°. The resultant voltage $\mathbf{V}_{in}$ makes an angle ϕ with $\mathbf{V}_R$ and $\mathbf{I}$. In Figure 18.11(b) the resistance R and reactance X_L form an impedance Z that also makes an angle of ϕ with respect to R. The power triangle shown in Figure 18.11(c) again is similar to the voltage phasor diagram and the impedance triangle. The apparent power VA makes an angle ϕ with respect to true power P.

Problem 18.3 Draw the power triangle for each circuit of Problem 18.1 and label the sides.

Circuit Application

Objective 4 Calculate the reactance, impedance, current, and voltage drops of an *RC* or *RL* series circuit when the resistor, reactive component, and applied input voltage are known.

In Objective 3 you learned how to construct the impedance triangle, voltage phasor diagram, and power triangle for an *RC* or *RL* series ac circuit. In this objective you will learn how to apply this to two different problems. Example 18.1 shows you how to work with a practical *RC* circuit, while Example 18.2 presents an *RL* circuit.

Theoretical Analysis

Example 18.1 Figure 18.12 shows the series *RC* circuit. Find: (a) reactance X_C, (b) impedance $\mathbf{Z}$, (c) $\mathbf{I}$, (d) $\mathbf{V}_R$, $\mathbf{V}_C$, (e) P, VAR_C, VA.

Solution

(a) $X_C = \dfrac{1}{2\pi f C} = \dfrac{1}{2\pi (60)(0.1 \times 10^{-6})} = \dfrac{10^6}{377(0.1)} = 26.525 \ k\Omega$

(b) Figure 18.13(a) shows the impedance triangle. Using Pythagoras' theorem, we have

$$Z = \sqrt{R^2 + X_C^2} = \sqrt{(33 \ k\Omega)^2 + (26.525 \ k\Omega)^2} = 42.34 \ k\Omega$$

$$\phi = -\arctan \frac{X_C}{R} = -\arctan \frac{26.525 \ k\Omega}{33 \ k\Omega} = -38.8°$$

$$\mathbf{Z} = 42.34 \ k\Omega \ \underline{/-38.8°}$$

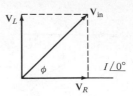

(a) Voltage phasor diagram

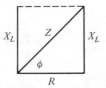

(b) Impedance triangle

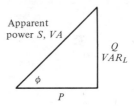

(c) Power triangle

Figure 18.11 Similarities between voltage phasor diagram, impedance triangle, and power triangle for an *RL* series circuit

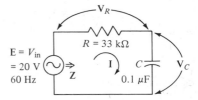

Figure 18.12 Circuit for Example 18.1

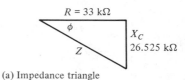

(a) Impedance triangle

Figure 18.13 Figures for solution of Example 18.1

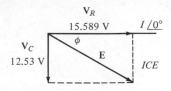

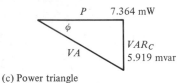

(b) Voltage phasor diagram

(c) Power triangle

Figure 18.13 (Continued)

(c) $\mathbf{I} = \dfrac{\mathbf{E}}{\mathbf{Z}} = \dfrac{20 \text{ V}}{42.34 \text{ k}\Omega} = 0.4724 \text{ mA } \underline{/0^\circ}$

(d) Figure 18.13(b) shows the voltage phasor diagram. Applying Ohm's law, we have

$$\mathbf{V}_R = \mathbf{I} \times \mathbf{R} = (0.4724 \text{ mA } \underline{/0^\circ})(33 \text{ k}\Omega \ \underline{/0^\circ}) = 15.589 \text{ V } \underline{/0^\circ}$$
$$\mathbf{V}_C = \mathbf{I} \times \mathbf{X}_C = (0.4724 \text{ mA } \underline{/0^\circ})(26.525 \text{ k}\Omega \ \underline{/-90^\circ})$$
$$= 12.53 \text{ V } \underline{/-90^\circ}$$

(e) Figure 18.13(c) shows the power triangle.

$$P = I^2 R = (0.4724 \text{ mA})^2 (33 \text{ k}\Omega) = 7.364 \text{ mW}$$

or

$$P = IV_R = (0.4724 \text{ mA})(15.589 \text{ V}) = 7.364 \text{ mW}$$
$$VAR_C = I^2 X_C = (0.4724 \text{ mA})^2 (26.525 \text{ k}\Omega) = 5.919 \text{ mvar}$$

or

$$VAR_C = IV_C = (0.4724 \text{ mA})(12.53 \text{ V}) = 5.919 \text{ mvar}$$
$$VA = IV_{\text{in}} = (0.4724 \text{ mA})(20 \text{ V}) = 9.448 \text{ mVA}$$

Checking VA by using Pythagoras' theorem, we obtain

$$VA = \sqrt{P^2 + VAR_C^2} = \sqrt{7.364^2 + 5.919^2}$$
$$VA = 9.448 \text{ mVA}$$

Example 18.2 The inductor in Figure 18.14 has an inductance of 10 H and a dc resistance of 500 Ω. Find: (a) X_L, (b) $\mathbf{Z}_{\text{circuit}}$, (c) $\mathbf{I}$, (d) $\mathbf{V}_R$, $\mathbf{V}_L$. (For $\mathbf{V}_L$, do not neglect the resistance of the inductor.)

Solution

(a) Reactance $X_L = 2\pi f L = 377 \times 10 = 3770 \ \Omega$

(b) $\mathbf{Z}_{\text{circuit}} = R + R_{\text{coil}} + jX_L = 1000 + 500 + j3770$
$$= \sqrt{R_{\text{total}}^2 + X_L^2} = \sqrt{(1500)^2 + (3770)^2}$$
$$= 4057 \ \Omega \ \underline{/+68.3^\circ}$$

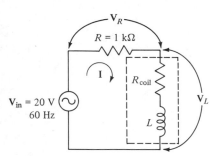

Figure 18.14 Circuit for Example 18.2

(c) $\mathbf{I} = \dfrac{\mathbf{V}_{\text{in}}}{\mathbf{Z}_{\text{circuit}}} = \dfrac{20 \text{ V}}{4057 \ \Omega} = 4.93 \text{ mA } \underline{/0^\circ}$

(d) $\mathbf{V}_R = \mathbf{I} \times \mathbf{R} = 4.93 \text{ mA } \underline{/0^\circ} \times 1 \text{ k}\Omega \ \underline{/0^\circ} = 4.93 \text{ V } \underline{/0^\circ}$
$$\mathbf{V}_L = \mathbf{I} \times \mathbf{Z}_{\text{coil}}$$
Since $\mathbf{Z}_{\text{coil}} = R_{\text{coil}} + jX_L = 500 + j3770 = 3803 \ \Omega \ \underline{/82.45^\circ}$
$$\therefore \quad \mathbf{V}_L = \mathbf{I} \times \mathbf{Z}_{\text{coil}} = (4.93 \text{ mA } \underline{/0^\circ})(3803 \ \Omega \ \underline{/82.45^\circ})$$
$$= 18.75 \text{ V } \underline{/82.45^\circ}$$

Problem 18.4 Figure 18.15 shows a series RC circuit. Complete Table 18.1 and draw the voltage phasor diagrams.

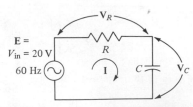

Figure 18.15 Circuit for Problem 18.4

Table 18.1 Table for Problem 18.4

C, μF	X_C, Ω	R, Ω	Z, Ω	I, mA	V_R, V	V_C, V	P, mW	VAR, mvar	VA, mVA
0.1		5 kΩ							
0.1		27 kΩ							
1		5 kΩ							
1		27 kΩ							

Table 18.2 Table for Problem 8.5

L, H	X_L, Ω	R, Ω	Z, Ω	I, mA	V_R, V	V_L, V	P, mW	VAR, mvar	VA, mVA
14		5 kΩ							
14		10 kΩ							

Problem 18.5 Figure 18.16 shows a series *RL* circuit. Complete Table 18.2 and draw the voltage phasor diagram.

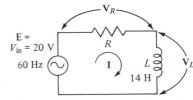

Figure 18.16 Circuit for Problem 18.5

Analysis of Laboratory Data

Objective 5 Calculate the current, reactance, and impedance of an *RC* or *RL* series circuit when the voltage drops across the resistor and the reactive component are known.

In this objective you will learn how to use given data on laboratory voltage to find other characteristics of a series *RC* or *RL* circuit. The following three examples show this.

Example 18.3 The voltage drop $\mathbf{V}_R$ across a 33-kΩ resistor is 15.6 V, and the voltage drop $\mathbf{V}_C$ across an unknown capacitor is 12.5 V. (See Figure 18.17.) Find: (a) current **I**, (b) reactance X_C, (c) capacitance *C*, (d) circuit impedance **Z**.

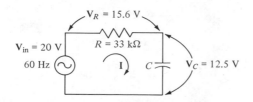

Figure 18.17 Circuit for Example 18.3

Solution Ohm's law for an ac circuit is $\mathbf{I} = \mathbf{V}/\mathbf{Z}$, where $\mathbf{Z}$ could be a resistor, a reactance, or a vector combination of both, depending on the circuit and data given. If you know a resistor and the voltage across that resistor, then

(a) $\mathbf{I} = \dfrac{\mathbf{V}_R}{\mathbf{R}} = \dfrac{15.6 \text{ V } \underline{/0°}}{33 \text{ k}\Omega \underline{/0°}} = 0.473 \text{ mA } \underline{/0°}$

(b) $X_C = \dfrac{V_C}{I} = \dfrac{12.5 \text{ V}}{0.473 \text{ mA}} = 26.44 \text{ k}\Omega$

(c) $X_C = \dfrac{1}{2\pi f C}$ $\therefore$ $C = \dfrac{1}{2\pi f X_C} = \dfrac{1}{377(26.44 \text{ k}\Omega)} = 0.1 \ \mu\text{F}$

(d) $Z = \sqrt{R^2 + X_C^2} = \sqrt{(33 \text{ k}\Omega)^2 + (26.4 \text{ k}\Omega)^2} = 42.3 \text{ k}\Omega$

$$\phi = -\arctan \dfrac{X_C}{R} = -\tan^{-1}\left(\dfrac{26.4 \text{ k}\Omega}{33 \text{ k}\Omega}\right) = -38.7°$$

or $\sin \phi = -\dfrac{X_C}{Z}$

$$\therefore \quad \phi = -\arcsin \dfrac{X_C}{Z} = -\sin^{-1}\left(\dfrac{26.44 \text{ k}\Omega}{42.3 \text{ k}\Omega}\right) = -38.7°$$

$$\mathbf{Z} = 42.3 \text{ k}\Omega \ \underline{/-38.7°}$$

Example 18.4 The voltage drop $\mathbf{V}_R$ across a 20-kΩ resistor is 19.6 V and the voltage drop $\mathbf{V}_L$ across an unknown inductor is 3.7 V. (See Figure 18.18.) Assume that the resistance of the coil L is negligible compared with the circuit resistance. Find: (a) current $\mathbf{I}$, (b) reactance X_L, (c) L, (d) circuit impedance $\mathbf{Z}$. (e) Draw the impedance triangle and label the sides.

Solution

(a) $\mathbf{I} = \dfrac{\mathbf{V}_R}{\mathbf{R}} = \dfrac{19.6 \text{ V } \underline{/0°}}{20 \text{ k}\Omega \underline{/0°}} = 0.98 \text{ mA } \underline{/0°}$

(b) $X_L = \dfrac{V_L}{I} = \dfrac{3.7 \text{ V}}{0.98 \text{ mA}} = 3.775 \text{ k}\Omega$

(c) $L = \dfrac{X_L}{2\pi f} = \dfrac{3.775 \text{ k}\Omega}{377} = 10.015 \text{ H} \approx 10 \text{ H}$

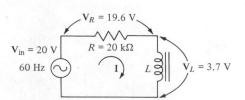

Figure 18.18   Circuit for Example 18.4

(d) $\phi = \arctan \dfrac{X_L}{R}$

$\phi = \tan^{-1}\left(\dfrac{3775\ \Omega}{20\ \text{k}\Omega}\right) = 10.7°$

$Z = \dfrac{X_L}{\sin \phi} = \dfrac{3.775\ \text{k}\Omega}{\sin 10.7°}$

$Z = 20.35\ \text{k}\Omega$

$\mathbf{Z} = 20.35\ \text{k}\Omega\ \underline{/10.7°}$

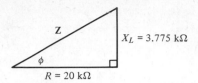

Figure 18.19 Figure for solution e of Example 18.4

(e) Figure 18.19 shows the impedance triangle.

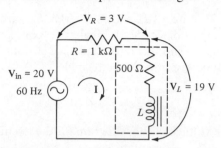

Figure 18.20 Circuit for Example 18.5

Example 18.5 The inductance of the inductor L in Figure 18.20 is not known. However, its resistance is equal to 500 Ω; therefore do not neglect it in your calculations. Find (a) current $\mathbf{I}$, (b) Z_{coil}, (c) X_L, (d) L, (e) $\mathbf{Z}_{\text{circuit}}$.

Solution

(a) Applying Ohm's law, we find that

$$\mathbf{I} = \frac{\mathbf{V}_R}{\mathbf{R}} = \frac{3\ \text{V}\ \underline{/0°}}{1\ \text{k}\Omega\ \underline{/0°}} = 3\ \text{mA}\ \underline{/0°}$$

(b) We cannot find X_L of the inductor by using $\mathbf{V}_L/\mathbf{I}$ because $\mathbf{V}_L$ is *not* across a *pure* inductance. The $\mathbf{V}_L$ reading is across an R_{coil} in series with L; therefore it is across an impedance.

$$Z_{\text{coil}} = \frac{\mathbf{V}_L}{\mathbf{I}} = \frac{19\ \text{V}}{3\ \text{mA}} = 6.333\ \text{k}\Omega$$

(c) To find X_L, we can make use of the Pythagorean theorem:

$$Z_{\text{coil}}^2 = R_{\text{coil}}^2 + X_{L,\ \text{coil}}^2$$
$$\therefore \quad X_L = \sqrt{Z_{\text{coil}}^2 - R_{\text{coil}}^2} = \sqrt{(6.333\ \text{k}\Omega)^2 - (500\ \Omega)^2}$$
$$X_L = 6.313\ \text{k}\Omega$$

Note that, even with 500 Ω of resistance in the coil, $X_L \approx Z_{\text{coil}}$.

(d) $L = \dfrac{X_L}{2\pi f} = \dfrac{6.313\ \text{k}\Omega}{377} = 16.75\ \text{H}$

(e) $\mathbf{Z}_{\text{circuit}} = R + (R_{\text{coil}} + jX_L) = 1000 + (500 + j6313)$

$\qquad = 1500 + j6313 = 6.489\ \text{k}\Omega\ \underline{/76.64°}$

Problem 18.6 Figure 18.21 shows a series *RC* circuit. Find: (a) $\mathbf{I}$, (b) reactance X_C, (c) C, (d) impedance $\mathbf{Z}$. (e) Draw the impedance triangle.

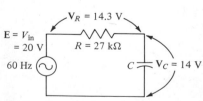

Figure 18.21 Circuit for Problem 18.6

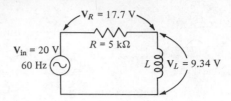

Figure 18.22 Circuit for Problem 18.7

Problem 18.7 Figure 18.22 shows a series *RL* circuit. Assume that the resistance of coil *L* is negligible compared with the circuit resistance. Find: (a) **I**, (b) reactance X_L, (c) *L*, (d) impedance **Z**. (e) Draw the impedance triangle.

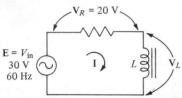

Figure 18.23 Circuit for Test Problem 1

Test

1. Draw the voltage phasor diagram for the circuit shown in Figure 18.23 and calculate V_L.

$$V_L = \underline{\hspace{2cm}} \quad \underline{\big\lfloor\hspace{1cm}}$$

2. Figure 18.24 shows a series *RL* circuit. Calculate the following.

$$X_L = \underline{\hspace{3cm}}$$

$$Z = \underline{\hspace{2cm}} \quad \underline{\big\lfloor\hspace{1cm}}$$

$$I = \underline{\hspace{2cm}} \quad \underline{\big\lfloor\hspace{1cm}}$$

$$V_R = \underline{\hspace{2cm}} \quad \underline{\big\lfloor\hspace{1cm}}$$

$$V_C = \underline{\hspace{2cm}} \quad \underline{\big\lfloor\hspace{1cm}}$$

$$P = \underline{\hspace{3cm}}$$

$$VAR_L = \underline{\hspace{3cm}}$$

$$VA = \underline{\hspace{3cm}}$$

Draw the impedance triangle and the power triangle.

3. Figure 18.25 shows a series *RC* circuit. Calculate the following.

$$I = \underline{\hspace{2cm}} \quad \underline{\big\lfloor\hspace{1cm}}$$

$$X_C = \underline{\hspace{3cm}}$$

$$C = \underline{\hspace{3cm}}$$

$$Z = \underline{\hspace{2cm}} \quad \underline{\big\lfloor\hspace{1cm}}$$

Draw the impedance triangle.

Figure 18.24 Circuit for Test Problem 2

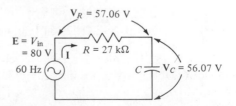

Figure 18.25 Circuit for Test Problem 3

4. Figure 18.26 shows a series *RC* circuit. Calculate the following.

$$Z = \underline{\hspace{2cm}} \, \underline{/\hspace{1cm}}$$

$$I = \underline{\hspace{2cm}} \, \underline{/\hspace{1cm}}$$

$$V_R = \underline{\hspace{2cm}} \, \underline{/\hspace{1cm}}$$

$$V_C = \underline{\hspace{2cm}} \, \underline{/\hspace{1cm}}$$

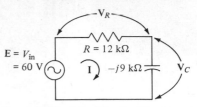

Figure 18.26 Circuit for Test Problem 4

Draw the voltage phasor diagram.

5. The series *RL* circuit shown in Figure 18.27 has 20 V dropped across the 10-kΩ resistor. Given that the inductance has a voltage drop of 45.83 V, find the value of *L*. (Assume that the resistance of coil *L* is small compared with circuit impedance.)

$$L = \underline{\hspace{3cm}}$$

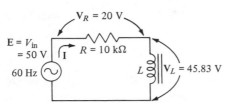

Figure 18.27 Circuit for Test Problem 5

Unit 19 *Series RLC Circuits*

In Unit 18 you learned how to apply Kirchhoff's voltage law and Ohm's law to a simple *RL* or *RC* series ac circuit. Now you will learn how to apply the same laws to a series circuit consisting of a resistor, an inductor, and a capacitor. In reading this unit you will learn more about a series *RLC* ac circuit.

Objectives

After completing all the work associated with this unit, you should be able to:

1. Draw an impedance triangle, the voltage phasor diagram, and the power triangle for any specified *RLC* series circuit. Label the sides with name and magnitude.
2. Calculate the current, voltage drop, power, and impedance (magnitude and angle) in any series ac circuit when R, X_L or L, X_C or C, and the applied voltage and frequency are known.
3. Calculate the current $\mathbf{I}$, X_C or C, and X_L or L in a series *RLC* circuit when the voltage drops across each component are known.

Series *RLC* Circuit

Objective 1 Draw an impedance triangle, the voltage phasor diagram, and the power triangle for any specified *RLC* series circuit. Label the sides with name and magnitude.

Figure 19.1 shows a typical *RLC* series circuit. Since current $\mathbf{I}$ is common in a series circuit, we shall use $I\ \underline{/0^\circ}$ or $\mathbf{I}$ as our refer-

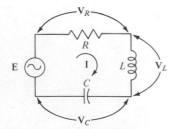

Figure 19.1 Sketch of a typical series *RLC* circuit

368

ence. Let us express the voltage drop across each individual element in terms of **I**.

$$\mathbf{V}_R = I\ \underline{/0^\circ} \times R\ \underline{/0^\circ} \qquad \text{(in phase)} \qquad (19.1)$$

$$\mathbf{V}_L = I\ \underline{/0^\circ} \times X_L\ \underline{/+90^\circ} \qquad (ELI) \qquad (19.2)$$

$$\mathbf{V}_C = I\ \underline{/0^\circ} \times X_C\ \underline{/-90^\circ} \qquad (ICE) \qquad (19.3)$$

The voltage phasor diagram in Figure 19.2 shows a sketch of the voltages across each individual component.

Since current **I** or $I\ \underline{/0^\circ}$ is used as a reference in the series circuit of Figure 19.1, it is drawn along the horizontal axis in Figure 19.2. And since, in Equation (19.1), the voltage $\mathbf{V}_R$ across the pure resistance R is in phase with the current **I**, it is also drawn along the horizontal axis. The voltage $\mathbf{V}_L$ in Equation (19.2) is leading the reference current **I** by 90° (*ELI*), and so it is drawn upward on the voltage phasor diagram. The voltage $\mathbf{V}_C$ lags the reference current **I** by 90° (*ICE*), as is drawn downward on the diagram. Note that the voltages $\mathbf{V}_L$ and $\mathbf{V}_C$ are directly opposite, or 180° out of phase. To simplify the voltage phasor diagram, we define $\mathbf{V}_X$ as the *net voltage difference* between $\mathbf{V}_L$ and $\mathbf{V}_C$. As shown in Figure 19.2, we have assumed that $\mathbf{V}_L$ is larger than $\mathbf{V}_C$.

In constructing the voltage phasor diagram, let us check and see whether KVL still applies, that is, whether the applied voltage **E** is equal to the sum of all the voltage drops around any closed path. KVL must still hold. The three voltage drops $\mathbf{V}_R$, $\mathbf{V}_L$, and $\mathbf{V}_C$ are written in Equations (19.1), (19.2), and (19.3). Since these voltages are all out of phase with one another, we cannot add them algebraically. But we can write KVL in vector notation. It is

$$\mathbf{E} = V_R + jV_L - jV_C = V_R + j(V_L - V_C) \qquad (19.4)$$

$$\mathbf{E} = V_R + jV_X \qquad (19.5)$$

where $V_X = (V_L - V_C)$

Therefore the applied voltage **E** is equal to the vector sum of the voltage drops around any closed path. In Equation (19.5), the applied voltage **E** is equal to the vector sum of $\mathbf{V}_R$ and $\mathbf{V}_X$.

The voltage $\mathbf{V}_X$ has three values: positive, zero, and negative. In this unit, we shall discuss only two conditions—$\mathbf{V}_X$ positive and $\mathbf{V}_X$ negative. The condition of $\mathbf{V}_X$ equal to zero is the condition of *resonance*. We shall discuss this condition of resonance in a later unit.

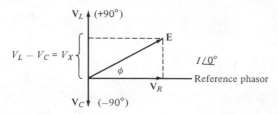

Figure 19.2 Voltage phasor diagram of a series *RLC* circuit (V_X positive)

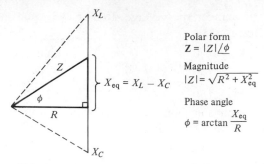

Figure 19.3 Impedance triangle for series *RLC* circuit ($X_L > X_C$)

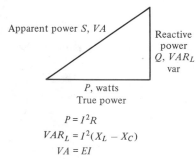

Figure 19.4 Power triangle for series *RLC* circuit ($X_L > X_C$)

V_X Positive

When the reactance $\mathbf{X}_L > \mathbf{X}_C$, then $\mathbf{V}_L > \mathbf{V}_C$. Since $\mathbf{V}_X = \mathbf{V}_L - \mathbf{V}_C$, then $\mathbf{V}_X$ is positive. The condition of $\mathbf{V}_X$ positive was drawn as a voltage phasor diagram in Figure 19.2. Figure 19.3 shows the corresponding impedance triangle. Note that the phase angle ϕ is positive. Figure 19.3 also shows the polar form of writing and evaluating impedance.

Figure 19.4 shows the corresponding power triangle for $\mathbf{X}_L > \mathbf{X}_C$ or $\mathbf{V}_L > \mathbf{V}_C$.

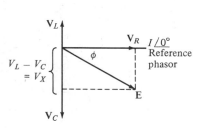

Figure 19.5 Voltage phasor diagram for series *RLC* circuit ($X_L < X_C$)

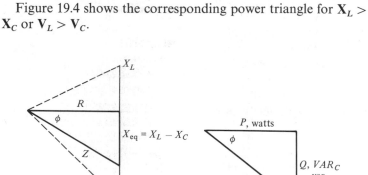

(a) Impedance triangle   (b) Power triangle

Figure 19.6 Impedance and power triangles for series *RLC* circuit ($X_L < X_C$)

V_X Negative

When the reactance $\mathbf{X}_L < \mathbf{X}_C$, then $\mathbf{V}_L < \mathbf{V}_C$. Since $\mathbf{V}_X = \mathbf{V}_L - \mathbf{V}_C$, then $\mathbf{V}_X$ is negative. Figure 19.5 shows the voltage phasor diagram corresponding to this condition.

Note that the phase angle ϕ is negative. Figure 19.6 shows the corresponding impedance and power triangles.

Problem 19.1 Figure 19.7 shows a series *RLC* circuit. Assume that $\mathbf{X}_L > \mathbf{X}_C$. Draw: (a) an impedance triangle, (b) a voltage phasor diagram, (c) a power triangle. Label all the sides.

Figure 19.7 Circuit for Problem 19.1

Circuit Applications

Objective 2   Calculate the current, voltage drop, power, and impedance (magnitude and angle) in any series ac circuit when R, X_L or L, X_C or C, and the applied voltage and frequency are known.

In this objective you apply the material you learned in Objective 1 to a practical *RLC* circuit. You do this by applying two well-known laws—Kirchhoff's voltage law (KVL) and Ohm's law.

To calculate the current **I** flowing in any circuit, we must first find the total series circuit impedance. We do this by using Equation (19.4), which is

$$\mathbf{E} = V_R + jV_L - jV_C \tag{19.6}$$

Since current **I** is the same in a series circuit, we can rewrite Equation (19.6) by using Ohm's law:

$$\mathbf{E} = \mathbf{I}(R + jX_L - jX_C) \tag{19.7}$$

or

$$\mathbf{E} = \mathbf{I}(\mathbf{Z}) \tag{19.8}$$

where **Z** is the series impedance of the circuit, which is equal to $R + jX_L - jX_C$.

Therefore the *total series circuit* impedance is

$$\mathbf{Z} = R + jX_L - jX_C \tag{19.9}$$

or, in vector form,

$$|Z| = \sqrt{R^2 + X_{eq}^2} \qquad \text{where } X_{eq} = X_L - X_C \tag{19.10}$$

and

$$\phi = \arctan \frac{X_{eq}}{R}$$

Once the impedance **Z** is known and the applied voltage is given, Ohm's law gives the resultant current **I**, by

$$\mathbf{I} = \frac{\mathbf{E}}{\mathbf{Z}} \tag{19.11}$$

We shall now use the equations for impedance **Z** [Equation (19.9)] and current **I** [Equation (19.11)] in an example to illustrate how these two fundamental laws are all that are needed to solve a practical *RLC* series circuit.

Theoretical Analysis

Example 19.1   Figure 19.8 shows a series *RLC* circuit. Assume that R_{coil} is small (negligible). (a) Find the voltage drops $\mathbf{V}_R$, $\mathbf{V}_L$, and $\mathbf{V}_C$; draw the voltage phasor diagram. (b) Draw the power triangle; include all calculated values.

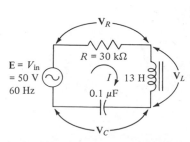

Figure 19.8   Circuit for Example 19.1

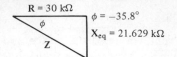

(a) Impedance triangle

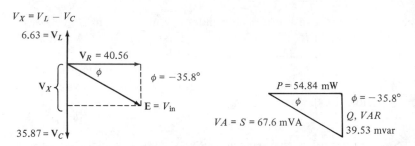

$V_X = V_L - V_C$

(b) Voltage phasor diagram   (c) Power triangle

Figure 19.9   Diagrams for solution of Example 19.1

Solution

(a) $X_L = 2\pi f L = 2\pi(60)(13)$ $X_C = \dfrac{1}{2\pi f C} = \dfrac{1}{377(0.1 \times 10^{-6})}$

$\qquad = 377(13) = 4901\ \Omega$

$\qquad\qquad\qquad\qquad\qquad\qquad\qquad = 26{,}530\ \Omega$

$\mathbf{X}_{eq} = jX_L - jX_C = j4901 - j26{,}530$

$\qquad\quad = -j21{,}629 \qquad \text{or capacitive reactance}$

$Z = \sqrt{R^2 + X_{eq}^2} = \sqrt{(30\text{ k}\Omega)^2 + (21.6\text{ k}\Omega)^2} = 36.98\text{ k}\Omega$

$\quad \approx 37\text{ k}\Omega$

$\phi = \arctan\dfrac{21{,}629}{30{,}000} = -35.8°$

$\therefore \quad \mathbf{I} = \dfrac{\mathbf{E}}{\mathbf{Z}} = \dfrac{50\text{ V}}{36.98\text{ k}\Omega} = 1.352\text{ mA }\underline{/0°}$

$\mathbf{V}_R = \mathbf{I} \cdot \mathbf{R} = 1.352\text{ mA }\underline{/0°} \times 30\text{ k}\Omega\ \underline{/0°} = 40.56\text{ V }\underline{/0°}$

$\mathbf{V}_L = \mathbf{I} \cdot \mathbf{X}_L = 1.352\text{ mA }\underline{/0°} \times 4901\ \Omega\ \underline{/90°} = 6.63\text{ V }\underline{/90°}$

$\mathbf{V}_C = \mathbf{I} \cdot \mathbf{X}_C = 1.352\text{ mA }\underline{/0°} \times 26.53\text{ k}\Omega\ \underline{/-90°}$

$\qquad\qquad = 35.87\text{ V }\underline{/-90°}$

The voltage phasor diagram is shown in Figure 19.9(b).

(b) $P = I^2 R = IV_R = (1.352\text{ mA})(40.56\text{ V}) = 54.84\text{ mW}$

$VAR = I^2 X_{eq} = IV_X = (1.352\text{ mA})(29.24\text{ V}) = 39.53\text{ mvar}$

$VA = I^2 Z = IV_{in} = (1.352\text{ mA})(50\text{ V}) = 67.6\text{ mVA}$

Note: The angle ϕ is the same for the impedance triangle, the voltage phasor diagram, and the power triangle.

Problem 19.2   Figure 19.10 shows a series *RLC* circuit. Assume that the R_{coil} of the inductor is negligible.

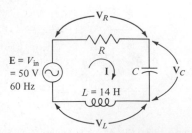

Figure 19.10   Circuit for Problem 19.2

Table 19.1 Table for Problem 19.2

	$C, \mu F$	R, Ω	X_C, Ω	X_L, Ω	$\mathbf{Z}, \Omega$	$\mathbf{I}, mA$	$\mathbf{V}_R, V$	$\mathbf{V}_L, V$	$\mathbf{V}_C, V$
1	1	1 kΩ							
2	1	5 kΩ							
3	1	10 kΩ							
4	0.1	10 kΩ							
5	0.1	33 kΩ							
	$C, \mu F$	R					P	VAR	VA
6	1	1 kΩ							

Complete Table 19.1. Draw the voltage phasor diagrams for rows 1 and 4. Draw the power triangle for row 6.

Analysis of Laboratory Data

Objective 3   Calculate the current **I**, X_C or C, and X_L or L in a series RLC circuit when the voltage drops across each component are known.

This objective is very similar to Objective 2 in the sense that two fundamental laws—KVL and Ohm's law—are all that are needed to solve an RLC series circuit. The information given in this objective represents typical laboratory data collected in an experimental setup. Here the components L and C are not known, but the measured voltages across each element with a known applied voltage are all the information that is needed to find X_L, L, X_C and C. This objective is best illustrated by the following example.

Example 19.2   The voltage drops across each component in Figure 19.11 are given. Assume that the coil resistance is negligible compared with the circuit resistance. Find: (a) current **I**, (b) capacitive reactance X_C, (c) C, (d) inductive reactance X_L, (e) L. (f) Draw the voltage phasor diagram to scale.

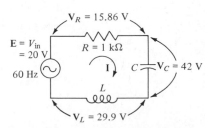

Figure 19.11   Circuit for Example 19.2

Solution Since we know the voltage drop across a known resistor, we can find the circuit current **I** without finding Z of the circuit.

(a) $\mathbf{I} = \dfrac{\mathbf{V}_R}{\mathbf{R}} = \dfrac{15.86 \text{ V } \underline{/0°}}{1 \text{ k}\Omega \underline{/0°}} = 15.86 \text{ mA } \underline{/0°}$

(b) $X_C = \dfrac{V_C}{I} = \dfrac{42 \text{ V}}{15.86 \text{ mA}} = 2.648 \text{ k}\Omega$

(c) $X_C = \dfrac{1}{2\pi f C}$ $\therefore$ $C = \dfrac{1}{2\pi f X_C} = \dfrac{1}{377(2.648 \text{ k}\Omega)}$

      $= 1.0017 \ \mu\text{F} \approx 1 \ \mu\text{F}$

(d) $X_L = \dfrac{V_L}{I} = \dfrac{29.9 \text{ V}}{15.86 \text{ mA}} = 1.885 \text{ k}\Omega$

(e) $L = \dfrac{X_L}{2\pi f} = \dfrac{1885}{377} = 5 \text{ H}$

(f) The voltage phasor diagram is shown in Figure 19.12.

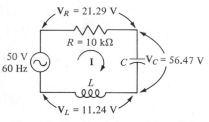

Figure 19.12   Circuit for solution of Example 19.2

Problem 19.3   The voltage drops across each component in the series *RLC* circuit shown in Figure 19.13 are known. Find: (a) Circuit current **I**, (b) X_C, (c) C, (d) X_L, (e) L. (f) Draw the voltage phasor diagram to scale.

Figure 19.13   Circuit for Problem 19.3

Test

1. Determine **I**, X_C, C, X_L, and L for the circuit shown in Figure 19.14. Draw the voltage phasor diagram, showing the resultant.

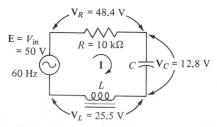

Figure 19.14   Circuit for Test Problem 1

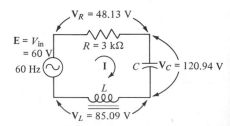

Figure 19.15   Circuit for Test Problem 2

2. Determine **I**, X_C, C, X_L, L, VAR_L, VAR_C, and VA for the circuit shown in Figure 19.15. Draw the voltage phasor diagram and the power triangle.

3. Calculate $\mathbf{Z}$, $\mathbf{I}$, $\mathbf{V}_R$, $\mathbf{V}_L$, and $\mathbf{V}_C$ for the circuit shown in Figure 19.16. Draw the impedance triangle.

$\mathbf{Z} = $ _____ $\angle$ _____

$\mathbf{I} = $ _____ $\angle$ _____

$\mathbf{V}_R = $ _____ $\angle$ _____

$\mathbf{V}_L = $ _____ $\angle$ _____

$\mathbf{V}_C = $ _____ $\angle$ _____

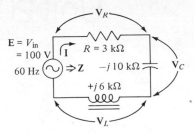

Figure 19.16 Circuit for Test Problem 3

4. Calculate X_L, X_C, $\mathbf{Z}$, $\mathbf{I}$, $\mathbf{V}_R$, $\mathbf{V}_L$, and $\mathbf{V}_C$ for the circuit shown in Figure 19.17. Draw the impedance triangle and the voltage phasor diagram.

$X_L = $ _____

$X_C = $ _____

$\mathbf{Z} = $ _____ $\angle$ _____

$\mathbf{I} = $ _____ $\angle$ _____

$\mathbf{V}_R = $ _____ $\angle$ _____

$\mathbf{V}_L = $ _____ $\angle$ _____

$\mathbf{V}_C = $ _____ $\angle$ _____

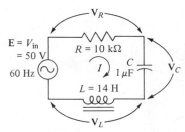

Figure 19.17 Circuit for Test Problem 4

Unit 20 *Parallel Circuits*

In Unit 19 you learned how to analyze a series *RLC* circuit. In this unit you will learn the characteristics of a parallel circuit.

Objectives

After completing all the work associated with this unit, you should be able to:

1. Draw the current phasor diagram and calculate the impedance (magnitude and angle) and the total current **I** for an *RC* or *RL* parallel circuit at a given frequency.

2. Draw the current phasor diagram for a parallel *RLC* circuit. Label all the phasors including the resultant.

3. Recall the definition, letter symbol, and units for the following: conductance, susceptance, and admittance.

4. Draw an admittance triangle for a parallel *RLC* circuit.

5. Calculate the impedance for a parallel *RC* or *RL* circuit using conductance, susceptance, and admittance.

6. (*Advanced*) Calculate the impedance (magnitude and angle) and the current through each component in a parallel *RLC* circuit when the input voltage and frequency are known.

Impedance and Current Phasor Diagrams for *RL*, *RC*, and *RLC* Parallel Circuits

Objective 1 Draw the current phasor diagram and calculate the impedance (magnitude and angle) and the total current **I** for an *RC* or *RL* parallel circuit at a given frequency.

This first objective is broken down into two parts: *RC* parallel circuits and *RL* parallel circuits. We shall describe these two types of circuits in the following paragraphs.

RC Parallel Circuit

Figure 20.1 shows a parallel *RC* circuit. The parallel combination of *R* and *C* forms a parallel impedance $\mathbf{Z}_{\parallel}$ to the ac voltage source **E**. The total current $\mathbf{I}_T$ drawn from the source is split into a resistive component $\mathbf{I}_R$ and a capacitive component $\mathbf{I}_C$. The applied voltage **E** is across the resistor *R* and also across capacitor *C*. The parallel impedance $\mathbf{Z}_{\parallel}$ of two parallel elements is determined by taking the *product divided by the sum*. However, we are now working with vector quantities. The parallel impedance is

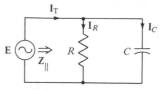

Figure 20.1   Parallel *RC* circuit

$$\mathbf{Z}_{\parallel} = \mathbf{R}\|\mathbf{X}_C = \frac{R\,\underline{/0^\circ} \times X_C\,\underline{/-90^\circ}}{R - jX_C} \quad \text{where } X_C = 1/(2\pi f C) \quad (20.1)$$

Notice that the parallel impedance $\mathbf{Z}_{\parallel}$ depends on the value of the capacitive reactance X_C, which depends on frequency. Moreover, in Equation (20.1), the numerator could very easily be written in polar notation, while the denominator is written in vector rectangular form. The denominator must be changed to polar form so that the division of two vector quantities can be performed easily.

Important note: Parallel impedance $\mathbf{Z}_{\parallel}$ is different from series impedance **Z**, and cannot be calculated by using the impedance triangle for series circuits. This will be shown in Objective 6. The impedance triangle can be used only to solve series circuits, because current in a series circuit is common. Therefore series impedance is found by adding the resistances and reactances vectorially. In parallel circuits, the voltage across each component is the same, and thus a different technique must be used.

Before we try to calculate the total current $\mathbf{I}_T$, let us look at the parallel *RC* circuit in Figure 20.1 and see what is common to each element. In a series circuit, the current is the same through all the elements, while in a parallel circuit, the voltage is the same across each of the elements. Since the voltage is common in parallel circuits, we shall use it as reference and let it be $E\,\underline{/0^\circ}$, that is, **E** with a phase angle of 0°. Now the current through the resistor *R* still satisfies Ohm's law and is

$$\mathbf{I}_R = \frac{E\,\underline{/0^\circ}}{R\,\underline{/0^\circ}} \quad \text{in phase} \quad (20.2)$$

while the current $\mathbf{I}_C$ through the capacitor is

$$\mathbf{I}_C = \frac{E\,\underline{/0^\circ}}{X_C\,\underline{/-90^\circ}} = \left|\frac{E}{X_C}\right|\,\underline{/+90^\circ} \quad ICE \quad (20.3)$$

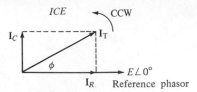

Figure 20.2 Current phasor diagram for parallel *RC* circuit

Note that the current $\mathbf{I}_C$ is leading the applied voltage $\mathbf{E}$. Figure 20.2 shows a sketch of the current phasor diagram. Notice that for parallel circuits the reference phasor $E\underline{/0°}$ is $\mathbf{E}$, and it is drawn on the horizontal axis. The current $\mathbf{I}_R$ through resistor R is in phase with the applied voltage, while the current $\mathbf{I}_C$ through capacitor C is leading the applied voltage $\mathbf{E}$ (*ICE*). Kirchhoff's current law must hold for parallel circuits. That is, the vector sum of all the branch currents must add up to the total line current $\mathbf{I}_T$.

$$\mathbf{I}_T = I_R + jI_C \tag{20.4}$$

or

$$\mathbf{I}_T = \sqrt{I_R^2 + I_C^2}\,\Big/\tan^{-1}\frac{I_C}{I_R} \tag{20.5}$$

We can also calculate the magnitude of the total current $\mathbf{I}_T$ by applying Ohm's Law to the total circuit.

$$\mathbf{I}_T = \frac{\mathbf{E}}{\mathbf{Z}_{\|}} \tag{20.6}$$

where $\mathbf{E}$ = applied voltage and $\mathbf{Z}_{\|}$ = parallel impedance of overall circuit

The following example illustrates the concepts learned so far.

Example 20.1 Figure 20.3 shows a parallel *RC* circuit. Find: (a) parallel impedance $\mathbf{Z}_{\|}$, (b) current $\mathbf{I}_R$, (c) current $\mathbf{I}_C$, (d) total current $\mathbf{I}_T$. (e) Draw the current phasor diagram.

Solution

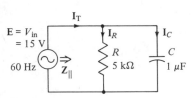

Figure 20.3 Circuit for Example 20.1

(a) $X_C = \dfrac{1}{2\pi f C} = \dfrac{1}{(377)(1 \times 10^{-6})} = 2653\ \Omega$

$R = 5\ \text{k}\Omega$

Using Equation (20.1), we obtain

$$\mathbf{Z}_{\|} = \mathbf{R}\|\mathbf{X}_C = \frac{R\underline{/0°} \times X_C\underline{/-90°}}{R - jX_C} = \frac{5\ \text{k}\Omega\underline{/0°} \times 2653\ \Omega\underline{/-90°}}{5\ \text{k}\Omega - j2653\ \Omega}$$

$$= \frac{13.265 \times 10^6\ \Omega\underline{/-90°}}{5.66\ \text{k}\Omega\underline{/-27.9°}} = 2.34\ \text{k}\Omega\underline{/-62.1°}$$

(See Figure 20.4.)

(b) Using Equation (20.2), we find that

$$\mathbf{I}_R = \frac{E\underline{/0°}}{R\underline{/0°}} = \frac{15\ \text{V}\underline{/0°}}{5\ \text{k}\Omega\underline{/0°}} = 3\ \text{mA}\underline{/0°} = 3\ \text{mA}$$

(c) From Equation (20.3),

$$\mathbf{I}_C = \frac{E\underline{/0°}}{X_C\underline{/-90°}} = \frac{15\ \text{V}\underline{/0°}}{2653\ \Omega\underline{/-90°}} = 5.654\ \text{mA}\underline{/+90°} = j5.654\ \text{mA}$$

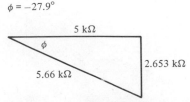

$\phi = -27.9°$

Figure 20.4 Vector diagram for denominator term in solution (a) of Example 20.1

(d) $\mathbf{I}_T = I_R + jI_C$

$\qquad = 3 \text{ mA} + j5.654 \text{ mA}$

Since $\mathbf{I}_T$ is a vector sum of $\mathbf{I}_R$ and $\mathbf{I}_C$,

$$|I_T| = \sqrt{I_R^2 + I_C^2} \qquad\qquad \phi = \tan^{-1}\left(\frac{5.654}{3}\right)$$

$$\qquad = \sqrt{3^2 + (5.654)^2} \text{ mA}$$

$$|I_T| = 6.41 \text{ mA} \qquad\qquad \phi = 62.1°$$

We can find $\mathbf{I}_T$ another way by using $\mathbf{Z}_{||}$ from part (a).

$$\mathbf{I}_T = \frac{E \underline{/0°}}{Z_{||} \underline{/\phi}} = \frac{15 \text{ V} \underline{/0°}}{2.34 \text{ k}\Omega \underline{/-62.1°}} = 6.41 \text{ mA} \underline{/+62.1°}$$

(e) The current phasor diagram is shown in Figure 20.5.

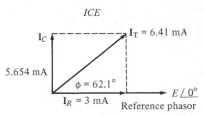

Figure 20.5 Diagram for solution (e) of Example 20.1

RL Parallel Circuit

Figure 20.6 shows a parallel *RL* circuit. We are now neglecting the resistance of the inductor. Unit 21 will discuss the condition in which the inductor has a dc component of resistance. The parallel combination of *R* and *L* forms a parallel impedance $\mathbf{Z}_{||}$ to the ac voltage source $\mathbf{E}$. The total current $\mathbf{I}_T$ drawn from the source is split into a resistive component $\mathbf{I}_R$ and an inductive component $\mathbf{I}_L$. We can compute the parallel impedance $\mathbf{Z}_{||}$ the same way we computed parallel *RC* circuits:

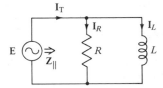

Figure 20.6 Parallel *RL* circuit

$$\mathbf{Z}_{||} = \mathbf{R}\|\mathbf{X}_L = \frac{(R \underline{/0°})(X_L \underline{/+90°})}{R + jX_L} \qquad \text{where } X_L = 2\pi f L \quad (20.7)$$

The parallel impedance $\mathbf{Z}_{||}$ depends on the resistance *R* and on the inductive reactance of *L* at a given frequency. This impedance, as in parallel *RC* circuits, cannot be calculated by using the impedance triangle of series circuits.

Since the applied voltage $\mathbf{E}$ is common to both *R* and *L*, it is used as a reference phasor $E \underline{/0°}$. The current $\mathbf{I}_R$ through resistor *R*, by Ohm's law, is

$$\mathbf{I}_R = \frac{E \underline{/0°}}{R \underline{/0°}} \qquad \text{in phase} \qquad\qquad (20.8)$$

$$\mathbf{I}_L = \frac{E \underline{/0°}}{X_L \underline{/+90°}} = \left|\frac{E}{X_L}\right| \underline{/-90°} \qquad ELI \qquad (20.9)$$

Note that the current $\mathbf{I}_L$ is lagging the applied voltage $\mathbf{E}$. Figure 20.7 shows a sketch of the current phasor diagram. Again, for parallel circuits, the applied voltage $\mathbf{E}$ or $E \underline{/0°}$ is drawn on the horizontal axis. The current $\mathbf{I}_R$ through resistor *R* is in phase with the applied voltage, while the current $\mathbf{I}_L$ through inductor *L* is lagging the applied voltage $\mathbf{E}$. This is shown in Equation (20.9).

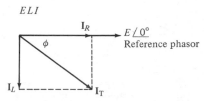

Figure 20.7 Current phasor diagram for parallel *RL* circuit

Kirchhoff's current law must hold for parallel *RL* circuits. That is, the vector sum of all the branch currents must add up to the total line current I_T. In mathematical terms,

$$I_T = I_R - jI_L \qquad (20.10)$$

or

$$I_T = \sqrt{I_R^2 + I_L^2} \; \left| -\tan^{-1}\left(\frac{I_L}{I_R}\right)\right. \qquad (20.11)$$

We can also calculate the magnitude of the total current I_T by applying Ohm's law to the total circuit.

$$I_T = \frac{E}{Z_{\parallel}} \qquad (20.12)$$

where E = applied voltage and $Z_{\parallel}$ = parallel impedance of overall circuit.

The following example illustrates the theory you have learned about parallel *RL* circuits.

Example 20.2 Figure 20.8 shows a parallel *RL* circuit. Find: (a) Parallel impedance $Z_{\parallel}$, (b) current I_R, (c) current I_L, (d) total current I_T. (e) Draw the current phasor diagram.

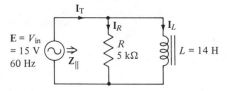

Figure 20.8 Circuit for Example 20.2

Solution
(a) $X_L = 2\pi f L = 377(14) = 5278 \; \Omega$
with $R = 5 \; k\Omega$ and using Equation (20.7),

$$Z_{\parallel} = R\|X_L = \frac{R\,\underline{/0°} \cdot X_L\,\underline{/+90°}}{R + jX_L}$$

$$= \frac{5\;k\Omega\,\underline{/0°} \cdot 5278\;\Omega\,\underline{/+90°}}{5\;k\Omega + j5278\;\Omega}$$

> Denom. evaluation is shown in Figure 20.9

$$= \frac{26.39 \times 10^6 \; \Omega \; \underline{/90°}}{7.27\;k\Omega \; \underline{/46.6°}}$$

$$Z_{\parallel} = 3.63 \; k\Omega \; \underline{/+43.4°} \qquad \text{Polar notation}$$

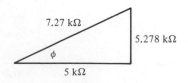

$\phi = 46.6°$

Figure 20.9 Diagram for denominator evaluation for solution (a) of Example 20.2

(b) $I_R = \dfrac{E \underline{/0°}}{R \underline{/0°}} = \dfrac{15 \text{ V} \underline{/0°}}{5 \text{ k}\Omega \underline{/0°}} = 3 \text{ mA} \underline{/0°} = 3 \text{ mA}$

(c) $I_L = \dfrac{E \underline{/0°}}{X_L \underline{/+90°}} = \dfrac{15 \text{ V} \underline{/0°}}{5278 \text{ }\Omega \underline{/+90°}} = 2.842 \text{ mA} \underline{/-90°}$

$\qquad = -j2.842 \text{ mA}$

(d) $I_T = I_R - jI_L = 3 \text{ mA} - j2.842 \text{ mA}$

Converting from rectangular notation to polar notation, and using Pythagoras' theorem, we obtain:

$|I_T| = \sqrt{I_R^2 + I_L^2}$
$\qquad = \sqrt{3^2 + 2.842^2} \text{ mA} = 4.13 \text{ mA}$

$\phi = -\tan^{-1}\left(\dfrac{2.842}{3}\right)$
$\qquad = -43.4°$

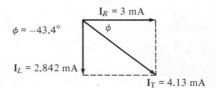

Figure 20.10 Diagram for solution e of Example 20.2

This may be found another way by using $Z_{\parallel}$ from part (a):

$$I_T = \dfrac{E \underline{/0°}}{Z_{\parallel} \underline{/0°}} = \dfrac{15 \text{ V} \underline{/0°}}{3.63 \text{ k}\Omega \underline{/+43.4°}}$$

$$I_T = 4.13 \text{ mA} \underline{/-43.4°}$$

(e) Figure 20.10 shows the current phasor diagram.

Problem 20.1 Figure 20.11 shows a parallel *RC* circuit. Complete Table 20.1, and draw the current phasor diagram for each circuit.

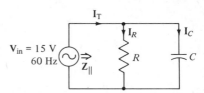

Figure 20.11 Circuit for Problem 20.1

Table 20.1 Table for Problem 20.1

$C, \mu F$	R, Ω	X_C, Ω	I_C, mA	I_R, mA	$I_T = (I_R + jI_C), \text{mA}$	$Z_{\parallel}, \Omega$	$I_T = \left(\dfrac{V_{in}}{Z_{\parallel}}\right), \text{mA}$
1	10 kΩ						
1	33 kΩ						
0.1	5 kΩ						
0.1	10 kΩ						
0.1	33 kΩ						

Table 20.2 Table for Problem 20.2

R, Ω	X_L, Ω	$\mathbf{I}_L$, mA	$\mathbf{I}_R$, mA	$\mathbf{I}_T = (I_R - jI_L)$, mA	$\mathbf{Z}_{\parallel}, \Omega$	$\mathbf{I}_T = \left(\dfrac{\mathbf{V}_{in}}{\mathbf{Z}_{\parallel}}\right)$, mA
$5\,\text{k}\Omega$						
$10\,\text{k}\Omega$						
$33\,\text{k}\Omega$						

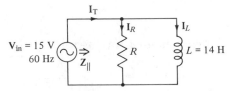

Figure 20.12   Circuit for Problem 20.2

Problem 20.2   Figure 20.12 gives a parallel *RL* circuit. Complete Table 20.2 and draw the current phasor diagram for each circuit.

RLC Parallel Circuit

Objective 2   Draw the current phasor diagram for a parallel *RLC* circuit. Label all the phasors, including the resultant.

Figure 20.13 shows a parallel *RLC* circuit. The total current $\mathbf{I}_T$ is divided into three branches, $\mathbf{I}_R$, $\mathbf{I}_L$, and $\mathbf{I}_C$. Again we are neglecting any coil resistance. Since the applied voltage $\mathbf{E}$ is common across each parallel element, we shall use it as a reference.

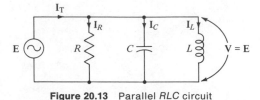

Figure 20.13   Parallel *RLC* circuit

Using $E \underline{/0°}$ as a reference and applying Ohm's law, we calculate each branch current as

$$\mathbf{I}_R = \frac{E \underline{/0°}}{R \underline{/0°}} \quad \text{in phase} \tag{20.13}$$

$$\mathbf{I}_C = \frac{E \underline{/0°}}{X_C \underline{/-90°}} \quad \text{or} \quad \left|\frac{E}{X_C}\right| \underline{/+90°} \quad ICE \tag{20.14}$$

$$\mathbf{I}_L = \frac{E \underline{/0°}}{X_L \underline{/+90°}} \quad \text{or} \quad \left|\frac{E}{X_L}\right| \underline{/-90°} \quad ELI \tag{20.15}$$

Since KCL must still hold, the vector sum of these three branch currents $\mathbf{I}_R$, $\mathbf{I}_L$, and $\mathbf{I}_C$ must equal the total current $\mathbf{I}_T$. This is shown by the equation

$$\mathbf{I}_T = I_R + jI_C - jI_L \tag{20.16}$$

or

$$\mathbf{I}_T = I_R + j(I_C - I_L) \tag{20.17}$$

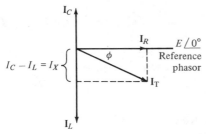

Figure 20.14 Current phasor diagram of a parallel *RLC* circuit ($I_L > I_C$)

We can draw Equation (20.17) as a current phasor diagram, as shown in Figure 20.14. The applied voltage **E**, being common to R, L, and C, is used as a reference phasor and is drawn on the horizontal axis as $E \underline{/0°}$. The current $\mathbf{I}_R$ through resistor R is also in phase with the parallel reference voltage **E**. From Equation (20.14), the current $\mathbf{I}_C$ is leading the applied reference voltage **E** by 90°, while from Equation (20.15), the current $\mathbf{I}_L$ is lagging the applied reference voltage **E** by 90°. In Figure 20.14, we are assuming that the current $\mathbf{I}_L > \mathbf{I}_C$. This results in a net current $\mathbf{I}_X$ (defined as $\mathbf{I}_C - \mathbf{I}_L$) in the $\mathbf{I}_L$ direction. As a result, the total current $\mathbf{I}_T$ is the vector sum of $\mathbf{I}_R$ and $\mathbf{I}_X$. The phase angle ϕ is negative.

We can easily calculate the overall parallel impedance $\mathbf{Z}_{\parallel}$ of the parallel *RLC* circuit by applying Ohm's law to the applied voltage $\mathbf{E}$ and the total current $\mathbf{I}_T$:

$$\mathbf{Z}_{\parallel} = |Z|\,\underline{/+\phi} = \frac{E\,\underline{/0°}}{I_T\,\underline{/-\phi}} \qquad (20.18)$$

where $E\,\underline{/0°}$ = the applied reference voltage and $I_T\,\underline{/-\phi}$ = total phasor current.

This method can be applied to any number of components in parallel. Note that it is very difficult to determine the impedance by the method of "product divided by sum" when there are more than two components in parallel in an ac circuit. This difficult method is presented in Objective 6.

Problem 20.3 Draw a current phasor diagram for the circuit shown in Figure 20.15. Label all the sides.

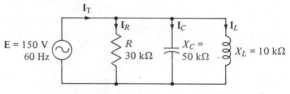

Figure 20.15 Circuit for Problem 20.3

Conductance, Susceptance, and Admittance

Objective 3 Recall the definition, letter symbol, and units for the following: conductance, susceptance, and admittance.

Definitions

The terms conductance, susceptance, and admittance are used most often in parallel circuits. The definition of each term is as follows.

Conductance G Units: siemens = S (or sometimes mhos = ohms spelled backward, ℧). Conductance is a measure of the ability of a resistor to pass current,

$$G = \frac{1}{R(\Omega)} \qquad (20.19)$$

Susceptance B   Units: siemens (mhos)—S(℧). Susceptance is a measure of the ability of a pure capacitor or inductor to pass ac current.

$$B = \frac{1}{X(\Omega)}$$ (20.20)

where **X** is capacitive or inductive reactance.

Admittance Y   Units: siemens (mhos)—S(℧). Admittance is a measure of the ability of a complete circuit (containing *R* and/or *C* and/or *L*) to pass ac current.

$$Y = \frac{1}{Z(\Omega)}$$ (20.21)

Problem 20.4   Define the three terms: conductance, susceptance, and admittance. Include symbols and units for each term.

Admittance Triangle

Objective 4   Draw an admittance triangle for a parallel *RLC* circuit.

In Objective 3 you learned the definition of conductance, susceptance, and admittance. In this objective you will learn to apply these three terms to the solving of parallel *RLC* circuits.

An admittance triangle is as important to parallel *RLC* circuits as an impedance triangle is to series *RLC* circuits. Recall that for a series circuit the current **I** is the same through each component, and is used as a reference. This is shown in Table 20.3, which summarizes what you have learned about series *RLC* circuits and also compares and summarizes the concepts used in solving parallel *RLC* circuits. (See page 386.)

The series impedance **Z** in Table 20.3 is given as

$$Z = R + jX_L - jX_C$$ (20.22)

The impedance triangle is drawn in Figure 20.16 for a condition of $X_L > X_C$. Having found the series impedance **Z**, we can prove KVL:

$$E = I(Z)$$ (20.23)

$$E = I(R + jX_L - jX_C)$$ (20.24)

$$E = V_R + jV_L - jV_C$$ (20.25)

A voltage phasor diagram is drawn below the impedance triangle of Figure 20.16.

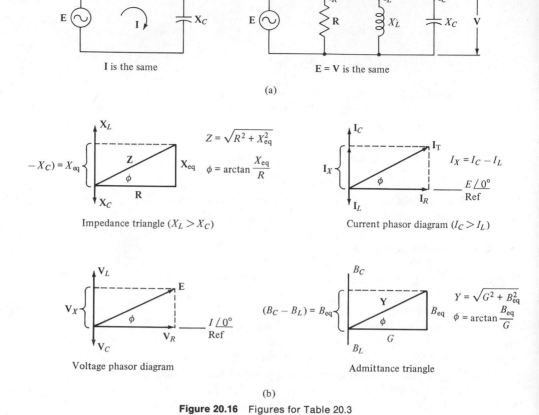

Figure 20.16 Figures for Table 20.3

Table 20.3 Summary of Series and Parallel *RLC* Circuits

	Series	Parallel
	$\mathbf{Z} = R + jX_L - jX_C$	$\mathbf{Z}_{\|} = R\|(jX_L)\|(-jX_C)$
	$\mathbf{Z} = R + j(X_L - X_C)$	KCL holds
	$\mathbf{Z} = R + jX_{eq}$ where $X_{eq} = X_L - X_C$	$\mathbf{I}_T = I_R + jI_C - jI_L$
	$\mathbf{E} = \mathbf{I}[\mathbf{Z}]$	$\mathbf{I}_T = VG + jVB_C - jVB_L$
	$\mathbf{E} = \mathbf{I}[R + jX_L - jX_C]$	$\mathbf{I}_T = V[G + jB_C - jB_L]$
	$\mathbf{E} = V_R + jV_L - jV_C$	$\mathbf{I}_T = \mathbf{V}[\mathbf{Y}]$
	KVL holds	

For parallel circuits, the voltage **V** or **E** is the same across each element, so it is used as a reference. The parallel impedance $\mathbf{Z}_{\|}$ is very messy to calculate by conventional techniques of "product over sum." An easier way is to use the currents flowing through each leg, then apply Ohm's law.

Referring first to Equation (20.16), we know that

$$\mathbf{I_T} = I_R + jI_C - jI_L \qquad (20.26)$$

which is actually KCL. Now applying the definitions of conductance, susceptance, and admittance, we have

$$\mathbf{I}_R = \frac{\mathbf{V}}{\mathbf{R}} = \mathbf{V}\left(\frac{1}{\mathbf{R}}\right) = \mathbf{VG} \qquad (20.27)$$

$$\mathbf{I}_C = \frac{\mathbf{V}}{\mathbf{X}_C} = \mathbf{V}\left(\frac{1}{\mathbf{X}_C}\right) = \mathbf{VB}_C \qquad (20.28)$$

$$\mathbf{I}_L = \frac{\mathbf{V}}{\mathbf{X}_L} = \mathbf{V}\left(\frac{1}{\mathbf{X}_L}\right) = \mathbf{VB}_L \qquad (20.29)$$

Substituting these equations into Equation (20.26), we get

$$\mathbf{I_T} = I_R + jI_C - jI_L \qquad (20.30)$$
$$\mathbf{I_T} = VG + jVB_C - jVB_L \qquad (20.31)$$

or

$$\mathbf{I_T} = V(G + jB_C - jB_L) \qquad (20.32)$$
$$\mathbf{I_T} = V(\mathbf{Y}) \qquad (20.33)$$

where admittance $\mathbf{Y} = G + jB_C - jB_L$

Notice that Equation (20.24) shows that impedance can be added vectorially in a series *RLC* circuit, because current **I** is the same through each component. On the other hand, Equation (20.32) shows that admittance can be added vectorially in a parallel *RLC* circuit, because the voltage across each component is common. The current phasor diagram and the admittance triangle are drawn in Figure 20.16.

We can draw the admittance triangles for two conditions: $\mathbf{B}_L > \mathbf{B}_C$ and $\mathbf{B}_L < \mathbf{B}_C$. These two conditions are shown in Figure 20.17 and 20.18.

Problem 20.5 Draw an admittance triangle for a parallel *RLC* circuit where $\mathbf{B}_C > \mathbf{B}_L$. Label all sides: conductance **G**, susceptance **B**, and admittance **Y**.

Impedance Calculation for *RC* and *RL* Parallel Circuit

Objective 5 Calculate the impedance for a parallel *RC* or *RL* circuit, using conductance, susceptance, and admittance.

In this objective, we present two examples showing how to calculate the parallel impedance of a circuit using conductance, susceptance, and admittance.

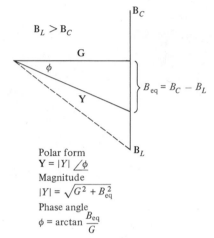

Polar form
$\mathbf{Y} = |Y| \underline{/\phi}$
Magnitude
$|Y| = \sqrt{G^2 + B_{eq}^2}$
Phase angle
$\phi = \arctan \dfrac{B_{eq}}{G}$

Figure 20.17 Admittance triangle for $B_L > B_C$

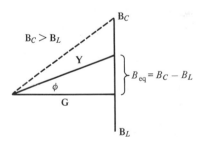

Polar form: $\mathbf{Y} = |Y| \underline{/\phi}$
(ϕ positive angle)

$\mathbf{I_T} = \mathbf{EY}$ or $\mathbf{I_T} = \mathbf{E}(\frac{1}{\mathbf{Z}})$

Figure 20.18 Admittance triangle for $B_C > B_L$

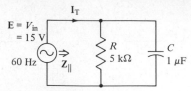

Figure 20.19   Circuit for Example 20.3

Example 20.3   Figure 20.19 shows the parallel RC circuit. Calculate: (a) conductance $\mathbf{G}$, (b) susceptance B_C, (c) admittance $\mathbf{Y}$. (d) Draw the admittance triangle. (e) Find total current $\mathbf{I}_T$. (f) Find impedance $\mathbf{Z}_{\parallel}$.

Solution

(a) $\mathbf{G} = \dfrac{1}{\mathbf{R}} = \dfrac{1}{5\text{ k}\Omega\ \underline{/0^\circ}} = 0.2 \times 10^{-3}$ S (siemens) $\underline{/0^\circ}$

(b) $B_C = 2\pi f C = 377(1 \times 10^{-6}) = 0.377 \times 10^{-3}$ S

(c) Admittance $Y = \sqrt{G^2 + B_C^2}$
$$= \sqrt{(0.2 \times 10^{-3})^2 + (0.377 \times 10^{-3})^2}$$

$$Y = 4.268 \times 10^{-4} \text{ S}$$

$$\phi = \arctan \frac{B_C}{G} = \tan^{-1}\left(\frac{0.377 \times 10^{-3}}{0.2 \times 10^{-3}}\right)$$

$$= \tan^{-1}(1.885) = 62.05^\circ$$

$$\mathbf{Y} = 4.268 \times 10^{-4} \text{ S } \underline{/+62.05^\circ}$$

(d) Figure 20.20 shows the admittance triangle.

(e) $\mathbf{I}_T$ from admittance calculations:

$$\mathbf{I}_T = \mathbf{E} \cdot \mathbf{Y} = (15 \text{ V } \underline{/0^\circ})(4.268 \times 10^{-4} \text{ S } \underline{/+62.05^\circ})$$
$$= 6.4 \text{ mA } \underline{/62.05^\circ}$$

(f) Impedance from admittance calculations:

$$\mathbf{Z}_{\parallel} = \frac{1}{\mathbf{Y}} = \frac{1}{4.268 \times 10^{-4}\ \underline{/62.05^\circ}} = 2.34 \text{ k}\Omega\ \underline{/-62.05^\circ}$$

Impedance check by Ohm's law:

$$\mathbf{Z}_{\parallel} = \frac{\mathbf{E}}{\mathbf{I}_T} = \frac{15 \text{ V } \underline{/0^\circ}}{6.4 \text{ mA } \underline{/+62.05^\circ}} = 2.34 \text{ k}\Omega\ \underline{/-62.05^\circ}$$

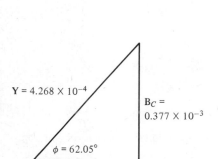

Figure 20.20   Admittance triangle for solution (d) of Example 20.3

Example 20.4   Figure 20.21 shows a parallel RL circuit. Find: (a) conductance $\mathbf{G}$, (b) susceptance B_L, (c) admittance $\mathbf{Y}$. (d) Draw the admittance triangle. (e) Find the total current $\mathbf{I}_T$. (f) Find the parallel impedance $\mathbf{Z}_{\parallel}$.

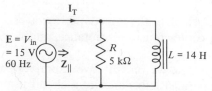

Figure 20.21   Circuit for Example 20.4

Solution

(a) $G = \dfrac{1}{R} = \dfrac{1}{5 \text{ k}\Omega \ /0^\circ} = 0.2 \times 10^{-3} \text{ S} \ /0^\circ$

(b) $B_L = \dfrac{1}{2\pi f L} = \dfrac{1}{(377)(14)} = 0.1894 \times 10^{-3} \text{ S}$

(c) Admittance $|Y| = \sqrt{G^2 + B_L^2}$

$\qquad = \sqrt{(0.2 \times 10^{-3})^2 + (0.1894 \times 10^{-3})^2}$

$\qquad |Y| = 2.755 \times 10^{-4} \text{ S}$

$\qquad \phi = -\arctan \dfrac{B_L}{G} = -\tan^{-1} \left(\dfrac{0.1894 \times 10^{-3}}{0.2 \times 10^{-3}} \right)$

$\qquad\qquad = -\tan^{-1}(0.947) = -43.4^\circ$

$\qquad Y = 2.755 \times 10^{-4} \text{ S} \ /-43.4^\circ$

(d) The admittance triangle is shown in Figure 20.22.

(e) Total current $\mathbf{I}_T = \mathbf{E} \cdot \mathbf{Y} = 15 \text{ V} \ /0^\circ \times 2.755 \times 10^{-4} \ /-43.4^\circ$

$\qquad\qquad = 4.13 \text{ mA} \ /-43.4^\circ$

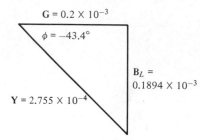

Figure 20.22 Diagram for solution (d) of Example 20.4

(f) $\mathbf{Z}_{||} = \dfrac{1}{\mathbf{Y}} = \dfrac{1}{2.755 \times 10^{-4} \ /-43.4^\circ} = 3.63 \text{ k}\Omega \ /+43.4^\circ$

Impedance check by Ohm's law:

$\qquad \mathbf{Z}_{||} = \dfrac{\mathbf{E}}{\mathbf{I}_T} = \dfrac{15 \ /0^\circ}{4.13 \text{ mA} \ /-43.4^\circ} = 3.63 \text{ k}\Omega \ /+43.4^\circ$

Problem 20.6 For the parallel circuit shown in Figure 20.23, solve for: (a) conductance **G**, (b) susceptance B_C, (c) admittance **Y**. (d) Draw the admittance triangle. (e) Find the total current $\mathbf{I}_T$. (f) Find the parallel impedance $\mathbf{Z}_{||}$.

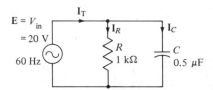

Figure 20.23 Circuit for Problem 20.6

Problem 20.7 For the parallel circuit shown in Figure 20.24, solve for (a) conductance **G**, (b) susceptance B_L, (c) admittance **Y**. (d) Draw the admittance triangle. (e) Find the total current $\mathbf{I}_T$. (f) Find the parallel impedance $\mathbf{Z}_{||}$.

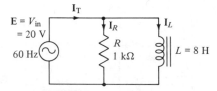

Figure 20.24 Circuit for Problem 20.7

Impedance Calculation for *RLC* Parallel Circuit

Objective 6 (*Advanced*) Calculate the impedance (magnitude and angle) and the current through each component in a parallel *RLC* circuit when the input voltage and frequency are known.

In this objective, you will learn how to calculate the impedance of a parallel *RLC* circuit by three different methods. This is illustrated in the following example.

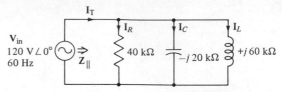

Figure 20.25   Circuit for Example 20.5

Example 20.5   Calculate the impedance (magnitude and angle), the total admittance, and the current through each component in the *RLC* parallel circuit shown in Figure 20.25. Draw the current phasor diagram.

Solution   Note that the inductive and capacitive reactances are given. This eliminates some calculations. We shall present three methods of solving this example here. The third method shows why the impedance triangle of series circuits cannot be used in solving the impedance of a parallel circuit.

Method 1   Since there are more than two circuit elements in parallel, and we have to find all the currents, we shall find $\mathbf{Z}_{||}$ by Ohm's law ($\mathbf{Z}_{||} = \mathbf{E}/\mathbf{I}_T$).

$$\mathbf{I}_R = \frac{\mathbf{V}_R}{\mathbf{R}} = \frac{120 \text{ V } \underline{/0^\circ}}{40 \text{ k}\Omega \ \underline{/0^\circ}} = 3 \text{ mA } \underline{/0^\circ} = 3 \text{ mA}$$

$$\mathbf{I}_L = \frac{\mathbf{V}_L}{\mathbf{X}_L} = \frac{120 \text{ V } \underline{/0^\circ}}{60 \text{ k}\Omega \ \underline{/+90^\circ}} = 2 \text{ mA } \underline{/-90^\circ} = -j2 \text{ mA}$$

$$\mathbf{I}_C = \frac{\mathbf{V}_C}{\mathbf{X}_C} = \frac{120 \text{ V } \underline{/0^\circ}}{20 \text{ k}\Omega \ \underline{/-90^\circ}} = 6 \text{ mA } \underline{/+90^\circ} = +j6 \text{ mA}$$

$$\therefore \quad \mathbf{I}_T = \mathbf{I}_R + \mathbf{I}_L + \mathbf{I}_C$$
$$= 3 \text{ mA} - j2 \text{ mA} + j6 \text{ mA} = 3 \text{ mA} + j4 \text{ mA}$$

The current phasor diagram is shown in Figure 20.26.

$$\mathbf{I}_T = \sqrt{3^2 + 4^2} \text{ mA} = 5 \text{ mA}, \qquad \phi = \tan^{-1}\left(\frac{4}{3}\right) = +53.1^\circ$$

$$\therefore \quad \mathbf{I}_T = 5 \text{ mA } \underline{/+53.1^\circ}$$

The parallel impedance $\mathbf{Z}_{||}$ can be calculated by Ohm's law:

$$\mathbf{Z}_{||} = \frac{\mathbf{E}}{\mathbf{I}_T} = \frac{120 \text{ V } \underline{/0^\circ}}{5 \text{ mA } \underline{/+53.1^\circ}} = 24 \text{ k}\Omega \ \underline{/-53.1^\circ}$$

The admittance is

$$\mathbf{Y}_T = \frac{\mathbf{I}_T}{\mathbf{E}} = \frac{5 \text{ mA } \underline{/+53.1^\circ}}{120 \text{ V } \underline{/0^\circ}} = 4.167 \times 10^{-5} \text{ S (siemens) } \underline{/+53.1^\circ}$$

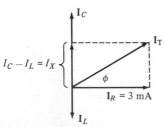

Figure 20.26   Current phasor diagram for solution to Example 20.5

Method 2 The parallel impedance can be calculated by using

$$\mathbf{Z}_\| = R \parallel -jX_C \parallel +jX_L = 40 \text{ k}\Omega \parallel -j20 \text{ k}\Omega \parallel +j60 \text{ k}\Omega$$

Applying the rule "product divided by the sum" (two components only), we can rewrite $(-j20 \text{ k}\Omega \parallel +j60 \text{ k}\Omega)$ as

$$(-j20 \text{ k}\Omega \parallel +j60 \text{ k}\Omega) = \frac{(-j20 \text{ k}\Omega)(+j60 \text{ k}\Omega)}{-j20 \text{ k}\Omega + j60 \text{ k}\Omega} = \frac{1200 \ (\text{k}\Omega)(\text{k}\Omega)}{+j40 \text{ k}\Omega}$$

$$= -j30 \text{ k}\Omega$$

$$\therefore \quad \mathbf{Z}_\| = (40 \text{ k}\Omega) \parallel (-j30 \text{ k}\Omega)$$

Again applying the same rule to two elements,

$$(40 \text{ k}\Omega) \parallel (-j30 \text{ k}\Omega) = \frac{(40 \text{ k}\Omega)(-j30 \text{ k}\Omega)}{40 \text{ k}\Omega - j30 \text{ k}\Omega} = \frac{-j1200 \ (\text{k}\Omega)(\text{k}\Omega)}{50 \text{ k}\Omega \ \underline{/-36.87°}}$$

$$= \frac{1200 \ (\text{k}\Omega)(\text{k}\Omega) \ \underline{/-90°}}{50 \text{ k}\Omega \ \underline{/-36.87°}}$$

$$\mathbf{Z}_\| = 24 \text{ k}\Omega \ \underline{/-53.13°} \qquad \text{Checks okay}$$

$$\mathbf{Y}_T = \frac{1}{\mathbf{Z}_\|} = \frac{1}{24 \text{ k}\Omega \ \underline{/-53.1°}} = 4.167 \times 10^{-5} \text{ S} \ \underline{/+53.1°}$$

Method 3 We can calculate the parallel impedance using the admittance triangle discussed in Objective 4 of this unit.

$$\mathbf{G} = \frac{1}{R} = \frac{1}{40 \text{ k}\Omega \ \underline{/0°}} = 0.025 \text{ mS} \ \underline{/0°}$$

$$\mathbf{B}_C = \frac{1}{\mathbf{X}_C} = \frac{1}{20 \text{ k}\Omega \ \underline{/-90°}} = 0.05 \text{ mS} \ \underline{/+90°}$$

$$\mathbf{B}_L = \frac{1}{\mathbf{X}_L} = \frac{1}{60 \text{ k}\Omega \ \underline{/+90°}} = 0.01667 \text{ mS} \ \underline{/-90°}$$

We can draw these vectors as the admittance triangle shown in Figure 20.27. Adding these vectors vectorially, we obtain the admittance **Y**:

$$Y = \sqrt{G^2 + (B_C - B_L)^2} = \sqrt{0.025^2 + (0.05 - 0.01667)^2}$$

$$Y = 0.04167 \text{ mS}$$

$$\phi = \arctan\left(\frac{B_{eq}}{G} = \frac{0.03333 \text{ mS}}{0.025 \text{ mS}} = 1.333\right) = +53.1°$$

$$\mathbf{Y} = 0.04167 \text{ mS} \ \underline{/+53.1°}$$

$$\mathbf{Z}_\| = \frac{1}{\mathbf{Y}} = \frac{1}{0.04167 \text{ mS} \ \underline{/+53.1°}} = 24 \text{ k}\Omega \ \underline{/-53.1°}$$

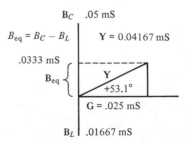

Figure 20.27   Admittance triangle for Example 20.5

Example 20.5 presented three methods of calculating the impedance of the circuits. Yet none of these methods involved the use of the impedance triangle. Let us see why.

In Objective 4 of this unit you were introduced to the admittance triangle for parallel circuits. Table 20.3 summarized the calculations used in series and parallel RLC circuits. Recall that for an ac series RLC circuit, the current is the same, and the total series impedance is a vector sum of all resistances and reactances connected in series. This vector addition requires the use of an impedance triangle. In dc parallel circuits which contained more than two resistances, we used the conductance calculation to obtain the overall parallel input resistance. Likewise, in ac parallel RLC circuits, the voltage is the same, and all the elements are computed in terms of their conductances. The vector sum of all the conductances and susceptances involves the use of an admittance triangle, not the impedance triangle.

In general, as more elements are placed in parallel, the total current I_T increases. This results in a total parallel impedance $Z_{||}$ smaller than that of any single element.

Problem 20.8 Refer to the circuit shown in Figure 20.28. Solve for all the currents, total impedance (magnitude and angle), and total admittance. Also draw the current phasor diagram.

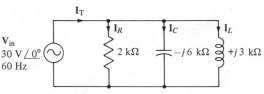

Figure 20.28   Circuit for Problem 20.8

Test

1. Figure 20.29 shows a parallel RC circuit. After completing the following blanks, draw the current phasor diagram.

$$X_C = \underline{\hspace{3cm}}$$

$$Z_{||} = \underline{\hspace{3cm}} \angle\underline{\hspace{1cm}}$$

$$I_R = \underline{\hspace{3cm}} \angle\underline{\hspace{1cm}}$$

$$I_C = \underline{\hspace{3cm}} \angle\underline{\hspace{1cm}}$$

$$I_T = \underline{\hspace{3cm}} \angle\underline{\hspace{1cm}}$$

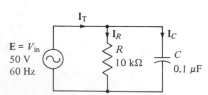

Figure 20.29   Circuit for Test Problem 1

2. Calculate the total current I_T for the circuit shown in Figure 20.30.

$I_T =$ _____ $\angle$

3. Calculate the impedance $Z_{||}$ and the admittance Y for the circuit shown in Figure 20.31. Draw an admittance triangle and label all the sides.

$Z_{||} =$ _____ $\angle$

$Y =$ _____ $\angle$

4. Calculate the impedance (magnitude and angle), the total admittance, and the current through each component in the *RLC* parallel circuit shown in Figure 20.32. Draw an admittance diagram and a current phasor diagram.

$I_T =$ _____ $\angle$

$Z_{||} =$ _____ $\angle$

$Y =$ _____ $\angle$

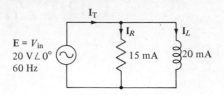

Figure 20.30 Circuit for Test Problem 2

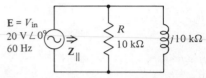

Figure 20.31 Circuit for Test Problem 3

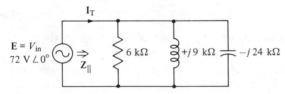

Figure 20.32 Circuit for Test Problem 4

Unit 21 *Power and Power-Factor Correction in ac Circuits*

In Units 18 and 19 you learned how to draw the power triangles for *RC*, *RL*, and *RLC* series circuits. In Unit 20 you learned about *RC*, *RL*, and *RLC* parallel circuits. In this unit you will learn more about power in series and parallel ac circuits. You will find out that most industrial electrical leads are inductive; that is, they operate at a lagging power factor. You will learn how to correct a lagging power factor so that you can operate a load at greater efficiency.

Objectives

After completing all the work associated with this unit, you should be able to:

1. Calculate the power (true, reactive, and apparent) developed in a resistor, capacitor, or perfect inductor.
2. Draw the power triangle for a capacitive or inductive series or parallel circuit. Label the sides of the power triangle.
3. Calculate the power factor of a circuit. Know what leading, lagging, and unity power factors are.
4. Calculate the amount of capacitance required to correct the power factor of an inductive circuit to unity.
5. Calculate the amount of capacitance required to correct the power factor of an inductive circuit to any other power factor.

394

Power in Components

Objective 1 Calculate the power (true, reactive, and apparent) developed in a resistor, capacitor, or perfect inductor.

One of the formulas you used in dc circuits to find power was $P = EI$. However, in ac circuits, voltage and current are not constant with respect to time, and the voltage across and current in inductors and capacitors are not in phase. Therefore we must use the *instantaneous voltage* and the *instantaneous current*. This gives an instantaneous power, or the power at a particular instant of time:

$$p = ei \qquad (21.1)$$

Lower-case letters are used in Equation (21.1) to designate instantaneous values. Equation (21.1), along with knowledge of the voltage and current phase relationships in a resistor, capacitor, and perfect inductor, will allow you to determine the power in these components.

Resistor

In a resistive ac circuit, the sine waves of applied voltage and current are in phase. Figure 21.1 shows this phase relationship of e and i. When we use Equation (21.1) to determine the instantaneous power, we obtain the p curve shown in Figure 21.1. Note that the instantaneous power curve is always positive, because when an instantaneous negative value of voltage is multiplied by an instantaneous negative value of current, we obtain an instantaneous positive value for power. Since the instantaneous power curve is always positive, the source is always delivering power to a resistor. The curve for instantaneous power in a resistor is also a sine curve. But the frequency of the instantaneous power curve is twice that of the applied voltage.

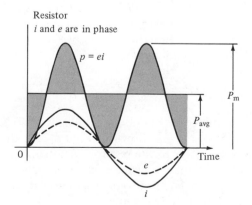

Figure 21.1 Instantaneous power in a resistor in an ac circuit

When we discuss power in ac circuits, we usually mean the average or effective power. That is, the true average is taken over a complete cycle. As shown in Figure 21.1, the average power P_{avg}, or effective power P, is equal to one-half the peak (or maximum) power P_m.

$$\text{Resistor} \qquad P_{avg} = P = \frac{P_m}{2} \qquad \text{(21.2)}$$

We can also prove this relationship between peak and average power in a resistor in an ac circuit mathematically, using the equations for a sine wave of voltage and current developed in Unit 15:

$$p = ei = (E_m \sin \omega t)(I_m \sin \omega t)$$
$$= E_m I_m \sin^2 \omega t = P_m \sin^2 \omega t \qquad \text{(21.3)}$$

From the half-angle relationship in trigonometry, Equation (21.3) becomes

$$p = \frac{P_m}{2}(1 - \cos 2 \omega t) \qquad \text{(21.4)}$$

The average of a cosine curve (or any sinusoidal wave) over a complete cycle is zero. Therefore Equation (21.4) becomes

$$P = \frac{P_m}{2} \qquad \text{(21.5)}$$

Using effective or rms values, Equation (21.5) becomes

$$P = \frac{P_m}{2} = \left(\frac{V_m}{\sqrt{2}}\right)\left(\frac{I_m}{\sqrt{2}}\right)$$
$$\therefore \quad P = V_R I_R \qquad \text{(21.6)}$$

where V_R is the effective voltage across the resistor and I_R is the effective current in the resistor.

The R subscript on the V and I is used to emphasize that we use Equation (21.6) for the power in the resistive part of the circuit.

Using Ohm's law in Equation (21.6), we can find two other forms of the power equation:

$$P = V_R I_R = I_R^2 R = \frac{V_R^2}{R} \qquad \text{(21.7)}$$

The power determined by Equations (21.6) and (21.7) is known as the *true* or *real power*. As shown in Figure 21.1, the source is always delivering power to the resistor. All the power delivered to the resistor is dissipated. The resistor converts the electrical energy into heat. The equations for true power in ac circuits are the same as for power in dc circuits.

Example 21.1 A 600-Ω resistive load is connected across a 120-V, 60-Hz ac source. What is the peak power and what is the average power drawn from the source?

Solution Remember, if p–p or p is not specified after an ac voltage, the voltage is understood to be rms.

$$\text{Resistive load} \qquad P = \frac{V_R^2}{R} = \frac{(120)^2}{600} = 24 \text{ W}$$

From

$$P = \frac{P_m}{2} \qquad \therefore \quad P_m = 2P = 48 \text{ W peak power}$$

Problem 21.1 A 500-Ω resistive load is connected across a 120-V, 60-Hz ac source. What is the peak and the average power drawn from the source?

Problem 21.2 A 100-Ω resistor is connected in an ac circuit in which a current of 50 mA peak exists. What is the true power dissipated in the resistor?

Capacitor

The current in a capacitor in an ac circuit leads the voltage across the capacitor by 90° (remember *ICE*). This phase relationship of *e* and *i* is shown in Figure 21.2. When we use Equation (21.1) to determine the instantaneous power curve, we obtain the *p* curve shown in Figure 21.2. The sine curve obtained is again twice the frequency of the applied voltage. However, it has equal positive and negative instantaneous power. Therefore the average power drawn from the source is zero. When the power curve is positive, the capacitor is absorbing power from the source. When the power curve is negative, the capacitor is returning power to the source. Therefore the average or true power in a capacitor is zero.

$$\text{Capacitor} \qquad P_{avg} = P = 0 \qquad\qquad (21.8)$$

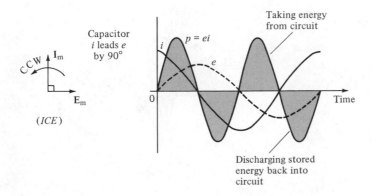

Capacitor
i leads *e*
by 90°

CCW I_m

E_m

(*ICE*)

Taking energy
from circuit

p = ei

i

e

0   Time

Discharging stored
energy back into
circuit

Figure 21.2 Instantaneous power in a capacitor in an ac circuit

Equation (21.8) can also be determined mathematically

$$p = ei = (E_m \sin \omega t)(I_m \cos \omega t)$$

Using the relationship from trigonometry, $\sin 2\alpha = 2 \sin \alpha \cos \alpha$, then

$$p = \frac{P_m}{2} \sin 2 \omega t = 0 \qquad \text{over a complete cycle}$$

The power developed in a capacitor is the product of the effective value of voltage across the capacitor and the effective value of current into the capacitor. However, since this power is not true power, it is designated as *reactive power*, represented by the symbol Q:

$$Q = V_C I_C \qquad (21.9)$$

Note: The unit used for reactive power is called the var. You may remember this as *v*olt *a*mperes *r*eactive.

By using Ohm's law in Equation (21.9), we obtain two other equations for reactive power.

$$Q = I_C^2 X_C = \frac{V_C^2}{X_C} \qquad (21.10)$$

where V_C is the effective voltage across the capacitor and I_C is the effective current in the capacitor.

Problem 21.3 A 100-μF capacitor is connected across a 35-V, 60-Hz source. What are the true power input to the capacitor, the reactive power input to the capacitor, and the apparent power input?

Inductor (Perfect)

The voltage across a perfect inductor (no resistance) in an ac circuit leads the current through it by 90° (remember *ELI*). Figure 21.3 shows the instantaneous voltage *e*, current *i*, and power *p* curves. Again, as with the capacitor, the instantaneous

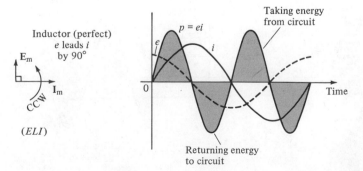

Figure 21.3 Instantaneous power in an inductor (perfect) in an ac circuit

power curve is a sine curve of twice the frequency of the applied voltage or current. The power curve above the horizontal axis is equal to the power curve below it. That is, the average power drawn by the inductor from the source is zero:

$$\text{Inductor} \qquad P_{\text{avg}} = P = 0 \qquad (21.11)$$

The reactive power in an inductor is

$$Q = V_L I_L = I_L^2 X_L = \frac{V_L^2}{X_L} \qquad (21.12)$$

where V_L is the effective voltage across the inductor and I_L is the effective current in the inductor.

Example 21.2   When connected to a 10-kHz source, a pure inductance of 10 mH passes a current of 15 mA. What are the reactive and true power inputs to the inductor?

Solution $X_L = 2\pi f L = 2\pi(10 \times 10^3)(10 \times 10^{-3}) = 628 \ \Omega$
$$Q = I_L^2 X_L = (15 \times 10^{-3})^2(628) = 141 \text{ mvar}$$
$$P = 0$$

Problem 21.4   When connected to a 1-kHz source, a pure inductance of 50 mH passes a current of 25 mA. What are the reactive, apparent, and true power inputs to the inductor?

Example 21.3   A solenoid has a resistance of 40 Ω and an impedance of 300 Ω when it is connected to a 120-V, 60-Hz source. What is the true power drawn by the solenoid when it is connected to the 120-V, 60-Hz source?

Solution   Refer to Figure 21.4:

$$I = \frac{E}{Z} = \frac{120 \text{ V}}{300 \ \Omega} = 0.4 \text{ A}$$

Since Figure 21.4 is a series circuit and current is common,
$$P = I^2 R = (0.4 \text{ A})^2(40 \ \Omega) = 6.4 \text{ W}$$

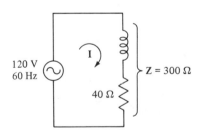

Figure 21.4   Circuit for Example 21.3

The Power Triangle

Objective 2   Draw the power triangle for a capacitive or inductive series or parallel circuit. Label the sides of the power triangle.

You have already drawn power triangles for capacitive and inductive series circuits in Units 18 and 19. When you drew these power triangles, you used current as a reference, because current is the common or reference phasor in series circuits.

When you worked with parallel circuits in Unit 20, you used voltage as the reference phasor. To standardize for all circuits—series, parallel, and series-parallel circuits—the international standards (SI) have chosen to use voltage E as the reference for all power triangles. You will use the power triangle a great deal in Objectives 3, 4, and 5 on power factor and power-factor correction. You will also find that a common way of correcting the power factor of a load is to add capacitance in parallel with the load. Since a parallel connection is used and the reference in parallel circuits is voltage, it should be easy for you to remember that *all power triangles use voltage as the reference axis.*

Figure 21.5 shows the power triangles for capacitive or inductive circuits. The actual circuit may contain resistors, capacitors,

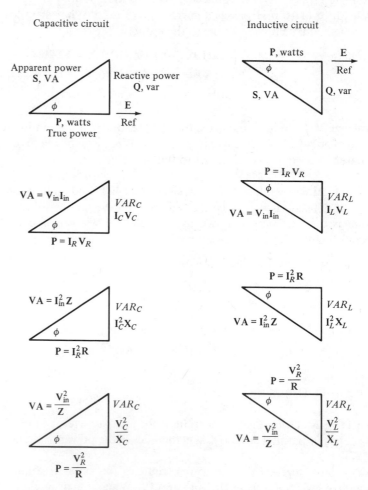

Figure 21.5 Power triangles for capacitive and inductive ac circuits using voltage as the reference phasor (note formulas given on triangles).

In a series circuit I_{in}, I_R, and I_X are the same value, because current is common.
In a parallel circuit V_{in}, V_R, and V_X are the same value, because voltage is common.

and inductors, but, as discussed in Units 18, 19, and 20, it appears either capacitive or inductive to the source. As shown in Figure 21.5, voltage is the reference phasor. Also shown on the power triangles in Figure 21.5 are the formulas you may use to find the true power, reactive power, and apparent power.

You will make considerable use of the formulas listed in Figure 21.5 to solve problems. Therefore, before we go on to the next objective, we want to discuss the use of the formulas for true power listed in Figure 21.5.

Be careful when you use the formula $P = V_R I_R$ for power in ac circuits. The formula applies only when V_R is the voltage drop across the resistor R and I_R is the current in resistor R.

In the formula $P = I_R^2 R$, I_R must be the current in resistor R. This form of the power equation is most commonly used when you have a series connection. Because if you know the current in the series connection, you know the current in R.

In the formula

$$P = \frac{V_R^2}{R}$$

V_R must be the voltage across the resistor R. This form of the power equation is most commonly used when you have a parallel connection. If you know the voltage across the parallel connection, you know the voltage across the resistor in the parallel connection. Do not use this form of the power equation in a series circuit when only the source voltage is known. It will not give you the correct answer, because you must know the voltage across the resistor R.

Comments similar to those listed above for true power could be made about the formulas for reactive power.

Problem 21.5 Draw the power triangles for the circuits shown in Figure 21.6. Use voltage as the reference phasor, and label the sides of the triangles. [*Hint*: For the series circuit, determine whether the circuit is capacitive or inductive. Then you can draw the power triangle.]

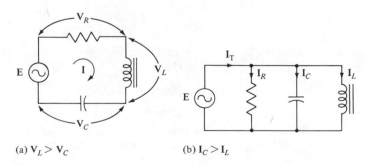

(a) $V_L > V_C$ (b) $I_C > I_L$

Figure 21.6 Circuits for Problem 21.5

Power Factor

Objective 3 Calculate the power factor of a circuit. Know what leading, lagging, and unity power factors are.

The power factor of a load is defined as

$$\text{Power factor of load} = \frac{\text{true power in load}}{\text{apparent power of load}} \quad (21.13)$$

Referring to Figure 21.5, you can see that the power factor is equal to

$$\text{Power factor (PF)} = \frac{P}{VA} = \frac{P}{S} \quad (21.14)$$

By trigonometry, the ratio of P to VA in a right triangle is equal to the cosine of the angle ϕ shown in Figure 21.5. Therefore

$$\text{PF} = \cos \phi \quad (21.15)$$

Some authors use $\cos \phi$ as the symbol for power factor, just as they use I as the symbol for current.

Remember that the angle ϕ in the power triangles in Figure 21.5 can also be found from the impedance triangle for series circuits or the admittance triangle for parallel circuits. Therefore you may be able to find the power factor of a load without actually determining the true and apparent power of the load. This is illustrated in the following example.

Example 21.4 A load whose impedance is $100 \ \Omega \ \underline{/+30°}$ is connected across a 120-V, 60-Hz voltage source. What is the power factor of the load?

Solution First we can solve this problem by using Equation (21.14). Using the value given for the impedance of the load, we can find the resistance by

$$R = Z \cos \phi = 100(0.866) = 86.6 \ \Omega$$

The magnitude of the current drawn by the load may be found by

$$I = \frac{E}{Z} = \frac{120 \ \text{V}}{100 \ \Omega} = 1.2 \ \text{A}$$

Therefore, using the equations from Figure 21.5, we can now find the values required by Equation (21.14):

$$P = I^2 R = (1.2)^2(86.6) = 124.7 \ \text{W}$$
$$VA = I^2 Z = (1.2)^2(100) = 144 \ \text{VA}$$

$$\text{PF} = \frac{P}{VA} = \frac{124.7 \ \text{W}}{144 \ \text{VA}} = 0.866$$

Second, we can solve the same problem using Equation (21.15). The angle and magnitude for impedance were given. Therefore $\phi = 30°$.

$$PF = \cos \phi = \cos 30° = 0.866$$

Example 21.5 A single-phase induction motor draws 6 A at 220 V, with the current lagging the voltage. A wattmeter, which measures true power, connected in the circuit indicates 720 W. What is the power factor of the circuit and what is the angle by which the current lags the voltage?

Solution Use Equation (21.14) and refer to Figure 21.5 (inductive circuit).

$$PF = \frac{P}{VA} = \frac{720 \text{ W}}{220(6) \text{ VA}} = 0.545$$

$$\cos \phi = 0.545, \qquad \phi = \cos^{-1}(0.545) = 56.9°$$

Problem 21.6 A load whose impedance is $80 \ \Omega \ \underline{/40°}$ is connected across a 120-V, 60-Hz voltage source. Find the power factor of the load by the two methods illustrated in Example 21.4.

By combining Equations (21.14) and (21.15), we can obtain another formula that may be useful for determining true power. By solving for P, we obtain

$$P = VA \cos \phi \qquad (21.16)$$

or

$$P = S \cos \phi$$

As discussed in Objective 2, voltage is used as the reference for all power triangles. With voltage as the reference, we may then define leading, lagging, and unity power factors. When we use E as the reference, the power factor leads or lags depending on whether the circuit is capacitive or inductive. This is shown in Figure 21.7. When a load appears capacitive in nature, we speak of a *leading power factor*; see Figure 21.7(a). When a load appears inductive in nature, we speak of a *lagging power factor*; see Figure 21.7(c).

When the load appears to the source as a pure resistance, the power factor is said to be unity. In this case, the circuit contains R, L, and C circuit elements. However, the amount of inductive and capacitive reactance is the same, and therefore they cancel each other. When this happens, the source sees only a resistive load. For unity power factor, as shown in Figure 21.7(b), $\phi = 0°$.

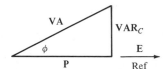

(a) Leading PF (circuit capacitive)

(b) Unity PF (circuit resistive)

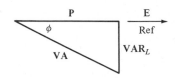

(c) Lagging PF (circuit inductive)

Figure 21.7 Power triangles for leading, unity, and lagging power-factor circuits

Therefore

$$PF = \cos \phi = \cos 0° = 1$$

When the power factor is unity, the apparent power drawn from the source is equal to the true power in the load. As shown in Figure 21.7(b), $P = S$ or $P = EI$.

A reactive factor is sometimes used. It is defined as

$$\text{Reactive factor} = \frac{\text{reactive power}}{\text{apparent power}}$$

$$\text{Reactive factor} = \frac{VAR}{VA} = \frac{Q}{S} = \sin \phi \qquad (21.17)$$

Example 21.6 Refer to the circuits shown in Figure 21.6. Do they have a leading or lagging power factor? What condition must exist in each circuit for there to be a unity power factor?

Solution In Figure 21.6(a): $V_L > V_C$; the circuit is inductive. Therefore we have a lagging power factor. For unity power factor, V_L must equal V_C.

In Figure 21.6(b): $I_C > I_L$; the circuit is capacitive. Therefore we have a leading power factor. For there to be a unity power factor, I_C must equal I_L.

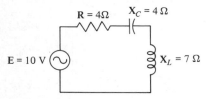

Figure 21.8 Circuit for Problem 21.7

Problem 21.7 For the circuit shown in Figure 21.8, determine the following: (a) Is the power factor of the circuit leading or lagging? (b) Calculate the true power. (c) Calculate the apparent power. (d) Find the power factor of the circuit by two methods. (e) Calculate the reactive power. (f) Find the reactive factor of the circuit.

Power-Factor Correction to Unity

Objective 4 Calculate the amount of capacitance required to correct the power factor of an inductive circuit to unity.

Before we show you how to master Objective 4, let us discuss why you might want to correct a power factor and how you can do it. For most domestic loads, such as your home, the power factor is near unity. Because the heavy load components—such as the electric stove, the electric clothes dryer, and other electrical appliances—are generally resistive. However, most industrial loads consist of electric motors, fluorescent lights, air conditioners, transformers, controls, and other equipment with low and lagging power factors.

The total current drawn by an industrial load may be several hundred or even thousands of amperes. This means that for a given supply voltage, the current drawn from the supply is greater than the current needed for the same useful power at unity power factor. This is illustrated in Example 21.7. The more current a system has to carry, the larger the copper losses, because the copper losses in a system depend on the square of the current ($P = I^2R$).

Power companies pay close attention to the power factors of their customers' circuits. If the power factor of a customer's load falls below some amount, for example 0.80, then a higher rate may be charged.

Another area in which capacitors are commonly used for power-factor correction is induction heating. Figure 21.9 shows a circuit diagram of an induction heating application. The name *induction* is used because that is how the energy from the heating coil gets to the load. There is no direct electrical connection between them. Induction heating may be used for hardening, forging, hot-forming, extruding, brazing, melting, mixing metals, and many other applications. The load shown in Figure 21.9— which is some type of metal—is inductive. Therefore, to correct the circuit to near unity power factor, capacitors are used. During the process of hardening or melting, the power factor of the load may change. Therefore some capacitors may have to be added to or removed from the circuit. This is why switches are shown in Figure 21.9.

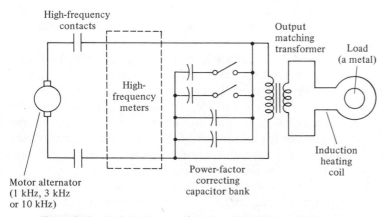

Figure 21.9   Circuit diagram of an induction heating application

Figure 21.10 Currents that flow in a typical induction heating application that requires 50 kW of power

The induction heating coil and load have a very low and lagging power factor. The coil usually draws several times more apparent power in kilovoltamperes than it does in true power in kilowatts. Figure 21.10 shows the currents that flow in a typical induction heating application that requires 50 kW of power. The motor-alternator only supplies 227 A, which is found by 50 kW/220 V. However, the load circuit requires 1458 A. This gives an apparent power of

$$VA = (220 \text{ V})(1458 \text{ A}) = 321 \text{ kVA}$$

This apparent power is many times the true power. As you can see, the power-factor-correcting capacitors make it possible to do this job by this process.

If you studied Figures 21.9 and 21.10, you would see the power-factor-correcting capacitors placed in parallel with the total load on the circuit. Consider a capacitor connected in series, Figure 21.11(a), to correct the power factor and reduce the current drawn from the source. The capacitor does reduce the net reactive power; however, it also increases the current drawn from the source. The current is greater because the circuit impedance is less. Also the source and the load voltage are not the same.

Now consider the capacitor connected in parallel, in Figure 21.11(b). In a parallel circuit, a change in current in one leg of the circuit does not affect the current in the other leg. However, the current in one leg can affect and actually reduce the total current for the two legs. The current in the capacitor cancels the reactive component of the current in the load. Therefore the current drawn from the source is small compared with the current in the capacitor and the inductive load circuit.

For example, if the inductive load of Figure 21.11(b) is a motor, it is connected across the source. The source and the voltage across the motor terminals are the same. Whatever current flows

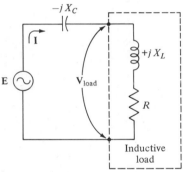

(a) Series capacitor

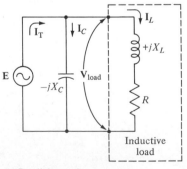

(b) Parallel capacitor

Figure 21.11 Consideration of where to place power-factor correcting capacitors

is current that is required by the motor. However, when we add capacitors in parallel, the current drawn from the source is less than the motor current. If the capacitors are connected at the load end of the circuit, the copper cable or bus bar connecting the source to the load may be physically smaller because it does not have to carry the high load current.

The conclusion that you should draw from the last few paragraphs is that—to correct power factor of an inductive load—one must insert capacitors in *parallel* with the load.

Example 21.7 An industrial load draws 600 A from a 440-V, 60-Hz supply at a lagging power factor of 0.6. How much current is drawn from the source if the power factor PF is corrected to unity? How much capacitive reactance is required to correct the PF to unity?

Solution Refer to Figure 21.11(b). The apparent power of the load is

$$VA = S = VI_L = (440 \text{ V})(600 \text{ A}) = 264 \text{ kVA}$$

The true power in the load is

$$P = VA \cos \phi = 264 \text{ kVA}(0.6) = 158.4 \text{ kW}$$

If the load is corrected to unity power factor, the current drawn from the source can be found by

$$P = VA \qquad \text{(unity power factor)}$$

$$I = \frac{P}{V} = \frac{158.4 \times 10^3 \text{ W}}{440 \text{ V}} = 360 \text{ A}$$

The reactive power of the load is

$$Q_L = \sqrt{S^2 - P_L^2} = \sqrt{(264 \text{ kVA})^2 - (158.4 \text{ kW})^2} = 211.2 \text{ kvar}$$

We may find the reactive power another way:

$$\sin \phi = \frac{Q_L}{S} \qquad \text{where } \phi = \cos^{-1}(0.6) = 53.13°$$

$$\therefore \quad Q_L = S \sin \phi = 264 \text{ kVA}(\sin 53.13°) = 211.2 \text{ kvar}$$

Therefore, to correct the power factor to unity, 211.2 kvar of capacitive reactance should be inserted in parallel.

As you can see by this example, there can be a considerable difference in the amount of current drawn from the source at some lagging power factor. In Example 21.7, one should use 211.2 kvar of capacitive reactance to correct to unity power factor. However, most power companies require only that the power factor remain within a certain range, say 0.8 to unity. If we were to correct to 0.8 lagging power factor, fewer capacitors would be required. This would save money.

Correcting to some power factor other than unity is discussed in the next objective. But before we go on to the next objective, we want to show you more examples and have you work some problems.

Example 21.8 A 1-kW, 120-V, 60-Hz motor runs at 75% lagging power factor. The power factor is corrected to unity. Find the motor line currents.

Solution To correct the power factor to unity, we insert capacitors in parallel with the motor; see Figure 21.12(a).

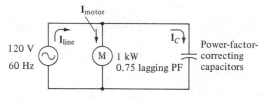

(a) Circuit

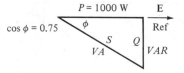

(b) Power triangle for motor

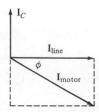

(c) Current phasor diagram

Figure 21.12 Diagrams for Example 21.8

Referring to Figure 21.12(b), the apparent power is

$$VA = \frac{P}{\cos \phi} = \frac{1000 \text{ W}}{0.75} = 1333 \text{ VA}$$

(Remember that $75\% = 0.75$)

If the power factor were not corrected (no capacitors), $\mathbf{I}_{line}$ from the source would equal $\mathbf{I}_{motor}$ because of the series circuit.

$$\mathbf{I}_{line} = \mathbf{I}_{motor} = \frac{\mathbf{VA}}{\mathbf{E}} = \frac{1333 \text{ VA}}{120 \text{ V}} = 11.11 \text{ A } \underline{/\phi}$$

When the power-factor-correcting capacitors are placed in the circuit, the motor current $\mathbf{I}_{motor}$ is still equal to 11.11 A. The motor must still have the 120-V source voltage across its terminals to make it run.

With the capacitors inserted into the circuit so that unity power factor is obtained, the current drawn from the source is a minimum. Refer to Figure 21.12(c). The current in the capacitors $\mathbf{I}_C$ cancels the effect of the reactive component of the motor current.

We determine $\mathbf{I}_{line}$ at unity power factor from the fact that

$$VA = P \qquad \text{(unity PF)}$$

$$\mathbf{I}_{line} = \frac{P}{E} = \frac{1000}{120} = 8.33 \text{ A } \underline{/0°}$$

Example 21.9 An induction motor draws 3.0 A at 0.8 lagging power factor from a 120-V, 60-Hz source. Find the value of C that must be placed in parallel with the motor to raise the PF to unity.

Solution Calculate the values for the power triangle, Figure 21.13.

$$\cos \phi = 0.8 \qquad \phi = 36.9° \qquad \text{and} \qquad \sin \phi = 0.6$$

The apparent power is

$$S = VA = (120 \text{ V})(3 \text{ A}) = 360 \text{ VA}$$

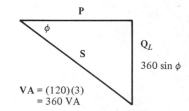

Figure 21.13 Power triangle for Example 21.9

The reactive power is

$$Q_L = VA \sin \phi = (360 \text{ VA})(0.6) = 216 \text{ var}$$

For unity power factor, Q_C must equal Q_L. Therefore, to correct to unity PF, we need to add 216 capacitive vars.

Since capacitance is placed in parallel, we may use the following form of Equation (21.10):

$$Q_C = \frac{V_C^2}{X_C}$$

Solving for X_C, we get

$$X_C = \frac{V_C^2}{Q_C} = \frac{(120 \text{ V})^2}{216 \text{ var}} = 66.67 \text{ }\Omega$$

From X_C we can now find C:

$$X_C = \frac{1}{2\pi fC}$$

$$C = \frac{1}{2\pi fX_C} = \frac{1}{377(66.67)} = 39.8 \text{ }\mu\text{F}$$

Instead of taking two steps to find the capacitance, as we did in Example 21.9, we could do it in one. Combine the formulas

$$Q_C = \frac{V_C^2}{X_C} \qquad \text{and} \qquad X_C = \frac{1}{2\pi fC}$$

by substituting the second formula into the first. Then

$$Q_C = \frac{V_C^2(2\pi fC)}{1}$$

or

$$C = \frac{Q_C}{2\pi fV_C^2} \tag{21.18}$$

Using Equation (21.18) to solve for C in Example 21.9, we obtain

$$C = \frac{216 \text{ var}}{(377)(120 \text{ V})^2} = 39.8 \text{ }\mu\text{F}$$

Example 21.10 In Example 21.9, after the power factor is corrected to unity, what are the magnitudes of the motor, capacitor, and line currents?

Solution Remember that the phase angle ϕ is the same in the power triangle and the current phasor diagram. Refer to Figure 21.12(c) for a sketch of a current phasor diagram. Therefore, to find the currents, we use

$$\mathbf{I}_{\text{motor}} = 3 \text{ A } \underline{/36.9°} \qquad \text{(this was given)}$$

$\mathbf{I}_C$ must equal the *vertical* component of $\mathbf{I}_{\text{motor}}$ for unity PF.

$$\mathbf{I}_C = \mathbf{I}_{\text{motor}} \sin \phi = 3 \sin 36.9° = 3(0.6) = 1.8 \text{ A } \underline{/90°}$$

With $\mathbf{I}_C$ canceling out the vertical component of $\mathbf{I}_{\text{motor}}$, all we have left is the horizontal component of $\mathbf{I}_{\text{motor}}$. $\mathbf{I}_{\text{line}}$ must equal this horizontal component:

$$\mathbf{I}_{\text{line}} = \mathbf{I}_{\text{motor}} \cos \phi = 3 \cos 36.9° = 3(0.8) = 2.4 \text{ A } \underline{/0°}$$

Problem 21.8 When a 1-hp (746-W) motor is connected to a 120-V, 60-Hz source, it operates at a 65% lagging power factor. How much capacitance is required to correct the power factor to unity? Once the power factor is corrected to unity, what are the motor and line currents?

Problem 21.9 An industrial load draws 800 A from a 440-V, 60-Hz supply at a lagging power factor of 0.7. How much current is drawn from the source if the power factor is corrected to unity? How much capacitance is required to correct the PF to unity?

Problem 21.10 An induction heating coil and load are connected to a 25-kW, 220-V, 10-kHz motor alternator. To correct to unity power factor, 150 kvar of capacitive reactance is required. What was the power factor of the coil and the load? At unity PF, what is the current drawn from the motor alternator? How much current do the capacitors and coil/load draw?

Other Power-Factor Corrections

Objective 5 Calculate the amount of capacitance required to correct the power factor of an inductive circuit to any other power factor.

As mentioned earlier, power companies usually allow a power factor to lag by some amount, say 0.8. If this is the case, then you do not have to add capacitors to bring the power factor up to unity, but only to bring it up to the specified power factor. The following examples show you how to determine the amount of capacitance required for any power factor.

Example 21.11 An induction motor that is connected across a 120-V, 60-Hz source has the power triangle shown in Figure 21.14(a). How many vars must be added to the circuit to produce an overall power factor of 0.9 lagging?

Solution The power (true power P) in the circuit remains the same when the power factor is corrected. The amount of capacitance in the circuit does not affect true power. However, it does affect the apparent power and the reactive power.

The angle ϕ' shown in Figure 21.14(b) is equal to

$$\phi' = \cos^{-1}(0.9) = 25.8°$$

To determine the amount of reactive power for $P = 800$ W and $\phi' = 25.8°$, we make use of the relation

$$\tan \phi' = \frac{Q'}{P}$$

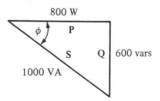

(a) Power triangle of an induction motor

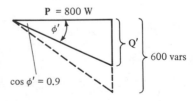

(b) Corrected power triangle

Figure 21.14 Power triangles for Example 21.11

or
$$Q' = P \tan \phi'$$
$$= 800(\tan 25.8°) = 800(0.484) = 387.2 \text{ var}$$

Therefore, to change the power factor from what was given in the problem (800 W/1000 VA = 0.8 lagging) to 0.9 lagging, the following number of vars of capacitive reactance must be added:

$$Q_C \text{ added} = 600 \text{ var} - 387 \text{ var} = 213 \text{ var}$$

From this value of vars, we can determine the actual value of capacitance; see Example 21.9.

Example 21.12   A 100-μF capacitor is connected across a load that is connected to a 120-V, 60-Hz source. The power factor of the circuit is raised from 0.6 to 0.8 lagging. What is the true power of the load?

Solution   Refer to Figure 21.15. The amount of reactive power the 100 μF added to the circuit is

$$X_C = \frac{1}{2\pi f C} = \frac{10^6}{377(100)} = 26.5 \text{ } \Omega$$

$$Q_C = \frac{E^2}{X_C} = \frac{(120 \text{ V})^2}{26.5 \text{ } \Omega} = 543 \text{ var}$$

The actual angles of ϕ and ϕ' shown on Figure 21.15 are equal to

$$\phi = \cos^{-1}(0.6) = 53.1°$$
$$\phi' = \cos^{-1}(0.8) = 36.9°$$

From the triangles shown in Figure 21.15,

$$\tan \phi = \frac{Q}{P} \quad \text{and} \quad \tan \phi' = \frac{Q'}{P}$$

and

$$Q = Q' + Q_C = Q' + 543 \text{ var}$$

Therefore

$$P(\tan \phi) = P(\tan 53.1°) = Q = Q' + 543 \text{ var}$$

or

$$1.333P = Q' + 543 \text{ var}$$

From
$$P(\tan \phi') = P(\tan 36.9°) = Q'$$

we obtain
$$0.75P = Q'$$

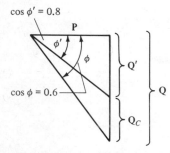

cos ϕ' = 0.8

cos ϕ = 0.6

Figure 21.15   Power triangles for Example 21.12

We now have two simultaneous equations,

$$1.333P = Q' + 543 \text{ var}$$
$$0.75P = Q'$$

Subtracting, we have $0.583P = 543 \text{ var}$

or

$$P = \frac{543 \text{ var}}{0.583} = 931 \text{ W}$$

Did you notice the similarities between Examples 21.11 and 21.12? In Example 21.11 you knew the true power, and were asked for the reactive power to change the power factor from one value to another. In Example 21.12, you knew the reactive power (you were given the value of the capacitor) that changed the power factor from one value to another, and you were asked the true power.

Example 21.13 In Example 21.8, what are the motor and line currents if the power factor is corrected to 90% lagging?

Solution The motor current I_{motor} is still equal to 11.11 A. The motor still has the 120-V source voltage across its terminals.

The power factor is corrected to 90% lagging. [*Note*: 90% = 0.90.] The apparent power is now

$$S = VA = \frac{P}{\cos \phi} = \frac{1000}{0.9} = 1111 \text{ VA}$$

The line current is equal to

$$I_{\text{line}} = \frac{VA}{E} = \frac{1111 \text{ VA}}{120 \text{ V}} = 9.26 \text{ A}$$

The line current is not as low as at unity power factor. However, not as many vars of capacitive reactance are required to correct to 90% lagging.

Problem 21.11 When connected to a 120-V, 60-Hz source, a bank of fluorescent lamps and its ballast inductance draws 3 A at a 60% lagging power factor. What capacitance should be connected across the lamps to bring the power factor up to 85% lagging?

Problem 21.12 Determine how much capacitance is required for the induction motor of Example 21.9 if the power factor is corrected only to 0.9 lagging.

Test

1. An electric toaster has a resistance of 13 Ω. When it is connected to a 120-V, 60-Hz ac line, what are the peak and average power drawn from the line?

2. When connected to a 60-Hz source, a 200-Ω resistor passes a current of 0.6 A. What are the true, reactive, and apparent power drawn by the resistor?

3. A 25-μF capacitor is connected across a 20-V, 60-Hz source. What are the true, reactive, and apparent power drawn by the capacitor?

4. When connected to a 20-kHz source, a pure inductance of 25 mH passes a current of 10 mA. What are the true, reactive, and apparent power drawn by the inductor?

5. Draw the power triangles for Figure 21.16. Use voltage as the reference phasor, and label the sides.

6. A single-phase induction motor draws 3 A at 120 V, with the current lagging the voltage. A wattmeter, which measures true power, connected in the circuit indicates 200 W. What is the power factor of the circuit, and what is the angle by which the current lags the voltage?

7. An 8-H choke has a resistance of 500 Ω. When connected to a 120-V, 60-Hz source, what is the true power drawn by the choke? What is the power factor, and is it leading or lagging?

8. A single-phase 60-Hz ac generator is rated to deliver 40 kW at 220 V and 0.8 power factor. Determine its current and voltampere rating (VA).

9. A single-phase 60-Hz induction motor draws 800 W at 120 V and 9 A lagging current. Determine the power factor, power factor angle, and voltampere rating (VA) of the motor.

10. A transformer draws 20 A at 0.9 lagging power factor from a 2300-V, 60-Hz power line. Determine the power and voltamperes (VA) taken by the transformer and the angle of lag of the circuit.

11. An inductive load takes 41.7 A at 240 V, 25 Hz. The current lags the voltage by an angle of 36.9°. What are the power factor, the power, and the voltamperes (VA) drawn by the inductive load?

12. 100 kW of power at 440 V, 3 kHz is required for an induction heating application. When corrected to unity power factor, 700 kvar is required. Determine the original power factor of the circuit. What are the values for I_T, I_C, and I_L at unity power factor?

13. An induction motor draws 2 A at 0.85 lagging power factor from a 120-V, 60-Hz source. Find the value of C

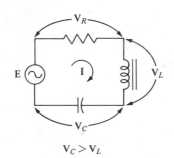

$V_C > V_L$

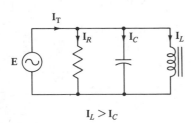

$I_L > I_C$

Figure 21.16   Circuits for Test Problem 5

that must be placed in parallel with the motor to raise the power factor to unity.

14. For the motor in Test Problem 13, what are the magnitudes of the final motor current, capacitor current, and line current?

15. How much capacitance is required to correct the power factor of the motor in Test Problem 11 to 0.9 lagging?

16. An induction motor that is connected across a 120-V, 60-Hz source has a power triangle as shown in Figure 21.17. How many vars must be added to the circuit to produce an overall power factor of 0.95 lagging? How must the capacitors be connected to produce this power factor?

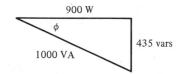

Figure 21.17 Power triangle for Test Problem 16

Unit 22 *Series-Parallel ac Circuits*

In the schematic of the phonograph amplifier given in Unit 1 (Figure 1.5) and in the transmission line represented in Figure 22.1, you see several series-parallel circuits. You will find many series-parallel ac circuits when you study advanced topics in electronics, measurements, and communications. As you found in your study of series and parallel ac circuits, it doesn't take much for the mathematics to become very complicated. With complicated series-parallel ac circuits, computers are usually required for an exact solution. However, a complicated series-parallel circuit can usually be reduced to a simple series, parallel, or series-parallel circuit by making assumptions about some of the circuit

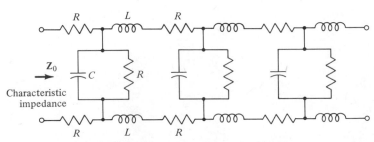

Figure 22.1　Equivalent circuit for a transmission line

elements or sections of the circuit. Then you may use the methods given in this unit to solve the circuit.

Objectives

After completing all the work associated with this unit, you should be able to:

1. Calculate the total impedance when two or more impedances are connected in series. Then find the current and/or voltages in the circuit.
2. Calculate the equivalent impedance when two impedances are connected in parallel. Then find the voltage and/or currents in the circuit.
3. Calculate the total admittance when two or more impedances are connected in parallel. Then, from the admittance, determine the total impedance.
4. Apply the voltage-divider and current-divider principles to ac circuits.
5. Convert a two-element parallel ac circuit to an equivalent two-element series circuit, and vice versa.
6. Calculate the total impedance of series-parallel ac circuits. Then find the currents and/or voltages in the circuits.
7. Apply Thévenin's theorem to series-parallel ac circuits.

The first two objectives of this unit are a review of previous units on *RC*, *RL*, and *RLC* series and parallel circuits. However, the approach used to solve the circuits is different. We shall consider an impedance as a circuit element instead of just a resistor, capacitor, or inductor. We shall develop a general formula for the solution of the circuit. Since this formula is general, it may be used for any impedance. The impedance may consist of just a resistor, or it may have both resistive and reactive components. This procedure may be considered a systems approach to the solution of series and parallel ac circuits. It will make it easier for you to complete the later objectives of this unit.

Impedances in Series

Objective 1   Calculate the total impedance when two or more impedances are connected in series. Then find the current and/or voltages in the circuit.

Figure 22.2 shows a series ac circuit with two impedances. The **Z**'s shown in the boxes in Figure 22.2 may represent any one

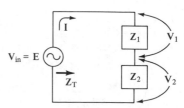

Figure 22.2   Series impedances

of the following:

$$\mathbf{Z} = R \qquad \text{resistance only} \qquad (22.1)$$

$$\mathbf{Z} = +jX_L \quad \text{inductive reactance only} \qquad (22.2)$$

$$\mathbf{Z} = -jX_C \quad \text{capacitive reactance only} \qquad (22.3)$$

$$\mathbf{Z} = R + jX_L \quad \text{resistance and inductive reactance} \qquad (22.4)$$

$$\mathbf{Z} = R - jX_C \quad \text{resistance and capacitive reactance} \qquad (22.5)$$

Equations (22.1) through (22.5) are written in rectangular notation; however, they may be given in polar notation.

Writing KVL around the loop in the circuit shown in Figure 22.2, we obtain

$$\mathbf{E} = \mathbf{V}_1 + \mathbf{V}_2 \qquad (22.6)$$

Remember that Equation (22.6) is a vector sum. Dividing both sides of Equation (22.6) by current $\mathbf{I}$, we obtain

$$\frac{\mathbf{E}}{\mathbf{I}} = \frac{\mathbf{V}_1}{\mathbf{I}} + \frac{\mathbf{V}_2}{\mathbf{I}} \qquad (22.7)$$

However, since this is a series circuit, the circuit current is the same current that is flowing through elements $\mathbf{Z}_1$ and $\mathbf{Z}_2$. Or $\mathbf{I} = \mathbf{I}_1 = \mathbf{I}_2$. Therefore we may modify Equation (22.7):

$$\frac{\mathbf{E}}{\mathbf{I}} = \frac{\mathbf{V}_1}{\mathbf{I}_1} + \frac{\mathbf{V}_2}{\mathbf{I}_2}$$

We then obtain

$$\mathbf{Z}_T = \mathbf{Z}_1 + \mathbf{Z}_2 \qquad (22.8)$$

For more than two impedances in series, the total impedance becomes

$$\mathbf{Z}_T = \mathbf{Z}_1 + \mathbf{Z}_2 + \mathbf{Z}_3 + \cdots + \mathbf{Z}_N \qquad (22.9)$$

Therefore, in a series circuit, we add impedances vectorially. If the impedances are expressed in polar notation, they must be changed into rectangular notation before they are added. Example 22.1 will show you how to work with impedances in series.

Example 22.1 Refer to the circuit with two series impedances shown in Figure 22.2. With V_{in} equal to 120 V at 60 Hz and $\mathbf{Z}_1$ equal to 10 kΩ $\underline{/20°}$ and $\mathbf{Z}_2$ equal to 6 kΩ $\underline{/42°}$, solve for $\mathbf{Z}_T$, $\mathbf{I}$, $\mathbf{V}_1$, and $\mathbf{V}_2$. (Use $\mathbf{I}$ as the reference phasor.)

Solution

$$\mathbf{Z}_T = 10 \text{ k}\Omega \; \underline{/20°} + 6 \text{ k}\Omega \; \underline{/42°}$$

$$= 10 \text{ k}\Omega \cos 20° + j10 \text{ k}\Omega \sin 20° + 6 \text{ k}\Omega \cos 42°$$
$$\quad + j6 \text{ k}\Omega \sin 42°$$

$$= 10 \text{ k}\Omega \, (0.9397) + j10 \text{ k}\Omega \, (0.3420) + 6 \text{ k}\Omega \, (0.7431)$$
$$\quad + j6 \text{ k}\Omega \, (0.6691)$$

$$= 9.397 \text{ k}\Omega + j3.42 \text{ k}\Omega + 4.459 \text{ k}\Omega + j4.015 \text{ k}\Omega$$

$$= 13.856 \text{ k}\Omega + j7.435 \text{ k}\Omega = 15.725 \text{ k}\Omega \; \underline{/28.2°}$$

We can find ϕ by using the arc tangent

$$\phi = \tan^{-1} \left(\frac{7.435 \text{ k}\Omega}{13.856 \text{ k}\Omega} \right) = \tan^{-1} (0.5366) = 28.2°$$

If you have a calculator, you can eliminate many of the steps shown above.

$$\mathbf{I} = \frac{\mathbf{E}}{\mathbf{Z}_T} = \frac{120 \text{ V} \; \underline{/28.2°}}{15.725 \text{ k}\Omega \; \underline{/28.2°}} = 7.631 \text{ mA} \; \underline{/0°}$$

Since this is a series circuit, $\mathbf{I}$ is used as the reference phasor ($I \; \underline{/0°}$). For $\mathbf{I}$ to have an angle of zero degrees, the input voltage must have the same angle as the total impedance. If you were to choose $\mathbf{E}$ as the reference phasor, the $\mathbf{I}$ would lag $\mathbf{E}$ by 28.2°.

$$\mathbf{V}_1 = \mathbf{IZ}_1 = (7.631 \text{ mA} \; \underline{/0°})(10 \text{ k}\Omega \; \underline{/20°}) = 76.31 \text{ V} \; \underline{/20°}$$

$$\mathbf{V}_2 = \mathbf{IZ}_2 = (7.631 \text{ mA} \; \underline{/0°})(6 \text{ k}\Omega \; \underline{/42°}) = 45.79 \text{ V} \; \underline{/42°}$$

We check to see whether KVL is satisfied.

$$\mathbf{V}_1 + \mathbf{V}_2 = 76.31 \text{ V} \; \underline{/20°} + 45.79 \text{ V} \; \underline{/42°}$$
$$= 71.71 + j26.10 + 34.03 + j30.64$$
$$= 105.74 + j56.74 = 120 \text{ V} \; \underline{/28.2°} = \mathbf{E}$$

Therefore the vector sum of $\mathbf{V}_1$ plus $\mathbf{V}_2$ does equal the input voltage.

Problem 22.1 Use the circuit with two series impedances shown in Figure 22.2. The input voltage V_{in} is 120 V, 60 Hz, with $\mathbf{Z}_1$ equal to 20 kΩ $\underline{/60°}$ and $\mathbf{Z}_2$ equal to 15 kΩ $\underline{/-20°}$. Solve for $\mathbf{Z}_T$, $\mathbf{I}$, $\mathbf{V}_1$, and $\mathbf{V}_2$. (Use $\mathbf{I}$ as the reference phasor.)

Two Impedances in Parallel

Objective 2 Calculate the equivalent impedance when two impedances are connected in parallel. Then find the voltage and/or currents in the circuit.

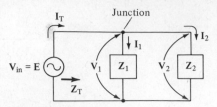

Figure 22.3 Parallel impedances

Figure 22.3 shows a parallel ac circuit with two impedances. As with the series ac circuit discussed in Objective 1, the **Z**'s shown in the boxes in Figure 22.3 may represent any one of Equations (22.1) through (22.5).

When only two impedances are connected in parallel, you can find the total impedance in a manner similar to the way you found the total resistance with two resistors in parallel. You may use the product over the sum. However, remember that you are working with vector quantities.

$$\mathbf{Z}_T = \mathbf{Z}_1 \| \mathbf{Z}_2 = \frac{\mathbf{Z}_1 \mathbf{Z}_2}{\mathbf{Z}_1 + \mathbf{Z}_2} \qquad (22.10)$$

Example 22.2 illustrates the use of Equation (22.10) and also shows you what you need to know to complete Objective 2.

Example 22.2 Refer to the circuit with two parallel impedances shown in Figure 22.3. The input voltage V_{in} is 120 V, 60 Hz, with $\mathbf{Z}_1$ equal to 10 kΩ $\underline{/20°}$ and $\mathbf{Z}_2$ equal to 6 kΩ $\underline{/42°}$. Solve for $\mathbf{Z}_T$, $\mathbf{I}_T$, $\mathbf{I}_1$, and $\mathbf{I}_2$.

Solution

$$\mathbf{Z}_T = \mathbf{Z}_1 \| \mathbf{Z}_2$$

Express the numerator in polar notation so that you can multiply. Express the denominator in rectangular notation so that you can add vectorally.

$$\mathbf{Z}_T = \frac{(10 \text{ kΩ } \underline{/20°})(6 \text{ kΩ } \underline{/42°})}{(9.397 \text{ kΩ} + j3.420 \text{ kΩ}) + (4.459 \text{ kΩ} + j4.015 \text{ kΩ})}$$

For the work for the denominator, see Example 22.1 for 10 kΩ $\underline{/20°}$ + 6 kΩ $\underline{/42°}$.

$$\mathbf{Z}_T = \frac{60 \times 10^6 \ \underline{/62°}}{15.725 \text{ k } \underline{/28.2°}} = 3.816 \text{ kΩ } \underline{/33.8°}$$

$$\mathbf{I}_T = \frac{\mathbf{E}}{\mathbf{Z}_T} = \frac{120 \text{ V } \underline{/0°}}{3.816 \text{ kΩ } \underline{/33.8°}} = 31.45 \text{ mA } \underline{/-33.8°}$$

This is a parallel circuit; therefore you use the voltage as the reference phasor.

$$\mathbf{I}_1 = \frac{\mathbf{V}_1}{\mathbf{Z}_1} = \frac{120 \text{ V } \underline{/0°}}{10 \text{ kΩ } \underline{/20°}} = 12 \text{ mA } \underline{/-20°}$$

$$\mathbf{I}_2 = \frac{\mathbf{V}_2}{\mathbf{Z}_2} = \frac{120 \text{ V } \underline{/0°}}{6 \text{ kΩ } \underline{/42°}} = 20 \text{ mA } \underline{/-42°}$$

Let us check to see whether KCL is satisfied.

$$\mathbf{I}_1 + \mathbf{I}_2 = 12 \text{ mA } \underline{/-20°} + 20 \text{ mA } \underline{/-42°}$$
$$= 11.28 \text{ mA} - j4.10 \text{ mA} + 14.86 \text{ mA} - j13.38 \text{ mA}$$
$$= 26.14 \text{ mA} - j17.49 \text{ mA} = 31.45 \text{ mA } \underline{/-33.8°} = \mathbf{I}_T$$

Therefore the vector sum of $\mathbf{I}_1$ plus $\mathbf{I}_2$ does equal the total current $\mathbf{I}_T$.

Problem 22.2   Use the circuit with two parallel impedances shown in Figure 22.3. The input voltage is 120 V, 60 Hz, with $\mathbf{Z}_1$ equal to 20 kΩ $\underline{/60°}$ and $\mathbf{Z}_2$ equal to 15 kΩ $\underline{/-20°}$. Solve for $\mathbf{Z}_T$, $\mathbf{I}_T$, $\mathbf{I}_1$, and $\mathbf{I}_2$.

The circuit shown in Figure 22.3 may be considered a simple series-parallel circuit. If both $\mathbf{Z}$'s have positive angles (inductive reactance), then the circuit may be drawn as shown in Figure 22.4. In the first leg, $\mathbf{R}_1$ and $\mathbf{X}_{L_1}$ are connected in series. In the second leg, $\mathbf{R}_2$ and $\mathbf{X}_{L_2}$ are connected in series, while the second leg is connected in parallel with the first leg.

When more than two impedances are connected in parallel, we could find the total impedance by repeated application of Equation (22.10). However, the math becomes difficult. Objective 3 shows you how to work with more than two impedances in parallel.

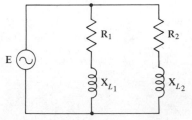

Figure 22.4   Simple series-parallel circuit—both $\mathbf{Z}_1$ and $\mathbf{Z}_2$ have a positiv phase angle

Total Admittance

Objective 3   Calculate the total admittance when two or more impedances are connected in parallel. Then, from the admittance, determine the total impedance.

Writing KCL at the junction shown in Figure 22.3, we obtain

$$\mathbf{I}_T = \mathbf{I}_1 + \mathbf{I}_2 \qquad (22.11)$$

Dividing both sides of Equation (22.11) by the voltage $\mathbf{E}$, we obtain

$$\frac{\mathbf{I}_T}{\mathbf{E}} = \frac{\mathbf{I}_1}{\mathbf{E}} + \frac{\mathbf{I}_2}{\mathbf{E}} \qquad (22.12)$$

This is a parallel circuit; therefore the voltages across each impedance must be the same, or $\mathbf{E} = \mathbf{V}_1 = \mathbf{V}_2$. Therefore we may modify Equation (22.12):

$$\frac{\mathbf{I}_T}{\mathbf{E}} = \frac{\mathbf{I}_1}{\mathbf{V}_1} + \frac{\mathbf{I}_2}{\mathbf{V}_2}$$

We then obtain the total admittance of the circuit,

$$\mathbf{Y}_T = \mathbf{Y}_1 + \mathbf{Y}_2 \qquad (22.13)$$

For more than two impedances in parallel, the total admittance becomes

$$\mathbf{Y}_T = \mathbf{Y}_1 + \mathbf{Y}_2 + \mathbf{Y}_3 + \cdots + \mathbf{Y}_N \qquad (22.14)$$

We can find the total impedance of the circuit from the total admittance,

$$\mathbf{Z}_T = \frac{1}{\mathbf{Y}_T} \qquad (22.15)$$

Therefore, in a parallel circuit, we add admittances vectorially. If the total impedance of the circuit is required, we take the reciprocal of the admittance. Example 22.3 shows you how to work with impedances connected in parallel by using admittances.

Example 22.3 Solve the circuit given in Example 22.2 by using admittances. Solve for $\mathbf{Y}_T$ and $\mathbf{Z}_T$ only.

Solution

$$\mathbf{Y}_1 = \frac{1}{\mathbf{Z}_1} = \frac{1}{10 \text{ k}\Omega \; \underline{/20°}} = 100 \times 10^{-6} \text{ S} \; \underline{/-20°}$$

$$= 100 \times 10^{-6} \text{ S} \cos 20° - j100 \times 10^{-6} \text{ S} \sin 20°$$

$$= 93.97 \times 10^{-6} \text{ S} - j34.20 \times 10^{-6} \text{ S}$$

$$\mathbf{Y}_2 = \frac{1}{\mathbf{Z}_2} = \frac{1}{6 \text{ k}\Omega \; \underline{/42°}}$$

$$= 166.7 \times 10^{-6} \text{ S} \cos 42° - j166.7 \times 10^{-6} \text{ S} \sin 42°$$

$$= 123.87 \times 10^{-6} \text{ S} - j111.54 \times 10^{-6} \text{ S}$$

$$\mathbf{Y}_T = \mathbf{Y}_1 + \mathbf{Y}_2 = (217.84 \text{ S} - j145.74 \text{ S}) \times 10^{-6}$$

$$= 262.1 \times 10^{-6} \text{ S} \; \underline{\left| \tan^{-1} -\left(\frac{145.74}{217.84}\right) \right.} = 262.1 \; \mu\text{S} \; \underline{/-33.8°}$$

$$\mathbf{Z}_T = \frac{1}{\mathbf{Y}_T} = \frac{1}{262.1 \times 10^{-6} \text{ S} \; \underline{/-33.8°}} = 3.815 \text{ k}\Omega \; \underline{/33.8°}$$

Problem 22.3 Solve Problem 22.2 for $\mathbf{Z}_T$ using admittances.

Problem 22.4 Solve the circuit shown in Figure 22.5 for $\mathbf{Z}_T$ using admittances. The input voltage is 120 V, 60 Hz, with $\mathbf{Z}_1$ equal to 10 kΩ $\underline{/30°}$, $\mathbf{Z}_2$ equal to 14.14 kΩ $\underline{/45°}$, and $\mathbf{Z}_3$ equal to 15 kΩ $\underline{/-36.87°}$.

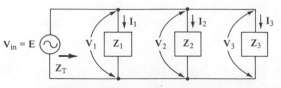

Figure 22.5   Three parallel impedances

Voltage- and Current-Divider Principles

Objective 4 Apply the voltage-divider and current-divider principles to ac circuits.

Voltage-Divider Principle

The voltage-divider principle that you learned for dc circuits also applies in ac circuits. However, in ac circuits, you must remember that you are working with vector quantities. To find the voltage across a particular impedance in a series circuit, take the ratio of that particular impedance to the total impedance in the circuit and multiply this ratio by the voltage that is across the total impedance. For two impedances in series, Figure 22.2, you can find $\mathbf{V}_2$ from

$$\mathbf{V}_2 = \left(\frac{\mathbf{Z}_2}{\mathbf{Z}_1 + \mathbf{Z}_2}\right)\mathbf{E} \tag{22.16}$$

For more than two impedances in series, you can find the voltage across one or more elements of the circuit from

$$\mathbf{V}_X = \left(\frac{\mathbf{Z}_X}{\mathbf{Z}_T}\right)\mathbf{E} \tag{22.17}$$

where $\mathbf{Z}_X$ is the impedance of the element or elements called X

$\mathbf{V}_X$ is the unknown voltage across impedance $\mathbf{Z}_X$

$\mathbf{Z}_T$ is the total impedance of the series circuit

Example 22.4 Find $\mathbf{V}_2$ in the circuit given in Example 22.1 using the voltage-divider principle.

Solution

$$\mathbf{V}_2 = \left(\frac{\mathbf{Z}_2}{\mathbf{Z}_1 + \mathbf{Z}_2}\right)\mathbf{E} = \left(\frac{6\ \text{k}\Omega\ \underline{/42°}}{10\ \text{k}\Omega\ \underline{/20°} + 6\ \text{k}\Omega\ \underline{/42°}}\right)120\ \text{V}\ \underline{/28.2°}$$

The angle for the voltage across an impedance must be the same as the angle associated with that impedance. This results from using current as the reference phasor (0°) in series circuits. Therefore the phase angle for the applied voltage $\mathbf{V}_{in}$ must be the same as the angle for the total impedance.

From Example 22.1, the denominator is equal to 15.725 kΩ $\underline{/28.2°}$; therefore

$$\mathbf{V}_2 = \frac{(6\ \text{k}\Omega\ \underline{/42°})(120\ \text{V}\ \underline{/28.2°})}{15.725\ \text{k}\Omega\ \underline{/28.2°}} = 45.79\ \text{V}\ \underline{/42°}$$

Problem 22.5 Find V_2 and V_1 in the circuit given in Problem 22.1 using the voltage-divider principle.

Current-Divider Principle

As with the voltage-divider principle, the current-divider principle that you learned in dc circuits also applies in ac circuits. For only two impedances in parallel (Fig. 22.3), to find the current in one impedance, take the ratio of the opposite impedance divided by the sum (vector) of the two impedances and multiply this ratio by the total current into the junction. To solve for the current I_2 in Figure 22.3 using the current-divider principle, you use

$$I_2 = \left(\frac{Z_1}{Z_1 + Z_2}\right) I_T \tag{22.18}$$

Example 22.5 Find I_2 in the circuit given in Example 22.2 using the current-divider principle. Assume that I_T is equal to 31.45 mA $\underline{/-33.8°}$.

Solution

$$I_2 = \left(\frac{Z_1}{Z_1 + Z_2}\right) I_T$$

$$= \left(\frac{10 \text{ k}\Omega \ \underline{/20°}}{10 \text{ k}\Omega \ \underline{/20°} + 6 \text{ k}\Omega \ \underline{/42°}}\right) 31.45 \text{ mA} \ \underline{/-33.8°}$$

Using 15.725 kΩ $\underline{/28.2°}$ from Example 22.1 for the denominator, we get

$$I_2 = \frac{(10 \text{ k}\Omega \ \underline{/20°})(31.45 \text{ mA} \ \underline{/-33.8°})}{15.725 \text{ k}\Omega \ \underline{/28.2°}} = 20 \text{ mA} \ \underline{/-42°}$$

Problem 22.6 Find I_2 in the circuit given in Problem 22.2 using the current-divider principle. Assume that I_T is equal to 10.80 mA $\underline{/-13.16°}$.

If there are more than *two* impedances in parallel, then we cannot use the current-divider principle, but must work with admittances as we did in Objective 3.

$$I_X = \left(\frac{Y_X}{Y_T}\right) I_T \tag{22.19}$$

where I_X is the current in element X whose admittance is Y_X.

Problem 22.7 Solve for I_1 in the circuit given in Example 22.2 using Equation (22.19). Use the admittances found in Example 22.3, and assume that I_T is equal to 31.45 mA $\underline{/-33.8°}$.

Conversion Between Circuits

Objective 5 Convert a two-element parallel ac circuit to an equivalent two-element series circuit, and vice versa.

Did you notice that when you worked with parallel dc or ac circuits, you always obtained a series circuit when you found the equivalent (total) resistance or impedance of the parallel circuit? It is also possible to replace a series circuit with an equivalent parallel circuit. For this objective we develop the conversion equations that you may use for ac circuits. Even though we shall show you how the conversion equations were derived, you will probably use only the equations.

Figure 22.6 shows the circuits and the conversion equations that you may use to change a parallel circuit to an equivalent series circuit. Figure 22.7 shows how to convert from a series to an equivalent parallel circuit.

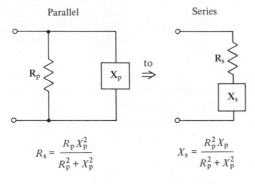

$$R_s = \frac{R_p X_p^2}{R_p^2 + X_p^2} \qquad X_s = \frac{R_p^2 X_p}{R_p^2 + X_p^2}$$

(Magnitudes only are used in these equations)

Figure 22.6 Conversion from a parallel circuit to an equivalent series circuit

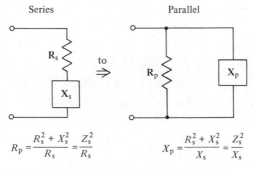

$$R_p = \frac{R_s^2 + X_s^2}{R_s} = \frac{Z_s^2}{R_s} \qquad X_p = \frac{R_s^2 + X_s^2}{X_s} = \frac{Z_s^2}{X_s}$$

(Magnitudes only are used in these equations)

Figure 22.7 Conversion from a series circuit to an equivalent parallel circuit

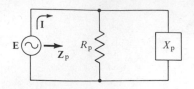

Parallel: $\mathbf{Z} = R_p \| (\pm j X_p)$

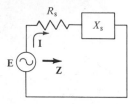

Series: $\mathbf{Z} = R_s \pm j X_s$

Figure 22.8 Circuits for the derivation of the conversion from a parallel to an equivalent series circuit

Parallel Circuit to an Equivalent Series Circuit

To derive the conversion equations listed in Figure 22.6, from a parallel to an equivalent series circuit, is straightforward. The two-element parallel circuit is shown in Figure 22.8. The subscript p denotes a *parallel* element. The X element may be a capacitor or a perfect inductor. The impedance of the parallel circuit shown in Figure 22.8 is equal to

$$\mathbf{Z}_p = \frac{\text{product}}{\text{sum}} = \frac{R_p(\pm j X_p)}{R_p \pm j X_p} \tag{22.20}$$

where the sign on the reactive component is $+$ if it is inductive or $-$ if it is capacitive.

To solve for $\mathbf{Z}_p$, we multiply both the numerator and denominator by the conjugate of the denominator.

$$\mathbf{Z}_p = \frac{R_p(\pm j X_p)}{R_p \pm j X_p} \times \frac{R_p \mp j X_p}{R_p \mp j X_p} = \frac{-j^2 R_p X_p^2 \pm j R_p^2 X_p}{R_p^2 - j^2 X_p^2}$$

$$= \frac{R_p X_p^2}{R_p^2 + X_p^2} \pm j \frac{R_p^2 X_p}{R_p^2 + X_p^2} = R_s \pm j X_s$$

where
$$R_s = \frac{R_p X_p^2}{R_p^2 + X_p^2} \tag{22.21}$$

$$X_s = \frac{R_p^2 X_p}{R_p^2 + X_p^2} \tag{22.22}$$

The subscript s denotes a *series* element.

To use Equations (22.21) and (22.22), work with the *magnitude* of X_p. Don't worry about the $\pm j$ that is usually attached. Also note in the derivation that an inductor in parallel converts to an inductor in series. A capacitor in parallel converts to a capacitor in series.

Example 22.6   An ac circuit consists of an inductor in parallel with a resistor. The inductor has an inductive reactance at the frequency of operation of 5 kΩ. The resistor is 10 kΩ. Find the equivalent series circuit.

Solution

$$R_s = \frac{R_p X_p^2}{R_p^2 + X_p^2} = \frac{10 \text{ k}\Omega(5 \text{ k}\Omega)^2}{(10 \text{ k}\Omega)^2 + (5 \text{ k}\Omega)^2} = \frac{250 \times 10^9}{125 \times 10^6} = 2 \text{ k}\Omega$$

$$X_s = \frac{R_p^2 X_p}{R_p^2 + X_p^2} = \frac{(10 \text{ k}\Omega)^2(5 \text{ k}\Omega)}{125 \times 10^6} = \frac{500 \times 10^9}{125 \times 10^6}$$

$$= 4 \text{ k}\Omega \qquad \text{(inductive)}$$

The equivalent series circuit consists of a 2-kΩ resistor in series with a 4-kΩ inductive reactance.

Problem 22.8   An ac circuit consists of an inductor in parallel with a resistor. The inductor has an inductive reactance of 100 Ω at the frequency of operation. The resistor is equal to 1 kΩ. Find the equivalent series circuit.

Problem 22.9   An ac circuit consists of a capacitor in parallel with a resistor. The capacitor has a capacitive reactance of 2 kΩ at the frequency of operation, and the resistor is equal to 20 kΩ. Find the equivalent series circuit.

Series Circuit to an Equivalent Parallel Circuit

To derive the conversion equations listed in Figure 22.7, from a series to an equivalent parallel circuit, we must first solve for the admittance of the series circuit of Figure 22.9. Again we use the subscript s to denote a series element.

$$\mathbf{Z}_s = R_s \pm jX_s$$

Admittance is the reciprocal of impedance:

$$\mathbf{Y}_s = \frac{1}{R_s \pm jX_s} \tag{22.23}$$

To solve for $\mathbf{Y}_s$, we do not take the reciprocal of R_s and add it to the reciprocal of X_s. We must work with the whole $(R_s \pm jX_s)$ term. To solve for $\mathbf{Y}_s$ in Equation (22.23), we multiply both the numerator and the denominator by the conjugate of the denominator.

$$\mathbf{Y}_s = \frac{1}{R_s \pm jX_s} = \frac{1}{R_s \pm jX_s} \times \frac{R_s \mp jX_s}{R_s \mp jX_s} = \frac{R_s \mp jX_s}{R_s^2 - j^2 X_s^2}$$

$$\mathbf{Y}_s = \frac{R_s}{R_s^2 + X_s^2} \mp j \frac{X_s}{R_s^2 + X_s^2} \tag{22.24}$$

Writing the equation for the admittance of the parallel circuit in Figure 22.9, we obtain

$$\mathbf{Y}_p = G_p \mp jB_p$$

$$\mathbf{Y}_p = \frac{1}{R_p} \mp j \frac{1}{X_p} \tag{22.25}$$

Comparing Equation (22.24) with Equation (22.25), we find that

$$R_p = \frac{R_s^2 + X_s^2}{R_s} \tag{22.26}$$

$$X_p = \frac{R_s^2 + X_s^2}{X_s} \tag{22.27}$$

As we did with Equations (22.21) and (22.22), we substitute the magnitude of X_s into Equations (22.26) and (22.27). An inductor in series converts to an inductor in parallel. And a capacitor in series converts to a capacitor in parallel.

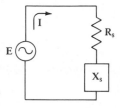

Series: $\mathbf{I} = \dfrac{E}{R_s \pm jX_s}$

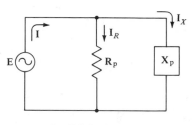

Parallel: $\mathbf{I} = I_R \mp jI_X$

$$= \frac{E}{R_p} \mp j \frac{E}{X_p}$$

Figure 22.9   Circuits for the derivation of the conversion from a series to an equivalent parallel circuit

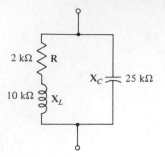

Figure 22.10   Circuit for Example 22.7

Example 22.7   Find a two-element equivalent parallel circuit for the circuit shown in Figure 22.10.

Solution   First we find the equivalent parallel circuit for the 2-kΩ resistor in series with the 10-kΩ inductive reactance.

$$R_p = \frac{R_s^2 + X_s^2}{R_s} = \frac{(2 \text{ k}\Omega)^2 + (10 \text{ k}\Omega)^2}{2 \text{ k}\Omega}$$

$$= \frac{104 \times 10^6}{2 \text{ k}} = 52 \text{ k}\Omega$$

$$X_p \text{ (inductive)} = \frac{R_s^2 + X_s^2}{X_s} = \frac{104 \times 10^6}{10 \text{ k}} = 10.4 \text{ k}\Omega$$

To find the equivalent X in parallel with the R_p, we must work with the equivalent susceptance.

$$B_{eq} = B_C - B_L = \frac{1}{X_C} - \frac{1}{X_L}$$

$$= \frac{1}{25 \text{ k}\Omega} - \frac{1}{10.4 \text{ k}\Omega} = 40 \times 10^{-6} \text{ S} - 96.15 \times 10^{-6} \text{ S}$$

$$= 56.15 \text{ } \mu\text{S} \qquad \text{(inductive)}$$

$$X_L = \frac{1}{B_{eq}} = \frac{1}{56.15 \times 10^{-6}} = 17.8 \text{ k}\Omega$$

The equivalent parallel circuit consists of a 52-kΩ resistor in parallel with a 17.8-kΩ inductive reactance.

Problem 22.10   An ac circuit consists of an inductor in series with a resistor. The inductor has an inductive reactance of 5 kΩ at the frequency of operation, and the resistor is equal to 500 Ω. Find the equivalent parallel circuit.

Problem 22.11   An ac circuit consists of a capacitor in series with a resistor. The capacitor has a capacitive reactance of 1 kΩ at the frequency of operation and the resistor is equal to 1 kΩ. Find the equivalent parallel circuit.

Problem 22.12   Find a two-element equivalent parallel circuit for the circuit shown in Figure 22.11.

Approximations

Examining the parallel-to-series conversion equations, Equations (22.21) and (22.22), you can see that if the ratio of R_p/X_p is 10 or greater, then X_s is approximately equal to X_p. However, R_s is much smaller than R_p. The series resistor R_s will approximately equal R_p divided by the *square* of the ratio of R_p/X_p.

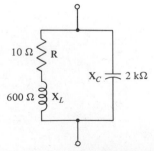

Figure 22.11   Circuit for Problem 22.10

Similar statements could be made about Equations (22.26) and (22.27). However, now the ratio X_s/R_s must be 10 or greater for the following statements to hold. The parallel reactance X_p is approximately equal to X_s. The parallel resistor R_p is many times greater than R_s, approximately the square of the ratio X_s/R_s times greater. A practical inductor or coil has series resistance in addition to the inductance. You will learn in the unit on Resonance that the ratio of the reactance of the coil to the resistance of the coil (X_s/R_s) is known as the *Q of the coil*.

Total Impedance

Objective 6   Calculate the total impedance of series-parallel ac circuits. Then find the currents and/or voltages in the circuits.

In this objective you will not learn any new concepts. You will see how you can use the first five objectives to solve various series-parallel ac circuits. You will do this by studying the examples that follow. Then you will be asked to solve various series-parallel ac circuits.

Example 22.8   Refer to the circuit shown in Figure 22.12(a). Find the total impedance $\mathbf{Z}_{ab}$ (both magnitude and direction) looking into terminals (a) and (b). Also find the total impedance $\mathbf{Z}_{cd}$ looking into terminals (c) and (d).

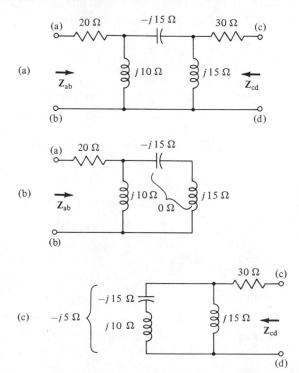

Figure 22.12   Circuits for Example 22.8

Solution: *Impedance* $\mathbf{Z}_{ab}$ To determine $\mathbf{Z}_{ab}$ in the circuit shown in Figure 22.12(a), we look to the *right* from terminals (a) and (b). There is no current in the 30-Ω resistor in the circuit, because terminals (c) and (d) are open. Therefore, to determine $\mathbf{Z}_{ab}$, we can reduce the circuit shown in Figure 22.12(a) to that shown in Figure 22.12(b).

Note that the capacitor ($-j15$ Ω) and the inductor ($+j15$ Ω) have reactance equal in magnitude but opposite in sign. Therefore this capacitor and inductor act as a short circuit. This short circuit is in parallel with the $+j10$ Ω inductor, and therefore shorts it out.

$$\therefore \quad \mathbf{Z}_{ab} = 20 \text{ Ω } \underline{/0^\circ}$$

Impedance $\mathbf{Z}_{cd}$ To determine $\mathbf{Z}_{cd}$ in the circuit shown in Figure 22.12(a), we must look to the *left* from terminals (c) and (d). Terminals (a) and (b) are open, and therefore the 20-Ω resistor is not in the circuit. The circuit in Figure 22.12(a) reduces to the circuit shown in Figure 22.12(c). The 10-Ω inductive reactance cancels all but 5 Ω of the 15-Ω capacitive reactance.

$$\mathbf{Z}_{cd} = 30 + (j15\|-j5) = 30 + \frac{j15(-j5)}{j15 - j5}$$

$$= 30 + \frac{75 \underline{/0^\circ}}{10 \underline{/90^\circ}} = 30 + 7.5 \underline{/-90^\circ} = 30.92 \text{ Ω } \underline{/-14^\circ}$$

Problem 22.13   Refer to the circuit shown in Figure 22.13. Find the total impedance $\mathbf{Z}_{ab}$ (both magnitude and direction) looking into terminals (a) and (b). Also find the total impedance $\mathbf{Z}_{cd}$ looking into terminals (c) and (d).

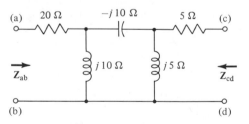

Figure 22.13   Circuit for Problem 22.13

Example 22.9   Refer to the circuit shown in Figure 22.14. Find $\mathbf{Z}_T$, $\mathbf{V}_1$, and $\mathbf{V}_2$.

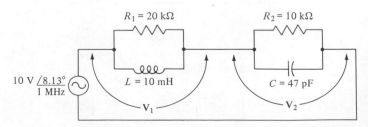

Figure 22.14   Circuit for Example 22.9

Solution

$$X_L = 2\pi f L = 2\pi(1 \times 10^6)(10 \times 10^{-3}) = 62.83 \text{ k}\Omega$$

$$X_C = \frac{1}{2\pi f C} = \frac{1}{2\pi(1 \times 10^6)(47 \times 10^{-12})} = 3.386 \text{ k}\Omega$$

Using the conversions from a parallel to an equivalent series circuit, we obtain

$$R_{s1} = \frac{R_p X_p^2}{R_p^2 + X_p^2} = \frac{20 \text{ k}\Omega(62.83 \text{ k}\Omega)^2}{(20 \text{ k}\Omega)^2 + (62.83 \text{ k}\Omega)^2} = 18.16 \text{ k}\Omega$$

$$X_{s1} = \frac{R_p^2 X_p}{R_p^2 + X_p^2} = \frac{(20 \text{ k}\Omega)^2(62.83 \text{ k}\Omega)}{4.348 \times 10^9} = 5.78 \text{ k}\Omega \text{ (inductive)}$$

$$R_{s2} = \frac{R_p X_p^2}{R_p^2 + X_p^2} = \frac{10 \text{ k}\Omega(3.386 \text{ k}\Omega)^2}{(10 \text{ k}\Omega)^2 + (3.386 \text{ k}\Omega)^2} = 1.029 \text{ k}\Omega$$

$$X_{s2} = \frac{R_p^2 X_p}{R_p^2 + X_p^2} = \frac{(10 \text{ k}\Omega)^2(3.386 \text{ k}\Omega)}{1.115 \times 10^8} = 3.038 \text{ k}\Omega \text{ (capacitive)}$$

$$R_{s \text{ total}} = R_{s1} + R_{s2} = 18.16 \text{ k}\Omega + 1.029 \text{ k}\Omega = 19.19 \text{ k}\Omega$$

$$\mathbf{X}_{s \text{ total}} = \mathbf{X}_{s1} + \mathbf{X}_{s2} = 5.78 \text{ k}\Omega \text{ (inductive)} - 3.038 \text{ k}\Omega \text{ (capacitive)}$$
$$= 2.742 \text{ k}\Omega \text{ (inductive)}$$

$$\therefore \quad \mathbf{Z}_T = 19.19 \text{ k}\Omega + j2.742 \text{ k}\Omega = 19.38 \text{ k}\Omega \underline{/8.13°}$$

The voltage across each of the parallel branches can be found by using the voltage-divider principle.

$$\mathbf{V}_1 = \frac{\mathbf{Z}_1}{\mathbf{Z}_T} \mathbf{E} = \frac{(18.16 \text{ k}\Omega + j5.87 \text{ k}\Omega)}{19.38 \text{ k}\Omega \underline{/8.13°}} 10 \text{ V} \underline{/8.13°}$$

$$= \left(\frac{19.08 \text{ k}\Omega \underline{/17.9°}}{19.38 \text{ k}\Omega \underline{/8.13°}} \right) 10 \text{ V} \underline{/8.13°}$$

$$= 9.85 \text{ V} \underline{/17.9°}$$

$$\mathbf{V}_2 = \frac{\mathbf{Z}_2}{\mathbf{Z}_T} \mathbf{E} = \frac{(1.029 \text{ k}\Omega - j3.038 \text{ k}\Omega)}{19.38 \text{ k}\Omega \underline{/8.13°}} 10 \text{ V} \underline{/8.13°}$$

$$= \left(\frac{3.208 \text{ k}\Omega \underline{/-71.3°}}{19.38 \text{ k}\Omega \underline{/8.13°}} \right) 10 \text{ V} \underline{/8.13°}$$

$$= 1.65 \text{ V} \underline{/-71.3°}$$

Problem 22.14 Refer to the circuit given in Example 22.9 (Figure 22.14). Find the impedance of each of the parallel branches 1 and 2 by the product over the sum. Then find $\mathbf{Z}_T$. Determine $\mathbf{I}$ from $\mathbf{Z}_T$, then find $\mathbf{V}_1$ and $\mathbf{V}_2$ by using Ohm's law.

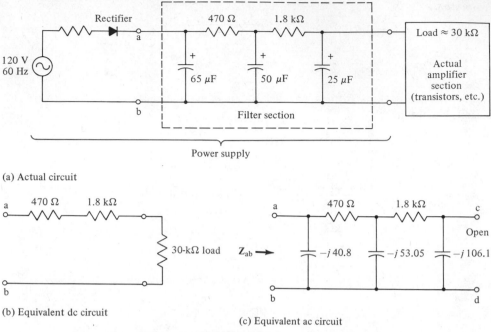

Figure 22.15 Phonograph amplifier power-supply circuit

Example 22.10 (*Advanced*) Figure 22.15(a) shows a phonograph amplifier power-supply circuit. Determine the ac input impedance at a given frequency (60 Hz).

Discussion The purpose of the power-supply circuit in Figure 22.15(a) is to change the 120-V ac voltage to a dc voltage. The rectifier section changes the input 60-Hz ac voltage to a pulsating dc voltage. By this we mean that we can represent the input to the filter part of the circuit in Figure 22.15(a) as a combination of a dc voltage and an ac voltage. The transistors in the amplifier section need a dc voltage applied to them to make them function correctly. The purpose of this filter network (series-parallel ac circuit) is to filter out the ac and let only the dc through.

Almost all the input dc voltage gets to the load. The capacitors act as open circuits for dc, and therefore the equivalent dc circuit appears as shown in Figure 22.15(b). Figure 22.15(b) is a simple voltage-divider circuit; very little dc voltage is lost in the 470-Ω and 1.8-kΩ series resistors.

The parallel capacitors provide a very low impedance for the ac voltage. Therefore most of the ac current flows through the capacitors back to the input circuit, with no (or very little) ac voltage getting to the load.

Solution To solve for the ac input impedance, we shall simplify the circuit given in Figure 22.15(a). We shall consider the 30-kΩ load to be very large when compared with the impedance of the other components in the filter section. Therefore the load appears as an infinite impedance or an open circuit, as shown in Figure 22.15(c). The circuit shown in Figure 22.15(c) is like the ladder networks we worked with in dc circuits. However, now the math becomes difficult because of the vector algebra.

First we find the capacitive reactance of each of the capacitors shown in Figure 22.15(a).

$$X_C = \frac{1}{2\pi f C} \quad \text{at 60 Hz}$$

$$65 \ \mu F: \quad X_C = \frac{1}{377(65 \times 10^{-6})} = 40.8 \ \Omega$$

$$50 \ \mu F: \quad X_C = \frac{1}{377(50 \times 10^{-6})} = 53.05 \ \Omega$$

$$25 \ \mu F: \quad X_C = \frac{1}{377(25 \times 10^{-6})} = 106.1 \ \Omega$$

Then, using the equivalent ac circuit shown in Figure 22.15(c), we find the impedance $\mathbf{Z}_{ab}$. With terminals (c) and (d) open, the 1.8-kΩ resistor is in series with the 106.1-Ω capacitive reactance. This branch is connected in parallel with the 53.05-Ω capacitive reactance. This is then connected in series with the 470-Ω resistor. And finally, this is all connected in parallel with the 40.8-Ω capacitive reactance. This is mathematically written as

$$\mathbf{Z}_{ab} = (40.8 \ \Omega \ \underline{/-90°})$$
$$\times \ \|[470 \ \Omega \ \underline{/0°} + 53.05 \ \Omega \ \underline{/-90°} \| (1.8 \ k\Omega - j106.1 \ \Omega)]$$

Let us now solve this mathematical expression section by section. The 1.8-kΩ resistor in series with the 106.1-Ω capacitive reactance is

$$1.8 \ k\Omega - j106.1 = 1803 \ \Omega \ \underline{/-3.37°}$$

As you see, the 25-μF capacitor with its 106.1-Ω capacitive reactance is almost negligible when compared with the 1.8-kΩ series resistor.

The parallel combination of the 53.05-Ω capacitive reactance and the two series components is equal to

$$53.05 \ \Omega \ \underline{/-90°} \| 1803 \ \Omega \ \underline{/-3.37°}$$

$$= \frac{(53.05 \ \Omega \ \underline{/-90°})(1803 \ \Omega \ \underline{/-3.37°})}{53.05 \ \Omega \ \underline{/-90°} + 1803 \ \Omega \ \underline{/-3.37°}}$$

$$= \frac{(53.05 \ \Omega \ \underline{/-90°})(1803 \ \Omega \ \underline{/-3.37°})}{-j53.05 + 1800 - j106}$$

$$= \frac{95,649.15 \ \underline{/-93.37°}}{1800 - j159.05} = \frac{95,649.15 \ \underline{/-93.37°}}{1807 \ \underline{/-5.05°}}$$

$$= 52.93 \ \underline{/-88.32°} = 1.55 - j52.91$$

The series combination of the 470-Ω resistor and the previous parallel combination is equal to

$$470 \ \Omega \ \underline{/0°} + 52.93 \ \Omega \ \underline{/-88.32°} = 470 + 1.55 - j52.91$$

$$= 471.55 - j52.91$$

$$= 474.5 \ \Omega \ \underline{/-6.4°}$$

And finally the total impedance is equal to

$$\mathbf{Z}_{ab} = 40.8 \ \Omega \ \underline{/-90°} \| 474.5 \ \Omega \ \underline{/-6.4°}$$

$$= \frac{(40.8 \ \Omega \ \underline{/-90°})(474.5 \ \Omega \ \underline{/-6.4°})}{40.8 \ \Omega \ \underline{/-90°} + 474.5 \ \Omega \ \underline{/-6.4°}}$$

$$= \frac{(40.8 \ \Omega \ \underline{/-90°})(474.5 \ \Omega \ \underline{/-6.4°})}{-j40.8 + 471.55 - j52.91}$$

$$= \frac{19,359.6 \ \underline{/-96.4°}}{471.55 - j93.71} = \frac{19,359.6 \ \underline{/-96.4°}}{480.8 \ \underline{/-11.24°}}$$

$$= 40.27 \ \Omega \ \underline{/-85.16°}$$

The answer for the total ac input impedance is very close to the reactance of the first parallel capacitor. This is reasonable, because it has the lowest impedance of any of the elements, and it is in parallel with a resistor plus other elements whose impedance is more than 10 times that of the capacitor.

It took considerable time to figure out this total impedance. Since this is a filter network, it would also be necessary to figure out how much of the input ac voltage gets to the filter output terminals, terminals (c) and (d) of Figure 22.15(c). This would also take time; see Example 22.12. This is why approximate methods or procedures using charts and graphs have been developed so that you can design filter networks for electronic circuits without all the difficulty with vector algebra.

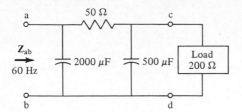

Figure 22.16 Circuit for Problem 22.15

Problem 22.15 (*Advanced*) Figure 22.16 shows a filter network for a high-power class A phonograph amplifier. Determine the ac input impedance. (Consider the 200-Ω load to be very large compared with the impedance of the other components of the filter section.)

Thévenin's Theorem

Objective 7 Apply Thévenin's theorem to series-parallel ac circuits.

You have already learned how to apply Thévenin's theorem to various series-parallel dc circuits. You apply Thévenin's theorem to ac circuits in a similar manner; however, you must remember that you are working with vector quantities.

We can apply Thévenin's theorem to the series-parallel ac circuit shown in Figure 22.17(a). As a result, we obtain the *Thévenin's equivalent circuit (TEC)* with load attached shown in Figure 22.18. The load for the TEC may be a single circuit element, or it may be several elements, including transistors, power supplies, or even complete circuits.

To find the open-circuit voltage $\mathbf{V}_{oc}$, remove the load of the circuit shown in Figure 22.17(a). Then determine the open-circuit voltage that exists across the load terminals [Figure 22.17(b)]. No current flows through $\mathbf{Z}_3$; therefore the $\mathbf{V}_{oc}$ is the voltage across $\mathbf{Z}_2$. We may use the voltage-divider principle to determine $\mathbf{V}_{oc}$:

$$\mathbf{V}_{oc} = \left(\frac{\mathbf{Z}_2}{\mathbf{Z}_1 + \mathbf{Z}_2}\right)\mathbf{E} \qquad (22.28)$$

To find $\mathbf{Z}_{oc}$, replace the voltage source by its internal impedance. Most of the time the internal impedance is considered to be zero; see Figure 22.17(c). Then determine the impedance $\mathbf{Z}_{oc}$ by looking back into the circuit from the load terminals.

$$\mathbf{Z}_{oc} = \mathbf{Z}_3 + (\mathbf{Z}_2 \| \mathbf{Z}_1) \qquad (22.29)$$

Remember that all the terms in Equation (22.29) are vector quantities.

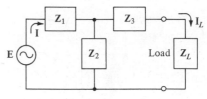

(a) Original circuit

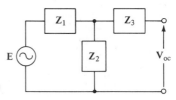

(b) Circuit to determine $\mathbf{V}_{oc}$

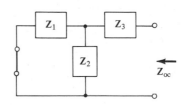

(c) Circuit to determine $\mathbf{Z}_{oc}$

Figure 22.17 Applying TEC to series-parallel ac circuits

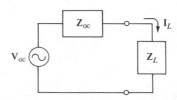

Figure 22.18 TEC of circuit in Figure 22.17(a) with load attached

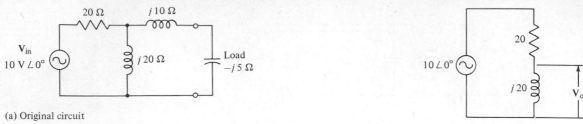

(a) Original circuit

(b) Circuit to determine $\mathbf{V}_{oc}$

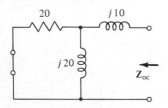

(c) Circuit to determine $\mathbf{Z}_{oc}$

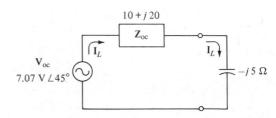

(d) TEC with load attached

Figure 22.19 Circuit for Example 22.11

Example 22.11   Refer to the circuit given in Figure 22.19(a). Consider the $-j5$-Ω capacitive reactance as the load, and determine the TEC. Solve for the voltage across and the current in the load.

Solution   Refer to Figure 22.19(b):

$$\mathbf{V}_{oc} = \frac{20\ \underline{/90^\circ}}{20 + j20}(10\ \underline{/0^\circ}) = \frac{20\ \underline{/90^\circ}}{28.28\ \underline{/45^\circ}}(10\ \underline{/0^\circ}) = 7.07\ \text{V}\ \underline{/45^\circ}$$

Refer to Figure 22.19(c):

$$\mathbf{Z}_{oc} = j10 + \frac{20\ \underline{/0^\circ}(20\ \underline{/90^\circ})}{20 + j20} = j10 + \frac{400\ \underline{/90^\circ}}{28.28\ \underline{/45^\circ}}$$

$$= j10 + 14.14\ \underline{/45^\circ} = j10 + 10 + j10 = 10\ \Omega + j20\ \Omega$$

Refer to Figure 22.19(d):

$$\mathbf{I}_L = \frac{\mathbf{V}_{oc}}{\mathbf{Z}_{\text{total}}} = \frac{7.07\ \underline{/45^\circ}}{10 + j20 - j5} = \frac{7.07\ \underline{/45^\circ}}{18.03\ \underline{/56.3^\circ}} = 0.392\ \text{A}\ \underline{/-11.3^\circ}$$

$$\mathbf{V}_L = \mathbf{I}_L\mathbf{Z}_L = 0.392\ \underline{/-11.3^\circ}\ (5\ \underline{/-90^\circ}) = 1.96\ \text{V}\ \underline{/-101.3^\circ}$$

If we wanted only the voltage across the load, we would use the voltage-divider principle.

$$\mathbf{V}_L = \frac{5\ \underline{/-90^\circ}}{18.03\ \underline{/56.3^\circ}}(7.07\ \text{V}\ \underline{/45^\circ}) = 1.96\ \text{V}\ \underline{/-101.3^\circ}$$

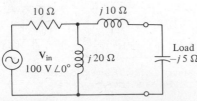

Figure 22.20 Circuit for Problem 22.16

Problem 22.16   Refer to the circuit given in Figure 22.20. Consider the $-j5$-Ω capacitive reactance as the load, and determine the TEC. Solve for the voltage across and the current in the load.

Example 22.12 (*Advanced*) Refer to the filter section of the phonograph amplifier power-supply circuit given in Figure 22.15(a). Use the TEC and determine how much of the input ac voltage actually gets to the load.

Solution Refer to Example 22.10 for some of the numbers that are to be used to solve this example. We shall apply Thévenin's theorem twice to solve this problem.

First application of Thévenin's theorem We call everything enclosed in the box in Figure 22.21(a) the load, and replace the other part of the circuit by the TEC. This gives the $\mathbf{V}_{oc}$ and $\mathbf{Z}_{oc}$ shown in Figure 22.21(b).
Refer to Figure 22.21(c):

$$\mathbf{V}_{oc} = \frac{-j53.05}{470 - j53.05}\,\mathbf{V}_{in} = \frac{53.05\,\underline{/-90°}}{473\,\underline{/-6.4°}}\,\mathbf{V}_{in}$$
$$= (0.112\,\underline{/-83.6°})\mathbf{V}_{in}$$

At this point in the circuit the input ac voltage is reduced considerably (about one-tenth).

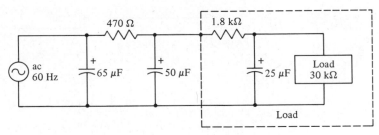

(a) Load for first application of TEC of circuit in Figure 22.15(a)

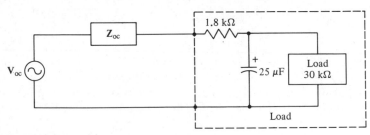

(b) TEC of Figure 22.15(a)

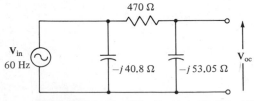

Figure 22.21 Circuit for Example 22.12 (*continued on next page*)

(c) Circuit to determine $\mathbf{V}_{oc}$

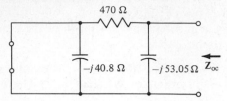

(d) Circuit to determine $\mathbf{Z}_{oc}$

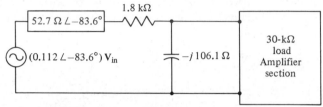

(e) Second application of TEC to circuit in Figure 22.15(a)

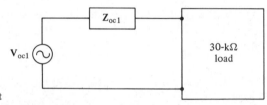

(f) Final TEC of circuit

Figure 22.21   *(continued)*

Refer to Figure 22.21(d):

$$\mathbf{Z}_{oc} = (53.05 \underline{/-90^\circ}) \| (470 \underline{/0^\circ}) = \frac{(53.05 \underline{/-90^\circ})(470 \underline{/0^\circ})}{470 - j53.05}$$

$$= \frac{(53.05)(470) \underline{/-90^\circ + 0^\circ}}{473 \underline{/-6.4^\circ}} = 52.7 \ \Omega \ \underline{/-83.6^\circ}$$

$$= 5.876 \ \Omega - j52.37 \ \Omega$$

Second application of Thévenin's theorem   Now we can call the actual load we started with (30 kΩ) the load for the second application of Thévenin's theorem. Refer to the circuit given in Figure 22.21(e). The new open-circuit voltage—call it $\mathbf{V}_{oc\,1}$—can be calculated by the voltage-divider principle:

$$\mathbf{V}_{oc1} = \left(\frac{106.1 \underline{/-90^\circ}}{(5.876 - j52.37) + 1.8 \ \text{k}\Omega - j106.1} \right)(0.112 \underline{/-83.6^\circ})\mathbf{V}_{in}$$

$$= \left(\frac{106.1 \underline{/-90^\circ}}{1806 - j158} \right)(0.112 \underline{/-83.6^\circ})\mathbf{V}_{in}$$

$$= 0.00655 \underline{/-168.6^\circ} \ \mathbf{V}_{in}$$

$$\mathbf{Z}_{oc1} = (106.1 \: \underline{/-90°})\|(1.8 \text{ k}\Omega \: \underline{/0°} + 52.7 \: \underline{/-83.6°})$$

$$= \frac{(106.1 \: \underline{/-90°})(1.8 \text{ k}\Omega \: \underline{/0°} + 52.7 \: \underline{/-83.6°})}{(106.1 \: \underline{/-90°}) + (1.8 \text{ k}\Omega \: \underline{/0°} + 52.7 \: \underline{/-83.6°})}$$

$$= \frac{106.1 \: \underline{/-90°} \: (1807 \: \underline{/-1.66°})}{1813 \: \underline{/-5°}} = 105.7 \: \Omega \: \underline{/-86.7°}$$

The open-circuit impedance $\mathbf{Z}_{oc1}$ is approximately equal to the 106.1-Ω capacitive reactance. We should expect this because this leg of the parallel circuit is considerably smaller than the other leg.

By the voltage-divider principle, the voltage across the 30-kΩ load is equal to

$$\mathbf{V}_L = \frac{\mathbf{Z}_L}{\mathbf{Z}_T} (\mathbf{V}_{oc1}) = \frac{30 \text{ k}\Omega \: \underline{/0°}}{\mathbf{Z}_{oc1} + \mathbf{Z}_L} (0.00655 \: \underline{/-168.6°}) \: \mathbf{V}_{in}$$

$$= \frac{30 \text{ k}\Omega \: \underline{/0°}}{105.7 \: \Omega \: \underline{/-86.7°} + 30 \text{ k}\Omega \: \underline{/0°}} (0.00655 \: \underline{/-168.6°}) \mathbf{V}_{in}$$

$$= (0.00655 \: \underline{/-168.6°}) \mathbf{V}_{in}$$

As you see, the input ac that actually gets to the load is very, very small (actually $0.00655 \mathbf{V}_{in}$). Therefore this circuit has prevented almost all the ac voltage from getting to the load (the actual amplifier section).

Test

1. Refer to the circuit with two series impedances shown in Figure 22.2. With V_{in} equal to 120 V at 60 Hz, $\mathbf{Z}_1$ equal to 14.14 kΩ $\underline{/45°}$, and $\mathbf{Z}_2$ equal to 30 kΩ $\underline{/-53.13°}$, solve for $\mathbf{Z}_T$, $\mathbf{I}$, $\mathbf{V}_1$, and $\mathbf{V}_2$. (Use $\mathbf{I}$ as the reference phasor.)

2. Use the circuit with two parallel impedances shown in Figure 22.3. The input voltage is 120 V, 60 Hz, with $\mathbf{Z}_1$ equal to 14.14 kΩ $\underline{/45°}$ and $\mathbf{Z}_2$ equal to 30 kΩ $\underline{/-53.13°}$. Solve for $\mathbf{Z}_T$, $\mathbf{I}_T$, $\mathbf{I}_1$, $\mathbf{I}_2$.

3. Refer to the circuit shown in Figure 22.5 for three parallel impedances. The input voltage is 120 V, 60 Hz, with $\mathbf{Z}_1$ equal to 2 kΩ $\underline{/60°}$, $\mathbf{Z}_2$ equal to 1 kΩ $\underline{/0°}$, and $\mathbf{Z}_3$ equal to 5 kΩ $\underline{/-20°}$. Solve for $\mathbf{Z}_T$, using admittances.

4. Refer to the circuit shown in Figure 22.5 for three parallel impedances. The input voltage is 20 V, 1 kHz, with $\mathbf{Z}_1$ made up of a 3900-Ω resistor, $\mathbf{Z}_2$ made up of a 10-kΩ resistor in series with a 500-mH inductor, and $\mathbf{Z}_3$ made up of a 5-kΩ resistor in series with a 0.05-μF capacitor. Solve for $\mathbf{Z}_T$, $\mathbf{I}_T$, $\mathbf{I}_1$, $\mathbf{I}_2$, and $\mathbf{I}_3$.

5. Solve for $\mathbf{V}_2$ in Test Problem 1, using the voltage-divider principle.

6. A 10-kΩ resistor and a 6-H inductor are connected in series across a 120-V $\underline{/12.74°}$, 60-Hz source. Find the voltage across the inductor by the voltage-divider principle.

7. Solve for $\mathbf{I}_2$ in Test Problem 2, using the current-divider principle. Use the $\mathbf{I}_T$ found in the problem.

8. Solve for $\mathbf{I}_3$ in Test Problem 4, using Equation (22.19).

9. An ac circuit consists of an inductor in parallel with a resistor. The inductor has an inductive reactance of 500 Ω at the frequency of operation. The resistor is equal to 8 kΩ. Find the equivalent series circuit.

10. An ac circuit consists of a capacitor in parallel with a resistor. The capacitor has a capacitive reactance of 3 kΩ at the frequency of operation, and the resistor is equal to 300 Ω. Find the equivalent series circuit.

11. An ac circuit consists of an inductor in series with a resistor. The inductor has an inductive reactance of 1200 Ω at the frequency of operation, and the resistor is equal to 600 Ω. Find the equivalent parallel circuit.

12. An ac circuit consists of a capacitor in series with a resistor. The capacitor has a capacitive reactance of 3 kΩ at the frequency of operation, and the resistor is equal to 9 kΩ. Find the equivalent parallel circuit.

13. Find a two-element equivalent parallel circuit for the circuit shown in Figure 22.22. The frequency of operation is 10 kHz.

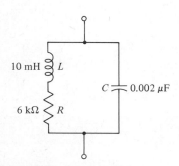

Figure 22.22 Circuit for Test Problem 13

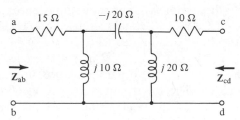

Figure 22.23 Circuit for Test Problem 14

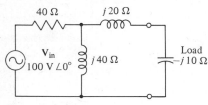

Figure 22.24 Circuit for Test Problem 15

14. Refer to the circuit shown in Figure 22.23. Find the total impedance $\mathbf{Z}_{ab}$ (both magnitude and direction) looking into terminals (a) and (b). Also find the total impedance $\mathbf{Z}_{cd}$ looking into terminals (c) and (d).

15. Refer to the circuit given in Figure 22.24. Consider the $-j10$-Ω capacitive reactance as the load, and determine the TEC. Then solve for the voltage across and the current in the load.

16. Determine the total impedance of the circuit shown in Figure 22.25. Express the answer in polar notation.

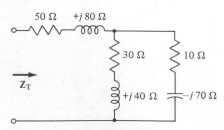

Figure 22.25 Circuit for Test Problem 16

Resonance Unit 23

The topic of resonance became very important in the early 1900s when the radio began to emerge. An understanding of resonant circuits was required for the production and reception of radio waves. Since that time, resonance has found many applications in aircraft and navigational systems; depth, range and altitude finders; telemetering systems; induction heating; and even welding systems. Resonance is not limited to the electrical field. It can also be found in mechanical systems, such as in the design of bridges and other structures, and in the architecture of buildings.

In this unit you will learn about two types of resonant electric circuits: series and parallel. Each of these is discussed in some detail in the following sections.

Objectives

After completing all the work associated with this unit, you should be able to:

1. Determine when a series ac circuit is resonant. List four characteristics of a series resonant circuit.
2. Calculate the resonant frequency of a series ac circuit. Also determine the current in the circuit and the voltage across the circuit elements at resonance.

3. Calculate the Q of a coil and the Q of a series resonant circuit.
4. Determine the bandwidth BW of a series resonant circuit.
5. Determine when a parallel ac circuit is resonant. List four characteristics of a parallel resonant circuit.
6. Calculate the resonant frequency, impedance, and bandwidth of a practical parallel resonant circuit (a capacitor in parallel with a coil that has resistance and reactance). Given the supply voltage, calculate the line and tank current.
7. Calculate the circuit impedance, circuit Q, and bandwidth of a circuit with a resistor in parallel with a practical resonant circuit.
8. Determine the loading resistor required for a specified bandwidth of a practical parallel resonant circuit.

The preceding eight objectives are broken down into two main areas. The first four objectives cover series resonant *RLC* circuits, while the remaining four objectives cover parallel resonant circuits.

Characteristics of Series Resonant Circuits

Objective 1   Determine when a series ac circuit is resonant. List four characteristics of a series resonant circuit.

In Unit 19, in the study of series *RLC* circuits, you learned about the circuit characteristics for the conditions $X_L > X_C$ and $X_L < X_C$. In this objective, you will learn about the condition $X_L = X_C$. This condition is called *resonance*. In order for resonance to occur, only two circuit elements are required, inductance L and capacitance C. Usually in a practical circuit some resistance is also included to represent the effective dc resistance of the coil L. A sketch of a practical series resonant circuit is shown in Figure 23.1.

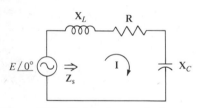

Figure 23.1   Sketch of a practical series resonant circuit

The inductance L is represented by its reactance X_L and resistance R, while the capacitance C is represented by its reactance X_C. The *resonant frequency* f_0 can be computed from the condition $X_L = X_C$. Since

$$X_L = 2\pi f_0 L \qquad (23.1)$$

and

$$X_C = \frac{1}{2\pi f_0 C} \qquad (23.2)$$

then

$$2\pi f_0 L = \frac{1}{2\pi f_0 C} \tag{23.3}$$

$$4\pi^2 f_0^2 LC = 1 \tag{23.4}$$

$$f_0^2 = \frac{1}{4\pi^2 LC} \tag{23.5}$$

$$f_0 = \frac{1}{2\pi\sqrt{LC}} \tag{23.6}$$

where f_0 = resonant frequency in hertz

   L = inductance in henries

   C = capacitance in farads

Note that the resonant frequency f_0 in Equation (23.6) is independent of resistance R, but depends only on the product of L and C.
   We can write the series impedance $\mathbf{Z}_s$ in rectangular form as

$$\mathbf{Z}_s = R + jX_L - jX_C = R + j(X_L - X_C) \tag{23.7}$$

or

$$\mathbf{Z}_s = R + jX_0 \tag{23.8}$$

where $X_0 = X_L - X_C$

   The series impedance $\mathbf{Z}_s$ can be written in polar form as

$$Z_s \, \underline{/\phi} = |Z_s| \, \underline{/\phi} \tag{23.9}$$

where $|Z_s| = \sqrt{R^2 + X_0^2}$ = magnitude of $\mathbf{Z}_s$

$$\underline{/\phi} = \arctan\frac{X_0}{R} = \text{angle of } \mathbf{Z}_s$$

Since, at resonance, $X_L = X_C$, so

$$X_0 = X_L - X_C = 0 \tag{23.10}$$

then

$$\underline{/\phi} = \arctan\left(\frac{X_0}{R}\right) = \arctan\left(\frac{0}{R}\right) = 0° \tag{23.11}$$

   Thus at resonance the polar form of the series impedance $\mathbf{Z}_s$ [Equation (23.9)] can be written as

$$|Z_s| = \sqrt{R^2 + 0^2} = R \tag{23.12}$$

and

$$\underline{/\phi} = 0° \tag{23.13}$$

From Equations (23.12) and (23.13), we find that the series impedance $\mathbf{Z}_s$ is a pure resistance whose value is R. Furthermore, the series impedance has an angle of $0°$ and does not depend on inductance L or capacitance C. In summary, the series impedance

$$\mathbf{Z}_s = R \underline{/0°} \qquad \text{at resonance} \qquad (23.14)$$

But the following question may arise: What is the form of the series impedance for frequencies off resonance?

By referring to Figure 23.1 and recalling Equation (23.9), you notice that at resonance $X_L = X_C$, so $X_0 = (X_L - X_C) = 0$. Since the inductive reactance X_L cancels out with the capacitive reactance X_C, the magnitude of $\mathbf{Z}_s$ depends only on $\sqrt{R^2}$. For frequencies above resonance, the inductive reactance $X_L > X_C$, so that $X_0 > 0$. This nonzero value of X_0 affects the total impedance of the series circuit. Now for frequencies above resonance, the series impedance

$$Z_s = \sqrt{R^2 + X_0^2} \qquad (23.15)$$

Likewise for frequencies below resonance, the capacitive reactance $X_C > X_L$, so that $X_0 < 0$. Again X_0 is nonzero, and the magnitude of series impedance $\mathbf{Z}_s$ is calculated by Equation (23.15). From the above observations, we can conclude that at resonance the series impedance $\mathbf{Z}_s$ is at a minimum. Figure 23.2 shows a sketch of the frequency variation of series impedance Z_s. Notice that Z_s has a minimum value at resonant frequency f_0. We can calculate the current $\mathbf{I}$ in the circuit of Figure 23.1, at resonance, by applying Ohm's law as

$$\mathbf{I} = \frac{\mathbf{E}}{\mathbf{Z}_s} = \frac{E \underline{/0°}}{R \underline{/0°}} \qquad (23.16)$$

$$\therefore \quad \mathbf{I} = \frac{E}{R} \underline{/0°} \qquad \text{at resonance} \qquad (23.17)$$

From Equation (23.17) we can see that if the series impedance $\mathbf{Z}_s$ is at a minimum, then the current I is a maximum. Figure 23.2 also includes a variation of the current I versus frequency f. Current I is drawn as a dashed line; it reaches its maximum value at resonant frequency f_0.

We can summarize the four characteristics of a series resonant circuit at resonance as follows.

1. $X_L = X_C$ or $f_0 = \dfrac{1}{2\pi\sqrt{LC}}$

2. $\mathbf{Z}_s = R \underline{/0°}$ minimum resistance

3. $\mathbf{I} = \dfrac{E}{R} \underline{/0°}$ maximum current

4. $\phi = 0°$ (Since phase angle of series impedance $\mathbf{Z}_s$ is zero, current $\mathbf{I}$ is in phase with applied voltage $\mathbf{E}$.)

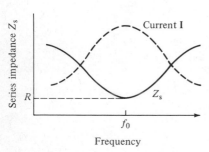

Figure 23.2 Graph showing variation of current *I* and series impedance Z_s versus frequency *f*.

Problem 23.1 List the four characteristics of a series resonant circuit.

Objective 2 Calculate the resonant frequency of a series ac circuit. Also determine the current in the circuit and the voltage across the circuit elements at resonance.

In this objective you will learn how to apply the resonant conditions you studied in Objective 1 to a practical circuit.

Example 23.1 The series *RLC* circuit shown in Figure 23.3 has $L = 100 \ \mu H$, $C = 100$ pF, and $R = 10 \ \Omega$. The applied voltage **E** is 100 V $\underline{/0°}$ and its frequency is the same as the resonant frequency of the circuit. Find: (a) Resonant frequency of circuit f_0, (b) input impedance $\mathbf{Z}_s$, (c) current **I**, (d) voltage across *R*, *L*, and *C*.

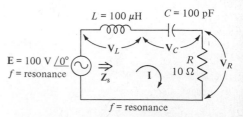

Figure 23.3 Figure for Example 23.1

Solution

(a) The resonant circuit frequency is

$$f_0 = \frac{1}{2\pi\sqrt{LC}} = \frac{1}{6.28\sqrt{100 \times 10^{-6} \times 100 \times 10^{-12}}}$$

$$= \frac{1}{6.28 \times 100 \times 10^{-9}} = 0.159 \times 10^7 = 1590 \text{ kHz}$$

(b) The circuit impedance at resonance is

$$\mathbf{Z}_s = R = 10 \ \Omega \ \underline{/0°}$$

(c) Using Ohm's law, we find the current

$$\mathbf{I} = \frac{\mathbf{E}}{\mathbf{Z}_s} = \frac{100 \text{ V} \ \underline{/0°}}{10 \ \Omega \ \underline{/0°}} = 10 \text{ A} \ \underline{/0°}$$

(d) The voltage across each element is

$$\mathbf{V}_R = \mathbf{I} \times \mathbf{R} = (10 \text{ A} \ \underline{/0°})(10 \ \Omega \ \underline{/0°}) = 100 \text{ V} \ \underline{/0°} = 100 \text{ V}$$

$$X_L = X_C = 2\pi f L = 6.28 \times 1590 \times 10^3 \times 100 \times 10^{-6} = 998.5 \ \Omega$$

$$\mathbf{V}_L = \mathbf{I} \times \mathbf{X}_L = 10 \text{ A} \ \underline{/0°} \times (998.5 \ \underline{/90°}) = 9985 \text{ V} \ \underline{/+90°}$$
$$= +j9985 \text{ V}$$

$$\mathbf{V}_C = \mathbf{I} \times \mathbf{X}_C = 10 \text{ A} \ \underline{/0°} \times 998.5 \ \underline{/-90°} = 9985 \text{ V} \ \underline{/-90°}$$
$$= -j9985 \text{ V}$$

Note that even though the voltages across the inductor ($\mathbf{V}_L$) and capacitor ($\mathbf{V}_C$) are larger than the applied voltage **E**, Kirchhoff's voltage law (KVL) must still hold. That is,

$$\mathbf{E} = \mathbf{V}_R + \mathbf{V}_L + \mathbf{V}_C$$
$$100 \text{ V} = 100 \text{ V} + j9985 \text{ V} - j9985 \text{ V}$$
$$100 \text{ V} = 100 \text{ V}$$

Problem 23.2 A series resonant circuit consists of $L = 5$ mH, $C = 50$ pF, $R = 20 \ \Omega$ and an applied voltage $\mathbf{E} = 100 \text{ V} \ \underline{/0°}$. Find the resonant frequency, the current at this frequency, and the voltage drop across each component.

From the previous example, you learned that the resonant frequency depends only on the product of inductance L and capacitance C. It is independent of the resistance R. If the product of L and C is large, then the resonant frequency is low. Let us present here another very important requirement in electric circuits. If you know only one component L (or C) and you are given the desired resonant frequency, you can find the value of the other component C (or L). This is shown by the following equations. Recalling Equation (23.4), we have

$$4\pi^2 f_0^2 LC = 1 \qquad (23.18)$$

Solving for L, we get

$$L = \frac{1}{4\pi^2 f_0^2 C} \qquad (23.19)$$

or if you know L, the value of C is

$$C = \frac{1}{4\pi^2 f_0^2 L} \qquad (23.20)$$

The following example demonstrates the use of these two equations.

Example 23.2 What value of capacitor must be in series with an inductor of 50 μH to have a resonant frequency of 5 kHz?

Solution   Since L is given as 50 μH, we can use Equation (23.20), which is

$$C = \frac{1}{4\pi^2 f_0^2 L} = \frac{1}{4 \times (3.14)^2 \times (25 \times 10^6) \times (50 \times 10^{-6})}$$

$$= \frac{1}{5000 \times (3.14)^2} = 20.3 \ \mu\text{F}$$

Problem 23.3   What value of inductance L must be in series with a capacitor C of 10 μF to have a resonant frequency of 10 kHz?

The Quality Q

Objective 3   Calculate the Q of a coil and the Q of a series resonant circuit.

In Units 11 and 13, you learned that an inductor stores energy in a magnetic field, while a capacitor stores energy on its plates. But what happens to this energy at resonance? In Unit 19, you learned that in a series RLC circuit, the voltage across an inductor is 180° out of phase with the voltage across a capacitor. Since the current I is the same everywhere in a series circuit, then power P ($P = V \times I$) in an inductor is $180°$ out of phase with the power in a capacitor. Furthermore, since energy W is related to power P by

$$W = P \times t \tag{23.21}$$

where $P =$ power in watts and $t =$ time in seconds, then the energy stored in an inductor is also $180°$ out of phase with the energy stored by a capacitor, which results in an exchange of energy between the inductor and capacitor at resonance. In an ideal case, if there is *no* resistance R between the inductor and the capacitor, then energy is transferred between inductor and capacitor forever, with no loss in energy. This is an *ideal oscillator*.

But in a practical sense, inductors do have a resistive component R. Therefore some energy is always dissipated (or lost) in this resistive element. Because of this loss, it is often desirable to relate the energy stored in a reactive component (L or C) to the energy dissipated in a resistive component (R). This ratio is called the Q *of the circuit*. It is written as

$$Q = \frac{W_Q}{W_{diss}} \tag{23.22}$$

where $W_Q =$ energy stored by reactive element

$W_{diss} =$ energy dissipated by resistive element

In Equation (23.21), energy W is defined as

$$W = P \times t \tag{23.23}$$

and since current I is the same everywhere in a series circuit, then power P is

$$P_Q = I^2 \times X_L \tag{23.24}$$
$$P_{diss} = I^2 \times R \tag{23.25}$$

where $I =$ current in series circuit

$X_L =$ reactance of inductor

$R =$ resistance of dissipative element

Then we can write Equation (23.22) as

$$Q = \frac{W_Q}{W_{diss}} = \frac{P_Q \times t}{P_{diss} \times t} = \frac{I^2 X_L t}{I^2 R t} \tag{23.26}$$

or

$$Q = \frac{X_L}{R} \tag{23.27}$$

Since at resonance $X_L = X_C$, then

$$Q = \frac{X_C}{R} \qquad (23.28)$$

This Q is called the *quality of the circuit* if R is the resistance of the circuit. On the other hand, if the resistance R is associated only with the resistance of the coil, then Q is often referred to as the *Q of the coil*. Thus we can express the quality of any inductor L as

$$Q = \frac{X_L}{R} = \frac{\omega_0 L}{R} = \frac{2\pi f_0 L}{R} \qquad (23.29)$$

where f_0 = resonant frequency in hertz

$\omega_0 = 2\pi f_0$ (radian frequency)

R = resistance of coil in ohms

L = inductance of coil in henries

We can also express the quality Q of a capacitor at resonance as

$$Q = \frac{X_C}{R} = \frac{1}{\omega_0 CR} = \frac{1}{2\pi f_0 CR} \qquad (23.30)$$

where C = capacitance in farads and R = equivalent series resistance of capacitor.

Practically speaking, the Q of a coil is usually about $Q \geq 50$. The following example illustrates how to calculate the Q of a coil or of a circuit.

Example 23.3   The series *RLC* circuit shown in Figure 23.4 is made up of a 30-mH inductor, a 100-Ω resistance, and a capacitor of 1000 pF. Find: (a) resonant frequency f_0 of circuit, (b) circuit impedance $\mathbf{Z}_s$ at resonance, (c) current $\mathbf{I}$, (d) circuit Q, (e) voltage $\mathbf{V}_L$ across inductor, (f) voltage $\mathbf{V}_C$ across capacitor, (g) power dissipated in resistance.

Solution

(a) $f_0 = \dfrac{1}{2\pi\sqrt{LC}} = \dfrac{1}{6.28\sqrt{30 \times 10^{-3} \times 1000 \times 10^{-12}}}$

$\quad = 29.07$ kHz

(b) The circuit impedance $\mathbf{Z}_s$ at resonance is

$$\mathbf{Z}_s = R = 100\ \Omega\ \underline{/0^\circ}$$

(c) The circuit current $\mathbf{I}$ is

$$\mathbf{I} = \frac{\mathbf{E}}{\mathbf{Z}_s} = \frac{10\ \text{V}\ \underline{/0^\circ}}{100\ \Omega\ \underline{/0^\circ}} = 100\ \text{mA}\ \underline{/0^\circ}$$

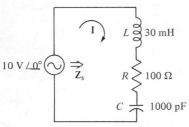

Figure 23.4   Figure for Example 23.3

(d) $Q_{circuit} = Q = \dfrac{X_L}{R} = \dfrac{2\pi f_0 L}{R} = \dfrac{2\pi(29.07 \times 10^3)(30 \times 10^{-3})}{100}$

$$= 54.77$$

(e) $\mathbf{V}_L = \mathbf{I} \times \mathbf{X}_L$

and since $\mathbf{I} = \mathbf{E}/\mathbf{R}$, we have

$$\mathbf{V}_L = \left(\dfrac{\mathbf{E}}{\mathbf{R}}\right) \times \mathbf{X}_L = \mathbf{E} \times \dfrac{\mathbf{X}_L}{\mathbf{R}} = \mathbf{E} \times Q$$

$$= 10 \text{ V} \times 54.77 = 547.7 \text{ V} \; \underline{/90^\circ}$$

(f) Since $X_L = X_C$ at resonance and current $\mathbf{I}$ is the same, then

$$V_C = V_L = 547.7 \text{ V} \qquad [\textit{Note:} \; \mathbf{V}_C = 547.7 \text{ V} \; \underline{/-90^\circ}]$$

(g) The power dissipated in resistor R is

$$P_{\text{diss}} = I^2 R = (0.1)^2(100) = 1 \text{ W}$$

In Example 23.3 you saw that you can easily calculate the voltage across a reactive component at resonance if you know Q. This is shown in the following proof. Since the voltage $\mathbf{V}_L$ across an ideal inductor (an inductor with no dc resistance) is

$$\mathbf{V}_L = \mathbf{I} \times \mathbf{X}_L \qquad (23.31)$$

and the current $\mathbf{I}$ in a series resonant circuit is

$$\mathbf{I} = \left(\dfrac{\mathbf{E}}{\mathbf{R}}\right) = \dfrac{E \; \underline{/0^\circ}}{R \; \underline{/0^\circ}} \qquad (23.32)$$

then, substituting Equation (23.32) into Equation (23.31), you get

$$\mathbf{V}_L = \left(\dfrac{\mathbf{E}}{\mathbf{R}}\right) \times \mathbf{X}_L = \dfrac{E \; \underline{/0^\circ}}{R \; \underline{/0^\circ}} \times X_L \; \underline{/90^\circ} = \left(E \times \dfrac{X_L}{R}\right) \underline{/90^\circ} \quad (23.33)$$

But

$$\dfrac{X_L}{R} = Q \qquad (23.34)$$

$$\therefore \quad \mathbf{V}_L = (E \times Q) \; \underline{/90^\circ} = j(E \times Q) \qquad (23.35)$$

Since, at resonance, $X_L = X_C$, we can likewise find the voltage V_C across a capacitor to be

$$\boxed{V_C = V_L = E \times Q} \qquad (23.36)$$

Problem 23.4   A series resonant circuit has $L = 200$ μH, $C = 200$ pF, and a resistor of 20 Ω. The applied voltage $\mathbf{E}$ is 120 V $\underline{/0°}$. Find the resonant frequency f_0, circuit Q, and voltage drop across each element.

Half-Power Points

Objective 4   Determine the bandwidth BW of a series resonant circuit.

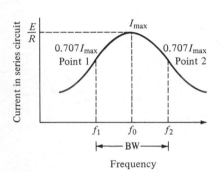

Figure 23.5   Variation of current *I* versus frequency in a series *RLC* circuit

In many amplifiers or electric circuits, we often want to find out how the output of a given circuit behaves as the frequency of the input signal is varied around resonance. Referring to Figure 23.5, note that the current I is a maximum at resonance and drops toward zero as the frequency is lowered or raised above resonance.

There is a range of frequencies between f_1 and f_2 at which the curve I is close to its maximum value. At frequencies f_1 and f_2, the current drops down to $0.707I_{\max}$. The lower frequency f_1 is below resonance, and the higher frequency f_2 is above resonance. These frequencies are sometimes called *cutoff* frequencies, or *half-power* frequencies. The following example illustrates the term half-power frequencies.

Example 23.4   Calculate the power dissipated in a resistor at frequencies f_0 and f_1. Refer to Figure 23.5.

Solution   Since power is defined as $P = I^2R$, then at f_0,

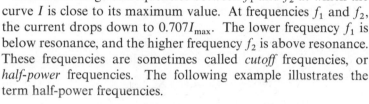

$$P_0 = I_{\max}^2 R$$

At f_1,

$$P_1 = (0.707I_{\max})^2R = 0.5I_{\max}^2R$$

Relating P_1 to P_0, we have $P_1 = 0.5P_0$.
Since the power at point 1 (frequency f_1) is equal to half the power at the point of resonance (f_0), point 1 is called the *half-power* point. The same holds true for point 2 (frequency f_2).

Bandwidth

The band of frequencies between f_2 and f_1 is defined as the *bandwidth* (**BW**) of the circuit. It can be found as

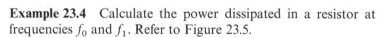

$$\mathrm{BW} = f_2 - f_1 \tag{23.37}$$

The bandwidth of a series resonant circuit could also be calculated in terms of resonant frequency f_0 and Q. It is

$$BW = \frac{f_0}{Q} \qquad (23.38)$$

where $f_0 = $ resonant frequency and $Q = $ circuit Q.

Equation (23.38) states that if we have a high circuit Q, then the bandwidth BW is small or narrow. In many communications circuits it is necessary to have narrow bandwidths (high Q) so that only certain frequencies can be passed through, while those frequencies outside the passband are greatly attenuated (reduced). This selection of desired frequencies is called *selectivity*. The following two examples illustrate how you can determine the bandwidth of a series resonant circuit.

Example 23.5 For the circuit in Example 23.3, calculate the bandwidth BW and specify the cutoff frequencies f_1 and f_2.

Solution From Example 23.3, we have the resonant frequency

$$f_0 = 29.07 \text{ kHz} \qquad \text{and} \qquad Q = 54.77$$

Using Equation (23.38) we have

$$BW = \frac{f_0}{Q} = \frac{29.07 \text{ kHz}}{54.77} = 531 \text{ Hz}$$

Referring to the sketch in Figure 23.6, we obtain

$$f_1 = f_0 - \frac{BW}{2} = 29{,}070 - \frac{531}{2} = 29{,}070 - 266 = 28.8 \text{ kHz}$$

Likewise,

$$f_2 = f_0 + \frac{BW}{2} = 29{,}070 + \frac{531}{2} = 29{,}070 + 266 = 29.34 \text{ kHz}$$

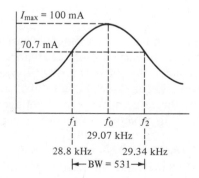

Figure 23.6 Sketch for solution of Example 23.5

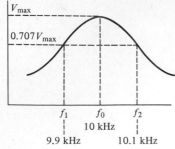

Figure 23.7   Figure for Example 23.6

Example 23.6   The variation of voltage versus frequency is shown in Figure 23.7. The circuit has a dc resistance of 50 Ω. Find the circuit Q, the inductance L, and the capacitance C.

Solution

(a) From the sketch in Figure 23.7, we know that bandwidth BW is

$$\text{BW} = f_2 - f_1 = 10.1 \text{ kHz} - 9.9 \text{ kHz} = 0.2 \text{ kHz} = 200 \text{ Hz}$$

Since bandwidth BW is related to circuit Q by $\text{BW} = f_0/Q$, then

$$Q = \frac{f_0}{\text{BW}} = \frac{10 \text{ kHz}}{200 \text{ Hz}} = 50$$

(b) From Equation (23.29), we have

$$Q = \frac{X_L}{R} = \frac{X_C}{R} \quad \text{or} \quad X_L = X_C = Q \times R = 50 \times 50 = 2500 \text{ } \Omega$$

Since $X_L = 2\pi f_0 L$, then

$$L = \frac{X_L}{2\pi f_0} = \frac{2500}{6.28 \times 10 \times 10^3} = 39.8 \text{ mH}$$

(c) Since $X_L = X_C = 1/(2\pi f_0 C)$, then

$$C = \frac{1}{2\pi f_0 X_C} = \frac{1}{6.28 \times 10 \times 10^3 \times 2500}$$

$$= 0.00637 \text{ } \mu\text{F} \quad \text{or} \quad 6370 \text{ pF}$$

Problem 23.5   The resonant frequency of a series RLC circuit is 1200 kHz. Find the bandwidth of the circuit, given that the circuit Q is 60. Also calculate the half-power frequencies f_1 and f_2.

Problem 23.6   Figure 23.8 shows the variation of the current I versus frequency. The circuit has a dc resistance of 50 Ω. Find the circuit Q, the inductance L, and the capacitance C.

In this objective you have learned about the variation of voltage (or current) versus frequency around resonance. But what is the value of the voltage (or current) further above or below the resonant frequency? Let us take the data of Example 23.6 and see what would be the value of the output voltage $\mathbf{V}_R$ across the circuit resistance. This is shown in the following example.

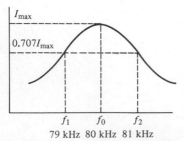

Figure 23.8   Sketch for Problem 23.6

Example 23.7   A series RLC circuit is resonant at 10 kHz. The inductance L is 39.8 mH and the capacitor C is 6370 pF. The applied input voltage is 100 V $\underline{/0°}$. Calculate the voltage $\mathbf{V}_R$ across a 50-Ω resistor at the resonant frequency, 10 kHz, and at a frequency of 9.8 kHz.

Solution

(a) At resonance $f_0 = 10$ kHz. Also

$$X_L = X_C \qquad \text{and} \qquad \mathbf{Z} = R = 50 \ \Omega \ \underline{/0^\circ}$$

$$\therefore \quad \mathbf{I} = \frac{\mathbf{V}_{\text{in}}}{\mathbf{R}} = \frac{100 \text{ V } \underline{/0^\circ}}{50 \ \Omega \ \underline{/0^\circ}} = 2 \text{ A } \underline{/0^\circ}$$

$$\mathbf{V}_R = \mathbf{I} \cdot \mathbf{R} = (2 \text{ A } \underline{/0^\circ})(50 \ \Omega \ \underline{/0^\circ}) = 100 \text{ V } \underline{/0^\circ}$$

Thus at resonance the voltage $\mathbf{V}_R$ is equal to the applied input voltage $\mathbf{V}_{\text{in}}$.

(b) At a frequency $f = 9.8$ kHz,

$$\mathbf{X}_L = +j2\pi f L = j6.28(9.8 \times 10^3)(39.8 \times 10^{-3}) = +j2.449 \text{ k}\Omega$$

$$\mathbf{X}_C = -j\frac{1}{2\pi f C} = \frac{-j}{6.28 \times 9.8 \times 10^3 \times 6.37 \times 10^{-9}}$$

$$= -j2.55079 \text{ k}\Omega$$

$$\mathbf{X}_0 = \mathbf{X}_L - \mathbf{X}_C = -j101.79 \ \Omega$$

$$Z = \sqrt{X_0^2 + R^2} = \sqrt{101.79^2 + 50^2} = 113.408 \ \Omega$$

$$\mathbf{I} = \frac{\mathbf{V}_{\text{in}}}{\mathbf{Z}} = \frac{100 \text{ V } \underline{/0^\circ}}{113.408 \ \Omega \ \underline{/-63.8^\circ}} = 0.88177 \text{ A } \underline{/+63.8^\circ}$$

$$\mathbf{V}_R = \mathbf{I} \cdot \mathbf{R} = (0.88177 \text{ A } \underline{/+63.8^\circ})(50 \ \Omega \ \underline{/0^\circ}) = 44.09 \text{ V } \underline{/+63.8^\circ}$$

Thus the voltage $\mathbf{V}_R$ at a frequency of 9.8 kHz is 44.1 V, or approximately 44.1% of the maximum input voltage $\mathbf{V}_{\text{in}}$. Furthermore, note that the phase angle ϕ is no longer zero.

Applying the technique described in Example 23.7, we can calculate the output voltage V_R versus frequency. Figure 23.9

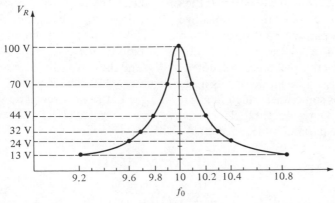

Frequency, kHz

Figure 23.9 A plot of the frequency response of V_R versus frequency for Example 23.7

Table 23.1   Frequency Response of a Series *RLC* Circuit
BW = 200 Hz,   resonant frequency f_0 = 10 kHz

Frequency f	Output voltage $\mathbf{V}_R$
10 kHz	100%
9.9 kHz and 10.1 kHz ($\pm\frac{1}{2}$ BW)	70%
9.8 kHz and 10.2 kHz (± 1 BW)	44%
9.7 kHz and 10.3 kHz ($\pm 1\frac{1}{2}$ BW)	32%
9.6 kHz and 10.4 kHz (± 2 BW)	24%
9.2 kHz and 10.8 kHz (± 4 BW)	13%

shows a plot of the frequency response of V_R versus frequency. The approximate voltage V_R versus frequency is also tabulated in Table 23.1.

Referring to Figure 23.9, we find that the maximum voltage V_R is 100 V (or 100% of V_{in}) at the resonant frequency of 10 kHz. Since the bandwidth BW is 200 Hz, the output voltage is 70 V (or 70%) at a frequency of 9.9 kHz or 10.1 kHz. These two frequencies are $\pm\frac{1}{2}$ bandwidth away from resonance, 10 kHz. At a frequency of 9.2 kHz, the output voltage V_R drops to 13% of the input voltage V_{in}. The information shown in Figure 23.9 and Table 23.1 may help you to calculate bandwidth and find the frequency response of any general series *RLC* circuit.

Characteristics of Parallel Resonant Circuits

Objective 5   Determine when a parallel ac circuit is resonant. List four characteristics of a parallel resonant circuit.

In Unit 20, you studied the general characteristics of a parallel *RLC* circuit. In this objective you will study the characteristics of a parallel *RLC* resonant circuit. This occurs for the condition $X_L = X_C$. Figure 23.10 shows a parallel *RLC* circuit. The inductive reactance is denoted by X_L, and the capacitive reactance is X_C. The inductor has a dc resistance of R. The total current $\mathbf{I}_T$ is sometimes called *line current* $\mathbf{I}_{line}$. Let us now recall how to calculate the parallel impedance $\mathbf{Z}_{||}$ of a parallel *RLC* circuit. The parallel impedance is

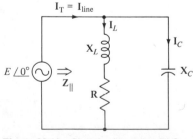

Figure 23.10   Sketch of parallel *RLC* circuit

$$\mathbf{Z}_{||} = (R + jX_L)\|(-jX_C) \tag{23.39}$$

$$\mathbf{Z}_{||} = \frac{(R + jX_L)(-jX_C)}{R + jX_L - jX_C} \tag{23.40}$$

Since at resonance $X_L = X_C$, we can simplify the denominator, resulting in

$$\mathbf{Z}_{\|} = \frac{(R + jX_L)(-jX_C)}{R} \tag{23.41}$$

In most practical resonant circuits, the circuit $Q \geq 10$. Since

$$Q = \frac{X_L}{R} \geq 10 \qquad \text{then} \qquad X_L \geq 10R \tag{23.42}$$

Since X_L is much larger than R, we can approximate the numerator of Equation (23.41) by

$$\mathbf{Z}_{\|} = \frac{(jX_L)(-jX_C)}{R} = \frac{-j^2 X_L X_C}{R} = \frac{X_L X_C}{R} \tag{23.43}$$

$$\therefore \quad \mathbf{Z}_{\|} = \left(\frac{X_L X_C}{R} = \frac{X_L^2}{R} = \frac{X_C^2}{R} \right) \underline{/0^\circ} \qquad \text{at resonance for } Q \geq 10 \tag{23.44}$$

Since $Q = X_L/R$, we can rewrite Equation (23.44) as

$$\boxed{Z_{\|} = Q \cdot X_L = Q \cdot X_C} \tag{23.45}$$

We can rewrite Equation (23.44) in terms of L, C, and R by replacing X_L by $2\pi f L$ and X_C by $1/(2\pi f C)$, yielding

$$\mathbf{Z}_{\|} = \frac{2\pi f L}{2\pi f C R} \underline{/0^\circ} = \frac{L}{CR} \underline{/0^\circ} \tag{23.46}$$

Equations (23.44), (23.45), and (23.46) are the same equation for parallel impedance, written in different forms. Moreover note that the parallel impedance at resonance is a pure resistance with a phase angle $\underline{/\phi} = 0^\circ$. It can readily be observed that in our development we included the vector notation for all the circuit elements. With Equation (23.43), the vector angles $(-j^2)$ are simplified, resulting in a phase angle of 0°.

Let us now see how the parallel impedance varies as the frequency is changed. At resonance the parallel impedance is a maximum and a pure resistance. This is shown in Equation (23.46) and in the following discussion. Recall that in the development, we wrote the impedance of each leg of Figure 23.10 in vector form so as to include the angles. Carrying these angles through the development and assuming that $Q \geq 10$, we found that the resultant angle [Equation (23.44)] became zero.

Let us now refer to Figure 23.10 and look at the reactance X_C. At frequency $f = \infty$, the reactance $X_C = 0$. Applying this to Equation (23.39), we obtain

$$\mathbf{Z}_{\|} = (R + jX_L) \| (-j0) = -j0 \qquad \text{at } f = \infty \tag{23.47}$$

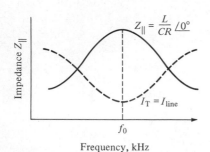

Likewise, at frequency $f = 0$, $X_L = 0$, and since $X_L \geq 10R$, we get

$$\mathbf{Z}_{\|} = (j0)\|(X_C) = j0 \qquad \text{at } f = 0 \qquad (23.48)$$

From Equations (23.47) and (23.48), we can conclude that the parallel impedance $\mathbf{Z}_{\|}$ at resonance is a maximum. Figure 23.11 shows this impedance versus frequency, with the line current I_T drawn as a dashed line. Note that the line current is a minimum at the resonant frequency f_0. This can be seen by applying Ohm's law as

$$\mathbf{I}_{T\min} = \frac{\mathbf{E}}{\mathbf{Z}_{\|\max}} \qquad \text{at resonance} \qquad (23.49)$$

Figure 23.11 Variation of impedance and current versus frequency in a parallel *RLC* circuit

Referring to Figures 23.10 and 23.11, let us consider two cases: the ideal perfect coil and the practical coil.

Perfect Coil

In a perfect coil the dc resistance $R = 0$. With $R = 0$ and using Equation (23.46), we obtain

$$Z_{\|} = \frac{L}{CR} = \frac{L}{C(0)} = \infty \qquad (23.50)$$

Here we have the condition that the impedance of a parallel circuit is infinite. This results in the line current $\mathbf{I}_T$ being zero. Ideally it means that the energy in the inductor and capacitor is continuously being exchanged, with no losses. This oscillation can continue forever with no additional line current $\mathbf{I}_T$ needed from the source.

Practical Coil

A practical coil has some dc resistance associated with it. This results in a parallel impedance $\mathbf{Z}_{\|}$ that may be very high at resonance and a line current $\mathbf{I}_T$ that is small. Since resonance for a parallel circuit occurs when $X_L = X_C$, the resultant frequency f_0 is the same as that for a series resonant circuit. It is

$$f_0 = \frac{1}{2\pi\sqrt{LC}} \qquad (23.51)$$

Likewise the quality of the coil Q for a parallel circuit is the same as that for a series circuit, and we can again calculate the bandwidth BW from Q. It is again

$$\text{BW} = \frac{f_0}{Q} \qquad (23.52)$$

In summary, the four characteristics of a parallel resonant circuit are as follows.

1. $X_L = X_C$ $\quad f_0 = \dfrac{1}{2\pi\sqrt{LC}}$

2. $Z_{||} = Q \cdot X_L = \dfrac{L}{CR} = \text{maximum}$

3. The parallel impedance $\mathbf{Z}_{||}$ is a pure resistance, since $\underline{/\phi} = 0°$.

4. The line current $\mathbf{I}_T$ is in phase with the applied voltage and is a minimum.

Problem 23.7 List the four characteristics of a parallel resonant circuit.

Simple Parallel *RLC* Circuits

Objective 6 Calculate the resonant frequency, impedance, and bandwidth of a practical parallel resonant circuit (a capacitor in parallel with a coil that has resistance and reactance). Given the supply voltage, calculate the line and tank current.

In this objective you will learn how to apply the theory you learned in Objective 5 to a practical parallel resonant circuit.

Example 23.8 A 2500-pF capacitor is placed in parallel with a coil that has a resistance of 40 Ω and an inductance of 40 mH. The supply voltage is 100 V $\underline{/0°}$. Find the resonant frequency, Q of the coil, the parallel impedance $\mathbf{Z}_{||}$, the circuit bandwidth, the line current, and the tank current.

Solution The circuit for this example looks like Figure 23.12.
(a) Assuming that $Q > 10$, the resonant frequency

$$f_0 = \frac{1}{2\pi\sqrt{LC}} = \frac{1}{6.28\sqrt{40 \times 10^{-3} \times 2500 \times 10^{-12}}} = 15.923 \text{ kHz}$$

(b) The reactance of the inductor is

$$X_L = 2\pi f_0 L = 6.28(15.923 \times 10^3)(40 \times 10^{-3}) = 4\text{k}\Omega$$
$$X_C = X_L = 4 \text{ k}\Omega$$

(c) $$Q_{\text{coil}} = Q = \frac{X_L}{R} = \frac{4000}{40} = 100$$

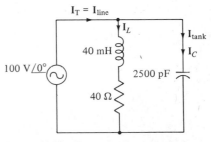

Figure 23.12 Figure for Example 23.8

Since $Q = 100$, then the assumption in part (a) that $Q > 10$ is valid, and the resonant frequency is correct.

The quality of the circuit Q_P is the same as coil Q because 40 Ω is the only resistance in the circuit.

(d) The parallel circuit impedance at resonance is

$$\mathbf{Z}_{\parallel} = \frac{L}{CR} \: \underline{/0°} = \frac{40 \times 10^{-3}}{40 \times 2500 \times 10^{-12}} = 400 \text{ k}\Omega \: \underline{/0°}$$

Another way of calculating the parallel impedance is by using Equation (23.45):

$$\mathbf{Z}_{\parallel} = Q \times X_L = 100 \times 4 \text{ k}\Omega = 400 \text{ k}\Omega$$

(e) The bandwidth BW is

$$\text{BW} = \frac{f_0}{Q} = \frac{15.923 \times 10^3}{100} = 159 \text{ Hz}$$

(f) The line current $\mathbf{I}_T$ can be calculated as

$$\mathbf{I}_T = \frac{\mathbf{E}}{\mathbf{Z}_{\parallel}} = \frac{100 \text{ V} \: \underline{/0°}}{400 \text{ k}\Omega \: \underline{/0°}} = 0.25 \text{ mA} \: \underline{/0°} = 250 \text{ μA} \: \underline{/0°}$$

(g) The tank current $\mathbf{I}_{\text{tank}} = \mathbf{I}_C$ at resonance and we can find it in two ways. First, by Ohm's law, we have

$$\mathbf{I}_C = \frac{\mathbf{E}}{\mathbf{X}_C} = \frac{100 \text{ V} \: \underline{/0°}}{4 \text{ k}\Omega \: \underline{/-90°}} = 25 \text{ mA} \: \underline{/+90°} = +j25 \text{ mA}$$

The second way, by using Q, is

$$\mathbf{I}_C = Q \times \mathbf{I}_T = 100 \times 0.25 \text{ mA} = 25 \text{ mA}$$

In Example 23.8 you learned how to calculate the tank current $\mathbf{I}_{\text{tank}}$ ($\mathbf{I}_L$ or $\mathbf{I}_C$) and the line current $\mathbf{I}_{\text{line}}$ ($\mathbf{I}_T$) by using the value of Q. The relationship between the line current $\mathbf{I}_T$ and the tank current $\mathbf{I}_C$ (or $\mathbf{I}_L$) is

$$I_C = I_L = I_{\text{tank}} = Q \times I_T \qquad (23.53)$$

Let's see how we can derive this equation. Referring to Figure 23.10 and using Ohm's law, we have

$$\mathbf{E} = \mathbf{I}_T \times \mathbf{Z}_{\parallel} \qquad (23.54)$$

But from Equation (23.43) we know that

$$\mathbf{Z}_{\parallel} = \frac{X_L X_C}{R} \: \underline{/0°} \qquad (23.55)$$

Substituting Equation (23.55) into Equation (23.54), we obtain

$$E = I_T \left[\frac{X_L X_C}{R} \right] = (I_T \times Q) \times X_C \qquad (23.56)$$

Dividing both sides of Equation (23.56) by X_C, we have

$$\frac{E}{X_C} = I_T \times Q \qquad (23.57)$$

and since $I_C = I_L$ and $I_C = E/X_C$, then

$$I_C = I_L = \frac{E}{X_C} = I_T \times Q \qquad (23.58)$$

To summarize all that you have learned up to this point, let us list all the equations for series and parallel circuits in tabular form in Table 23.2.

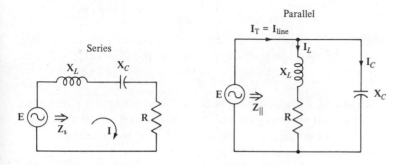

Figure 23.13 Sketch for Table 23.2

Table 23.2 Comparison of Series and Parallel Resonance (see Figure 23.13)

Series	Parallel
Current **I** same	Voltage **E** same
$X_L = X_C$	$X_L = X_C$
$\mathbf{Z}_s = \mathbf{R} = \text{minimum}$	$\mathbf{Z}_{\parallel} = \left(\dfrac{X_L^2}{R} = \dfrac{X_C^2}{R} = \dfrac{L}{CR} \right) \underline{/0^\circ} = \text{maximum}$
$\mathbf{I} = \dfrac{\mathbf{E}}{\mathbf{R}} = \text{maximum}$	$I_T = \dfrac{\mathbf{E}}{\mathbf{Z}_{\parallel}} = \text{minimum}$
$f_0 = \dfrac{1}{2\pi\sqrt{LC}}$	$f_0 = \dfrac{1}{2\pi\sqrt{LC}}$
$Q = \dfrac{\omega_0 L}{R} = \dfrac{2\pi f_0 L}{R}$	$Q = \dfrac{\omega_0 L}{R} = \dfrac{2\pi f_0 L}{R}$
$\mathbf{V}_R = \mathbf{I} \times \mathbf{R} = \mathbf{E}$	
$\mathbf{V}_L = \mathbf{I} \times \mathbf{X}_L = Q \times E$	$I_L = \dfrac{\mathbf{E}}{\mathbf{X}_L} = Q \times I_T$
$\mathbf{V}_C = \mathbf{I} \times \mathbf{X}_C = Q \times E$	$I_C = \dfrac{\mathbf{E}}{\mathbf{X}_C} = Q \times I_T$
$V_L = V_C$	$I_L = I_C$
$\text{BW} = \dfrac{f_0}{Q}$	$\text{BW} = \dfrac{f_0}{Q}$

Problem 23.8   A 1000-pF capacitor is placed in parallel with a coil of 50 Ω dc resistance and an inductance of 4 mH. The power-supply voltage is 160 V. Find the resonant frequency, Q of the circuit, the parallel impedance $\mathbf{Z}_\parallel$, the circuit bandwidth, the tank current, and the line current.

Parallel *RLC* Circuit Loaded with External Resistor

Objective 7   Calculate the circuit impedance, circuit Q, and bandwidth of a circuit with a resistor in parallel with a practical resonant circuit.

In practical applications, such as communications circuits, we often have a resistor placed in parallel with a parallel resonant circuit. In this objective you will learn how to calculate the new circuit Q and the new bandwidth.

Figure 23.14 shows the circuit diagram representing this. The original parallel *RLC* resonant circuit is now loaded by the load resistor R_{load}, drawn dashed in Figure 23.14. Since the bandwidth of any circuit depends on the circuit Q, any changes in circuit Q affect the bandwidth BW. Recall that the circuit Q is a ratio of the energy stored in a reactive component to the energy dissipated in a resistive component. Since an additional dissipative element R_{load} has been added to the original parallel *RLC* resonant circuit, the circuit Q is reduced. A new calculation of circuit Q becomes extremely difficult. Furthermore, since bandwidth

$$\text{BW} = \frac{f_0}{Q} \qquad (23.59)$$

then bandwidth is also affected. In the following section you will learn how to calculate the new circuit Q and the new bandwidth using the equations developed so far.

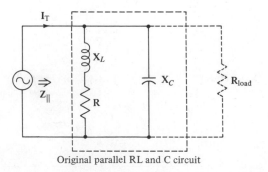

Figure 23.14   Practical parallel *RLC* circuit loaded by R_{load}

Original parallel RL and C circuit

Loading effect on parallel resonant circuit

Recall from Equation (23.46) that the parallel impedance of only the original parallel *RLC* resonant circuit is a pure resistance whose value is

$$\mathbf{Z}_{\|} = \frac{L}{CR} \underline{/0^\circ} \tag{23.60}$$

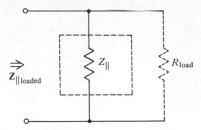

Figure 23.15 Equivalent circuit of practical parallel *RLC* circuit loaded by R_{load} shown in Figure 23.14

If we replaced the original parallel *RLC* resonant circuit of Figure 23.14 by a resistance $Z_{\|}$, then the equivalent circuit would look like Figure 23.15. Here we have two resistances, $Z_{\|}$ and R_{load}, in parallel. We can compute the new parallel impedance $\mathbf{Z}_{\| \text{ loaded}}$ as

$$\mathbf{Z}_{\| \text{ loaded}} = Z_{\|} \| R_{load} \underline{/0^\circ} \tag{23.61}$$

where $\mathbf{Z}_{\| \text{ loaded}}$ = equivalent parallel resistance of *loaded* parallel *RLC* resonant circuit

$Z_{\|}$ = equivalent parallel resistance of original parallel *RLC* resonant circuit

R_{load} = load resistance added

The new parallel impedance $\mathbf{Z}_{\| \text{ loaded}}$ is lowered when R_{load} is connected. But how does this affect the circuit Q? Recall from Equation (23.45) that the circuit impedance $Z_{\|}$ is related to circuit Q by the relationship

$$Z_{\|} = Q \times X_L \tag{23.62}$$

Furthermore the resonant frequency f_0 does not change when a load resistor R_{load} is connected to the parallel *RLC* resonant circuit. Consequently the reactance X_L of the inductor also does not change. In conclusion, since X_L is a constant, we can state from Equation (23.62) that the parallel impedance $Z_{\|}$ is directly related to circuit Q. Reducing parallel impedance $Z_{\|}$ by a factor of 2 reduces the circuit Q by a factor of 2. This relationship of $Z_{\|}$ and Q can be mathematically expressed as

$$Z_{\| \text{ loaded}} = Q_{loaded} \times X_L \tag{23.63}$$

where $Z_{\| \text{ loaded}}$ = impedance of loaded resonant circuit

$Z_{\|}$ = impedance of original parallel *RLC* resonant circuit

Q_{loaded} = the new circuit Q with loaded resistor R_{load}

Note that Equation (23.62) is similar to Equation (23.63) except for the term *loaded* attached to $Z_{\|}$ and Q. Relating Equations (23.62) and (23.63), we have

$$\frac{Z_{\|}}{Z_{\| \text{ loaded}}} = \frac{Q \times X_L}{Q_{loaded} \times X_L} \tag{23.64}$$

Since X_L does *not* change, Equation (23.64) can be reduced to the form

$$\frac{Z_{||}}{Z_{|| \text{ loaded}}} = \frac{Q}{Q_{\text{loaded}}} \tag{23.65}$$

Since bandwidth BW $= f_0/Q$, we can also include bandwidth in Equation (23.65). Since

$$\text{BW}_{\text{loaded}} = \frac{f_0}{Q_{\text{loaded}}} \quad \text{and} \quad \text{BW} = \frac{f_0}{Q}$$

then Equation (23.65) can be written as

$$\frac{Z_{||}}{Z_{|| \text{ loaded}}} = \frac{Q}{Q_{\text{loaded}}} = \frac{\text{BW}_{\text{loaded}}}{\text{BW}} \tag{23.66}$$

The following example illustrates the approach you take to calculate the bandwidth of a loaded parallel RLC resonant circuit.

Example 23.9 The parallel RLC resonant circuit of Example 23.8 is loaded with an $R_{\text{load}} = 600 \text{ k}\Omega$. (See Figure 23.16.) Find the resonant frequency, the parallel impedance $Z_{|| \text{ loaded}}$, the circuit Q, and the overall circuit bandwidth.

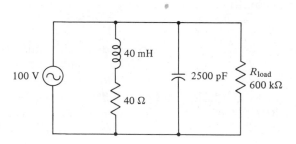

Figure 23.16   Figure for Example 23.9

Solution

(a) The resonant frequency is the same as in Example 23.8. It is

$$f_0 = \frac{1}{2\pi\sqrt{LC}} = 15.923 \text{ kHz}$$

(b) The parallel impedance of RLC (R_{load} excluded) is

$$Z_{||} = \frac{L}{CR} = 400 \text{ k}\Omega$$

We can compute the new parallel impedance $Z_{|| \text{ loaded}}$ from Equation (23.61) as

$$Z_{|| \text{ loaded}} = Z_{||} \| R_{\text{load}} = 400 \text{ k}\Omega \| 600 \text{ k}\Omega = 240 \text{ k}\Omega$$

(c) We can find the new circuit Q from the relation given in Equation (23.63):

$$Z_{||\,\text{loaded}} = Q_{\text{loaded}} \times X_L$$

Since

$$X_L = 2\pi f_0 L = (6.28)(15.923 \times 10^3)(40 \times 10^{-3}) = 4000 \ \Omega$$

$$\therefore \quad Q_{\text{loaded}} = \frac{Z_{||\,\text{loaded}}}{X_L} = \frac{240 \text{ k}\Omega}{4 \text{ k}\Omega} = 60$$

Recall that the Q of Example 23.7 without R_{load} was 100, while in Example 23.8, with R_{load} connected, the Q dropped to 60.
(d) We can calculate the overall bandwidth BW by

$$\text{BW}_{\text{loaded}} = \frac{f_0}{Q_{\text{loaded}}} = \frac{15.923 \text{ kHz}}{60} = 265 \text{ Hz}$$

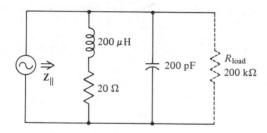

Figure 23.17   Sketch for Problem 23.9

Problem 23.9   The parallel RLC resonant circuit shown in Figure 23.17 is loaded with an $R_{\text{load}} = 200 \text{ k}\Omega$. Find the following. The resonant frequency f_0:

With R_{load} removed   With R_{load}

With R_{load} removed	With R_{load}
$Z_{||} = $ _____	$Z_{||\,\text{loaded}} = $ _____
Circuit $Q = $ _____	$Q_{\text{loaded}} = $ _____
BW $= $ _____	$\text{BW}_{\text{loaded}} = $ _____

Objective 8   Determine the loading resistor required for a specified bandwidth of a practical parallel resonant circuit.

In Objective 7 you learned that we can modify the bandwidth of a parallel resonant circuit by placing a resistance in parallel with the parallel resonant circuit. In this objective, you will learn how to calculate the value of the load resistance R_{load} for a desired bandwidth. This is illustrated in the following example.

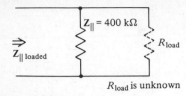

Figure 23.18 Sketch for solution of Example 23.10

Example 23.10 Determine the value of the loading resistor to produce a bandwidth of 796 Hz for the circuit given in Example 23.8.

Solution Figure 23.18 shows the circuit diagram for this example. The resonant frequency, from Example 23.8, is

$$f_0 = 15{,}923 \text{ Hz} \quad \text{and} \quad X_L = 4 \text{ k}\Omega$$

Let us define $Q_{\|\text{ loaded}} = Q_P$. Since the bandwidth

$$\text{BW} = \frac{f_0}{Q_P} \quad \text{then} \quad Q_P = \frac{f_0}{\text{BW}}$$

Therefore the Q_P required for the desired bandwidth is

$$Q_P = \frac{f_0}{\text{BW}} = \frac{15{,}923 \text{ Hz}}{796} = 20$$

One of the ways of changing the original Q of 100 of Example 23.8 without changing L or C is to add a resistor in parallel. But how much resistance must be placed in parallel with $Z_{\|} = 400 \text{ k}\Omega$ of Example 23.8? Recall that

$$Z_{\|\text{ loaded}} = Q_P \times X_L$$
$$= 20 \times 4 \text{ k}\Omega = 80 \text{ k}\Omega$$

Since

$$Z_{\|\text{ loaded}} = Z_{\|} \| R_{\text{load}} = \frac{Z_{\|} \times R_{\text{load}}}{Z_{\|} + R_{\text{load}}}$$

We rewrite the above equation, and have

$$Z_{\|\text{ loaded}} (Z_{\|} + R_{\text{load}}) = Z_{\|} \times R_{\text{load}}$$

$$R_{\text{load}} = \frac{Z_{\|\text{ loaded}} \times Z_{\|}}{Z_{\|} - Z_{\|\text{ loaded}}} = \frac{80 \text{ k}\Omega \times 400 \text{ k}\Omega}{400 \text{ k}\Omega - 80 \text{ k}\Omega} = 100 \text{ k}\Omega$$

Problem 23.10 Determine the value of the loading resistor to produce a bandwidth of 796 Hz for the circuit shown in Figure 23.19. Find the loading resistor R_{load}.

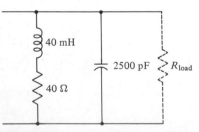

Figure 23.19 Circuit for Problem 23.10

Test

1. A series RLC circuit has the following components: $L = 40$ mH, $C = 400$ pF, and resistance $R = 125 \ \Omega$. The applied voltage $\mathbf{E}$ is 25 V $\underline{/0°}$. Find the resonant frequency f_0. Also at resonance, find circuit impedance $\mathbf{Z}_s$, current $\mathbf{I}$, and the voltage drops across R, L, and C.

2. What value of inductance must we place in series with a 15-μF capacitor to have a resonant frequency of 8 kHz?

3. A series resonant circuit has $L = 32$ mH, $C = 0.008$ μF and $R = 40$ Ω. Find the resonant frequency f_0, circuit Q, and bandwidth BW.

4. The following data were obtained from the circuit shown in Figure 23.20.

Frequency f, kHz	V_{out}, V
790	6.972
791	8.008
792	9.610
793	11.289
794	13.309
795	15.160
796	15.966
797	15.160
798	13.309
799	11.289
800	9.610

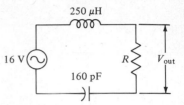

Figure 23.20 Sketch of circuit for Test Problem 4

From these data, find the resonant frequency f_0, BW, Q, and resistance R.

5. A 200-pF capacitor is placed in parallel with a 200-μH inductor. This inductor has a dc resistance of 50 Ω. At resonance find the circuit impedance $\mathbf{Z}_{||}$, circuit Q, and bandwidth.

6. Given that a 60-kΩ resistor is placed in parallel with the 200-pF capacitor of Test Problem 5, what are the new circuit impedance $\mathbf{Z}_{||}$, circuit Q, and bandwidth BW_{loaded}?

7. For Figure 23.21, find V_{out} at resonant frequency.

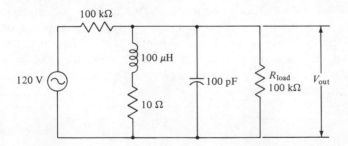

Figure 23.21 Circuit for Test Problem 7

8. Determine the value of the loading resistor needed to produce a bandwidth of 318 Hz for the circuit of Example 23.8.

Unit 24 *Transformers*

You have worked with ac circuits. And you are aware that the most common ac voltage is 120 V, 60 Hz, which is available out of the receptacle outlets in your home. But what if you want a higher or lower voltage?

You have also learned that for maximum power transfer, the load attached to a circuit should equal the internal resistance of the circuit. If the load is not equal to the internal resistance, is there any way that it can be made to appear equal?

How do power companies transfer power from the generating station to your home efficiently?

We have asked you several questions. In this unit you will find the answers because you will learn about an electric device called a *transformer*. A transformer is used in power supplies to step up or step down voltages. It may also be used to step up or step down current. It may be used for impedance and phase matching. And it can be used to electrically isolate one part of a circuit from another.

Objectives

After completing all the work associated with this unit, you should be able to:

1. Recall the electric symbol for a transformer, its definition, and a brief description of its operation.

2. Calculate the unknown voltage when the turns ratio and either the primary or secondary voltage are known, and vice versa.

3. Calculate the unknown current when the turns ratio and either the primary or secondary current are known or can be found, and vice versa.

4. Calculate the unknown impedance when the turns ratio and either the input or load impedance are known, and vice versa.

5. Briefly describe the construction of an iron-core transformer (both secondary windings and core).

6. Determine the phase relationship between the input and output voltage of a transformer, given the winding direction or sense dots.

7. Draw the complete equivalent circuit of an iron-core transformer and explain each of the elements in the circuit. For those elements in the equivalent circuit that represent losses, be able to calculate the transformer efficiency and copper and core losses when results of short-circuit and open-circuit tests are available.

8. Compare an autotransformer with a standard transformer with respect to isolation and typical turns ratio. Also be able to calculate the output voltage and currents when input values are given, and vice versa.

9. Compare an air-core transformer with a standard transformer with respect to coefficient of coupling and method of calculation.

Introduction to Transformers

Objective 1 Recall the electric symbol for a transformer, its definition, and a brief description of its operation.

Definition

A *transformer* is an electric device that can step up or step down ac voltage or current. It can also be used to match impedance, and it provides isolation. The transformer consists of a primary and a secondary winding linked by a mutual magnetic field. Transformers may have an air core, iron core, or variable core, depending on the operating frequency and application.

Transformers vary in size and shape depending on the application. You can find very small transformers in electronic circuits and very large transformers in circuits used by commercial power companies. You can find transformers used at power-line frequencies (between 60 and 400 Hz), audio frequencies (20 to 20,000 Hz), ultrasonic frequencies (20,000 to 100,000 Hz), and radio frequencies (over 30 kHz).

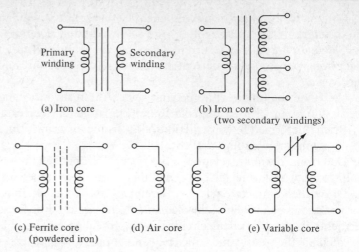

Figure 24.1 Electric symbols for transformers

(a) Iron core

(b) Iron core (two secondary windings)

(c) Ferrite core (powdered iron)

(d) Air core

(e) Variable core

Electric Symbol

The electric symbols used for transformers are shown in Figure 24.1. The primary winding is connected to an ac power source, and the secondary winding is connected to the load. Transformers used at power-line frequencies and audio frequencies usually have an iron core. At high frequencies, the power losses become excessive, and a special powdered iron core or solid ferromagnetic core is used. And at very high frequencies, no core (air core) is used.

Description of Operation

A transformer transfers electric energy from the primary circuit to the secondary circuit by electromagnetic mutual induction. As you can see in the electric symbols for the transformer (Figure 24.1), there is no electrical connection between the primary and second windings.

Figure 24.2 illustrates how a transformer depends on mutual induction. You learned about mutual induction in the unit on Inductance. The voltage source attached to the primary winding

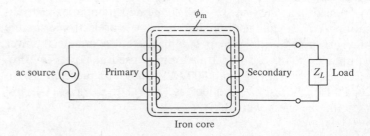

Figure 24.2 Transformer showing mutual induction

in Figure 24.2 causes an ac current in the primary. The current sets up a changing magnetic field. The iron core that the primary coil is wrapped around concentrates and intensifies the magnetic field. If another coil (secondary) is also wrapped around the iron core, as shown in Figure 24.2, the magnetic lines of force set up by the primary coil link the secondary coil. When this happens, an emf is induced across the secondary winding. If a load is attached to the secondary winding, then current is obtained in the load. For a voltage to be induced across the secondary winding, there must be a changing current in the primary winding. Therefore a transformer is an ac device. If a dc source were connected to the primary winding, the current would become so large that the transformer would burn out. On dc ($f = 0$ Hz), the primary winding acts like a low resistance because once the current reaches a steady value, the inductive reactance is equal to zero ohms.

Mathematically we can describe the above discussion as follows. From Faraday's law, the primary voltage is equal to

$$e_p = N_p \frac{d\phi_p}{dt} \tag{24.1}$$

where N_p is the number of turns in the primary and $d\phi_p/dt$ is the rate of change of the lines of flux in the primary. (Recall that we discussed the meaning of rate of change (d/dt) in Units 11, 13, 14, and 15.) The magnitude of the voltage induced across the secondary e_s, then, is

$$e_s = N_s \frac{d\phi_m}{dt} \tag{24.2}$$

where ϕ_m is the portion of the primary flux ϕ_p that links the secondary and N_s is the number of turns in the secondary. The coefficient of coupling between the primary and second coils is determined by the ratio of ϕ_m to ϕ_p.

$$k \text{ (coefficient of coupling)} = \frac{\phi_m}{\phi_p} \tag{24.3}$$

In an iron-core transformer operated at power-line frequencies, the coefficient of coupling is equal to 1. Therefore $\phi_m = \phi_p$ and

$$e_s = N_s \frac{d\phi_p}{dt} \tag{24.4}$$

The average emf induced in the second coil is then equal to

$$E_{ave} = N_s \frac{\phi_{m\,max}}{t}$$

where t is the time (in seconds) it takes the flux to rise from zero to ϕ_m. The flux in the secondary coil is a sine wave, and therefore rises from zero to ϕ_m in one quarter of a cycle ($t = 1/4f$).

$$E_{ave} = N_s \frac{\phi_{m\,max}}{\dfrac{1}{4f}} = 4fN_s\phi_m \qquad (24.5)$$

The relation between the effective value of a sine wave and the average value is

$$E = 1.11 E_{ave}$$

because the average value is 0.637 of the peak value and the effective value is 0.707 of the peak value.

Therefore the effective secondary voltage is equal to

$$E_s = 4.44fN_s\phi_{m\,max} \qquad (24.6)$$

When Equation (24.6) is used without subscripts, it is known as the *general transformer equation*.

$$E = 4.44fN\phi_{m\,max} \qquad (24.7)$$

At the beginning of this section we said that there is no electrical connection between the primary and secondary windings of a transformer. That is, energy is fed from one circuit to another by mutual induction, not by direct electrical connection. This characteristic is called *isolation*. The isolation feature is so important that some transformers are designed just to isolate one circuit from another. When a transformer is designed for the sole purpose of isolation, it usually has a one-to-one voltage ratio. One of the main purposes of isolation is to prevent electric shock. An example of this is the isolation of the electronic power supply in a TV set or radio from the house power line.

Problem 24.1   From memory, draw the various electric symbols that are used for transformers. Define and briefly describe the operation of a transformer.

Calculation of Voltage

Objective 2   Calculate the unknown voltage when the turns ratio and either the primary or secondary voltage are known, and vice versa.

To develop the relationship between the number of turns in the windings and the voltage on the primary and secondary, we shall assume that we have an ideal transformer. By ideal, we mean an iron-core transformer with no winding and core losses and a coefficient of coupling of 1. Most low-frequency transformers manufactured today are almost ideal. Therefore the equations that we develop are very close, if not exact.

From the general transformer equation, Equation (24.7), the effective value of the voltage across the primary winding is

$$E_p = 4.44 f N_p \phi_{p\,max} \tag{24.8}$$

Assuming that the coefficient of coupling is 1, the effective value of the voltage across the secondary winding is

$$E_s = 4.44 f N_s \phi_{p\,max} \tag{24.9}$$

Dividing Equation (24.8) by Equation (24.9), we obtain

$$\boxed{\frac{E_p}{E_s} = \frac{N_p}{N_s}} \tag{24.10}$$

Therefore the ratio of the primary voltage to the secondary voltage is the same as the ratio of the number of primary turns to the number of secondary turns. These ratios are called the turns ratio or the *transformation ratio*. As you study Equation (24.10), you see that if the number of turns in the secondary is less than the number of turns in the primary, we have a *step-down transformer*. That is, the secondary voltage E_s is smaller than the primary voltage E_p. Whereas, if the number of turns in the secondary is greater than the number of turns in the primary, we have a *step-up transformer*.

Example 24.1   A transformer with a transformation ratio (turns ratio) of 10 to 1 is connected to a 120-V, 60-Hz source of emf. What is the secondary voltage?

Solution   Using Equation (24.10) and solving for E_s, we obtain

$$E_s = E_p \frac{N_s}{N_p} = 120 \times \frac{1}{10} = 12 \text{ V}$$

Example 24.2   Suppose that you want a voltage of 35 V out of a transformer that is to be connected to the 120-V, 60-Hz line. What transformation ratio must there be?

Solution   From Equation (24.10)

$$\frac{N_p}{N_s} = \frac{E_p}{E_s} = \frac{120}{35} = \frac{24}{7}$$

Problem 24.2   A transformer with a turns ratio of 15 to 2 is connected to a 120-V, 60-Hz source. What is the secondary voltage?

Problem 24.3   A low-voltage power supply is being designed for a transistor radio. An ac voltage of 20 V is required for the input to the rectifier section of the power supply. A transformer is to be used to obtain the 20 V. What transformation ratio is required if the transformer is connected to the 120-V, 60-Hz line?

Calculation of Current

Objective 3   Calculate the unknown current when the turns ratio and either the primary or secondary current are known or can be found, and vice versa.

Again we are considering an ideal transformer, or a transformer that is 100% efficient. Therefore, by the law of the conservation of energy, the power in the primary P_p must equal the power in the secondary P_s.

$$P_p = P_s \tag{24.11}$$

From the power equation $P = EI$, we substitute into Equation (24.11):

$$E_p I_p = E_s I_s$$

Solving for the ratio of E_p to E_s, we obtain

$$\frac{E_p}{E_s} = \frac{I_s}{I_p} \tag{24.12}$$

Substituting Equation (24.10) into (24.12), we obtain

$$\frac{E_p}{E_s} = \frac{N_p}{N_s} = \frac{I_s}{I_p}$$

or

$$\boxed{\frac{N_p}{N_s} = \frac{I_s}{I_p}} \tag{24.13}$$

Examining Equations (24.10) and (24.13), we see that if we step up voltage, we step down current. Or if we step down voltage, we step up current.

Example 24.3   A 100-Ω resistive load is connected across the secondary of a transformer with a 10-to-1 turns ratio. When the transformer is connected to the 120-V, 60-Hz line, how much current is drawn from the line?

Solution   Using Equation (24.10), we can find the secondary voltage:

$$E_s = \frac{N_s}{N_p} E_p = \frac{1}{10} (120) = 12 \text{ V}$$

Using Ohm's law, we find that the secondary current is equal to

$$I_s = \frac{V_s}{R_L} = \frac{12}{100} = 120 \text{ mA}$$

Then we can find the primary current by using Equation (24.13)

$$I_p = \frac{N_s}{N_p} I_s = \frac{1}{10} (120) = 12 \text{ mA}$$

The higher the current a transmission line has to carry, the larger the I^2R losses. Therefore power companies step up the voltage for transmission (which steps down the current). Then, at the other end of the transmission line, they step down the voltage to what is required at the load. You will learn more about this when you study polyphase circuits in Unit 25. The transformer in the next example may be a transformer used at the end of a transmission line.

Example 24.4   A 2300/230-V, 60-Hz transformer is rated at 250 kVA. Determine the primary and secondary current when it is operating at 250 kVA.

Solution   The primary current drawn is

$$I_p = \frac{250 \text{ kVA}}{2300 \text{ V}} = 108.7 \text{ A}$$

We can find the secondary current in a similar manner, using the kVA rating and the secondary voltage. Or we can find the secondary current by using Equation (24.12).

$$I_s = \frac{E_p}{E_s} I_p = \frac{2300}{230} (108.7) = 1087 \text{ A}$$

Problem 24.4   A 20-Ω resistive load is connected across the secondary of a transformer with a 3-to-1 turns ratio. When the transformer is connected to the 120-V, 60-Hz line, how much current is drawn from the line? What is the minimum VA rating the transformer should have?

Problem 24.5 A transformer with 1200 primary turns and 150 secondary turns delivers 90 A at 230 V to a resistive load (unity power factor). Determine the power delivered to the load and the primary current drawn from the source.

Calculation of Impedance

Objective 4 Calculate the unknown impedance when the turns ratio and either the input or load impedance are known, and vice versa.

To determine how the primary and secondary impedance relate to the turns ratio or transformation ratio, we determine the ratio of primary impedance to secondary impedance.

$$\frac{Z_p}{Z_s} = \frac{E_p/I_p}{E_s/I_s} \tag{24.14}$$

Rearranging Equation (24.14), we obtain

$$\frac{Z_p}{Z_s} = \frac{E_p}{E_s} \times \frac{I_s}{I_p} \tag{24.15}$$

From Equation (24.10), we get

$$\frac{E_p}{E_s} = \frac{N_p}{N_s}$$

From Equation (24.13),

$$\frac{I_s}{I_p} = \frac{N_p}{N_s}$$

Therefore Equation (24.15) becomes

$$\frac{Z_p}{Z_s} = \frac{N_p}{N_s} \times \frac{N_p}{N_s}$$

or

$$\boxed{\frac{Z_p}{Z_s} = \left(\frac{N_p}{N_s}\right)^2} \tag{24.16}$$

To express this in words: The ratio of primary impedance to secondary impedance is equal to the *square* of the ratio of primary turns to secondary turns.

Solving Equation (24.16) for Z_p, we obtain

$$Z_p = \left(\frac{N_p}{N_s}\right)^2 Z_s$$

Assuming that the impedance of the secondary winding is small when compared with the impedance of the load, we may say that $Z_s = Z_L$. Then

$$Z_p = \left(\frac{N_p}{N_s}\right)^2 Z_L \qquad (24.17)$$

where Z_p is known as the *reflected impedance*. The reflected impedance is found by multiplying the load impedance by the square of the turns ratio.

Examining Equation (24.17), you can see that a transformer can be used as an impedance-matching device. The following examples illustrate this.

Example 24.5 Figure 24.3 shows a single-stage transistor amplifier circuit. The 8-Ω speaker is to be connected as a load; however, for maximum power transfer, the transistor should have a 1-kΩ load. Determine what turns ratio is required to match the speaker to the transistor. Assume that both impedances are resistive and that the transformer is ideal at the operating frequency.

Solution Using Equation (24.16), we may find the required turns ratio:

$$\left(\frac{N_p}{N_s}\right)^2 = \frac{Z_p}{Z_s}, \qquad \frac{N_p}{N_s} = \sqrt{\frac{Z_p}{Z_s}} = \sqrt{\frac{1000}{8}} = \frac{11.2}{1}$$

Example 24.6 A power amplifier has an output transformer in the circuit so that a 4-Ω, 8-Ω, or 16-Ω speaker may be used. Refer to Figure 24.4. Instead of an 8-Ω speaker being connected to

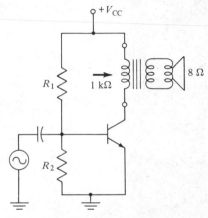

Figure 24.3 Circuit for Example 24.5

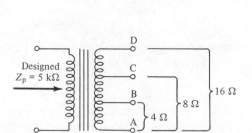

Figure 24.4 Output transformer that may take a 4-Ω, 8-Ω, or 16-Ω speaker

taps C and A, it was connected to taps D and C. What impedance does the primary of the transformer now see?

Solution The reflected impedance of each of the loads must be the same on the primary side of the transformer. Using Equation (24.17), we have

$$Z_p = \left(\frac{N_p}{N_{CA}}\right)^2 8 = \left(\frac{N_p}{N_{DA}}\right)^2 16 = \left(\frac{N_p}{N_{DA} - N_{CA}}\right)^2 Z_x$$

where Z_x is the impedance that should be placed between taps D and C to obtain the correct reflected impedance on the primary side. Working with the last three equalities, and taking the square root of each to simplify the expression, we obtain

$$\frac{N_p}{N_{CA}}\sqrt{8} = \frac{N_p}{N_{DA}}\sqrt{16} = \frac{N_p}{N_{DA} - N_{CA}}\sqrt{Z_x}$$

We can work with this relation to determine what value of Z_x gives the designed reflected impedance. After working with the first two equalities to find N_{DA} in terms of N_{CA}, we then substitute this into the last two equalities. As a result, we find that

$$\sqrt{Z_x} = \sqrt{16} - \sqrt{8} \quad \text{or} \quad Z_x = 1.37 \ \Omega$$

From the original equation, we find that

$$\left(\frac{N_p}{N_{DA} - N_{CA}}\right)^2 = \frac{Z_p}{Z_x}$$

We now modify the original equation to take any value for the impedance connected between taps D and C.

$$Z_p' = \left(\frac{N_p}{N_{DA} - N_{CA}}\right)^2 Z_x' = \left(\frac{Z_p}{Z_x}\right) Z_x' = \left(\frac{5000}{1.37}\right) 8 = 29.2 \ k\Omega$$

It does make a difference which taps you use.

Problem 24.6 A transformer is used to make a 16-Ω load appear as a 500-Ω load. Find the transformation ratio of the transformer.

Problem 24.7 A transformer with 2000 primary turns and 50 secondary turns is used as an impedance-matching device. What is the reflected value of a 100-Ω resistive load connected to the secondary?

Problem 24.8 Assume, in Example 24.6, that a 4-Ω speaker is connected between taps C and B. What impedance does the primary of the transformer see?

Construction of an Iron-Core Transformer

Objective 5 Briefly describe the construction of an iron-core transformer (both secondary windings and core).

Secondary Windings

A transformer may have multiple secondary windings, as shown in Figure 24.5. This enables the transformer to supply several different values of ac voltages from one input source. The transformer in Figure 24.5 has both step-up and step-down circuits.

The transformer shown in Figure 24.6 has a center-tap secondary winding. There are many applications that use a center-tap or multiple-tapped secondary winding. You saw one application of a multiple-tapped secondary winding in Example 24.6. And we shall briefly describe two applications for the center-tap secondary winding. Both applications are in electronic dc power supplies. An electronic dc power supply changes the input line voltage of 120 V, 60 Hz to a fixed dc or variable-output dc voltage.

The circuit in Figure 24.7 provides a positive dc output voltage. The circuit in Figure 24.8 provides both a positive and a negative dc voltage, which is required for some integrated circuits. In both these circuits, note the transformer with the center-tap secondary winding.

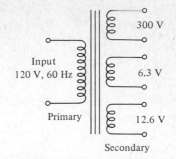

Figure 24.5 Multiple secondary windings

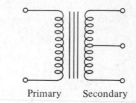

Figure 24.6 Center-tap secondary

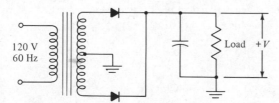

Figure 24.7 Full-wave rectifier power-supply circuit (positive dc output voltage)

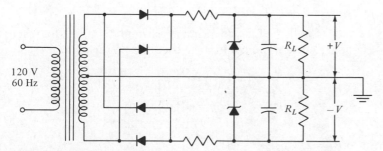

Figure 24.8 Electronic power-supply circuit (provides both positive and negative dc output voltages)

The transformer in Figure 24.6 has a single primary winding and a center-tap secondary. You may also find some transformers with a center-tap primary winding and a single secondary winding.

No new formulas are required to work with transformers that have multiple secondary windings. However, we do want to work one example and have you work one problem to show you that you cannot add currents or voltages in the secondaries to determine the current or voltage in the primary. You must calculate each one separately.

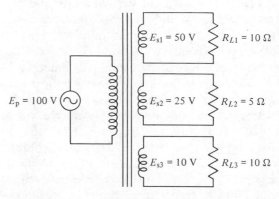

Figure 24.9　Circuit for Example 24.7

Example 24.7　Refer to the transformer shown in Figure 24.9. Find the primary current (primary fuse rating).

Solution　We solve for each secondary current using Ohm's law. Then we use Equation (24.12) to find the primary current.

$$I_{s1} = \frac{50}{10} = 5 \text{ A} \qquad I_{p1} = \frac{E_{s1}}{E_p} I_{s1} = \frac{50}{100}(5) = 2.5 \text{ A}$$

$$I_{s2} = \frac{25}{5} = 5 \text{ A} \qquad I_{p2} = \frac{E_{s2}}{E_p} I_{s2} = \frac{25}{100}(5) = 1.25 \text{ A}$$

$$I_{s3} = \frac{10}{10} = 1 \text{ A} \qquad I_{p3} = \frac{E_{s3}}{E_p} I_{s3} = \frac{10}{100}(1) = 0.1 \text{ A}$$

$$I_p = I_{p1} + I_{p2} + I_{p3} = 3.85 \text{ A}$$

Instead of solving for the primary current required for each secondary current, we could have used the power relations:

$$P_{s1} = I_{s1}E_{s1} = (5)(50) = 250 \text{ W}$$
$$P_{s2} = I_{s2}E_{s2} = (5)(25) = 125 \text{ W}$$
$$P_{s3} = I_{s3}E_{s3} = (1)(10) = \underline{10 \text{ W}}$$
$$\text{Total } P_s = \overline{385 \text{ W}}$$

Since $P_p = P_s$, then

$$I_p = \frac{P_p}{E_p} = \frac{385}{100} = 3.85 \text{ A}$$

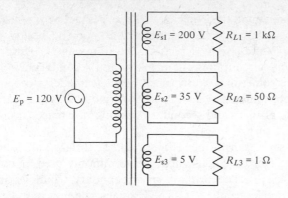

$E_p = 120$ V

$E_{s1} = 200$ V $R_{L1} = 1$ kΩ

$E_{s2} = 35$ V $R_{L2} = 50$ Ω

$E_{s3} = 5$ V $R_{L3} = 1$ Ω

Figure 24.10   Circuit for Problem 24.9

Problem 24.9   Refer to the transformer shown in Figure 24.10. Find the primary current.

Core Construction

Transformers used at power-line frequencies and audio frequencies usually have an iron core. These iron cores are not made of solid iron or steel. Instead the steel, known as *transformer steel*, is usually laminated; see Figures 24.11 and 24.12. By laminated we mean composed of thin slices, 0.007 to 0.014 inch thick, coated with a nonconducting, insulating varnish. The actual thickness used depends on the operating frequency of the transformer. If the cores were not laminated, considerable power would be lost in the core, because the lines of flux induce current, called *eddy current*, in the core. The thicker the steel, the larger the current, and therefore the larger the core losses. We'll discuss eddy-current losses in more detail in a later objective.

There are two main core shapes, shown in Figures 24.11 and 24.12. In a core-type transformer, Figure 24.11, the windings or

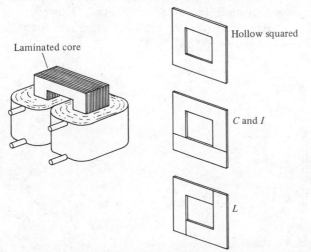

Laminated core

Hollow squared

C and I

L

Figure 24.11   Core-type transformer and laminations

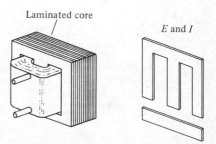

Laminated core

E and I

Figure 24.12   Shell transformer and laminations

copper surround the iron core. The core is in the form of a hollow square made up of laminations that may be hollow-square-shaped, C- and I-shaped, or L-shaped stampings. To improve the coefficient of coupling, the primary and secondary windings or coils are sometimes both placed on the same leg close together. Therefore sometimes, instead of two windings, as in Figure 24.11, both the primary and secondary windings are placed around one leg of the core.

In a shell transformer (Figure 24.12), the iron surrounds the copper. This core construction is commonly used, because it it very efficient. The primary and secondary windings are both placed on the center leg. The primary winding may be inside, with the secondary wound over the primary. The primary and secondary windings may be interleaved, or the coils may be wound in pancake form.

Problem 24.10   Briefly describe the construction of an iron-core transformer. Be sure to cover multiple secondary windings, tapped secondary windings, and core construction.

Phase Relationship

Objective 6   Determine the phase relationship between the input and output voltage of a transformer, given the winding direction or sense dots.

A transformer may be used for phase matching. Depending on the winding direction of the primary and secondary, the output voltage can be either in phase with or 180° out of phase with the input. Generally, for efficiency, the primary and secondary windings are wrapped around the same core area. However, in Figure 24.13(a) we have shown the windings separated for sim-

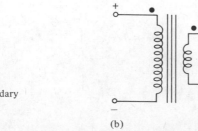

(a)   (b)   (c)

Figure 24.13   Transformer phase relationship—no phase shift

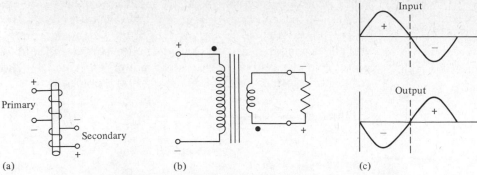

Figure 24.14 Transformer phase relationship—180° phase shift

plification. Note that the primary and secondary windings are both wound in the same direction. The *sense dots* (●) in Figure 24.13(b) indicate which ends of the windings have the same polarity at the same instant of time. Figure 24.13(c) shows the input and output voltage waveforms for the transformer in Figure 24.13(b). The input and output voltages are in phase.

If the secondary winding is wound in a direction opposite that of the primary [Figure 24.14(a)], then the output voltage is 180° out of phase with respect to the input voltage. Figure 24.14(b) shows the sense dots on the windings, and Figure 24.14(c) shows the waveforms.

Problem 24.11 For a transformer with sense dots as shown in Figure 24.15, sketch the output voltage waveform for the input voltage shown.

Equivalent Circuit and Losses

Objective 7 Draw the complete equivalent circuit of an iron-core transformer and explain each of the elements in the circuit. For those elements in the equivalent circuit that represent losses, be able to calculate the transformer efficiency and copper and core losses when results of short-circuit and open-circuit tests are available.

Transformer Efficiency

In our previous objectives we have assumed that a transformer was 100% efficient. The actual efficiency of a transformer can be determined by

$$\% \text{ Efficiency} = \frac{\text{power output}}{\text{power input}} \times 100 \quad (24.18)$$

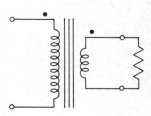

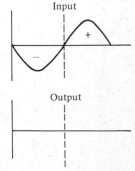

Figure 24.15 Transformer for Problem 24.11

When we assume a resistive load, the input power is equal to the product of the primary voltage and the primary current. The output power is equal to the product of the secondary voltage and the secondary current. The difference between the input power and the output power is the power consumed by the various transformer losses. Equation (24.18) becomes

$$\% \text{ Efficiency} = \frac{\text{power output}}{\text{power output} + \text{transformer losses}} \times 100 \quad (24.19)$$

Transformer Losses

We shall consider three types of transformer losses. When current flows in a conductor, power is dissipated. The losses in the copper windings of the transformer are called *copper losses*. The other two types of losses are *eddy-current losses* and *hysteresis losses*. Both are core losses.

Copper losses   Even though the windings in a transformer are made of copper, they still have a small amount of resistance. Copper losses, then, come from the I^2R loss that results from the resistance of the winding. The transformer must be able to dissipate the heat due to the copper losses.

Copper losses may be kept low if the windings are designed properly. One should use low-resistance copper and also wire that is the proper size. The shape of the cross section of the wire or copper is also important, particularly as frequency increases. Example 24.8 explains how to calculate the copper losses in a transformer.

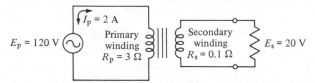

Figure 24.16   Transformer circuit for Example 24.8

Example 24.8   Refer to the transformer circuit in Figure 24.16. Calculate the copper losses in the transformer.

Solution   Copper losses are determined by taking I^2R. Therefore we must know both the primary and secondary current. Using Equation (24.12), we find I_s:

$$I_s = \frac{E_p}{E_s} I_p = \frac{120}{20} (2) = 12 \text{ A}$$

Therefore

Primary copper loss $I_p^2 R_p = \quad (2)^2(3) = 12 \quad$ W
Secondary copper loss $I_s^2 R_s = (12)^2(0.1) = 14.4$ W
Total copper losses $\overline{26.4 \text{ W}}$

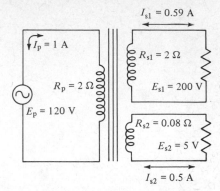

Figure 24.17   Circuit for Problem 24.12

One additional comment: The total power supplied to the primary is $(120 \text{ V})(2 \text{ A}) = 240$ VA. Of the 240 VA supplied, 26.4 W or 11% are lost in copper losses.

Problem 24.12   Refer to the transformer circuit in Figure 24.17. Calculate the copper losses in the transformer.

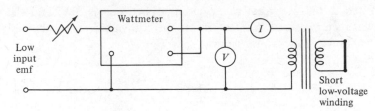

Figure 24.18   Short-circuit transformer test

We can run a *short-circuit* transformer test to determine the actual copper losses in a transformer. Figure 24.18 shows the setup for the short-circuit test. The low-voltage winding is shorted. Then an input voltage is applied. The input voltage must be such that the rated current flows in the transformer windings. Because one of the windings is shorted, the input voltage is greatly reduced from what the transformer normally has applied. We do not have to worry about the core losses affecting our determination of copper losses. The exciting current that flows is very low because the input emf is low. Therefore core losses are reduced by the square of the reduction of the exciting current. They are so small that they may be neglected.

We can determine the equivalent resistance by using the wattmeter and ammeter readings obtained in the test setup of Figure 24.18.

$$R_e = \frac{P}{I^2} \tag{24.20}$$

In our discussion of losses we have considered the transformer windings to have only resistance. But because the windings are wound in the form of a coil, they also have a certain amount of reactance. Since a voltmeter and ammeter were inserted in the test setup of Figure 24.18, we can determine the amount of

equivalent reactance. First we determine the equivalent impedance, then the equivalent reactance.

$$Z_e = \frac{V}{I} = \sqrt{R_e^2 + X_e^2} \qquad (24.21)$$

$$\therefore \quad X_e = \sqrt{Z_e^2 - R_e^2} \qquad (24.22)$$

Since we shorted the low-voltage winding, the values for R_e
and X_e are for the high-voltage side of the transformer. If the
transformer were a step-up transformer, then these values for R_e
and X_e could be reflected to the primary side by dividing by the
turns ratio squared.

Example 24.9 A short-circuit test is run on a transformer. For a
rated primary current of 2 A, the applied emf is 3 V, and the
wattmeter reads 4 W. Determine the equivalent resistance and
reactance of the transformer.

Solution

$$R_e = \frac{P}{I^2} = \frac{4}{2^2} = 1 \ \Omega$$

$$Z_e = \frac{V}{I} = \frac{3}{2} = 1.5 \ \Omega$$

$$X_e = \sqrt{Z_e^2 - R_e^2} = \sqrt{(1.5)^2 - (1)^2} = 1.12 \ \Omega$$

Core losses We mentioned in Objective 5 that the transformer
core was made of steel laminations to reduce the power loss due
to eddy currents. Whenever an alternating current flows in the
primary winding of a transformer, it produces a changing magnetic field. In addition to inducing a voltage across the secondary
winding, it also induces a voltage into the core material. This
induced voltage causes random currents, called *eddy currents*, to
flow in the core. Power loss due to eddy currents can be expressed
by

$$P_e = kB_m^2 f^2 t^2 \qquad (24.23)$$

where k = a constant that depends on the units for the other
quantities (it varies inversely with the resistivity ρ of
the core material)

B_m = maximum flux density

f = frequency

t = thickness of the lamination

Therefore, from Equation (24.23), you can see that the power loss
due to eddy currents increases directly with the square of the
thickness of the laminations. This is why transformer cores are
made from thin laminations that are insulated from one another
with a thin coat of varnish.

Equation (24.23) also shows that the power loss due to eddy currents increases directly with the square of the design frequency. This is why, for frequencies above 20 kHz, air cores or special ferrite cores must be used.

Another power loss that occurs in transformer cores is known as *hysteresis loss*. This is a characteristic of any iron or ferromagnetic material. When a magnetic field passes through a core, the core material becomes magnetized. To be magnetized, the magnetic domains (dipoles) within the material must be aligned with the external field. When the field reverses, the magnetic domains must realign themselves with the external field. However, some of the magnetic domains resist a change in position. As a result, flux density tends to lag the magnetic field intensity that created it. A definite amount of force is required to realign the magnetic domains. Therefore part of the magnetizing force must be used to overcome this internal friction, and this results in the production of heat and wasted energy. The higher the operating frequency of the transformer, the more rapidly the magnetic domains have to change their alignment, and therefore the greater the hysteresis loss.

An open-circuit transformer test may be run to determine the actual core losses in a transformer. Figure 24.19 shows the setup for the open-circuit test. The high-voltage winding is opened, so that only a low voltage is required for the input. The input voltage must be the exact rated voltage for that winding. Since the secondary of the transformer is open, the primary current is only that current necessary to establish the flux in the core. Therefore the wattmeter in Figure 24.19 shows the core losses. These core losses are due to the eddy currents and the hysteresis losses in the core. We can determine values for the core resistance R_c and mutual reactance X_m of the core from the meter readings in a manner similar to our determination of Equations (24.20), (24.21), and (24.22) from the short-circuit tests.

$$R_c = \frac{P}{I^2} \tag{24.24}$$

$$Z_c = \frac{V}{I} = \sqrt{R_c^2 + X_m^2} \tag{24.25}$$

$$X_m = \sqrt{Z_c^2 - R_c^2} \tag{24.26}$$

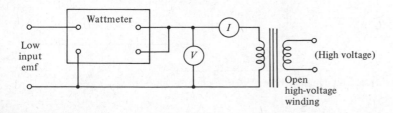

Figure 24.19 Open-circuit transformer test

Problem 24.13 The following short-circuit and open-circuit measurements were taken on a transformer.

$$\text{Short circuit}: I = 1 \text{ A}, V = 6 \text{ V}, \text{ and } P = 4 \text{ W}$$

$$\text{Open circuit}: V = 20 \text{ V}, I = 2 \text{ A}, \text{ and } P = 16 \text{ W}$$

Determine R_e, X_e, R_c, and X_m.

Equivalent Circuit of an Iron-Core Transformer

Up to this point in this objective, we have been discussing transformer losses and how to determine them. Now we want to put everything together and show you the actual equivalent circuit of a transformer. Figure 24.20 shows the equivalent circuit of an iron-core transformer. Note that within the equivalent circuit is the ideal transformer we usually consider that we have. The following features identify the nonideal characteristics of a transformer:

C_p and C_s = capacitance of primary and secondary,

C_w = equivalent capacitance between the windings

R_p and R_s = dc or winding resistance of primary and secondary

R_c = hysteresis and eddy-current (core) losses

X_p and X_s = winding leakage reactance of primary and secondary

X_m = inductive reactance associated with magnetization of the core (establishing flux ϕ_m in the core)

a = turns ratio of an ideal transformer N_p/N_s

Z_L = secondary load impedance.

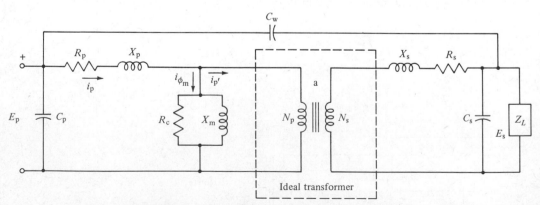

Figure 24.20 Equivalent circuit of an iron-core transformer

This equivalent circuit diagram would be almost impossible to work with unless you had a computer. You have already seen one simplification of it: That is, forget the losses and consider the transformer ideal. We'll give you a couple of other simplifications that you can use when results of the short-circuit and open-circuit tests are available. These simplifications do consider some of the effects of the reactance in the windings and core.

If the transformer is operated at low or medium frequency, the capacitance between turns and windings is low enough so that we can neglect its effect. Figure 24.21(a) shows this simplification. Figure 24.21(b) shows the elements on the secondary side of the ideal transformer being reflected to the primary side. The relations developed for Objective 4 are used, with a equal to N_p/N_s. Figure 24.21(c) is the same circuit as Figure 24.21(b) except that it uses electric symbols.

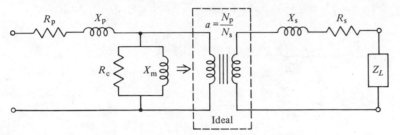

(a) Equivalent circuit assuming C small and frequency low
　(neglect effects of C_p, C_s, and C_w)

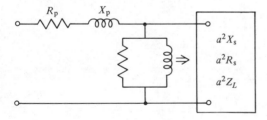

(b) Reflected secondary load

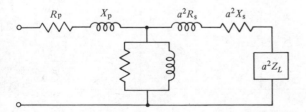

(c) Equivalent circuit showing reflected secondary load

Figure 24.21 Transformer equivalent circuit

Figure 24.22 shows how we can simplify the equivalent circuit further. Figure 24.22(a) results when core losses (R_c and X_m) are negligible. The circuit is very simple, and R_e and X_e can be determined from the short-circuit test. Figure 24.22(b), (c), and (d) consider the effect of frequency on the equivalent circuit. At low frequencies X_p and X_s are small and can be neglected; see Figure 24.22(b). At medium frequencies, the effect of X_m as well as

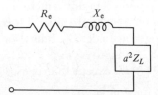

(a) Assuming core losses negligible, where $R_e = R_p + a^2 R_s$, $X_e = X_p + a^2 X_s$

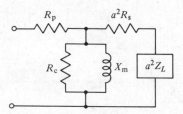

(b) Low-frequency equivalent, considering core losses

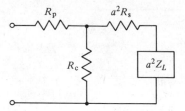

(c) Mid-frequency equivalent, considering core losses

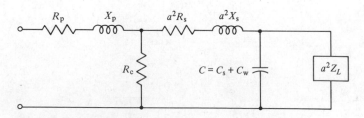

(d) High-frequency equivalent

Figure 24.22 Transformer equivalent circuits

X_p and X_s can be neglected; see Figure 24.22(c). At high frequencies, we must consider X_p and X_s because they are in series. Note that capacitance has been inserted in Figure 24.22(d). At high frequencies X_c becomes very small. Since it shunts some of the elements in the equivalent, it must be considered.

Problem 24.14 Draw the complete equivalent circuit of an iron-core transformer and explain each of the elements in the circuit.

Autotransformer

Objective 8 Compare an autotransformer with a standard transformer with respect to isolation and typical turns ratio. Also be able to calculate the output voltage and currents when input values are given, and vice versa.

In an earlier objective we mentioned that an important feature about transformers was isolation. If isolation is not required and a low turns ratio is needed, then we can use an autotransformer. We may still obtain the high efficiency and the voltage- and current-changing ability, but at reduced cost.

By using an autotransformer, we can transform ac energy without a separate primary and secondary winding. Figure 24.23 shows an autotransformer with a 2-to-1 voltage ratio. The winding designated AC in Figure 24.23 may consist of a single coil with a tap at point B. Or windings AB and BC may be separate coils, but connected so that they are in series and wound in the

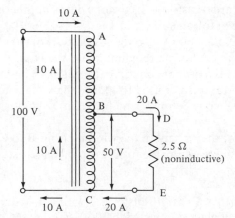

Figure 24.23   Autotransformer with a 2-to-1 voltage ratio

same direction. The turns between A and C are the primary turns. The turns between B and C are the secondary turns.

In the autotransformer in Figure 24.23, 100 V is applied across the primary winding AC. Since B is at the midpoint of the winding, the voltages from A to B and from B to C must each be 50 V. We can also find the voltage from B to C by using Equation (24.10) with a turns ratio of 2 to 1. With a 2.5-Ω resistive load connected across the 50-V secondary, 20 A must exist in the load. The power in the load is (50 V × 20 A), or 1000 W. Since the losses in this type of transformer are small and can usually be neglected, the input applied to the primary is also 1000 W. This means that the current entering winding AB is 1000 W/100 V or 10 A, which is flowing downward from A to B at the instant of consideration. Since the load current (from B to D through D to E and from E to C) is 20 A, then by KCL the current must be equal to 10 A, and it must flow upward from C to B. This 10 A combines with the 10 A flowing downward in AB to give the 20 A required by the load. The 10 A flowing upward from C to B is the transformed or secondary current. Notice that the winding from C to B has to carry only 10 A, whereas if the secondary winding had been isolated, as in a standard transformer, it would have had to carry the load current of 20 A.

Since there is no separate secondary winding in an autotransformer, there is a saving on the amount of copper required for the transformer. And, as shown in Figure 24.23, the secondary part of the winding, from B to C, carries the same value of current as the input. In a standard transformer with a 2-to-1 voltage ratio, the secondary would have to be large enough in cross-sectional area to carry twice the input current.

Another way of showing that an autotransformer is smaller than a standard transformer is to consider voltamperes (VA). The input to the autotransformer in Figure 24.23 is 100 V (10 A), or 1000 VA. Of the 1000 VA, only 50 V (10 A) or 500 VA is transformed to the load. The remaining 500 VA is conducted to the load through the primary winding (through AB and BC). Therefore an autotransformer used to reduce the voltage to a 1000-VA load can be considerably smaller than a standard 1000-VA transformer.

To see how the current in the winding from B to C of an autotransformer increases as the turns ratio increases, study Example 24.10.

Example 24.10 An autotransformer with a 5-to-1 voltage ratio draws 1000 W when connected to 100-V source; see Figure 24.24. Determine the secondary or load voltage and all the currents that flow in the windings.

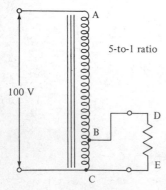

Figure 24.24 Autotransformer for Example 24.10

Solution From Equation (24.10), we have

$$E_s = \frac{N_s}{N_p} E_p = \frac{100}{5} = 20 \text{ V}$$

The input current, or current from A to B, is

$$I_{AB} = I_p = \frac{1000 \text{ W}}{100 \text{ V}} = 10 \text{ A}$$

The load current is equal to the current that flows in the secondary winding of a standard transformer. From Equation (24.13), we obtain

$$I_L = I_s = \frac{N_p}{N_s} I_p = 5(10) = 50 \text{ A}$$

Therefore, by KCL, the current that flows in the winding from B to C is equal to

$$I_{CB} = I_L - I_p = 50 - 10 = 40 \text{ A upward}$$

Did you notice that as the turns ratio of the autotransformer increased, the current in the BC part of the winding became more equal or closer to the load current and much greater than the primary current? This means that if only one winding with a tap is used, the whole winding has to be designed to handle the large current. The only other alternative is to design two separate windings (A to B and B to C). However, it would then be almost as economical to have separate primary and secondary windings. And, with two separate windings, we gain the isolation feature. This is why the autotransformer is more economical than the standard transformer only when the turns ratio is low.

Problem 24.15 How does an autotransformer compare with a standard transformer with respect to isolation and typical turns ratio?

Problem 24.16 An autotransformer with a 4-to-3 voltage ratio draws 1500 W when connected to a 100-V source; see Figure 24.25. Determine the secondary or load voltage and all the currents that flow in the windings.

Air-Core Transformers

Objective 9 Compare an air-core transformer with a standard transformer with respect to coefficient of coupling and method of calculation.

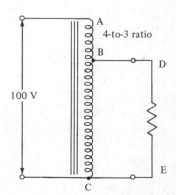

Figure 24.25 Autotransformer for Problem 24.16

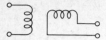

(a) Minimum coupling: k near 0

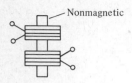

Nonmagnetic

(b) Loose coupling

(c) Interwound tight coupling: k near 1

Figure 24.26 Variations in transformer coupling

Transformers that operate at very high frequencies (radio frequency, or RF) have no cores (that is, they have air cores). The power losses in any kind of iron core, even powdered iron, are too excessive at these frequencies. As discussed in Objective 1, the purpose of the iron core is to improve the coefficient of coupling. The formulas developed for use in Objectives 2, 3, and 4 assumed an ideal transformer with no losses and a coefficient of coupling of 1. With no core (an air core), the coefficient of coupling can vary from near zero (the situation in which no primary flux cuts across the secondary) to unity, depending on the positioning of the coils. Figure 24.26 illustrates three variations in coupling.

Since most transformers that operate at RF have coefficients of coupling that are not equal to 1, the formulas developed for iron-core transformers do not apply. Transformers for RF circuits are generally tuned to resonance, and therefore the resonance factor is considered in the calculations. For example, capacitors C_1 and C_2 have been inserted in the air-core transformer circuit in Figure 24.27. The capacitors are adjusted so that both the primary and secondary circuits are resonant at the source frequency. When resonance occurs, there is a rise in current. This large current causes a large signal voltage to be induced in the secondary winding.

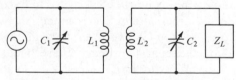

Figure 24.27 Air-core transformer circuit

We shall not develop the formulas that we must use to work with RF transformers because it is beyond the scope of this objective. However, most advanced electric circuits books cover this topic.

Problem 24.17 Compare an air-core transformer with a standard transformer with respect to coefficient of coupling and method of calculation.

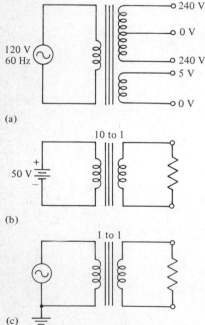

Figure 24.28 Circuits for Test Problem 1

Test

1. Discuss the operation of the transformers in Figure 24.28.
2. A transformer with a transformation ratio of 20 to 1 is connected to a 120-V, 60-Hz source of emf. What is the secondary voltage?

3. A voltage of 45 V is desired out of a transformer connected to the 120-V, 60-Hz line. What turns ratio is required?

4. A 15-Ω resistive load is connected across the secondary of a transformer with a 15-to-1 turns ratio. When the transformer is connected to the 120-V, 60-Hz line, how much current is drawn from the line?

5. A 120-kV/460-V, 60-Hz transformer connects the power lines from the city substation to the district station power lines. Find the primary and secondary current when it is operating at 3 kVA.

6. An audio transformer has 1500 primary turns and 40 secondary turns. What is the reflected value of a 4-Ω load connected to its secondary? A voltage of 10 V is applied to the primary; find the secondary voltage and current.

7. A transformer is used to make an 8-Ω load appear as a 2-kΩ load. If a 4-Ω load is connected instead of the 8-Ω load, what is the reflected primary impedance?

8. A transformer has 40 turns on the secondary winding and 760 turns on the primary winding.
   (a) What is the transformation ratio?
   (b) What voltage is available across the secondary winding when 120 V is applied across the primary?
   (c) A load of $5 + j0\ \Omega$ is connected to the secondary. What impedance is reflected into the primary circuit?

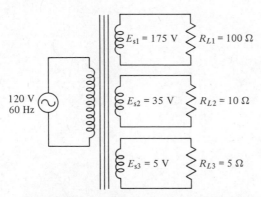

Figure 24.29 Transformer for Test Problem 9

9. Refer to the transformer shown in Figure 24.29. Find the primary current.

10. For a transformer with sense dots as shown in Figure 24.30, sketch the output-voltage waveform for the input voltage shown.

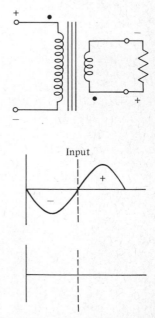

Figure 24.30 Transformer for Test Problem 10

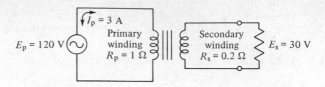

Figure 24.31 Transformer for Test
Problem 11

11. Refer to the transformer circuit in Figure 24.31. Calculate the copper losses in the transformer.

12. A short-circuit test is run on a 2300/208-V, 500-kVA, 60-Hz distribution transformer. For a rated primary current of 218 A, the applied emf is 90 V and the wattmeter reads 9.6 kW. Draw the simplified equivalent circuit, then calculate R_e and X_e.

13. An open-circuit test is run on the transformer of Test Problem 12. The high-voltage winding is opened. With 208 V applied to the low-voltage winding, the primary current is 80 A and the wattmeter reads 1.6 kW. Calculate R_c and X_m.

14. An autotransformer with a 3-to-2 voltage ratio draws 800 W when connected to a 120-V source; see Figure 24.32. Determine the secondary or load voltage and all the currents that flow in the windings.

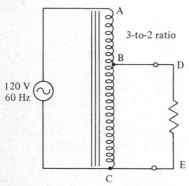

Figure 24.32 Autotransformer for Test
Problem 14

Polyphase Unit 25
Systems

Up to now you have been studying ac circuits that contained a single ac voltage source. These circuits are often called *single-phase ac*. Circuits that contain a number of ac voltage sources in which there is a phase difference are often called *polyphase ac*. In this unit you will study only the most common polyphase circuit—the three-phase (3ϕ) circuit.

Objectives

After completing all the work associated with this unit, you should be able to:

1. Describe how three-phase voltages are generated. Define the double-subscript notation used in three-phase systems.
2. Draw the wye (Y) and delta (Δ) connections for a three-phase system. Define and relate the line voltages and phase voltages in a wye or delta connection.
3. Calculate the power in a balanced wye and/or delta three-phase system.
4. Briefly describe the complete electrical system from the generating station to the final household outlet.
5. Calculate the power in an unbalanced wye and/or delta three-phase system.

Double-Subscript Notation

Objective 1 Describe how three-phase voltages are generated. Define the double-subscript notation used in three-phase systems.

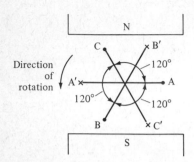

Figure 25.1   Three-phase generator

A three-phase ac generator can be considered to be made up of three separate ac generators, each producing a sine wave. Recall that in Unit 15 you studied the generation of a single-phase ac sine wave. The generation of three-phase (3ϕ) voltages can be accomplished in the same manner.

A magnetic field is set up in the air gap between the North and South poles of a magnet; see Figure 25.1. The North pole is on top, and the magnetic flux lines go from the North to the South pole. The generator of a single-phase ac sine wave is made up of a single conductor AA′. For three-phase generation, instead of a single conductor AA′, three sets of conductors, AA′, BB′, and CC′, are used. These three separate windings are mechanically fixed 120° apart. That is, between the A and B windings there is a 120° difference. Likewise the A and C windings are also 120° apart. Note that each winding is described by a double subscript AA′ or CC′. This *double-subscript notation* is used to describe precisely the voltage across the winding at any instant of time. For example, the notation $\mathbf{E}_{AA'}$ means the voltage at point A with respect to point A′. In general, the *unprimed* letters represent the point of voltage measurement, while the *primed* letters A′, B′, C′ represent the reference points.

If the three fixed windings AA′, BB′, and CC′ were rotated in a counterclockwise direction, they would cut the magnetic lines of flux developed by the magnetic field at different times. The three voltages $\mathbf{E}_{AA'}$, $\mathbf{E}_{BB'}$, and $\mathbf{E}_{CC'}$ that are generated by the rotating windings are shown in Figure 25.2(b). These three-phase (3ϕ) voltages $\mathbf{E}_{AA'}$, $\mathbf{E}_{BB'}$, and $\mathbf{E}_{CC'}$ represent the voltages at the unprimed variable (A, B, or C) with respect to the primed variable (A′, B′, or C′).

These voltages can also be represented as a voltage phasor diagram, as shown in Figure 25.2(a). As the three phasors $\mathbf{E}_{AA'}$, $\mathbf{E}_{BB'}$, and $\mathbf{E}_{CC'}$ rotate *counterclockwise*, they generate the 3ϕ ac voltage sine waves of Figure 25.2(b). The positions of the phasors in Figure 25.2(a) represent the instantaneous voltages of Figure 25.2(b) at the time $t = 0$. At this time, the voltage $\mathbf{E}_{AA'}$ is zero, $\mathbf{E}_{BB'}$ is negative, and $\mathbf{E}_{CC'}$ is positive. Note that voltage phasor $\mathbf{E}_{BB'}$ lags the voltage $\mathbf{E}_{AA'}$ by 120°. Furthermore voltage phasor $\mathbf{E}_{CC'}$ leads $\mathbf{E}_{AA'}$ by 120° (or lags $\mathbf{E}_{BB'}$ by 120°).

If we use the voltage phasor $\mathbf{E}_{AA'}$ as a reference whose magnitude E_p expresses the rms value, then we can express the voltage phasor $\mathbf{E}_{AA'}$ as

$$\mathbf{E}_{AA'} = E_p \underline{/0^\circ} \qquad (25.1)$$

The other two phasors can also be expressed as

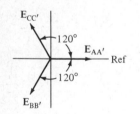

(a) Phasor voltages

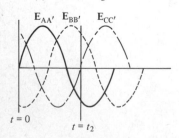

(b) Instantaneous voltages

Figure 25.2   Three-phase voltage waveforms

$$\mathbf{E}_{BB'} = E_p \underline{/-120^\circ} \qquad \text{or} \qquad E_p \underline{/+240^\circ} \qquad (25.2)$$

and

$$\mathbf{E}_{CC'} = E_p \underline{/+120°} \tag{25.3}$$

where E_p = magnitude of phasor expressing the rms value of the ac voltage.

Since the conductors AA', BB', and CC' are independent generators, they can be used as three separate single-phase ac generators. The only difference between the three ac generators is that one is lagging another generator voltage by 120°.

Problem 25.1 By referring to Figure 25.2(b) draw the three phasor voltages at the instant of time $t = t_2$.

Problem 25.2 Suppose that a polyphase system consisted of four generators, WW', XX', YY', and ZZ', separated by 90°. See Figure 25.3. Draw the output voltage.

Three-Phase Generators

Objective 2 Draw the wye (Y) and delta (Δ) connections for a three-phase system. Define and relate the line voltages and phase voltages in a wye or delta connection.

Three-phase systems generated in the first objective are used as a standard in the United States. The reason for their popularity is that they:

1. Provide less pulsating power
2. Provide a more constant torque
3. Provide 173% more power than a single-phase ac system

In this objective you study the two popular ways of interconnecting the three-phase generators: (1) the wye (Y) connection, (2) the delta (Δ) connection.

Wye Connections

If you look at the three voltage generators shown in Figure 25.1, you can see that each is separate. Drawing them separately and identifying them as AA', BB', and CC', we can connect them into the form of a Y (*wye connection*); see Figure 25.4. We did this by connecting all the primed variables (A', B', C') together. This common connection is called *neutral* and designated by the letter N.

The wye connection of the three-phase generators of Figure 25.4 can be redrawn in voltage phasor form as three separate

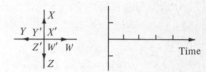

Figure 25.3 Sketch for Problem 25.2

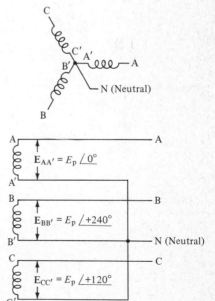

Figure 25.4 Three voltage generators connected in wye configuration

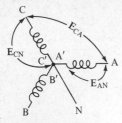

(a) Wye connection

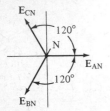

(b) Voltage phasor diagram

Figure 25.5 Wye connection of three-phase generators.

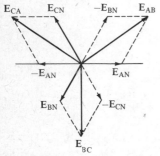

Figure 25.6 Determination of line voltage $\mathbf{E}_{CA}$

voltage generators $\mathbf{E}_{AN}$, $\mathbf{E}_{BN}$, and $\mathbf{E}_{CN}$. [See Figure 25.5(b).] The voltage $\mathbf{E}_{AN}$ represents the rms voltage drop across the winding AA′ and is considered to be a phase voltage. Any voltage from A, B, C to the neutral point N is defined as a *phase voltage*. Since most household appliances in the United States require single-phase 120-V ac, the magnitude E_p of any single-phase voltage will be assumed to be 120 V. Therefore the phase voltages $\mathbf{E}_{AN}$, $\mathbf{E}_{BN}$, and $\mathbf{E}_{CN}$ can be written in vector form as

$$\mathbf{E}_{AN} = E_p \; \underline{/0°} \qquad = 120 \text{ V } \underline{/0°} \tag{25.4}$$

$$\mathbf{E}_{BN} = E_p \; \underline{/+240°} = 120 \text{ V } \underline{/+240°} \tag{25.5}$$

$$\mathbf{E}_{CN} = E_p \; \underline{/+120°} = 120 \text{ V } \underline{/+120°} \tag{25.6}$$

In Figure 25.5(a) or 25.5(b), note that the wye connection is made up of four line terminals A, B, C, and N. Because of the four wires, the wye configuration is often referred to as the *four-wire, three-phase system*.

Suppose that, instead of measuring the voltage from terminal C to neutral N, we wanted to measure the voltage from terminal C to terminal A. What would be the magnitude of $\mathbf{E}_{CA}$? Refer to Figure 25.5(a). In general, the voltages $\mathbf{E}_{AB}$, $\mathbf{E}_{BC}$, or $\mathbf{E}_{CA}$ are called *line voltages* because they are measured across two lines feeding the wye connection. Since the line voltage $\mathbf{E}_{CA}$ is across two phase voltages, $\mathbf{E}_{CN}$ and $\mathbf{E}_{AN}$, we can redraw the voltage phasors $\mathbf{E}_{AN}$ and $\mathbf{E}_{CN}$ of Figure 25.5(a) and obtain a resultant voltage $\mathbf{E}_{CA}$, as shown in Figure 25.6. Since the resultant voltage phasor

$$\mathbf{E}_{CA} = \mathbf{E}_{CN} - \mathbf{E}_{AN} \tag{25.7}$$

then

$$\mathbf{E}_{CN} = 120 \text{ V } \underline{/+120°} = -120 \cos 60° + j120 \sin 60° \tag{25.8}$$

and

$$\mathbf{E}_{AN} = 120 \text{ V } \underline{/0°} = +120 \text{ V} \tag{25.9}$$

then

$$\mathbf{E}_{CA} = (-120 \cos 60° + j120 \sin 60°) - 120 \text{ V} \tag{25.10}$$

$$= -120\left(\frac{1}{2}\right) + j120\left(\frac{\sqrt{3}}{2}\right) - 120 \text{ V} \tag{25.11}$$

$$= -120\left(\frac{3}{2}\right) + j120\left(\frac{\sqrt{3}}{2}\right) = 120\left(-\frac{3}{2} + j\frac{\sqrt{3}}{2}\right) \tag{25.12}$$

$$\mathbf{E}_{CA} = 120(\sqrt{3}) \; \underline{/+150°} \tag{25.13}$$

$$\mathbf{E}_{CA} = 207.8 \text{ V } \underline{/+150°} \tag{25.14}$$

Since the phase voltage $E_{AN} = 120$ V, the line voltage E_{CA} can be related to it by

$$E_{CA} = \sqrt{3}\,E_{AN} \qquad (25.15)$$

where E_{CA} = line voltage (magnitude) and E_{AN} = phase voltage (magnitude).

In general, in a balanced (all phases alike) wye connection, the magnitude of the line voltage is related to the phase voltage by the following relationship:

$$E_{L-L} = \sqrt{3}\,E_{L-N} \qquad (25.16)$$

where E_{L-L} = voltage line-to-line and E_{L-N} = voltage line-to-neutral. Figure 25.6 also shows the remaining line voltages $\mathbf{E}_{AB}$ and $\mathbf{E}_{BC}$.

Problem 25.3 The line-to-line voltage of a wye connection is 4160 V. What is the phase voltage?

Problem 25.4 The phase voltage of a four-wire three-phase system is 550 V. What is the line voltage?

Delta Connections

Let us again refer to the three-phase generators shown in Figure 25.1. We can connect them into a delta (Δ) configuration. This is shown in Figure 25.7. Here terminal A′ is connected to B, B′ connected to C, and C′ is connected to A. We can redraw the sketch of Figure 25.7 in delta (Δ) form and also include the voltage phasors. This is shown in Figure 25.8.

In Figure 25.8(a), notice that in the delta (Δ) configuration there is no neutral wire N, but only three wires A, B, and C. Because of the three wires, the delta configuration is often referred to as the *three-wire three-phase system.*

In the wye (Y) configuration, we were not too concerned about miswiring the three-phase system. Here in the delta (Δ) configuration, you must take *care* in connecting the three-phase generator. Referring to Figure 25.8(b), the vector sum of the generators $\mathbf{E}_{AA'}$ and $\mathbf{E}_{BB'}$ results in a resultant voltage vector $\mathbf{E}_{AB'}$. Here

$$\mathbf{E}_{AB'} = \mathbf{E}_{AA'} + \mathbf{E}_{BB'} \qquad (25.17)$$

where

$$\mathbf{E}_{AA'} = 120 \text{ V } \underline{/0°} = +120 \qquad (25.18)$$

$$\mathbf{E}_{BB'} = 120 \text{ V } \underline{/-120°} = -120\cos 60° - j120\sin 60° \qquad (25.19)$$

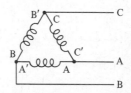

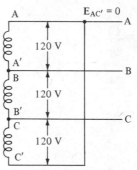

Figure 25.7 Three voltage generators connected in delta configuration.

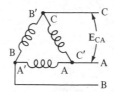

(a) Delta connection

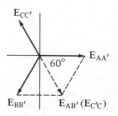

(b) Voltage phasor diagram

Figure 25.8 Delta connection of three-phase generators.

Substituting Equations (25.18) and (25.19) into Equation (25.17), we obtain

$$\mathbf{E}_{AB'} = 120 \text{ V} - 120 \text{ V} \cos 60° - j120 \text{ V} \sin 60° \quad (25.20)$$

$$= 120 \text{ V} - 120 \text{ V}\left(\frac{1}{2}\right) - j120 \text{ V}\left(\frac{\sqrt{3}}{2}\right) \quad (25.21)$$

$$= 120 \text{ V}\left(\frac{1}{2} - j\frac{\sqrt{3}}{2}\right) \quad (25.22)$$

$$\mathbf{E}_{AB'} = 120 \text{ V}\underline{/-60°} \quad (25.23)$$

Since point A is the same as point C′, and point B′ is the same as point C, then

$$\mathbf{E}_{AB'} = \mathbf{E}_{C'C} = 120 \text{ V}\underline{/-60°} \quad (25.24)$$

Adding the resultant voltage vector $\mathbf{E}_{C'C}$ to $\mathbf{E}_{CC'}$ results in zero. This means that the potential difference at A and at C′ is zero. In other words, point A and point C′ can be connected safely.

Let us now look at the situation in which one of the three-phase generators of Figure 25.1 is incorrectly connected. Suppose that the generator BB′ terminals are reversed, as shown in Figure 25.9. Then the resultant vector voltage $\mathbf{E}_{AC'}$ is written as

$$\mathbf{E}_{AC'} = \mathbf{E}_{AA'} + \mathbf{E}_{B'B} + \mathbf{E}_{CC'} \quad (25.25)$$

where $\mathbf{E}_{AA'} = 120 \text{ V}\underline{/0°} = 120$
$\quad \mathbf{E}_{B'B} = 120 \text{ V}\underline{/+60°} = 120 \cos 60° + j120 \sin 60°$
$\quad \mathbf{E}_{CC'} = 120 \text{ V}\underline{/+120°} = -120 \cos 60° + j120 \sin 60°$

$$\therefore \quad \mathbf{E}_{AC'} = 120 + (120 \cos 60° + j120 \sin 60°)$$
$$- 120 \cos 60° + j120 \sin 60° \quad (25.26)$$

$$= 120 + j120\left(\frac{\sqrt{3}}{2}\right) + j120\left(\frac{\sqrt{3}}{2}\right)$$

$$= 120 + j120\sqrt{3} = 120 + j207.84$$

$$\therefore \quad \mathbf{E}_{AC'} = 240 \text{ V}\underline{/+60°} \quad (25.27)$$

Here reversing the phase leads BB′ (Figure 25.9) gives a resultant voltage $E_{AC'} \cong 240$ V. This is summarized in Figure 25.10.

Now what is the relationship between the line voltages and the phase voltages in a delta configuration, shown in Figure 25.7 or 25.8(a). Since the phase voltages are across each separate winding, they are

$$E_{AA'} = E_{BB'} = E_{CC'} = 120 \text{ V} \quad (25.28)$$

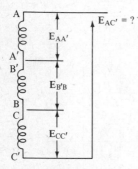

(a) Incorrect connection

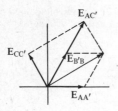

(b) Voltage phasor diagram

Figure 25.9 Three-phase delta generator connected incorrectly.

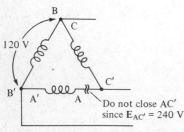

Figure 25.10 Summary of incorrect delta connection.

Since the points $C' = A$, $A' = B$, and $B' = C$, then the line voltages E_{CA}, E_{AB}, and E_{BC} can be easily found to be

$$E_{CA} = E_{CC'} = 120 \text{ V} \qquad (25.29)$$

$$E_{AB} = E_{AA'} = 120 \text{ V} \qquad (25.30)$$

$$E_{BC} = E_{BB'} = 120 \text{ V} \qquad (25.31)$$

Therefore, from Equations (25.29) through (25.31), we can see that in a delta configuration the phase voltages and the line voltages are the same.

Typically in power-distribution systems a Y-connected generator is connected to a step-up transformer whose primary is connected in a Δ connection. The secondary of the transformer is then used to feed high-voltage transmission lines.

The following problem illustrates the actual conditions.

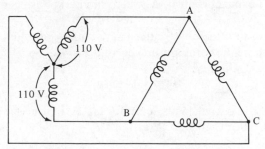

Figure 25.11 Sketch for Problem 25.5

Problem 25.5 The phase voltage of the Y connection of Figure 25.11 is 110 V. Find the phase and line voltage of the delta (Δ) connection.

Power Calculations in a Balanced Wye and/or Delta System

Objective 3 Calculate the power in a balanced wye and/or delta three-phase system.

Before we try to calculate the total power delivered to a Y- or Δ-connected three-phase system, let us first find the relationships between the line currents and the phase currents.

In a Y-connected system, the phase current is the same as the line current, since the current in the line goes into one individual phase.

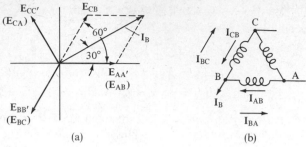

Figure 25.12   Determination of line current in delta system

In a Δ-connected system, the line currents and phase currents are related in a more complicated way. The current direction in Figure 25.12(b) from point A to point B is defined as I_{AB}. Using double-subscript notation, we can express the current from point B to point A as I_{BA}. The current I_{BA} or I_{AB} is defined as the *phase current*. The line current I_B is a vector sum of phase currents I_{AB} and I_{CB}.

Assuming that the phase current I_{AB} and I_{CB} is equal to 10 A, let us compute the resultant line current I_B.

$$I_B = I_{AB} + I_{CB} \qquad (25.32)$$

Assuming resistive loads, current I_{AB} is in the direction of $E_{AA'}$ (E_{AB}). [See Figure 25.12(a).] Thus

$$I_{AB} = 10 \text{ A } \underline{/0^\circ} = 10 \text{ A} \qquad (25.33)$$

Likewise phase current I_{CB} is in phase with voltage phasor E_{CB}. Thus

$$I_{CB} = 10 \text{ A } \underline{/+60^\circ} = 10 \text{ A} \cos 60^\circ + j10 \text{ A} \sin 60^\circ \quad (25.34)$$

Substituting Equations (25.33) and (25.34) into Equation (25.32), we obtain line current

$$I_B = 10 \text{ A} + (10 \text{ A} \cos 60^\circ + j10 \text{ A} \sin 60^\circ) \qquad (25.35)$$

$$I_B = 10 \text{ A} + 10(0.5) \text{ A} + j10 \text{ A} \left(\frac{\sqrt{3}}{2}\right) = 10 \text{ A} \left[\frac{3}{2} + j\frac{\sqrt{3}}{2}\right] \quad (25.36)$$

$$I_B = 10 \, (\sqrt{3}) \text{ A } \underline{/+30^\circ} \qquad (25.37)$$

Thus the *line current* I_B in a Δ configuration is related to the phase current by a factor of $\sqrt{3}$. Moreover the resultant line current I_B is 30° out of phase with one of the phase currents I_{AB}, as shown in Figure 25.12(a).

We can summarize the relationships between the line and phase voltages and line and phase currents of a Y or Δ system in tabular form, as we have done in Table 25.1. Note that, in a wye configuration, the line-to-line voltages are related to the phase voltage by a factor of $\sqrt{3}$, while the phase and line currents are the same. A sketch of the phase and line voltages is included in Figure 25.14 to show the phase relationship between them.

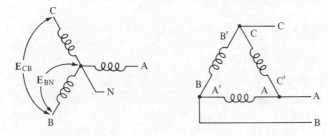

Figure 25.13 Wye and delta connections for Table 25.1

Table 25.1 Voltage and Current Relationships in a Y or Δ Three-Phase System

Wye connection	Delta connection
$E_{L-L} = \sqrt{3}E_{L-N}$	
where E_{L-L} = voltage line to line	
E_{L-N} = voltage line to neutral (phase voltage)	Line voltage = Phase voltage
$E_{AB} = \sqrt{3}E_{AN}$	$E_{BB'} = E_{BC}$
$E_{BC} = \sqrt{3}E_{BN}$	$E_{CC'} = E_{CA}$
$E_{CA} = \sqrt{3}E_{CN}$	$E_{AA'} = E_{BA}$
Line current = phase current	$I_{line} = (\sqrt{3})I_{phase}$
$I_{AB} = I_{AN} = I_{NB}$	$I_{B} = \sqrt{3}I_{BB'} = \sqrt{3}I_{AA'}$
$I_{BC} = I_{BN} = I_{NC}$	$I_{C} = \sqrt{3}I_{CC'} = \sqrt{3}I_{BB'}$
$I_{CA} = I_{CN} = I_{NA}$	$I_{A} = \sqrt{3}I_{AA'} = \sqrt{3}I_{CC'}$

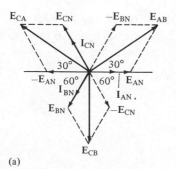

(a)

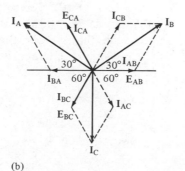

(b)

Figure 25.14 Current relationships in Y and Δ three-phase system

In a delta configuration, the line currents are related to phase currents by a factor of $\sqrt{3}$, while the phase and line-to-line voltages are the same. Figure 25.14(b) shows a similar sketch of the phase relationship of the line and phase currents.

Referring to Table 25.1, we are now ready to calculate the total power delivered to a wye- or delta-connected three-phase system. The following two examples illustrate the procedure we use to calculate the total power in a three-phase system.

Example 25.1 Three resistors of 10 Ω each are connected in delta to a 120-V, three-phase supply. Calculate (a) the current in each resistor, (b) the current in each line, and (c) the total power of the system.

Solution Figure 25.15 shows the circuit diagram for Example 25.1. (a) Since 120 V is across the 10-Ω resistor, we calculate the current through the resistor by Ohm's law to be

$$I_R = \frac{120 \text{ V}}{10 \text{ }\Omega} = 12 \text{ A}$$

(b) Since the system is balanced, the line current I_{line} is related to the resistor current (phase current) by a factor of $\sqrt{3}$.

$$\therefore \quad I_{\text{line}} = \sqrt{3}I_R = (\sqrt{3})(12 \text{ A}) = 20.78 \text{ A}$$

(c) The total power of the system can be calculated in a number of ways. One way is to calculate the phase power first, then add the three independent phases. Since power $P = IE$, then

$$P_{\text{phase}} = I_{\text{phase}}E_{\text{phase}} = I_pE_p$$

But

$$I_{\text{phase}} = I_R = 12 \text{ A}$$
$$E_{\text{phase}} = E_{\text{line}} = 120 \text{ V}$$
$$\therefore P_{\text{phase}} = (12 \text{ A})(120 \text{ V}) = 1440 \text{ W}$$

Since each phase is independent, then for three-phase power, the total power is

$$P_{\text{total}} = 3P_{\text{phase}} = 3(1440) = 4320 \text{ W}$$

In Example 25.1 we calculated the total power by first calculating the phase power, then multiplying each phase by 3 to get 3ϕ power. In general this is accomplished by

$$P_p = I_pE_p \tag{25.38}$$

where P_p = phase power

$\quad\quad I_p$ = phase current

$\quad\quad E_p$ = phase voltage

$$P_T = 3P_p = 3I_pE_p \tag{25.39}$$

where P_T = total power

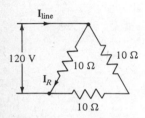

Figure 25.15 Sketch for Example 25.1

We can also calculate the total power of a delta configuration in terms of line voltages E_L and line currents I_L. Relating these to Equation (25.39), we have

$$P_T = 3I_p E_p \qquad (25.40)$$

But

$$E_L = E_p \qquad (25.41)$$

$$I_p = \frac{I_L}{\sqrt{3}} \qquad (25.42)$$

Substituting Equations (25.41) and (25.42) into Equation (25.40), we have

$$\therefore \quad P_T = 3\left(\frac{I_L}{\sqrt{3}}\right)(E_L) \qquad (25.43)$$

or

$$P_T = \sqrt{3}I_L E_L \qquad (25.44)$$

where I_L = line current and E_L = line voltage.

The following example illustrates the technique of calculating total power P_T for a wye-connected three-phase system.

Example 25.2 Three 12-Ω resistors are connected in a wye configuration to a 230-V three-phase supply. Calculate (a) the phase voltage, (b) the phase current, (c) the line current, (d) the total power, (e) the neutral current.

Solution Figure 25.16 shows the circuit diagram for this example. (a) The line-to-line voltage is given as 230 V. Since, for a wye connection,

$$E_p = \frac{E_{line}}{\sqrt{3}} = \frac{230}{1.732} = 132.8 \text{ V}$$

the phase voltage $E_p \cong 133$ V.
(b) The phase current I_p is calculated by Ohm's law to be

$$I_p = \frac{E_p}{R} = \frac{133 \text{ V}}{12 \ \Omega} \cong 11 \text{ A}$$

(c) Since the line current I_L is the same as the phase current I_P,

$$I_L = 11 \text{ A}$$

(d) The total power $P_T = 3P_p$ where P_p = phase power

But

$$P_p = I_p E_p$$
$$= (11A)(133 \text{ V}) = 1463 \text{ W}$$
$$P_T = 3P_p = 4389 \text{ W}$$

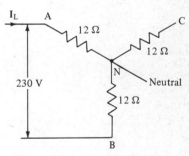

(a) Resistive Y connection

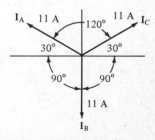

(b) Current phasor diagram

Figure 25.16 Sketch for Example 25.2

(e) The neutral current is the vector sum of all the phase currents I_A, I_B, and I_C. Referring to Figure 25.16(b),

$$I_A = 11 \underline{/+150^\circ} = -11 \cos 30^\circ + j11 \sin 30^\circ$$
$$= -9.53 + j5.5 \text{ A}$$
$$I_B = 11 \underline{/+270^\circ} = -j11 \text{ A}$$
$$I_C = 11 \underline{/+30^\circ} = 11 \cos 30^\circ + j11 \sin 30^\circ = +9.53 + j5.5\text{A}$$

Since $I_N = I_A + I_B + I_C$,

$$I_N = (-9.53 + j5.5) - j11 + (+9.53 + j5.5)$$
$$= (-9.53 + 9.53) + j(5.5 - 11 + 5.5)$$
$$= 0 + j0$$
$$I_N = 0 \text{ A}$$

Thus, in a balanced wye connection, the neutral current I_N is zero.

Problem 25.6   Given that three 5-kW resistive 120-V loads were connected in wye to a 208-V, four-wire, three-phase supply, calculate (a) the line current, and (b) the total power.

Problem 25.7   Three 20-Ω resistors are connected to a Δ-connected generator whose phase voltage is 240 V. Calculate (a) the phase currents, (b) the line currents, and (c) the total power.

Practical Distribution Systems

Objective 4   Briefly describe the complete electrical system from the generating station to the final household outlet.

Up to now you have been studying the generation of three-phase voltages and their configurations in a wye (Y) or delta (Δ) connection. In this objective you will learn how these three-phase circuits are applied and used in the practical sense.

Today generating plants can deliver large amounts of electric power. These plants can generate as much as 500,000 kW of power and are very efficient. They can produce about 1.5 kWh of electrical energy from one pound of coal. One kilowatt-hour can operate an electrical toaster for about one hour. Furthermore, because large amounts of coal or other fuels are burned to rotate the generators, large amounts of heat are generated. Because of the need to keep the generators cool, the generating plants are usually situated near large bodies of water.

All power generators are designed to operate with three-phase voltages and at a maximum possible speed of 60 hertz, or 3600 rpm. The output voltages of generators range from about

10,000 to 22,000 volts. As you can see, these voltages are extremely large! The reason they are so large is that, to obtain a given amount of power $P(= IE)$, raising the voltage means that less current is used. This lower current keeps the power line losses (I^2R) low.

Figure 25.17 shows a sketch of a typical generating station and substation. Here a power generator is producing 18,000 V at the output. To obtain this high voltage, the generator is connected in a Y connection. After it leaves the generator, the voltage is raised to a suitable level for transmission to various substations. For this raising there are three-phase step-up transformers that reduce the output current and still maintain a constant power. This reduced current keeps the line power losses (I^2R) to a minimum. A generally accepted transmission voltage can range from 66,000 to 110,000 volts. The voltage value depends on the distance the power is transmitted. A general rule of thumb for voltage drop in transmission is about 1000 V/mile. In Figure 25.17, the output of the transmission transformer is 110,000 V. Note that the connection is a wye configuration to keep the voltage high.

Once the three-phase high-voltage transmission lines reach their destination, they are fed to a substation. Substations are located close to small cities that have no generating station. Furthermore, the prime purpose of a substation is to step down the high transmission voltage. This is accomplished with the use of transformers. The stepped-down voltages out of the substation may vary from 2300 to 13,800 V, depending on the distance to be traveled. These voltages distributed out of the substation serve each individual customer. The stepped-down voltage out of the substation shown in Figure 25.17 is 4160 V.

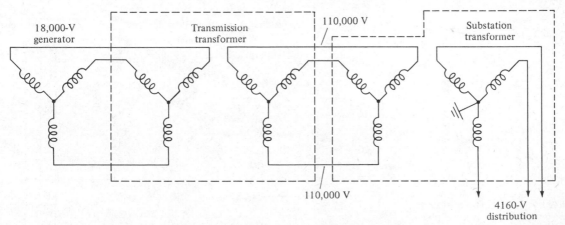

Figure 25.17 Sketch of typical generating station and substation

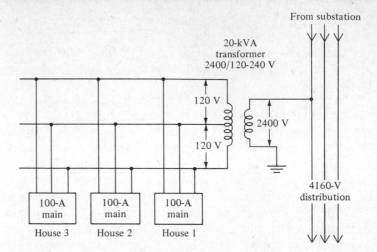

Figure 25.18 Distribution of electric power from substation to typical American household

Let us now take the output voltage of 4160 V of the substation and distribute it to a typical American household, as shown in Figure 25.18. One phase of the wye connection is fed to a 20-kVA transformer and stepped down to 120 V and 240 V. The reason that 240 V is also brought to a house is that such special appliances as electric ovens and electric dryers require 220 to 240 V. This single-phase (three wires) power is fed to the 100-A main switch of each home. Each 20-kVA transformer can handle a group of about four or five homes. Another group of homes could be powered by another 20-kVA transformer; they may be connected to another phase of the 4160-V distribution line. This is done to balance the loads in the three-phase 4160-V distribution line.

Another example of the 4160-V distribution line going to another customer is shown in Figure 25.19. Here a three-phase

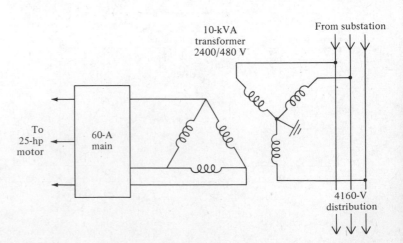

Figure 25.19 Distribution of electric power from substation to three-phase 25-hp customer

25-hp motor is connected to a 10-kVA 2400/4800 V transformer. The primary of the transformer is connected in a Y configuration while the secondary of the transformer is connected in a delta connection. Here the motor was designed to keep the line voltages low (reduce arcing) but the currents high.

Problem 25.8   What is the purpose of a substation?

Problem 25.9   Describe the electrical system from the generating station to the final household outlet.

Power Calculation in an Unbalanced Wye and/or Delta System

Objective 5   Calculate the power in an unbalanced wye and/or delta three-phase system.

In Example 25.2, you calculated the power in a balanced wye connection and found that the neutral current I_N was zero. In this objective, let us calculate the power delivered to an unbalanced load and find out whether the neutral current is still zero. The following example illustrates this.

Example 25.3   One of the three 12-Ω resistors of Example 25.2 is changed to 22 Ω to create an unbalanced load. The wye configuration is still connected to a 230-V three-phase supply. (See Figure 25.20.) Calculate (a) the phase voltage, (b) the phase current, (c) the line current, (d) the total power, and (e) the neutral current I_N.

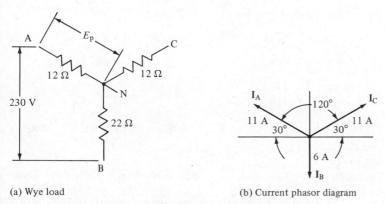

(a) Wye load

(b) Current phasor diagram

Figure 25.20   Unbalanced wye load of Example 25.3

Solution

(a) The phase voltage for a wye connection is again

$$E_p = \frac{E_{line}}{\sqrt{3}} = \frac{230 \text{ V}}{1.732} = 132.8 \text{ V} \approx 133 \text{ V}$$

(b) and (c) In a wye connection, the phase currents and the line currents are the same. We can write the line currents in vector form by referring to Figure 25.20(b). They are:

$$I_A = \frac{E_p}{R_A} = \frac{133 \text{ V} \underline{/+150^\circ}}{12 \text{ }\Omega} = 11 \text{ A } \underline{/+150^\circ}$$

$$I_B = \frac{E_p}{R_B} = \frac{133 \text{ V} \underline{/+270^\circ}}{22 \text{ }\Omega} = 6 \text{ A } \underline{/+270^\circ}$$

$$I_C = \frac{E_p}{R_C} = \frac{133 \text{ V} \underline{/+30^\circ}}{12 \text{ }\Omega} = 11 \text{ A } \underline{/+30^\circ}$$

(d) We cannot calculate the total power P_T of the wye-connected system by Equation (25.39) because the phase currents are different. The power must be calculated on a per-phase basis and then added.

$$\text{Phase power } P_A = I_A E_p = (11 \text{ A})(133 \text{ V}) \approx 1463 \text{ W}$$
$$\text{Phase power } P_B = I_B E_p = (6 \text{ A})(133 \text{ V}) \approx 798 \text{ W}$$
$$\text{Phase power } P_C = I_C E_p = (11 \text{ A})(133 \text{ V}) \approx 1463 \text{ W}$$

Thus the total power P_T is:

$$P_T = P_A + P_B + P_C = 1463 + 798 + 1463 \text{ W} = 3724 \text{ W}$$

(e) The neutral current I_N is the vector sum of the three phase currents I_A, I_B, and I_C. Referring to Figure 25.20(b), we have

$$I_A = 11 \text{ A } \underline{/+150^\circ} = -9.53 + j5.5 \text{ A}$$
$$I_B = 6 \text{ A } \underline{/+270^\circ} = -j6 \text{ A}$$
$$I_C = 11 \text{ A } \underline{/+30^\circ} = +9.53 + j5.5 \text{ A}$$

Since neutral current I_N is

$$I_N = I_A + I_B + I_C = (-9.53 + j5.5) + (-j6) + (+9.53 + j5.5)$$
$$I_N = +j5.5 - j6 + j5.5 = +j5 \text{ A}$$

Thus, in an unbalanced wye connection, the neutral current is *not* zero.

Example 25.4   Calculate the line currents and the total power delivered in the unbalanced delta-connected resistor loads in Figure 25.21. The line voltage is 120 V.

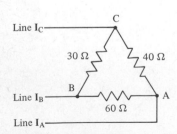

Figure 25.21   Unbalanced delta-connected loads of Example 25.4

Solution Refer to the delta current phasor diagram of Figure 25.14(b) and apply Ohm's law. The phase currents are:

$$\mathbf{I}_{CA} = \frac{\mathbf{E}_{CA}}{R} = \frac{120 \text{ V } \underline{/+120°}}{40 \text{ }\Omega} = 3 \text{ A } \underline{/+120°} = -1.5 \text{ A} + j2.6 \text{ A}$$

$$\mathbf{I}_{AB} = \frac{\mathbf{E}_{AB}}{R} = \frac{120 \text{ V } \underline{/0°}}{60 \text{ }\Omega} = 2 \text{ A } \underline{/0°} = +2 \text{ A}$$

$$\mathbf{I}_{BC} = \frac{\mathbf{E}_{BC}}{R} = \frac{120 \text{ V } \underline{/-120°}}{30 \text{ }\Omega} = 4 \text{ A } \underline{/-120°} = -2 \text{ A} - j3.46 \text{ A}$$

Figure 25.22 shows these phase currents.

The line currents $\mathbf{I}_A$, $\mathbf{I}_B$, and $\mathbf{I}_C$ are vector sums of the phase currents.

$$\mathbf{I}_A = \mathbf{I}_{BA} + \mathbf{I}_{CA} = -\mathbf{I}_{AB} + \mathbf{I}_{CA}$$
$$\mathbf{I}_A = (-2 \text{ A}) + (-1.5 \text{ A} + j2.6 \text{ A}) = -3.5 \text{ A} + j2.6 \text{ A}$$
$$= 4.36 \text{ A } \underline{/+143.4°}$$
$$\mathbf{I}_B = \mathbf{I}_{AB} + \mathbf{I}_{CB} = \mathbf{I}_{AB} - \mathbf{I}_{BC} = 2 \text{ A} - (-2 \text{ A} - j3.46 \text{ A})$$
$$= 4 \text{ A} + j3.46 \text{ A} = 5.29 \text{ A } \underline{/+40.9°}$$
$$\mathbf{I}_C = \mathbf{I}_{BC} + \mathbf{I}_{AC} = \mathbf{I}_{BC} - \mathbf{I}_{CA}$$
$$= (-2 \text{ A} - j3.46 \text{ A}) - (-1.5 \text{ A} + j2.6 \text{ A}) = -0.5 \text{ A} - j6.06 \text{ A}$$
$$= 6.08 \text{ A } \underline{/+265.3°}$$

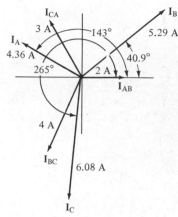

Figure 25.22 Sketch of line currents for solution of Example 25.4

We cannot calculate the power of the unbalanced delta-connected system using the three-phase equation (25.44) because the line currents are all different and are not 120° apart. (See Figure 25.22.) The power must be calculated on a per-phase basis and then added.

Power in Phase AC $= I_{CA}E = (3 \text{ A})(120 \text{ V}) = 360 \text{ W}$
Power in Phase AB $= I_{AB}E = (2 \text{ A})(120 \text{ V}) = 240 \text{ W}$
Power in Phase BC $= I_{BC}E = (4 \text{ A})(120 \text{ V}) = 480 \text{ W}$

Thus the total power P_T is

$$P_T = 360 + 240 + 480 = 1080 \text{ W}$$

Problem 25.10 The wye configuration in Figure 25.23 is connected to a 230-V three-phase supply. Calculate (a) the phase voltage, (b) phase current, (c) line currents, (d) total power, and (e) neutral current I_N.

Problem 25.11 The delta configuration in Figure 25.24 is connected to a line voltage of 180 V. Calculate the line currents and the total power delivered to the loads.

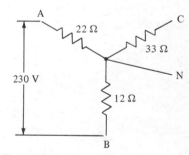

Figure 25.23 Circuit for Problem 25.10

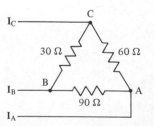

Figure 25.24 Circuit for Problem 25.11

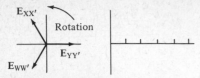

Figure 25.25   Sketch for Test Problem 1

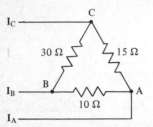

Figure 25.26   Delta connection for Test Problem 6

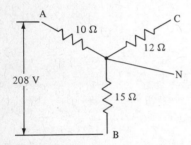

Figure 25.27   Sketch for Test Problem 7

Test

1. Figure 25.25 shows a three-phase generator at time $t = 0$. Draw the three-phase voltages generated $\mathbf{E}_{XX'}$ and $\mathbf{E}_{WW'}$ with respect to $\mathbf{E}_{YY'}$.

2. The line voltage of a four-wire, three-phase system is 4160 V. Find the phase voltage.

3. The phase voltage of a delta connection is 220 V. What is the line voltage?

4. Three 40-Ω resistors are connected in a delta arrangement to a 120-V three-phase supply. Calculate: (a) the current in each resistor, (b) the current in each line, (c) the total power of the system.

5. Three 20-Ω resistors are connected in a wye configuration to a 208-V three-phase supply. Calculate: (a) the phase voltage, (b) the phase current, (c) the line current, (d) the total power P_T, (e) the neutral current $\mathbf{I}_N$.

6. Three resistors of 10 Ω, 15 Ω, and 30 Ω are each connected in a delta configuration to a 300-V three-phase supply (Figure 25.26). Calculate the line currents $\mathbf{I}_A$, $\mathbf{I}_B$, and $\mathbf{I}_C$ and the total power of the system.

7. Three resistors of 10 Ω, 12 Ω, and 15 Ω are connected in a wye configuration as shown in Figure 25.27. The line-to-line voltage applied is 208 V. Calculate the three line currents, the neutral current $\mathbf{I}_N$, and the total power of the system.

Nonsinusoidal Unit 26
Circuits

In the earlier units, you learned about dc and ac sine waves. Recall that the ac sine wave is generated by rotating a loop of wire in a uniform magnetic field. Also remember that the ac sine wave is present in many forms of electric circuits, such as ac machines, or distributed ac electric power. In this unit, you will be introduced to other types of waveforms that are often observed at the output of wave signal generators or other communications equipment. For instance, the output of a radio or of a microphone is *not* a pure sine wave, but a complicated nonsinusoidal waveform. In this unit, we shall study these nonsinusoidal waveforms.

Objectives

After completing all the work associated with this unit, you should be able to:

1. Define an even or odd periodic function.
2. Define the term harmonic. Be able to identify the first, second, or nth harmonic of a wave and calculate the amplitude of each harmonic.
3. Interpret the Fourier-series representation of some elementary even and/or odd nonsinusoidal periodic functions. Be able to

513

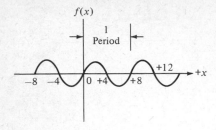

(a) ac sine wave

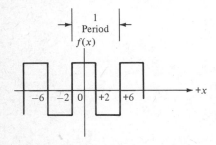

(b) Square wave

Figure 26.1 Examples of periodic functions

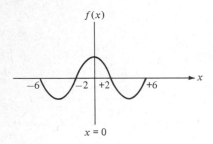

(a) Cosine wave

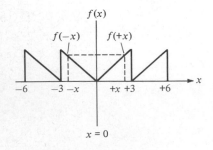

(b) Ramp function

Figure 26.2 Examples of even waveforms

evaluate the Fourier series at different instants of time and construct (draw) the complex function (waveform).

4. Calculate the effective voltage and power of a periodic nonsinusoidal wave.

Periodic Functions

Objective 1 Define an even or odd periodic function.

In this objective, we are interested only in waveforms that repeat over and over. These types of waveforms (functions) are said to be *periodic*. Figure 26.1 shows some examples of periodic functions.

Figure 26.1(a) shows a standard ac sine wave. The sine wave starts at point 0 and repeats its cycle at the point marked $+8$. Thus the *period* of the sine wave is 8 units. Figure 26.1(b) shows a square wave (nonsinusoidal wave). The square wave can be considered to start at point -2 and repeats again at point $+6$. Thus the period of the square wave is also 8 units.

In the study of nonsinusoidal circuits, we often examine two types of periodic waveforms—even waveforms and odd waveforms. These are defined in the following two paragraphs.

Even Waveforms

An *even waveform* can be defined as a waveform that has a mirror image about a certain given vertical axis. Figure 26.2 shows some examples of even functions. Figure 26.2(a) represents a cosine wave. Here the wave is symmetric around the line $x = 0$. That is, if one folds the waveform along the vertical axis $x = 0$, the waveform at point $+2$ falls on point -2. Point $+6$ lines up with point -6. Likewise Figures 26.2(b) and 26.1(b) represent even nonsinusoidal periodic waveforms. Again, folding the waveforms around line $x = 0$ results in all points lining up with one another. Mathematically, an even function can be expressed as

$$f(+x) = f(-x) \qquad \text{even function} \qquad (26.1)$$

Here Equation (26.1) states that for any positive value $+x$ the value of the function is the same as the function at the same negative value of x. (See Figure 26.2(b).) For instance, the value of the function at point $x = +6$ is the same as that at point $x = -6$, or mathematically

$$f(+6) = f(-6) \qquad \text{for even function}$$

Odd Waveforms

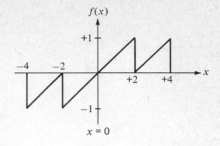

(a) Ramp function

An *odd waveform* can be defined as a waveform that satisfies the following two conditions:

Condition 1. $\quad f(x) = 0 \quad$ for $x = 0 \quad$ (26.2)

Condition 2. $\quad f(+x) = -f(-x) \quad$ (26.3)

Equation (26.2) states that when $x = 0$, the function must also be zero, and Equation (26.3) states that the value of the function at a positive point $(+x)$ is the same (except for sign) as at the negative point $(-x)$. Note that for the even waveform, condition 1 [Equation (26.2)] was not a requirement; the even waveform could be equal to zero at $x = 0$, or it could have some other value.

Figure 26.3 shows examples of odd functions. At $x = 0$ both functions have a value of zero, and at any other point the value of the function is the same (except for sign) as at the same negative point. In Figure 26.3(b), at the point $x = +4$, the function is $f(+4) = +3$, while at the point of $x = -4$, the function $f(-4) = -3$ or $-f(-4) = +3$. Mathematically this can be summarized as

At

$$x = +4 \qquad f(+4) = +3$$
$$x = -4 \qquad f(-4) = -3 \text{ or } -f(-4) = +3$$
$$\therefore \quad f(+4) = -f(-4)$$

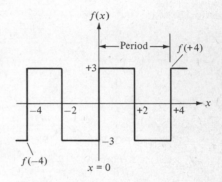

(b) Square wave

Figure 26.3 Examples of odd waveforms

Now that you know the definition of even and odd periodic nonsinusoidal waveforms, identify the waveforms given in the following problems.

Problem 26.1 Identify the waveforms in Figure 26.4 and specify their period.

(a) Type _____ Period _____

(b) Type _____ Period _____

(c) Type _____ Period _____

Harmonics

Objective 2 Define the term harmonic. Be able to identify the first, second, or nth harmonic of a wave and calculate the amplitude of each harmonic.

Baron Jean Baptiste Fourier in 1826 mathematically analyzed a periodic nonsinusoidal wave described in Objective 1. He

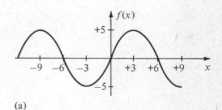

(a)

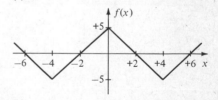

(b)

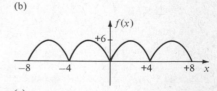

(c)

Figure 26.4 Sketches for Problem 26.1

found that any such wave can be constructed out of an infinite series of harmonically related pure sine waves. He stated that any such periodic wave can be mathematically expressed as:

$$f(x) = A_0 + A_1 \cos x + A_2 \cos 2x + A_3 \cos 3x + \cdots$$
$$+ A_n \cos nx + B_1 \sin x + B_2 \sin 2x + \cdots$$
$$+ B_n \sin nx + B_{n+1} \sin (n + 1)x + \cdots \qquad (26.4)$$

where $f(x)$ = function describing any periodic wave

A_0 = constant value (dc value)

A_1 = amplitude of cos x term

A_n = amplitude of cos nx term

B_1 = amplitude of sin x term

B_n = amplitude of sin nx term

An *infinite series* is a never-ending combination of terms. For example, Equation (26.4) is an infinite series of cos x and sin x terms of different frequencies and different amplitudes. Furthermore, the value for n is a very, very large number. Thus Fourier stated that any periodic wave described in Objective 1 can be described by the infinite series of Equation (26.4). In this objective you will be introduced to the term *harmonic*, and you will learn how to describe the different harmonically related pure sine waves used in Equation (26.4). In a later objective, you will learn how to apply Equation (26.4) to represent some elementary nonsinusoidal functions described earlier.

Definition

Figure 26.5 is a sketch of four harmonically related sine waves. Part (a) represents a pure sine wave of amplitude 1.0 and a fundamental frequency of 10 Hz. The waveform of Figure 26.5(b) has twice the fundamental frequency ($2f$, or 20 Hz) of the waveform of Part (a). A frequency twice the fundamental frequency is defined as a *second harmonic*. The amplitude associated with this second harmonic is 0.6. Likewise the waveform of Figure 26.5(c) is a sine wave of amplitude 0.7 and a frequency three times the fundamental frequency. This frequency is defined as the *third harmonic*. Figure 26.5(d) represents a sine wave of amplitude of 0.8 and a frequency four times the fundamental frequency. This frequency is defined as the *fourth harmonic*. In general, the nth harmonic is related to the fundamental frequency by the following equation:

$$n\text{th harmonic} = n \times (\text{fundamental frequency}) \qquad (26.5)$$

where n = an integer number.

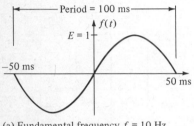

(a) Fundamental frequency f = 10 Hz
$f(t) = E \sin 2\pi ft = 1 \sin 62.8t$

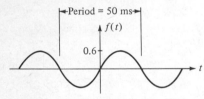

(b) Second harmonic = 2 × fund. = 20 Hz
$f(t) = 0.6 \sin 125.6t$

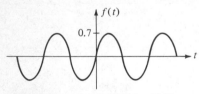

(c) Third harmonic = 3 × fund. = 30 Hz
$f(t) = 0.7 \sin 188.4t$

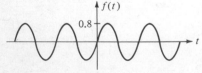

(d) Fourth harmonic = 4 × fund. = 40 Hz
$f(t) = 0.8 \sin 251.2t$

Figure 26.5 Harmonically related sine waves

As you see by Equation (26.5), the fundamental frequency and the first harmonic are the same. Referring back to Figure 26.5, we can write a mathematical expression for the fundamental (first harmonic), second harmonic, third harmonic, and fourth harmonic waves. Recall that a sine wave can be written as:

$$f(t) = E \sin(2\pi ft) = E \sin(6.28ft) \qquad (26.6)$$

where E = amplitude of the sine wave

$2\pi = 2(3.14) = 6.28$

$\quad f$ = frequency in hertz

$\quad t$ = time in seconds

Applying Equation (26.6) to the harmonically related sine waves of Figure 26.5, we obtain, for the fundamental frequency of 10 Hz,

First harmonic
$$f_1(t) = E \sin 2\pi ft = 1.0 \sin(6.28)(10t) = 1.0 \sin 62.8t \qquad (26.7)$$

Second harmonic
$$f_2(t) = 0.6 \sin(6.28)(20t) = 0.6 \sin(125.6t) \qquad (26.8)$$

Third harmonic
$$f_3(t) = 0.7 \sin(6.28)(30t) = 0.7 \sin(188.4t) \qquad (26.9)$$

Fourth harmonic
$$f_4(t) = 0.8 \sin(6.28)(40t) = 0.8 \sin(251.2t) \qquad (26.10)$$

If we replace the fundamental radian frequency ($62.8t$) by x, then we can rewrite Equation (26.7) as

$$f_1(t) = 1.0 \sin 62.8t = 1.0 \sin x \qquad (26.11)$$

Likewise we can rewrite Equations (26.8), (26.9), and (26.10) as:

$$f_2(t) = 0.6 \sin 125.6t = 0.6 \sin 2x \qquad (26.12)$$
$$f_3(t) = 0.7 \sin 188.4t = 0.7 \sin 3x \qquad (26.13)$$
$$f_4(t) = 0.8 \sin 251.2t = 0.8 \sin 4x \qquad (26.14)$$

Looking at the frequency components of the last four equations, you can observe that all of them are related by an integer number. For instance, the frequency of function f_4 is four times ($4x$) the fundamental frequency of function $f_1(t)$.

In the next objective, we'll be looking at complex periodic waveforms that may be made up of many harmonically related sine waves. For instance, a complex waveform may be of the form:

$$f(x) = f_1 + f_2 + f_3 + f_4 \qquad (26.15)$$
$$f(x) = \sin x + 0.6 \sin 2x + 0.7 \sin 3x + 0.8 \sin 4x \qquad (26.16)$$

Equation (26.16) represents a wave made up of a fundamental frequency x and three other harmonics, $2x$, $3x$, and $4x$.

We shall now show you how to take a complex waveform and find its fundamental frequency and other harmonics. This is illustrated in the following example.

Examples

Example 26.1   The Fourier series

$$f(t) = \frac{1}{\pi}\left(1 + \frac{\pi}{2}\cos 628t + \frac{2}{3}\cos 1256t - \frac{2}{15}\cos 2512t\right)$$

represents a complex periodic waveform. Find the amplitudes of the fundamental frequency and all its harmonics.

Solution   The first term, $(1/\pi)(1)$, is the dc component of the waveform, and has 0 frequency. The lowest frequency in the ac component is present in the wave $\cos 628t$. Since

$$\cos \omega_1 t = \cos 628t \qquad \text{and} \qquad \omega_1 = 2\pi f_1$$

then

$$f_1 = \frac{\omega_1}{2\pi} = \frac{628}{6.28} = 100 \text{ Hz}$$

The corresponding amplitude A_1 for the fundamental frequency is

$$A_1 = \frac{1}{\pi}\left(\frac{\pi}{2}\right) = \frac{1}{2}$$

The next-higher frequency term is $\cos 1256t$. Its frequency is twice the fundamental frequency $[2(628) = 1256]$. Thus the second harmonic is

$$f_2 = \frac{\omega_2}{2\pi} = \frac{1256}{6.28} = 200 \text{ Hz}$$

The corresponding amplitude A_2 for the second harmonic is

$$A_2 = \frac{1}{\pi}\left(\frac{2}{3}\right) = \frac{2}{3\pi}$$

The final frequency term is $\cos 2512t$. Its frequency is four times the fundamental $(2512/f_1 = 2512/628 = 4)$. Thus the fourth harmonic is

$$f_4 = \frac{\omega_4}{2\pi} = \frac{2512}{6.28} = 400 \text{ Hz}$$

The corresponding amplitude A_4 for the fourth harmonic is

$$A_4 = -\frac{1}{\pi}\left(\frac{2}{15}\right) = -\frac{2}{15\pi}$$

Note that this complex wave does not contain a third or a fifth harmonic.

Example 26.2 The Fourier series

$$f(t) = \frac{4V}{\pi}(\sin 62.8t + \tfrac{1}{3}\sin 188.4t + \tfrac{1}{5}\sin 314t)$$

where $V = 1.0$ V, represents a complex wave. Determine whether the complex wave is an even or an odd function. Also find the amplitudes of the fundamental frequency and all its harmonics.

Solution

(a) The first term of the Fourier series is a sine curve. If we draw a sine curve, you can see that the complex wave is an odd waveform. Since the other additional terms are also sine terms, the complex function is an odd function.

(b) The fundamental frequency f_1 is

$$f_1 = \frac{\omega_1}{2\pi} = \frac{62.8}{6.28} = 10 \text{ Hz}$$

and its amplitude B_1 is

$$B_1 = \frac{4V}{\pi} = \frac{4(1)}{\pi} = \frac{4}{\pi} = 1.27$$

Since $188.4/62.8 = 3$, then

$$f_3 = \frac{\omega_3}{2\pi} = \frac{188.4}{6.28} = 30 \text{ Hz} \qquad \text{(third harmonic)}$$

and its amplitude B_3 is

$$B_3 = \frac{4V}{\pi}\left(\frac{1}{3}\right) = \frac{4(1)}{\pi(3)} = \frac{4}{3\pi} = 0.425$$

Similarly, for the next frequency ratio of $314/62.8 = 5$,

$$f_5 = \frac{\omega_5}{2\pi} = \frac{314}{6.28} = 50 \text{ Hz (fifth harmonic)}$$

and its amplitude B_5 is

$$B_5 = \frac{4V}{\pi}\left(\frac{1}{5}\right) = \frac{4(1)}{\pi(5)} = \frac{4}{5\pi} = 0.255$$

A frequency spectrum analyzer is an instrument that displays the amplitudes of all the harmonics of a complex wave. If we

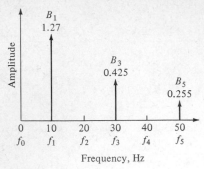

Figure 26.6 Amplitudes of harmonics displayed on frequency spectrum analyzer

injected the complex wave of Example 26.2 into a spectrum analyzer, the results would appear as shown in Figure 26.6.

Problem 26.2 Given the Fourier series:

$$f(t) = \frac{V}{2} - \frac{V}{\pi}(\sin 6.28t + \tfrac{1}{2}\sin 12.56t + \tfrac{1}{3}\sin 18.84t +$$
$$\tfrac{1}{4}\sin 25.12t + \tfrac{1}{5}\sin 31.4t)$$

where $V = 3$ V, find the amplitudes of the fundamental frequency and of all its harmonics.

Problem 26.3 The Fourier series

$$f(t) = \frac{8V}{\pi^2}\left(\sin \omega_0 t - \frac{1}{3^2}\sin 3\omega_0 t + \frac{1}{5^2}\sin 5\omega_0 t\right)$$

represents a complex wave. Determine whether the wave is even or odd.

Problem 26.4 The Fourier series

$$e(t) = 90\cos 754t - 30\cos 2262t + 18\cos 3770t$$

represents a complex wave. Determine whether the above wave is even or odd. Also find the amplitudes of the fundamental frequency and of all its harmonics. Plot the amplitudes as displayed on a frequency spectrum analyzer, as shown in Example 26.2.

Fourier Series Representation

Objective 3 Interpret the Fourier series representation of some common elementary even and/or odd nonsinusoidal periodic functions. Be able to evaluate the Fourier series at different instants of time and construct (draw) the complex function (waveform).

It is beyond the scope of this book to mathematically determine the infinite series of harmonic functions that represent a given complex nonsinusoidal periodic wave. Thus the purpose of this objective is to take a given infinite series representation and construct, point by point, the complex wave. By adding on additional harmonics, one can quickly conclude that Baron Fourier was correct in stating that all periodic waveforms can be expressed by an infinite series of cos x and sin x terms of different frequencies.

Before actually trying an evaluation of a specific function, let us first try to evaluate a single sine or cosine term at a specific time. This is illustrated in the following example.

Example 26.3 Evaluate the following functions at a time $t = 40$ ms. Let $f = 10$ Hz. (a) Function $f_1 = \sin \omega t$, (b) function $f_2 = \sin 5\omega t$, (c) function $f_3 = \cos 4\omega t$.

Solution

(a) Function f_1 at time $t = 40$ ms and $f = 10$ Hz:

$$f_1 = \sin \omega t = \sin 2\pi ft = \sin (6.28)(10)(40 \times 10^{-3}) = \sin 2.512 \text{ rad}$$

Converting radians to degrees, using the relation

$$\text{Degrees} = \frac{180°}{\pi} \times \text{radians} = \frac{180}{3.14}(2.512) = 144°$$

Thus

$$f_1 = \sin 2.512 \text{ rad} = \sin 144° = \sin 36° = 0.5878$$

(b) Function f_2 at time $t = 40$ ms and $f = 10$ Hz:

$$f_2 = \sin 5\omega t = \sin 5(6.28)(10)(40 \times 10^{-3})$$
$$= \sin 12.56 \text{ rad} = \sin 720° = \sin 360° = \sin 0°$$
$$= 0$$

(c) Function f_3 at $t = 40$ ms:

$$f_3 = \cos 4\omega t = \cos (4)(6.28)(10)(40 \times 10^{-3})$$
$$= \cos 10.048 \text{ rad} = \cos 576° = \cos 216° = -\cos 36°$$
$$= -0.8090$$

Example 26.3 demonstrated how to calculate a given function at a certain instant of time. Table 26.1 tabulates the evaluation of a 10-Hz sine function at ten different times in one cycle.

Table 26.1 Evaluation of Sine Function at Different Times

$$f = 10 \text{ Hz} \qquad \omega = 2\pi f$$

t, ms	$\sin \omega t$ $\sin 62.8t$	$\sin 2\omega t$ $\sin 125.6t$	$\sin 3\omega t$ $\sin 188.4t$	$\sin 4\omega t$ $\sin 251.2t$	$\sin 5\omega t$ $\sin 314t$	$\sin 6\omega t$ $\sin 376.8t$	$\sin 7\omega t$ $\sin 439.6t$
10	+0.5875	+0.9509	+0.9514	+0.5888	+0.0016	−0.5862	−0.9504
20	+0.9509	+0.5888	−0.5862	−0.9518	−0.0032	+0.9499	+0.5914
30	+0.9514	−0.5862	−0.5901	+0.9499	+0.0048	−0.9528	+0.5824
40	+0.5878	−0.9518	+0.9499	−0.5837	−0.0000	+0.5940	−0.9538
50	+0.0016	−0.0032	+0.0048	−0.0064	+0.0080	−0.0096	+0.0111
60	−0.5862	+0.9499	−0.9528	+0.5940	−0.0096	−0.5785	+0.9468
70	−0.9504	+0.5914	+0.5824	−0.9538	+0.0111	+0.9468	−0.6003
80	−0.9518	−0.5837	+0.5940	+0.9479	−0.0127	−0.9557	−0.5733
90	−0.5901	−0.9529	−0.9484	−0.5785	+0.0143	+0.6016	+0.9571
100	−0.0032	−0.0064	−0.0096	−0.0127	+0.0159	−0.0191	−0.0223

Table 26.2 Evaluation of Cosine Function at Different Times

$$f = 10 \text{ Hz} \qquad \omega = 2\pi f$$

t, ms	$\cos \omega t$ $\cos 62.8t$	$\cos 2\omega t$ $\cos 125.6t$	$\cos 3\omega t$ $\cos 188.4t$	$\cos 4\omega t$ $\cos 251.2t$	$\cos 5\omega t$ $\cos 314t$	$\cos 6\omega t$ $\cos 376.8t$	$\cos 7\omega t$ $\cos 439.6t$
10	+0.8092	+0.3096	−0.3081	−0.8083	−1	−0.8101	−0.3111
20	+0.3096	−0.8082	−0.8101	+0.3066	+1	+0.3127	−0.8064
30	−0.3081	−0.8101	+0.8073	+0.3127	−1	+0.3035	+0.8129
40	−0.8082	+0.3066	+0.3126	−0.8090	+1	−0.8045	+0.3005
50	−0.9999	+0.9999	−0.9999	+1.0000	−1	+1	−0.9999
60	−0.8101	+0.3127	+0.3035	−0.8045	+1	−0.8157	+0.3217
70	−0.3111	−0.8064	+0.8129	+0.3005	−1	+0.3217	+0.7997
80	+0.3066	−0.8120	−0.8045	+0.3187	+1	+0.2944	−0.8194
90	+0.8073	+0.3036	−0.3172	−0.8157	−1	−0.7988	−0.2899
100	+0.9999	1.0000	+0.9999	+0.9999	+1	+0.9998	+0.9998

Similarly, Table 26.2 represents the evaluation of a 10-Hz cosine function at ten different time intervals of one cycle. These tables also include the evaluation of other harmonic functions at precisely the same time. These evaluated tables are useful in calculating some of the functions presented in the following examples.

Example 26.4 The Fourier series representation of a square wave shown in Figure 26.7 is

$$v(t) = \frac{4V}{\pi}\left(\cos \omega_0 t - \tfrac{1}{3}\cos 3\omega_0 t + \tfrac{1}{5}\cos 5\omega_0 t - \cdots\right)$$

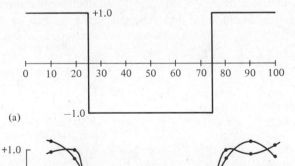

(a)

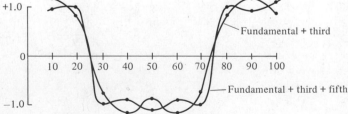

$$v(t) = \frac{4V}{\pi}\left[\cos \omega_0 t - \tfrac{1}{3}\cos 3\omega_0 t + \tfrac{1}{5}\cos 5\omega_0 t\right]$$

V = amplitude = 1.0

(b)

Figure 26.7 Sketch of square wave of Example 26.4

For $V = 1.0$ V and frequency $f = 10$ Hz, evaluate two terms of the series (fundamental and third harmonic). Draw this waveform. Reevaluate the three-term series (fundamental, third, and fifth harmonic) and compare it with the two-term series evaluated previously.

Solution   Let us first evaluate the two-term series for $V = 1.0$ V and $t = 10$ ms. Referring to Table 26.2 for time $t = 10$ ms, we find that the function is

$$v(t) = \frac{4(1)}{\pi} \left[\cos \omega_0 t - \tfrac{1}{3} \cos 3\omega_0 t \right]$$

$$= \frac{4}{\pi} [0.8092 - (0.333)(-0.3081)] = \frac{4}{\pi} (0.9118) = 1.1609$$

The remaining calculations are presented in tabular form in the first column of Table 26.3.

A sample calculation for the three-term series is as follows. Here, for time $t = 10$ ms,

$$v(t) = \frac{4(1)}{\pi} (\cos \omega_0 t - \tfrac{1}{3} \cos 3\omega_0 t + \tfrac{1}{5} \cos 5\omega_0 t)$$

$$= \frac{4}{\pi} [0.8092 - 0.333(-0.3081) + 0.2(-1)]$$

$$= \frac{4}{\pi} (0.7118) = 0.9063$$

A sketch of the two-term and the three-term series is shown in Figure 26.7. In comparing these sketches, you can see that, as we

Table 26.3

t, ms	Two-term series $\frac{4V}{\pi} (\cos \omega_0 t - \tfrac{1}{3} \cos 3\omega_0 t)$	Three-term series $\frac{4V}{\pi} (\cos \omega_0 t - \tfrac{1}{3} \cos 3\omega_0 t + \tfrac{1}{5} \cos 5\omega_0 t)$
10	1.1609	+0.9063
20	0.7377	+0.9923
30	−0.7346	−0.9892
40	−1.1616	−0.9070
50	−0.8492	−1.1038
60	−1.1601	−0.9055
70	−0.7408	−0.9954
80	+0.7315	+0.9861
90	+1.1624	+0.9078
100	+0.8492	+1.1038

add additional harmonics, the infinite series starts to approach the square-wave function.

Example 26.5   The Fourier series of a triangular wave is given as

$$v(t) = \frac{8V}{\pi^2}\left(\sin \omega_0 t - \frac{1}{3^2}\sin 3\omega_0 t + \frac{1}{5^2}\sin 5\omega_0 t - \cdots\right)$$

For $V = 1.0$ V and frequency $f = 10$ Hz, evaluate two terms of the series (fundamental + third harmonic) and draw them on the sketch of Figure 26.8(b). Reevaluate the three-term series (fundamental + third and fifth harmonics) and compare it with the two-term series evaluated previously.

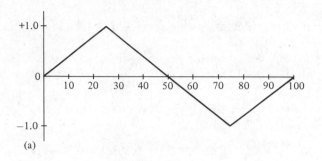

(a)

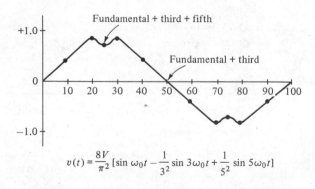

$$v(t) = \frac{8V}{\pi^2}[\sin \omega_0 t - \frac{1}{3^2}\sin 3\omega_0 t + \frac{1}{5^2}\sin 5\omega_0 t]$$

(b) $V =$ amplitude $= 1.0$

Figure 26.8   Sketch of triangular wave of Example 26.5

Solution   Referring to Table 26.1 for $V = 1.0$ and a time $t = 10$ ms, we can evaluate the two-term series as:

$$v(t) = \frac{8(1)}{\pi^2}(\sin \omega_0 t - \tfrac{1}{9}\sin 3\omega_0 t) = \frac{8}{\pi^2}[0.5875 - \tfrac{1}{9}(0.9514)]$$

$$= \frac{8}{\pi^2}(0.4818) = 0.3905 \qquad \text{at } t = 10 \text{ ms}$$

Table 26.4 Evaluation of Triangular Wave of Example 26.5

t, ms	$\frac{8}{\pi^2}(\sin \omega_0 t - \frac{1}{9}\sin 3\omega_0 t)$	$\frac{8}{\pi^2}(\sin \omega_0 t - \frac{1}{9}\sin 3\omega_0 t + \frac{1}{25}\sin 5\omega_0 t)$
10	+0.3905	+0.3906
20	+0.8236	+0.8235
25	+0.7205	+0.7531
30	+0.8243	+0.8245
40	+0.3917	+0.3915
50	+0.0009	+0.0012
60	−0.3893	−0.3896
70	−0.8228	−0.8224
80	−0.8250	−0.8254
90	−0.3929	−0.3924
100	−0.0017	−0.0022

A sample calculation of the three-term series is

$$v(t) = \frac{8(1)}{\pi^2}\left(\sin \omega_0 t - \tfrac{1}{9}\sin 3\omega_0 t + \tfrac{1}{25}\sin 5\omega_0 t\right)$$

$$= \frac{8}{\pi^2}\left[0.5875 - \tfrac{1}{9}(0.9514) + \tfrac{1}{25}(0.0016)\right]$$

$$= 0.3906 \qquad \text{at } t = 10 \text{ ms}$$

Table 26.4 shows the remaining calculations in tabular form.
A sketch of the two-term series and the three-term series is shown in Figure 26.8(b). Again, as more harmonics are added, the infinite series starts to approach the triangular-wave function.

In studying Examples 26.4 and 26.5, you found that only a few terms of the series were actually evaluated to determine the general shape of the curve. Let us briefly look at these two examples and see if we can quickly tell how many harmonics are really needed. If we plot the amplitude of each harmonic of the series presented in Examples 26.4 and 26.5 (see Figure 26.9), we can see that the amplitudes of the higher harmonics are smaller than the amplitude of the fundamental.

Figure 26.9(a) represents a square wave whose amplitude of the fundamental frequency is 1.27. The amplitude of the third harmonic is one-third of the fundamental frequency (0.4244), and the amplitude of the fifth harmonic is one-fifth of the fundamental (0.2546). The next-higher harmonics would be one-seventh, one-ninth, and so on, of the fundamental.

As one can readily see, the amplitudes of the higher-order harmonics are small, but still significant. In other words, harmonics of high frequency must be present in a square wave. This

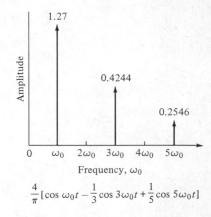

$$\frac{4}{\pi}[\cos \omega_0 t - \frac{1}{3}\cos 3\omega_0 t + \frac{1}{5}\cos 5\omega_0 t]$$

(a) Square wave; $V = 1.0$

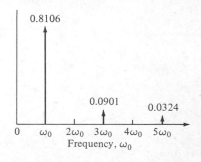

$$\frac{8}{\pi^2}[\sin \omega_0 t - \frac{1}{9}\sin 3\omega_0 t + \frac{1}{25}\sin 5\omega_0 t]$$

(b) Triangular wave; $V = 1.0$

Figure 26.9 Amplitudes of harmonics of square wave and triangular wave

is a logical conclusion, for if we look at the slope (sharpness or rise) of a square wave, we can see that only a high-frequency harmonic can contribute to this sharp slope.

Figure 26.9(b) represents the amplitudes of each term in the series representing a triangular wave. Here the amplitude of the fundamental frequency is 0.8106, while the amplitude of the third harmonic is one-ninth of the fundamental (0.0901) and the amplitude of the fifth harmonic is one twenty-fifth of the fundamental (0.0324). Note here that the amplitude terms of the harmonics of a triangular wave drop rapidly (as $1/3^2$, $1/5^2$, ...). Thus not many harmonic terms are needed to represent a triangular wave. The reasoning here is that a triangular wave has a small slope, and lower-frequency waves can describe its shape.

Problem 26.5   Using the data in Table 26.1, evaluate the Fourier series of the following function:

$$v(t) = \frac{4V}{\pi} \left[\sin \omega_0 t + \tfrac{1}{3} \sin 3\omega_0 t + \tfrac{1}{5} \sin 5\omega_0 t + \cdots \right]$$

at the times specified in Table 26.5. Let $V = 1.0$ V and frequency $f = 10$ Hz. Sketch the curves from the data in Table 26.5 and comment on the kind of waveform (square, triangular, or sine wave) and the type of waveform (even or odd). (See Figure 26.10.)

Table 26.5   Data for Problem 26.5

t, ms	Fundamental + third harmonic $\dfrac{4}{\pi}(\sin \omega_0 t + \tfrac{1}{3} \sin \omega_0 t)$	Fundamental + third + fifth harmonic $\dfrac{4}{\pi}(\sin \omega_0 t + \tfrac{1}{3} \sin 3\omega_0 t + \tfrac{1}{5} \sin 5\omega_0 t)$
10		+1.1528
20	+0.9624	
30		
40		
50		
60		
70		
80		
90		−1.1508
100	0	

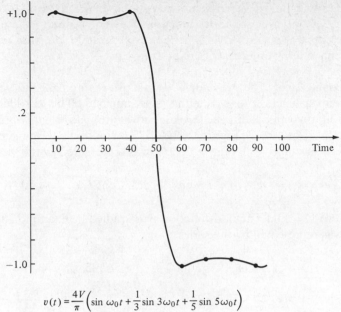

$$v(t) = \frac{4V}{\pi}\left(\sin \omega_0 t + \frac{1}{3}\sin 3\omega_0 t + \frac{1}{5}\sin 5\omega_0 t\right)$$

where V = amplitude = 1.0

Figure 26.10 Sketch of solution to Problem 26.5

Problem 26.6 Evaluate the Fourier series of the following function $(f = 10 \text{ Hz})$:

$$v(t) = \frac{4}{\pi}\left(\frac{1}{2} + \frac{1}{3}\cos \omega_0 t - \frac{1}{15}\cos 2\omega_0 t + \frac{1}{35}\cos 3\omega_0 t - \cdots\right)$$

Using the data from Table 26.2, complete Table 26.6 for the listed times. Comment on the type of function (even or odd).

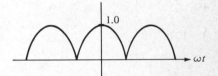

Figure 26.11 Sketch of solution to Problem 26.6

Table 26.6 Table for Problem 26.6

t, ms	$v(t) = \frac{4}{\pi}\left(\frac{1}{2} + \frac{1}{3}\cos \omega_0 t - \frac{1}{15}\cos 2\omega_0 t\right)$
10	
20	
30	
40	
50	
60	
70	
80	
90	
100	

Up to now, we have been studying periodic waveforms that were either even or odd functions. The following example presents a waveform that is neither even nor odd, but that may be made up of an even and an odd wave.

Example 26.6 The waveform shown in Figure 26.12(a) represents a function that is neither even nor odd. This particular waveform can readily be broken down into its even [Figure 26.12(b)] and odd [Figure 26.12(c)] functions. The even function was studied in Example 26.4; it was represented by the series

$$f_e = \frac{4}{\pi}(\cos \omega_0 t - \tfrac{1}{3} \cos 3\omega_0 t + \tfrac{1}{5} \cos 5\omega_0 t - \cdots)$$

where $V = 1.0$. The odd function was studied in Example 26.5; it was represented by the series

$$f_o = \frac{8}{\pi^2}\left(\sin \omega_0 t - \frac{1}{3^2} \sin 3\omega_0 t + \frac{1}{5^2} \sin 5\omega_0 t - \cdots\right)$$

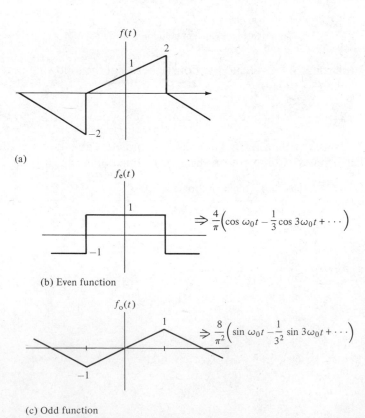

(a)

(b) Even function

(c) Odd function

Figure 26.12 Sketch of waveform made up of even and odd waveforms

The overall function (waveform) of Figure 26.12(a) would then be the *sum* of both even and odd functions. Thus

$$f(t) = f_e(t) + f_o(t)$$

or

$$f(t) = \frac{4}{\pi} \cos \omega_0 t + \frac{8}{\pi^2} \sin \omega_0 t - \frac{4}{3\pi} \cos 3\omega_0 t - \frac{8}{9\pi^2} \sin 3\omega_0 t + \cdots$$

Thus the overall function $f(t)$ is made up of even and odd terms. In general, recalling Equation (26.4), Example 26.6 illustrates the fact that we can express any periodic function in terms of even (cosine terms) and odd (sine terms) waveforms.

Effective Voltage and Power

Objective 4 Calculate the effective voltage and power of a periodic nonsinusoidal wave.

Before we try to calculate the effective value of a periodic nonsinusoidal wave, let us first review by example how to calculate the effective value of a pure ac sine wave. See Example 26.7.

Example 26.7 The ac sine wave $v(t) = 170 \sin 377t$ is applied across a 5-kΩ load resistor. Calculate the effective voltage and power dissipated in the load.

Solution
(a) Since $V_{peak} = 170$ V, then

$$V = V_{eff} = 0.707 V_{peak} = 0.707(170) = 120 \text{ V}$$

(b) The power dissipated P_{diss} can be calculated as

$$P_{diss} = \frac{V_{eff}^2}{R_L} = \frac{(120)^2}{5 \text{ k}\Omega} = 2.88 \text{ W}$$

In Example 26.7, the effective voltage or the power dissipated was independent of frequency. It depended only on the amplitude of the wave.

To find an equation to determine the effective voltage of a nonsinusoidal periodic wave, we must work with the power that is developed in a given load. In Objective 3, we learned that we can express any periodic nonsinusoidal wave as a series of ac sine and ac cosine waves of different frequencies. As a result, we can interpret this nonsinusoidal periodic wave as consisting of harmonically related sine waves that are all acting independently.

Since they are independent, we can apply the principle of super-position (Unit 10). The total power P_T is defined as

$$P_T = \frac{V_1^2}{R_L} + \frac{V_2^2}{R_L} + \cdots + \frac{V_n^2}{R_L} \tag{26.17}$$

where R_L = load resistor in ohms

$\qquad V_1$ = effective voltage of first harmonic

$\qquad V_n$ = effective voltage of nth harmonic

But the total power P_T is also related to effective voltage V_{eff} by

$$P_T = \frac{V_{eff}^2}{R_L} \tag{26.18}$$

Relating Equation (26.18) to (26.17), we obtain

$$V_{eff}^2 = V_1^2 + V_2^2 + V_3^2 + \cdots + V_n^2 \tag{26.19}$$

$$\therefore \quad V_{eff} = \sqrt{V_1^2 + V_2^2 + V_3^2 + \cdots + V_n^2} \tag{26.20}$$

We are now ready to calculate the effective value of a periodic nonsinusoidal wave. This is illustrated by Example 26.8.

Example 26.8 Calculate the effective voltage and power dissipated in a 5-Ω resistor if the square wave applied is of the form

$$v(t) = \frac{4V}{\pi} (\cos \omega_0 t - \tfrac{1}{3} \cos 3\omega_0 t + \tfrac{1}{5} \cos 5\omega_0 t - \tfrac{1}{7} \cos 7\omega_0 t)$$

where $V = 1.0$ V

Solution   Since

$$V_{eff} = \sqrt{V_1^2 + V_2^2 + \cdots + V_n^2}$$

where $V_1 \ldots V_n$ = effective value

we first must convert the peak values to effective values. Thus

$$V_1 = 0.707 \left[\frac{4(1)}{\pi} \right] 1 = 0.90018, \qquad V_3 = 0.707 \left[\frac{4(1)}{\pi} \right] \frac{1}{3} = 0.30006$$

$$V_5 = 0.707 \left[\frac{4(1)}{\pi} \right] \frac{1}{5} = 0.18004, \qquad V_7 = 0.707 \left[\frac{4(1)}{\pi} \right] \frac{1}{7} = 0.12860$$

$$\therefore \quad V_{eff} = \sqrt{V_1^2 + V_3^2 + V_5^2 + V_7^2}$$
$$= \sqrt{(0.90018)^2 + (0.30006)^2 + (0.18004)^2 + (0.12860)^2}$$
$$V_{eff} = \sqrt{0.94931} = 0.9743$$

Another way of calculating the V_{eff} is to use peak values V_p, such as:

$$V_{\text{eff}} = \sqrt{\frac{V_{1p}^2 + V_{2p}^2 + \cdots + V_{np}^2}{2}} = \frac{4(1)}{\pi}\sqrt{\frac{1^2 + \frac{1}{3^2} + \frac{1}{5^2} + \frac{1}{7^2}}{2}}$$

$$V_{\text{eff}} = \frac{4}{\pi}\sqrt{\frac{1.172}{2}} = 0.975 \text{ V}$$

The total power dissipated in a 5-Ω load is

$$P_T = \frac{V_{\text{eff}}^2}{R_L} = \frac{(0.975)^2}{5} = 0.190 \text{ W}$$

Problem 26.7 Calculate the effective voltage V_{eff} and power dissipated by the complex wave

$$v(t) = \frac{8}{\pi^2}\left(\sin \omega_0 t - \frac{1}{3^2}\sin 3\omega_0 t + \frac{1}{5^2}\sin 5\omega_0 t\right)$$

where $R_L = 7\ \Omega$

Example 26.9 Calculate the effective voltage V_{eff} of the saw-tooth waveform shown in Figure 26.13.

The equation of the complex wave for a sawtooth is given by

$$v(t) = V\left[\frac{1}{2} - \frac{1}{\pi}\left(\sin \omega_0 t + \frac{1}{2}\sin 2\omega_0 t + \frac{1}{3}\sin 3\omega_0 t + \cdots\right)\right]$$

where V = amplitude of sawtooth.

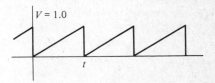

Figure 26.13 Sawtooth waveform of Example 26.9

Solution The equation for a sawtooth waveform is made up of a dc component $V(\frac{1}{2})$ and an ac component (V/π) $(\sin \omega_0 t + \frac{1}{2}\sin 2\omega_0 t + \cdots)$. The effective voltage V_{eff} is

$$V_{\text{eff}} = \sqrt{V_{dc}^2 + V_1^2 + V_2^2 + \cdots + V_n^2}$$

where V_{dc} = dc average voltage and $V_1 \ldots V_n$ = effective ac voltage of each harmonic component.

For instance, the amplitude of the first harmonic is given as $B_1 = V(1/\pi)$ volts peak. Thus the effective value of the first harmonic is

$$V_1 = 0.707 B_1 = 0.707\left(\frac{V}{\pi}\right)$$

Letting $V = 1.0$, we have

$$V_1 = 0.707\left(\frac{1}{\pi}\right) = 0.2250$$

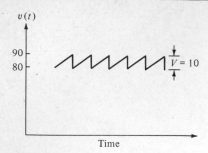

Figure 26.14 Output voltage of regulated power supply of Problem 26.8

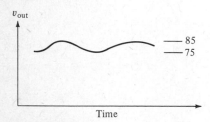

$v_{out} = 80 + 5 \sin \omega_0 t$

Figure 26.15 Output waveform of properly filtered power supply of Problem 26.9

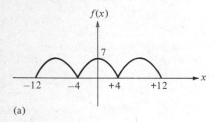

(a)

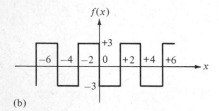

(b)

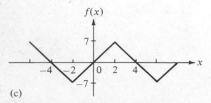

(c)

Figure 26.16 Waveforms for Test Problem 1

Likewise,

$$V_2 = 0.707\left(\frac{1}{2\pi}\right) = 0.1125, \qquad V_3 = 0.707\left(\frac{1}{3\pi}\right) = 0.0750$$

or in general,

$$V_n = 0.707\left(\frac{1}{n\pi}\right)$$

Since

$$V_{eff} = \sqrt{V_{dc}^2 + V_1^2 + V_2^2 + V_3^2 + \cdots + V_n^2}$$

and it consists of many terms, let us approximate the effective voltage V_{eff} by using only the first three terms of the harmonic series V_1, V_2, and V_3. Thus the effective voltage is approximately

$$V_{eff} \cong \sqrt{V_{dc}^2 + V_1^2 + V_2^2 + V_3^2}$$
$$\cong \sqrt{(\tfrac{1}{2})^2 + (0.2250)^2 + (0.1125)^2 + (0.075)^2}$$
$$\cong \sqrt{0.3189} \cong 0.5647$$

We observe the sawtooth waveform of Example 26.9 in many electric systems. It appears in many sweep generators and may appear as an output of a regulated power supply.

Problem 26.8   The waveform in Figure 26.14 appears at the output of a regulated power supply. Find the effective voltage V_{eff}.

Hint: $v(t) \cong 80 + 10\left[\dfrac{1}{2} - \dfrac{1}{\pi}\left(\sin \omega_0 t + \dfrac{1}{2}\sin 2\omega_0 t + \dfrac{1}{3}\sin 3\omega_0 t\right)\right]$

Problem 26.9   The output of a properly filtered power supply is shown in Figure 26.15. Calculate the effective voltage V_{eff}.

Test

1. Identify the waveforms in Figure 26.16 and specify their periods.

   (a) Type _____   Period _____

   (b) Type _____   Period _____

   (c) Type _____   Period _____

2. The series

   $$v(t) = E[1 + 2(\cos 125.6t + \cos 251.2t + \cos 376.8t)]$$

   represents a complex wave. Determine whether the waveform is even or odd. Calculate the amplitudes at each frequency value. List all higher-order harmonics.

3. Draw the complex wave

$$v(t) = 100 \cos 62.8t + 20 \cos 314t$$

Determine whether the waveform is even or odd.

4. Calculate the effective voltage and power dissipated in a 10-Ω resistor if the complex wave is

$$v(t) = \frac{4}{\pi}\left[\sin \omega_0 t + \frac{1}{3} \sin 3\omega_0 t + \frac{1}{5} \sin 5\omega_0 t \right]$$

5. The output of a power supply is shown in Figure 26.17. Write the equation of output v_{out}. Find the effective voltage V_{eff}.

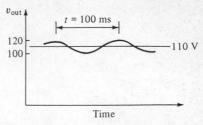

Figure 26.17 Output voltage of power supply for Test Problem 5

Unit 27 *Instruments*

In the preceding units we discussed the different electric-circuit parameters: voltage, current, resistance, and power. We showed ammeters, voltmeters, ohmmeters, and wattmeters in various circuits. However, we have not discussed the way that these quantities are actually measured. Furthermore, we have assumed that these instruments are perfect. That is, we have assumed that when the instruments are inserted into a circuit, they have no effect on the circuit operation. For example, an ammeter is considered to have no resistance. If it is inserted in a circuit, all the applied voltage is dropped across the load, and no voltage is dropped across the ammeter itself. Likewise a voltmeter is treated as an infinite resistance. When it is connected across a load resistor, all the current flows through the load resistor and none through the meter.

But are instruments perfect?

In this unit we shall first discuss the basic meter movement that is used in most instruments. The second, third, and fourth objectives will take this basic meter movement and use it to design a simple ammeter, voltmeter, and ohmmeter. We shall also discuss multifunction meters.

Objectives

After completing all the work associated with this unit, you should be able to:

1. Describe the basic D'Arsonval meter movement.
2. Briefly describe the way that a basic meter movement can

be used as an ammeter. Design an ammeter with several ampere ranges. Determine the loading effect of an ammeter in a given series circuit.

3. Briefly describe the way that a basic meter movement can be used as a voltmeter. Analyze the way that a voltmeter may load down a given circuit.

4. Briefly describe the way that the basic meter movement can be used as an ohmmeter. Design a simple series ohmmeter circuit.

5. Briefly describe the VOM. Determine whether the meter shows the proper deflection when connected as a voltmeter, ammeter, or ohmmeter.

6. Briefly describe the VTVM, TVM, and DVM. Compare their features with those of the VOM.

7. Briefly describe the wattmeter.

Basic Meter Movement

Objective 1   Describe the basic D'Arsonval meter movement.

One of the basic meter mechanisms involves the movement of a coil located in a permanent magnetic field. This movement is called the *D'Arsonval movement*, after its inventor, who developed this meter movement in 1881. Its principle is shown in Figure 27.1.

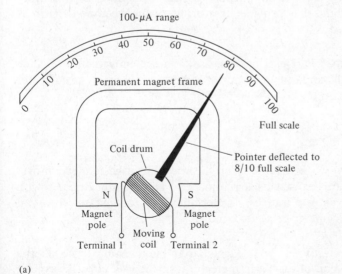

(a)

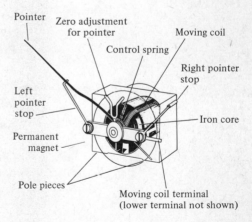

(b)

Figure 27.1   D'Arsonval meter movement

The permanent magnet located at the outside of the suspended moving coil establishes permanent magnetic lines of force (flux) in the air gap between the north and south poles. A suspended moving coil is placed between the North and South poles of the permanent magnet. The dc current applied at the inputs of the suspended moving coil creates a field around the moving coil that interacts with the permanent magnetic field. D'Arsonval found that the twisting force on the moving coil is *directly proportional* to the dc applied current. Therefore the scale is linear as shown.

For instance, a pointer mounted on the moving coil can be calibrated to indicate the actual dc current flowing in the coil. If the current to the moving coil is reduced to zero, the upper and lower control springs force the pointer of Figure 27.1(b) back to zero. Because of the dc current to the moving coil and the nature of construction of the overall meter movement, polarity must be observed at the input of the moving coil.

What would happen if such a meter were connected to an ac source? The pointer would not deflect, for the force on the moving coil would reverse alternately at a very fast rate. Since the meter could not move at such a fast rate, the pointer would indicate zero. Furthermore, a dc meter might burn out if it were connected to an ac source.

One of the variations of the *D'Arsonval galvanometer* (permanent-magnet moving-coil mechanism) is a moving coil supported with taut bands or ribbons of metal. These bands (ribbons) not only carry the dc current to the moving coil, but serve two additional purposes: They eliminate jewel-bearing friction, and they provide a restoring torque when no dc current is applied.

Problem 27.1   Describe the basic meter movement. Include a sketch containing the critical parts used in a basic meter movement.

Ammeters

Objective 2   Briefly describe the way that a basic meter movement can be used as an ammeter. Design an ammeter with several ampere ranges. Determine the loading effect of an ammeter in a given series circuit.

In Objective 1, you learned that the twisting force on the moving coil of the basic meter movement was directly proportional to the dc applied current. Since the scale is linear, the basic meter movement can be used directly as an ammeter. The typical electrical specifications of this meter movement might be 1 mA, 100 Ω. Therefore 1 mA of current would be necessary to produce a full-scale deflection on the scale.

But what about using this same meter to measure currents greater than 1 mA? To achieve this, a shunt resistor R_{sh} [see Figure 27.2(a)] must be placed in parallel with the meter movement. The excessive current above 1 mA must be shunted through this resistor. Example 27.1 illustrates the design of an ammeter with several ampere ranges.

Example 27.1 Using meter-movement characteristics of 50 μA at 0.25 V, design a dc ammeter with three full-scale ranges of 100 μA, 1 mA, and 10 mA. Specify the resistance of the ammeter for each of these ranges.

Solution Since the maximum allowable current I_m through the meter is 50 μA, the basic meter movement must employ a shunt (bypass resistor) in the design of ammeters with higher current ratings. This is shown in Figure 27.2(b). Since the voltage drop across the meter R_m is 0.25 V, the shunt voltage is also 0.25 V. The total full-scale current I_{in} of any ammeter is related to the meter current I_m and the shunt current I_{sh} by the relation

$$I_{in} = I_m + I_{sh}$$

(a) Referring to Figure 27.2(b), the total dc input current of 100 μA is divided between the meter and the shunt resistor. The shunt current is

$$I_{sh} = I_{in} - I_m = 100 \ \mu A - 50 \ \mu A = 50 \ \mu A$$

We can calculate the resistance of the meter R_m by Ohm's law:

$$R_m = \frac{0.25 \text{ V}}{I_m} = \frac{0.25 \text{ V}}{50 \ \mu A} = 5 \text{ k}\Omega$$

Since the voltage drop across the shunt resistor R_{sh} is 0.25 V, we can calculate the resistance of the shunt by Ohm's law to be

$$R_{sh} = \frac{0.25 \text{ V}}{I_{sh}} = \frac{0.25 \text{ V}}{50 \ \mu A} = 5 \text{ k}\Omega$$

The resistance R_A of the ammeter on the 100-μA range is the parallel combination of the meter resistance R_m and the shunt resistance R_{sh}. It is

$$R_A = R_m \| R_{sh} = 5 \text{ k}\Omega \| 5 \text{ k}\Omega = 2.5 \text{ k}\Omega$$

(b) The design of an ammeter for a full-scale deflection of 1 mA is similar to the design of step (a). Here with the total dc input

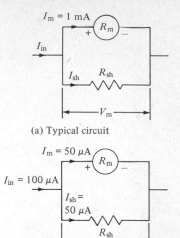

(a) Typical circuit

(b) Circuit for Example 27.1

Figure 27.2 Basic meter movement with shunt

current I_{in} of 1 mA, the maximum allowable current I_m through the meter is still 50 μA. The resulting shunt current I_{sh} [see Figure 27.2(a)] can be calculated to be

$$I_{sh} = I_{in} - I_m = 1 \text{ mA} - 50 \text{ } \mu\text{A} = 950 \text{ } \mu\text{A}$$

Since the voltage drop across the shunt resistor R_{sh} is still 0.25 V, the resistance of the shunt is

$$R_{sh} = \frac{0.25 \text{ V}}{I_{sh}} = \frac{0.25 \text{ V}}{950 \text{ } \mu\text{A}} = 263.16 \text{ } \Omega$$

The resistor R_A of the ammeter on the 1-mA range is the parallel combination of the meter resistance R_m and the shunt resistance R_{sh}. It is

$$R_A = R_m \| R_{sh} = 5 \text{ k}\Omega \| 263.16 \text{ } \Omega = 250 \text{ } \Omega$$

(c) The design of an ammeter for a 10-mA full-scale dc deflection is similar to the design of step (b).

$$I_{sh} = I_{in} - I_m = 10 \text{ mA} - 50 \text{ } \mu\text{A} = 9.950 \text{ mA}$$

$$R_{sh} = \frac{0.25 \text{ V}}{I_{sh}} = \frac{0.25 \text{ V}}{9.95 \text{ mA}} = 25.13 \text{ } \Omega$$

The resistance R_A of the ammeter on the 10-mA range is

$$R_A = R_m \| R_{sh} = 5 \text{ k}\Omega \| 25.13 \text{ } \Omega = 25 \text{ } \Omega$$

Figure 27.3 summarizes the design of the ammeter of Example 27.1.

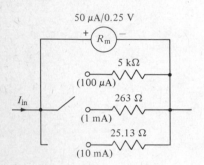

Figure 27.3 Design of ammeter of Example 27.1

Problem 27.2 The electrical specs of a meter movement are 50 μA at 0.25 V. Design a dc ammeter for a full scale of 100 mA. Calculate the overall resistance of the ammeter for a full scale of 100 mA.

Problem 27.3 A 50-μA meter movement has 2 kΩ meter resistance R_m. What shunt resistance R_{sh} is needed to extend the range to 500 μA?

The shunt design of the multiple-range ammeter in Example 27.1 is very poor. Refer to Figure 27.3 and let's see why. Suppose that an ammeter measurement was being taken on the 10-mA range and a range change was required. As the switch is being moved between current ranges, there is a point at which the shunt resistor is disconnected from the meter. At this point excessive current will flow through the meter movement and damage it.

A safe method (described in Example 27.2) is generally used for multiple current ranges in an ammeter. This design consists of a series-parallel circuit combination that provides a safe way of switching between current ranges. This circuit is called an *Ayrton shunt* or *universal shunt*. The same ammeter requirements of Example 27.1 are redesigned and shown in Example 27.2.

Example 27.2 Redesign the ammeter of Example 27.1 using the universal shunt or Ayrton shunt to provide a safe method of switching between current ranges. The meter movement is still 0.25 V at 50 μA. The current ranges are 100 μA, 1 mA, and 10 mA.

Solution Figure 27.4(a) shows the circuit configuration of the universal shunt, which consists of resistors R_1, R_2, and R_3. In position 1 (100 μA), all three resistors are in parallel with the meter movement R_m. In position 2 (1 mA), two of the resistors, R_2 and R_3, are in parallel with the series combination of R_1 and the meter movement R_m. In position 3 (10 mA), the resistor R_3 is in parallel with the series combination of resistors R_1 and R_2 and the meter.

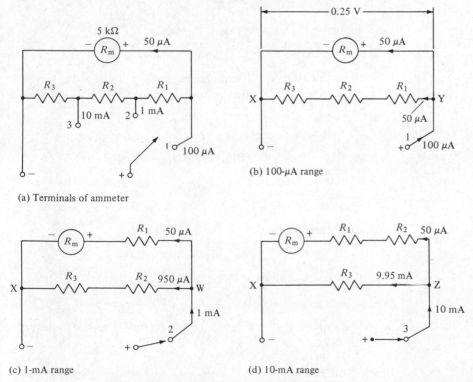

(a) Terminals of ammeter

(b) 100-μA range

(c) 1-mA range

(d) 10-mA range

Figure 27.4 Circuit configuration of universal shunt

(a) For the 100-μA full-scale range [see Figure 27.4(b)], 50 μA flows through the meter and the remaining 50 μA (100 μA − 50 μA) flows through resistors R_1, R_2, and R_3. Since the voltage across point XY is 0.25 V,

$$(50 \ \mu A)(5 \ k\Omega) = 0.25 \ V = 50 \ \mu A(R_1 + R_2 + R_3)$$
$$\therefore \quad R_1 + R_2 + R_3 = 5 \ k\Omega$$

Note that, since current is equally divided, the shunt resistor is the same as the meter movement resistor.

(b) For the 1-mA full-scale range, 50 μA still flows through resistor R_1 and meter R_m. [See Figure 27.4(c).] Since the voltage drop across points XW is the same as across resistors R_2 and R_3,

$$50 \ \mu A(R_m + R_1) = 950 \ \mu A(R_2 + R_3)$$
$$50 \ \mu A(5 \ k\Omega + R_1) = 950 \ \mu A(R_2 + R_3) \tag{a}$$

But from step (a), we have

$$R_1 + R_2 + R_3 = 5 \ k\Omega \quad \therefore \quad (R_2 + R_3) = 5 \ k\Omega - R_1$$

Substituting this into Equation (a), we obtain

$$50 \ \mu A(5 \ k\Omega + R_1) = 950 \ \mu A(5 \ k\Omega - R_1),$$
$$50(5 \ k\Omega + R_1) = 950(5 \ k\Omega - R_1)$$

Solving for R_1, we find that

$$R_1 = 4.5 \ k\Omega \quad \text{and} \quad R_2 + R_3 = 500 \ \Omega \tag{b}$$

(c) For the 10-mA full-scale range, see Figure 27.4(d). Here the voltage across XZ is the same as across resistor R_3; therefore

$$50 \ \mu A(R_m + R_1 + R_2) = 9.95 \ mA(R_3)$$
$$50 \ \mu A(5 \ k\Omega + 4.5 \ k\Omega + R_2) = 9.95 \ mA(R_3)$$
$$50(9.5 \ k\Omega + R_2) = 9950 R_3 \tag{c}$$

Recalling Equation (b), we obtain

$$R_2 + R_3 = 500 \ \Omega, \quad R_3 = 500 - R_2$$

Substituting this back into Equation (c), we get

$$50(9.5 \ k\Omega + R_2) = 9950(500 - R_2)$$

Solving for R_2, we have

$$R_2 = 450 \quad \therefore \quad R_3 = 500 - R_2 = 50 \ \Omega$$

The universal shunt design of the complete meter appears in Figure 27.5.

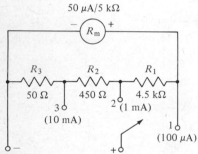

50 μA/5 kΩ

R_m

R_3 R_2 R_1

50 Ω 450 Ω 4.5 kΩ

3 (10 mA) 2 (1 mA)

1 (100 μA)

Figure 27.5 Universal shunt design of a meter for Example 27.2

100 Ω

R_m 100 μA

R_3 R_2 R_1

3 (10 mA) 2 (1 mA)

1 (200 μA)

Ammeter terminals

Figure 27.6 Universal shunt design for Problem 27.4

Problem 27.4 Design an ammeter using the universal shunt for the current ranges 200 μA, 1 mA, and 10 mA. Refer to Figure 27.6.

The meter resistance is 100 Ω, and the maximum meter current is 100 μA.

Up to this point you have learned how to design a multiple-range ammeter using a given meter movement. Example 27.3 presents the problems of loading that an ammeter may create in a series circuit.

Example 27.3 Determine the loading effect of an ammeter in the circuits shown in Figure 27.7. The internal resistance of the ammeter is 50 Ω.

Solution

(a) Without the ammeter, the theoretical current flowing in the circuit of Figure 27.7(a) is

$$I_{m1} = \frac{1.5 \text{ V}}{100 \text{ }\Omega} = 15 \text{ mA}$$

With the ammeter inserted in the circuit, the total resistance R_T of the circuit is

$$R_T = R + R_m$$

where R_m = resistance of ammeter = 50 Ω. Therefore the current now would be

$$I_{m1} = \frac{1.5 \text{ V}}{R + R_m} = \frac{1.5 \text{ V}}{100 \text{ }\Omega + 50 \text{ }\Omega} = 10 \text{ mA}$$

Comparing the theoretical current of 15 mA with the actual current through the ammeter of 10 mA, we can see that there is a large error (50%). Let's see if the error gets smaller for larger circuit resistances.

(b) Referring to Figure 27.7(b) the theoretical current without the ammeter is

$$I_{m2} = \frac{10 \text{ V}}{10 \text{ k}\Omega} = 1 \text{ mA}$$

With the ammeter inserted in series with the 10-kΩ resistor, the total circuit resistance R_T is now

$$R_T = 10 \text{ k}\Omega + R_m$$

where R_m = resistance of the ammeter = 50 Ω.

$$\therefore \quad R_T = 10 \text{ k}\Omega + 50 \text{ }\Omega = 10.05 \text{ k}\Omega$$

Thus, with the ammeter inserted, the current is

$$I_{m2} = \frac{10 \text{ V}}{R_T} = \frac{10 \text{ V}}{10.05 \text{ k}\Omega} \cong 1 \text{ mA}$$

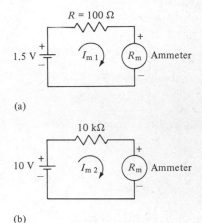

(a)

(b)

Figure 27.7   Loading effect of ammeter

Note that, in this case, the ammeter resistance had no effect on the large-value resistor in the circuit, while in the first circuit [Figure 27.7(a)], the ammeter resistance had an effect on the small-value resistance of the circuit.

In general, an ammeter will not load a circuit made up of large-value resistors. Large-value resistors are defined· as resistors whose value is *ten or more times greater* than the ammeter resistance R_m.

Problem 27.5 Determine the loading effect of an ammeter on the circuit shown in Figure 27.8.

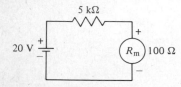

Figure 27.8 Circuit for Problem 27.5

Voltmeters

Objective 3 Briefly describe the way that a basic meter movement can be used as a voltmeter. Analyze the way that a voltmeter may load down a given circuit.

In Objective 2, you learned how to take the basic meter movement and design an ammeter. In this objective, we'll take the same meter movement and design a basic dc voltmeter, as shown in Example 27.4.

Example 27.4 Using a meter movement with electrical characteristics of 50 μA at 0.25 V drop, design a dc voltmeter of three full-scale ranges: 1 V, 2.5 V, and 10 V. Include the resistance of the voltmeter for these three ranges.

Solution
(a) Figure 27.9 shows the circuit configuration of a voltmeter. For a 1-V full-scale range, a series-dropping resistor R_s must be added in series with the basic meter. For a full-scale deflection, the meter R_m drops 0.25 V at a current level of 50 μA, leaving 0.75 V to be dropped across the series resistor R_s. Applying Ohm's law, the series resistor is

$$R_s = \frac{0.75 \text{ V}}{50 \text{ }\mu\text{A}} = 15 \text{ k}\Omega$$

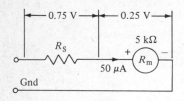

(a) 1-V range

(b) 2.5-V range

Figure 27.9 Basic meter movement used as voltmeter

The input resistance of the voltmeter for this 1-V range is the sum of the series-dropping resistor R_s and the meter resistance R_m. It is

$$R_T = R_m + R_s = 5 \text{ k}\Omega + 15 \text{ k}\Omega$$
$$R_T = 20 \text{ k}\Omega \qquad \text{for 1-V range}$$

(b) For the 2.5-V range, the design is similar to that of step (a). By referring to Figure 27.9(b), we see that the series dropping resistor R_s now has a voltage drop of 2.25 V, with 50 μA flowing through the meter. Thus

$$R_s = \frac{2.25 \text{ V}}{50 \text{ }\mu\text{A}} = 45 \text{ k}\Omega$$

The total resistance of the voltmeter on this range is

$$R_T = R_m + R_s = 5 \text{ k}\Omega + 45 \text{ k}\Omega = 50 \text{ k}\Omega \qquad \text{for 2.5-V range}$$

(c) For the 10-V range, the series-dropping resistor R_s drops $10 \text{ V} - 0.25 \text{ V} = 9.75 \text{ V}$ at a full-scale meter-deflection current of 50 μA. Thus

$$R_s = \frac{9.75 \text{ V}}{50 \text{ }\mu\text{A}} = 195 \text{ k}\Omega$$

The total resistance of the voltmeter is

$$R_T = R_m + R_s = 5 \text{ k}\Omega + 195 \text{ k}\Omega = 200 \text{ k}\Omega \qquad \text{for 10-V range}$$

In Example 27.4, you noticed that the input resistance of the voltmeter went up as the voltage range was increased from 1 V to 10 V. For instance, the input resistance of the meter is

$$R_T = 20 \text{ k}\Omega \text{ for 1-V range} \qquad (27.1)$$
$$R_T = 50 \text{ k}\Omega \text{ for 2.5-V range} \qquad (27.2)$$
$$R_T = 200 \text{ k}\Omega \text{ for 10-V range} \qquad (27.3)$$

The input resistance of any dc voltmeter can be easily computed using the following relation:

$$R_T = (\text{sensitivity in ohms per volt}) \times (\text{full-scale voltage range})$$
$$(27.4)$$

The *sensitivity* of a voltmeter may be defined as the reciprocal of the full-scale current reading of the meter movement. For instance, we can compute the sensitivity of the voltmeter of Example 27.4 by taking the full-scale current reading of the meter, 50 μA, and dividing this into 1.

$$\therefore \quad \text{Sensitivity} = \frac{1}{50 \text{ }\mu\text{A}} = 0.02 \text{ M}\Omega/\text{V} = 20 \text{ k}\Omega/\text{V}$$

Using this sensitivity, we find that input resistance of the dc voltmeter on the 10-V range is

$$R_T = (20 \text{ k}\Omega/\text{V}) \times (10\text{-V range}) = 200 \text{ k}\Omega$$

This agrees with Equation (27.3).

Problem 27.6 What series-dropping resistor R_s must be placed in series with the meter movement of Example 27.4 to increase the full-scale dc voltage range to 100 V?

By studying Example 27.4, you have learned how to design a dc voltmeter. Example 27.5 presents the loading problems a voltmeter may cause when one is measuring across certain circuit elements.

Example 27.5 Compute the voltage across resistor R_2 (see Figure 27.10), given that the measurement was made by a dc voltmeter whose sensitivity is 20 kΩ/V.

Solution
(a) Referring to Figure 27.10(a), we compute the theoretical voltage across R_2 by the voltage-divider principle as

$$V_{R2} = \frac{2 \text{ M}\Omega}{2 \text{ M}\Omega + 1 \text{ M}\Omega} 12 \text{ V} = 8 \text{ V} \qquad \text{theoretical}$$

With the dc voltmeter set to the 10-V range, using Equation (27.4) we calculate the input resistance of the meter to be

$$R_T = (20 \text{ k}\Omega/\text{V}) \times (10 \text{ V}) = 200 \text{ k}\Omega$$

Redrawing the circuit of Figure 27.10(a), we have the new equivalent circuit of Figure 27.11. The voltage

$$V_{\text{meter}} = \left(\frac{182 \text{ k}\Omega}{1 \text{ M}\Omega + 182 \text{ k}\Omega}\right) 12 \text{ V} = 1.848 \text{ V} \qquad \text{actual}$$

Thus the voltmeter with a resistance of 200 kΩ loads the 2-MΩ resistor and gives an erroneous reading (1.848 V versus 8 V).
(b) Referring to Figure 27.10(b), we see that the theoretical voltage across R_2 is

$$V_{R2} = \left(\frac{2 \text{ k}\Omega}{2 \text{ k}\Omega + 1 \text{ k}\Omega}\right) 12 \text{ V} = 8 \text{ V} \qquad \text{theoretical}$$

With the dc voltmeter again set on the 10-V range, the input resistance of the meter is again 200 kΩ. [Same as in part (a).]

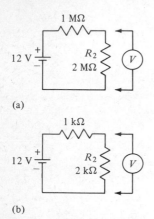

(a)

(b)

Figure 27.10 Effect of voltmeter loading on circuit measurement

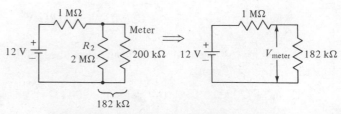

Figure 27.11 Equivalent circuit of Figure 27.10(a)

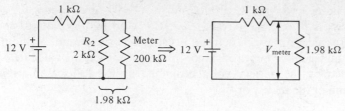

Figure 27.12 Equivalent circuit of Figure 27.10(b)

Redrawing the circuit of Figure 27.10(b), we obtain the new equivalent circuit of Figure 27.12.

The voltage measured across R_2 with the meter connected is

$$V_{meter} = \left(\frac{1.98 \text{ k}\Omega}{1 \text{ k}\Omega + 1.98 \text{ k}\Omega} \right) 12 \text{ V} = 7.97 \text{ V} \approx 8 \text{ V}$$

Thus the voltmeter with an input resistance of 200 kΩ does not load down a low-value resistance R_2. The actual voltage agrees very closely with the measured voltage (7.97 V versus 8 V).

Ohmmeters

Objective 4 Briefly describe the way that the basic meter movement can be used as an ohmmeter. Design a simple series ohmmeter circuit.

We can determine the amount of resistance in a circuit by using voltmeter and ammeter readings and Ohm's law. However, often it is more convenient to have an instrument that can measure resistance directly. This instrument is called an *ohmmeter*. We can construct an ohmmeter by adding to the basic moving-coil meter movement an internal battery and current-limiting resistance. Let us assume that we have a 1-mA basic meter movement that has 100 Ω internal resistance, a 3-V battery, and any resistors that are needed. The circuit we shall use is shown in Figure 27.13.

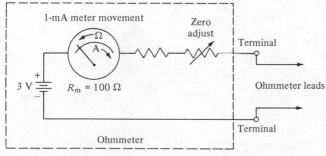

Figure 27.13 Ohmmeter circuit (series)

The purpose of the variable resistor marked *zero adjust* is to allow us to use the circuit even when the battery voltage has decreased with use. The total resistance in the circuit, including the meter movement resistance, must not exceed the value determined by the battery voltage divided by the full-scale current rating of the meter, or

$$R_{\text{meter total}} = \frac{V_{\text{battery}}}{I_{\text{fs}}} \qquad (27.5)$$

When the leads connected to the terminals of the ohmmeter circuit shown in Figure 27.13 are shorted (no resistance between them), the meter deflects to full scale. When the leads connected to the terminals are open (infinite resistance), the meter does not deflect at all. Note that when the meter scale is marked in ohms, it has 0 ohms on the right side and infinite ohms on the left side. The ohm scale increases from the right to the left.

Study Example 27.6; it reviews the design of the series ohmmeter circuit and shows you how to calibrate the scale.

Example 27.6 Using a 1-mA meter movement ($R_{\text{m}} = 100\ \Omega$) and a 3-V battery, design a simple series ohmmeter circuit. Determine the maximum meter circuit resistance and calibrate the meter scale so that it reads in ohms.

Solution The meter circuit of Figure 27.13 is redrawn showing a variable external resistance R_{ext} connected between the ohmmeter terminals (Figure 27.14).

The maximum resistance in the meter circuit is determined using Equation (27.5) when the ohmmeter leads are shorted ($R_{\text{ext}} = 0\ \Omega$).

$$R_{\text{meter total}} = \frac{V_{\text{battery}}}{I_{\text{fs}}} = \frac{3\ \text{V}}{1\ \text{mA}} = 3\ \text{k}\Omega$$

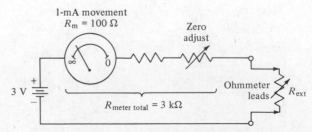

Figure 27.14 Ohmmeter circuit for Example 27.6

Table 27.1 Calculations for Marking Ohmmeter Scale

Scale marking R_{ext}, Ω	Total circuit Resistance, Ω	$I = \dfrac{V_{battery}}{R_T}$, mA	Meter deflection
0	3000	1.0	Full scale
1000	4000	0.75	3/4 scale
2000	5000	0.60	3/5 scale
3000	6000	0.50	1/2 scale
6000	9000	0.333	1/3 scale
9000	12,000	0.25	1/4 scale
15,000	18,000	0.167	1/6 scale
30,000	33,000	0.091	1/11 scale
60,000	63,000	0.048	1/21 scale
150,000	153,000	0.020	0
∞	∞	0	0

As the battery voltage decreases, the $R_{meter\ total}$ is decreased by the zero-adjust variable resistor. This makes it possible for the meter to still deflect to full scale, showing 0 Ω resistance, when the ohmmeter leads are shorted.

As stated above, when R_{ext} is equal to 0 Ω (ohmmeter leads shorted), the meter in the circuit of Figure 27.14 deflects to full scale. When R_{ext} is equal to infinite (∞) ohms (ohmmeter leads open), the meter does not deflect at all (0 mA). These are two points on the meter scale, and they are shown in Table 27.1.

Let us complete Table 27.1, and then we can mark our scale. To aid us in calibrating the meter scale in ohms, we temporarily mark the bottom of the scale in milliamperes. (Refer to Figure 27.15.) Using the values determined in Table 27.1, we write the values for ohms on the top of the scale. Once we have done this, we may erase the bottom or mA scale.

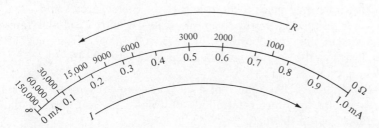

Figure 27.15 Scale marked first in current (0–1.0 mA) and then in ohms (∞ − 0 Ω)

As shown in Figure 27.14, the ohmmeter leads are connected *across* the resistance to be measured. However, the resistance to be measured must be removed from its circuit. The ohmmeter must never be inserted across a circuit with a power supply, since the power supply may cause extremely high currents and burn out the meter movement. Another reason for removing at least one end of the resistor to be measured from its circuit is that there may be some other paths in the circuit that are in parallel with the section to be measured. The parallel paths cause a low meter reading.

Did you notice that the ohmmeter scale in Figure 27.15 is not linear? It is expanded at the right near 0 Ω and crowded at the left near infinite (∞) ohms. As we were calibrating the ohmmeter scale in Example 27.6, you saw the reason for the nonlinear scale. The scale markings are determined by taking a reciprocal relation $1/X$ that represents a graph of a hyperbolic curve.

Problem 27.7   Using a 1-mA meter movement ($R_m = 100$ Ω) and a 1.5-V battery, design a simple series ohmmeter circuit. Determine the maximum meter circuit resistance and calibrate the meter scale so that it reads in ohms.

Did you notice that it is very difficult to measure low values of resistance (between 0 and 1000 Ω) with the meter shown in Figure 27.15? It is also difficult to measure actual resistor values above 15,000 Ω. Commercial ohmmeters provide for resistance measurements from less than 1 Ω up to several megohms. They do this by providing several ranges (for example, $R \times 1$, $R \times 100$, and $R \times 10$ k). The reading is obtained by multiplying the actual scale reading by the $R \times$ factor.

We must modify the simple series circuit shown in Figure 27.13 to obtain a meter that measures both low and high resistances. To obtain high resistance, we have to start with a lower-current meter movement, such as 50 μA. Also we may use a higher-voltage battery (for example, 7.5 V). Then, to obtain low resistance, we may place a shunt across the meter circuit and use a lower-voltage battery (for example, 1.5 V). Figure 27.16 illustrates a circuit that may be used for an $R \times 1$ range. Example 27.7 illustrates the operation of the circuit in Figure 27.16.

Example 27.7   Assume that a 12-Ω resistance is connected between the ohmmeter terminals of Figure 27.16. What is the deflection on the meter?

Solution   When 0 Ω resistance is connected between the ohmmeter terminals of Figure 27.16, I_m with the 1.5-V battery should

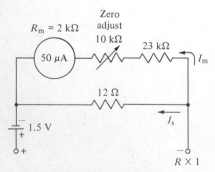

Figure 27.16   Low-resistance ohmmeter circuit

equal 50 μA. Therefore the value at which the zero-adjust potentiometer should be set can be determined by

$$R_{total} = \frac{E}{I_m} = \frac{1.5 \text{ V}}{50 \text{ } \mu\text{A}} = 30 \text{ k}\Omega$$

$$R_{zero \text{ } adjust} = 30 \text{ k}\Omega - R_m - 23 \text{ k}\Omega \qquad \text{where } R_m = 2 \text{ k}\Omega$$

$$\therefore \quad R_{zero \text{ } adjust} = 5 \text{ k}\Omega$$

The current in the 12-Ω resistor of Figure 27.16 is determined by Ohm's law (for $R_{ext} = 0 \text{ } \Omega$):

$$I_s = \frac{E}{R_s} = \frac{1.5 \text{ V}}{12 \text{ } \Omega} = 125 \text{ mA}$$

The meter current is equal to

$$I_m = \frac{1.5 \text{ V}}{30 \text{ k}\Omega} = 50 \text{ } \mu\text{A for } R_{ext} = 0 \text{ } \Omega$$

Figure 27.17 shows the ohmmeter connected across a 12-Ω load. The parallel branch of 12 Ω in parallel with 30 kΩ is approximately 12 Ω.

$$12 \text{ } \Omega \| 30 \text{ k}\Omega = 11.995 \text{ } \Omega \approx 12 \text{ } \Omega$$

Therefore the circuit of Figure 27.17 reduces to a simple circuit composed of two 12-Ω series resistors. Half the battery voltage is dropped across the 12-Ω R_{ext} resistor, leaving 0.75 V across the meter circuit.

$$I_m = \frac{0.75 \text{ V}}{30 \text{ k}\Omega} = 25 \text{ } \mu\text{A}$$

Therefore the meter deflects to center scale. Figure 27.18 gives a sketch of the scale for the low-resistance ohmmeter circuit of Figure 27.16.

VOM

Objective 5 Briefly describe the VOM. Determine whether the meter shows the proper deflection when connected as a voltmeter, ammeter, or ohmmeter.

The VOM is one of the most popular instruments because it is portable, versatile, and compact. The letters VOM stand for

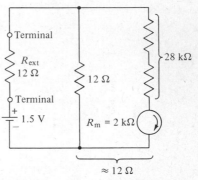

Figure 27.17 Circuit of Figure 27.16 redrawn showing the circuit when ohmmeter is connected across a 12-Ω resistance

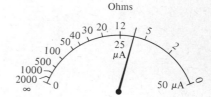

Figure 27.18 Ohmmeter scale for circuit of Figure 27.16

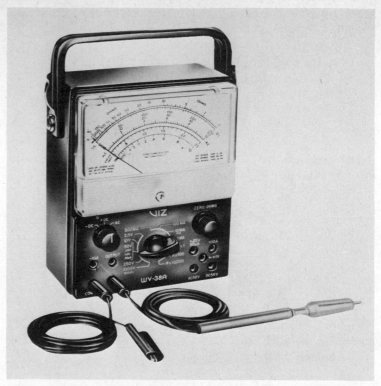

Figure 27.19   VIZ WV-38A VOM (Courtesy VIZ Manufacturing Company)

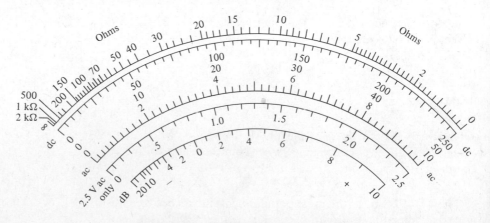

Figure 27.20   Various scales on typical VOM

voltmeter, *ohmmeter*, and *milliammeter*. Figure 27.19 shows the physical appearance of the VIZ WV-38A VOM. Figure 27.20 shows the various scales on a typical VOM. A typical VOM has a taut-band, permanent-magnet moving-coil meter movement. It has several dc voltage ranges, several dc current ranges, and three resistance ranges. The typical VOM has a modified full-wave bridge rectifier circuit that provides several ac voltage ranges. There is no provision for ac current measurement.

In this objective, you will be combining what you learned in Objectives 2, 3, and 4. As we mentioned above, a VOM is a combination of a voltmeter circuit, an ohmmeter circuit, and a milliammeter circuit using one meter movement.

Figure 27.21 shows the schematic for the VIZ WV-38A VOM.

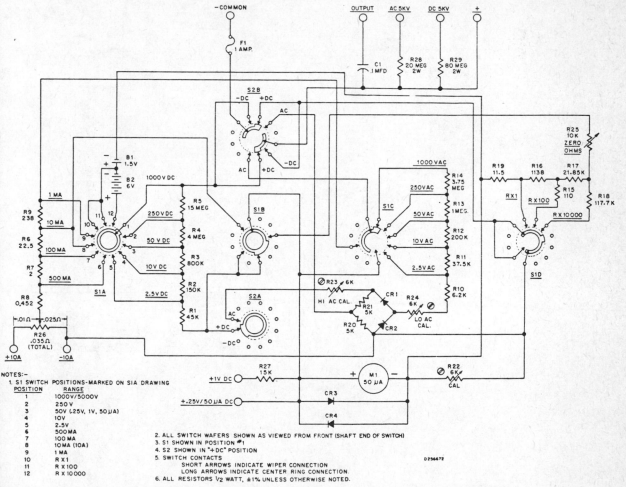

Figure 27.21 Schematic of VIZ WV-38A VOM (Courtesy VIZ Manufacturing Co.)

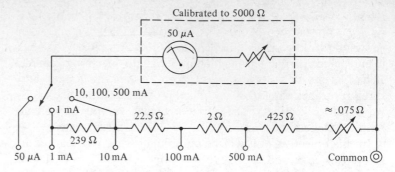

Figure 27.22   Typical VOM dc-current circuit

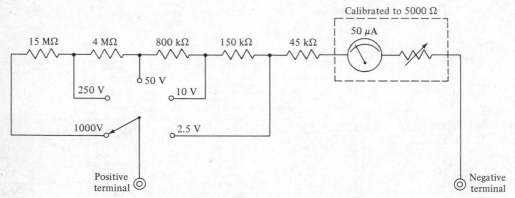

Figure 27.23   Typical VOM dc-voltage circuit

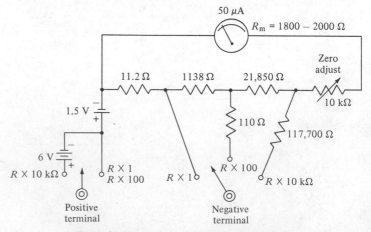

Figure 27.24   Typical VOM ohmmeter circuit

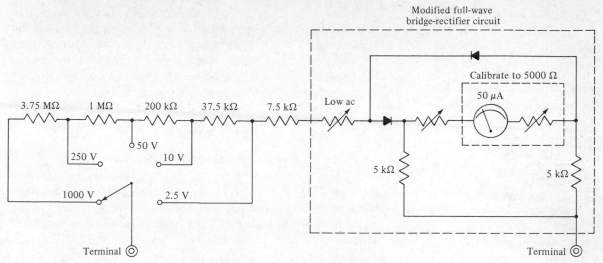

Figure 27.25 Typical VOM ac voltage circuit

Figures 27.22, 27.23, 27.24, and 27.25 show the dc current, dc voltage, ohmmeter, and ac voltage circuits of a typical VOM. You are already familiar with the circuits for dc voltage, dc current, and ohms.

Figure 27.25 shows a modified full-wave bridge rectifier circuit. This circuit converts the ac input so that current flows through the meter in only one direction.

To complete this objective, study Examples 27.8 and 27.9, then work the problems that follow the examples.

Example 27.8 A VOM is inserted into a circuit to measure 5 mA of current. Check to see whether the meter movement deflects to half scale on the 10-mA range. Assume that the meter contains the typical VOM dc current circuit shown in Figure 27.22.

Solution Figure 27.26 shows the section of Figure 27.22 for the 10-mA range. For the meter to deflect to half scale, 25 μA

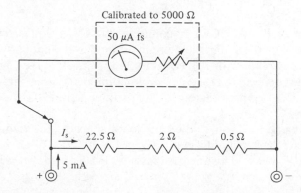

Figure 27.26 VOM dc circuit for 10 mA full scale

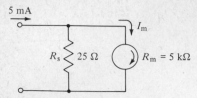

Figure 27.27 Simplified 10-mA full VOM circuit for Example 27.8

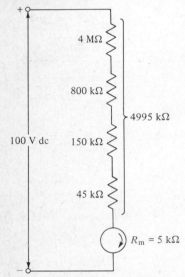

Figure 27.28 Simplified 250-V dc full-scale VOM circuit for Example 27.9

must flow through it. To solve this problem, we must determine how much current flows through the meter when 5 mA enters the meter circuit. We redraw the circuit of Figure 27.26, combining the resistors to simplify the circuit. Figure 27.27 shows this circuit.

Examining Figure 27.27, we see that we can use the current-divider principle to solve for the meter current I_m:

$$I_m = \left(\frac{R_s}{R_s + R_m}\right)I_{in} = \left(\frac{25\ \Omega}{5025\ \Omega}\right)5\ mA = 24.86 \approx 25\ \mu A$$

Therefore the meter does deflect to half scale.

Example 27.9 A VOM is connected across a 100-V dc source to check the voltage. The schematic of the dc voltage section of the VOM is shown in Figure 27.23. Check to see whether the meter deflects to 100/250 or two-fifths of full scale on the 250-V range.

Solution For the meter to deflect to two-fifths of full scale, $\frac{2}{5}(50\ \mu A) = 20\ \mu A$ must flow in the meter circuit. Figure 27.28 shows only the part of Figure 27.23 required to solve this problem.

As shown in Figure 27.28, the 100 V is across a total resistance of 5 MΩ. The meter current is determined by using Ohm's law.

$$I_m = \frac{E}{R_T} = \frac{100\ V}{5\ M\Omega} = 20\ \mu A$$

Therefore the meter does deflect to two-fifths of full scale.

Problem 27.8 A VOM is inserted into a circuit to measure 0.8 mA of current. Check to see whether the meter deflects to four-fifths of full scale on the 1-mA range. Assume that the meter contains the typical VOM dc current circuit shown in Figure 27.22.

Problem 27.9 A VOM is connected across a 25-V dc source. The schematic of the dc voltage section of the VOM is shown in Figure 27.23. Check to see whether the meter deflects to one-half of full scale on the 50-V range.

Problem 27.10 A VOM is connected across a 30-Ω resistor. Assume that the ohmmeter circuit is similar to that shown in Figure 27.17, but with R_{ext} equal to 30 Ω. Check to see whether the meter deflects the distance on the ohmmeter scale shown in Figure 27.20 (approximately 15/50, or three-tenths of full scale).

In Objective 3, you studied the loading effect of a voltmeter. The meter sensitivity of a typical VOM is 20,000 Ω/V on dc and 5000 Ω/V on ac. This means that a VOM loads down a high-resistance circuit. However, we can add an electronic circuit to the basic meter circuit to increase the input impedance of the meter to 10 MΩ or higher. The VTVM, or *vacuum-tube voltmeter*, has been a very popular high-impedance instrument. However, it is being replaced by the TVM (*transistorized voltmeter*). You will learn more about both the VTVM and the TVM in the next objective.

VTVM, TVM, and DVM

Objective 6 Briefly describe the VTVM, TVM, and DVM. Compare their features with those of the VOM.

VTVM

Figure 27.29 shows the physical appearance of a typical VTVM, or *vacuum-tube voltmeter*, with a line cord attached. The VTVM measures dc and ac voltage and resistance.

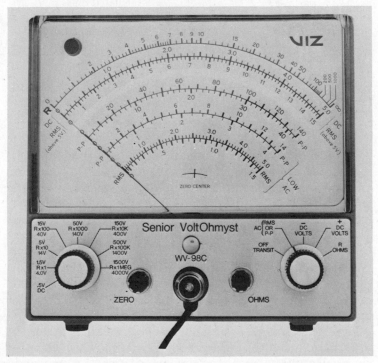

Figure 27.29 Physical appearance of a typical VTVM (Courtesy VIZ Manufacturing Co.)

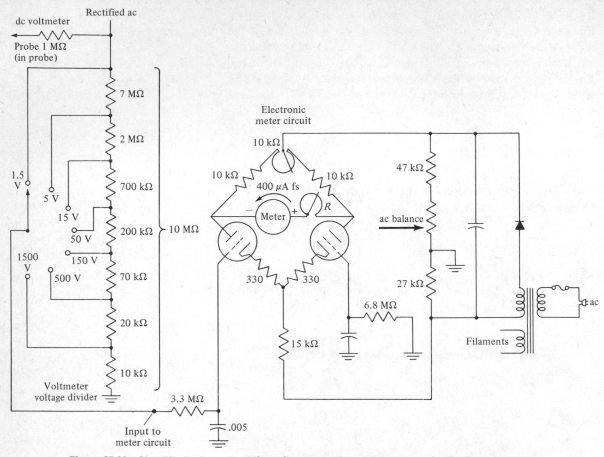

Figure 27.30 Simplified schematic of the voltmeter section and the meter circuit of a typical VTVM

Figure 27.30 shows a simplified schematic of the voltmeter section and the meter circuit of a typical VTVM. The meter circuit is very sensitive as a result of the bridge circuit and the amplifying ability of the vacuum triode tubes. The deflection of the pointer is proportional to the voltage across the voltage divider. The electronic circuit (vacuum-tube bridge circuit) has a very high input impedance, and therefore does not load down the voltage-divider circuit shown in Figure 27.30. Therefore, as shown in Figure 27.30, the input impedance for all dc voltage ranges is 11 MΩ, that is, 10 MΩ plus 1 MΩ in the probe. The input impedance for all ac voltage ranges is 10 MΩ.

The VTVM also has several ohm ranges, and can read resistors in the megohm range. Figure 27.30 does not show the ohmmeter circuit. However, the ohmmeter circuit contains a battery and a resistor for each ohm range. Therefore, when measuring resistance, use the same precautions for a VTVM that were recom-

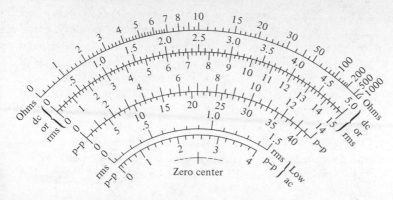

Figure 27.31 Scales on a VTVM

mended for a VOM. In the VTVM the output of the ohmmeter circuit is fed into the same electronic meter circuit as the output of the voltage circuit. As a result of using the electronic meter circuit, the ohmmeter readings increase from left to right, the same as the voltage readings.

Figure 27.31 shows the various scales on a typical VTVM. Note that the ohmmeter scale is the opposite of the ohmmeter scale for the VOM, which was shown in Figure 27.20.

One cannot use the VTVM to measure current because of its high input impedance. The VTVM must also be plugged into an ac power line for operation because it needs a power supply for its electronic circuit.

TVM

A TVM (*transistorized voltmeter*), which also measures dc and ac voltage and resistance, has several advantages over a VTVM. Since it uses transistors, it requires no warm-up time. It is comparatively lightweight and compact, and may be designed as a battery-operated instrument. Figure 27.32 shows one electronic

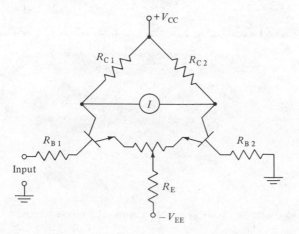

Figure 27.32 Basic TVM bridge configuration

circuit that has been used for a TVM; it is very similar to the electronic circuit used in the VTVM. Note that the triode vacuum tubes have been replaced by bipolar transistors. However, the input impedance of the bipolar transistors is not as high as that of the vacuum triodes. Therefore, when a high-impedance instrument is needed, a field-effect transistor (FET) may be used as a high-input-impedance amplifier in front of the bipolar transistor bridge circuit.

Once the TVM had been developed, other solid-state instruments soon became available. Figure 27.33 shows a solid-state VOM. Some solid-state VOMs measure ac current in addition

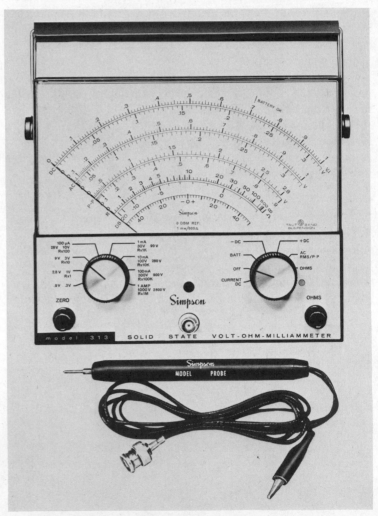

Figure 27.33   Solid-state VOM (Courtesy Simpson Electric Co.)

to dc current. Some manufacturers call their solid-state instrument an electronic multimeter or solid-state multimeter. You must read the meter specifications carefully. Some of the instruments have a very high input resistance; others have the old standard of 20,000 Ω/V on dc ranges and 5000 Ω/V on ac. Some instruments measure only dc and ac voltage and resistance, while others measure dc and ac voltage, dc and ac current, resistance, and decibels. Some instruments are portable (internal batteries), some have an optional battery pack available, and others require a 105–125-V ac 50/60-Hz power supply.

DVM

Up to this point we have been discussing *analog* instruments. That is, the angular deflection of the pointer across the scale of the meter corresponds *directly* to the measured quantity. All the instruments have used the basic meter movement we discussed in Objective 1. Remember that when we discussed the basic meter movement, we said that the deflection of the pointer was proportional to the dc current flowing in the meter. Even the meter in the ohmmeter circuit deflects in a way that is proportional to the amount of current. However, the scale is marked in ohms instead of in amount of current.

The letters DVM stand for *digital voltmeter*. A digital voltmeter—or any digital instrument—displays a digital readout of the reading. Figure 27.34 shows a digital panel meter. The readout is displayed automatically with numerical magnitude, decimal point, and polarity.

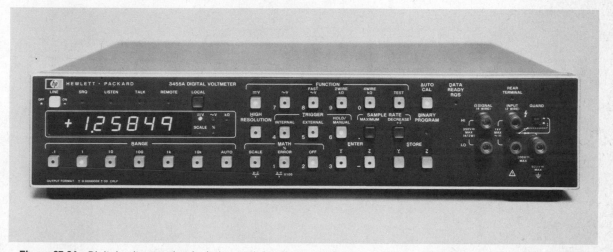

Figure 27.34 Digital voltmeter that includes math functions and remote-control operation (Courtesy Hewlett-Packard)

An analog-to-digital (A/D) converter is part of every DVM circuit. An A/D converter is a circuit that converts the dc voltage (analog) to an equivalent digital form. The output of the digital circuit is several discrete voltages (for example, several +5 V and/or 0 V). In a DVM, the output of the A/D converter is connected to some type of digital output display circuit.

There are several types of A/D converter circuits. Linear ramp, staircase ramp, integrator, and servo-balancing potentiometer are four circuits that may be used. The description of these circuits is beyond the scope of this text. However, we have included a block diagram of the counter-type or staircase ramp DVM in Figure 27.35. The dc input voltage is connected to a voltage conditioner that either decreases a very high voltage or increases a low voltage. The digital-to-analog (D/A) converter part of the circuit of Figure 27.35 provides a staircase ramp voltage that is continuously compared with the dc analog voltage. When the two voltages are the same, the comparator halts the counter and the output circuit displays the decimal readout.

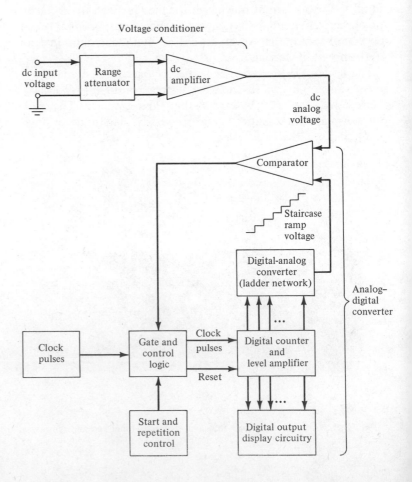

Figure 27.35 Block diagram of a counter-type (staircase ramp) DVM

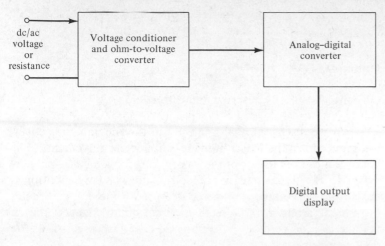

Figure 27.36 Digital voltmeter-ohmmeter block diagram

One functional block diagram of a digital volt-ohmmeter is shown in Figure 27.36. The input to the A/D converter is an analog dc voltage. The voltage conditioner and ohm-to-voltage converter provide the proper dc voltage. An input ac voltage is rectified so that it becomes a dc voltage. The dc voltage is increased if it is low or decreased if very high, so that it is the proper amplitude. The direct current that flows in the circuit when measuring resistance is also converted to a voltage.

As with the solid-state analog instruments, there are many digital instruments. They may have a single function and range. Or they may have many functions, like a multimeter. Figure 27.37 shows a digital multimeter that measures dc and ac voltage, dc and ac current, and resistance.

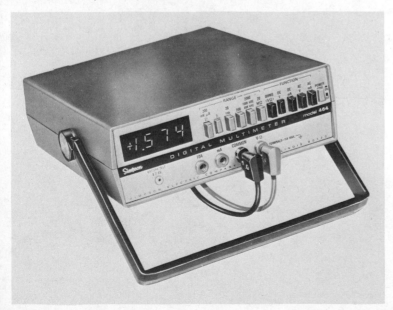

Figure 27.37 Digital multimeter (Courtesy Simpson Electric Co.)

Problem 27.11   Briefly describe the VTVM, TVM, and DVM. Compare the features of each with those of the VOM.

Wattmeters

Objective 7   Briefly describe the wattmeter.

When we discussed analog meters in the previous objectives, we referred to the basic moving-coil meter movement. The deflection of the pointer of the basic meter movement is proportional to the dc current in the coil. And when we mentioned measuring ac voltage, the meter still used a moving-coil meter movement along with a rectifier circuit that changed the ac voltage to a dc voltage. Therefore we have discussed only a dc meter movement.

There are instruments that measure ac voltage directly. That is, these instruments measure rms, not an average value calibrated to read rms. These instruments may contain a moving iron-vane mechanism, a thermocouple circuit, or an electrodynamometer mechanism. We shall not discuss the iron-vane mechanism or the thermocouple circuit meter. However, we shall discuss the electrodynamometer meter movement because it can be used to measure power.

The permanent magnet of the basic meter movement is replaced with an electromagnet to make an electrodynamometer meter

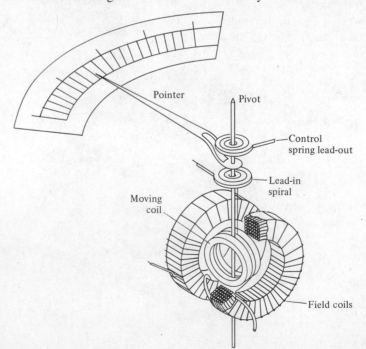

Figure 27.38 Electrodynamometer meter movement

movement. The electromagnet consists of two stationary coils, as shown in Figure 27.38, fixed in position, one on each side of the moving coil. The stationary and moving coils are connected in series to measure current, either dc or ac. The change in direction of current has no effect on the pointer, because the current in both coils reverses at the same instant, and so the change in direction of current has no effect on the reaction between the coils. The meter pointer indicates rms current. When the electrodynamometer meter movement is used in a voltmeter circuit, the meter reads rms voltage.

The instrument that measures power, a *wattmeter*, determines power by measuring the product of potential and current. The wattmeter must have the capacity to measure both current and potential at the same time. In the case of ac power, the phase angle between current and potential must also be considered. To measure current, the ammeter must be placed in series with the circuit components. To measure potential, a voltmeter is placed in parallel with the circuit element. Because of the stationary and moving coils, the electrodynamometer meter movement may be used for a wattmeter. The stationary coils are connected like the current coil, in series with the load current; see Figure 27.39.

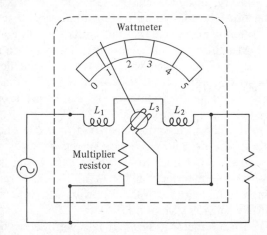

Figure 27.39 Electrodynamometer used as a wattmeter

Figure 27.39 also shows the moving coil with a multiplier in series connected across the load. The current in the moving coil is proportional to the load voltage. Therefore the deflection of the pointer is proportional to the product of current times potential. If the currents in the coils are out of phase, the pointer still indicates the true power because the deflection of the pointer depends on the interaction between coils.

Problem 27.12   Briefly describe the wattmeter.

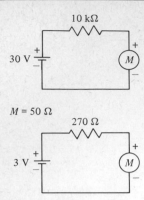

$M = 50\ \Omega$

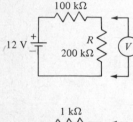

Figure 27.40 Figure for Test
Problem 4.

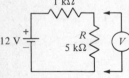

Figure 27.41 Figure for Test
Problem 6.

Test

1. The electrical characteristics of a given D'Arsonval meter movement are 50 Ω at a full-scale deflection current $I_m = 1$ mA. Calculate the voltage drop across the meter.

2. Using the electrical specs of the meter movement of Test Problem 1, design a dc ammeter for a full scale of 100 mA. Calculate the overall resistance of the ammeter for this full scale.

3. Design an ammeter using the universal shunt for the current ranges of 2 mA, 10 mA, and 100 mA. The meter resistance is 50 Ω, and the maximum meter current is 1 mA.

4. Determine the loading effect of an ammeter on the circuits shown in Figure 27.40.

5. Using the meter movement with electrical characteristics of 1 mA and a resistance of 50 Ω, design a dc voltmeter of three full-scale ranges 1 V, 2.5 V and 10 V. Include the resistance of the voltmeter for these three ranges.

6. Compute the voltage across resistor *R* (see Figure 27.41) if the measurement was made by a voltmeter whose sensitivity is 5 kΩ/V. Voltmeter range is 10 V.

7. Using a 1-mA meter movement ($R_m = 200\ \Omega$) and a 6-V battery, design a simple series ohmmeter circuit. Determine the maximum meter circuit resistance, and calibrate the meter scale so that it reads in ohms.

8. A VOM is inserted into a circuit to measure 0.6 mA of current. Check to see whether the meter deflects to 6/10 = three-fifths of full scale on the 1-mA range. Assume that the meter contains the typical VOM dc current circuit shown in Figure 27.22.

9. A VOM is connected across a 40-V dc source. The schematic of the dc voltage section of the VOM is shown in Figure 27.23. Check to see whether the meter deflects to 40/50 = four-fifths of full scale on the 50-V range.

10. Compare the features of the following instruments: the VOM, VTVM, TVM, and DVM.

11. Compare the meter movement in a wattmeter with the basic D'Arsonval meter movement.

Objectives

After completing this Appendix, you should be able to:

1. Read any single-scale meter.
2. Read any single-scale meter that has a multiple-range switch.
3. Read any multiscale voltmeter or ammeter.
4. Read any ohmmeter.

The following seven steps will help you complete the first three objectives. Study the examples listed under the objectives. Then work the problems.

1. Find the meter scale you are to read. Of course, there is no question when there is only one scale. On multiscale meters there is usually a function-selector switch that tells you which scale to read.
2. Determine the full-scale reading. If there is more than one range, a range switch will tell you the full-scale (fs) reading.
3. On the actual meter scale, find the full-scale reading.
4. Look at the whole scale and determine the major divisions, keeping in mind the fs reading.
5. Locate the major divisions on each side of the pointer.
6. Count the number of spaces between the major divisions, then determine the value per space.
7. Determine the actual reading.

Objective 1 Read any single-scale meter.

Example A.1 The meter shown in Figure A.1 has a full-scale value of 1.0 A. What is the meter reading?

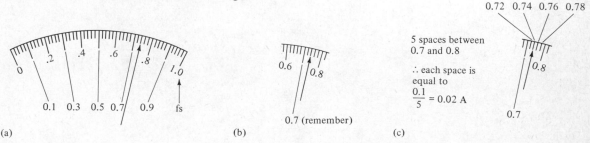

Figure A.1 Meter scale for Example A.1

565

Solution

Step 1 There is only one scale. It is shown in Figure A.1(a).

Step 2 The full-scale reading given is 1.0 A. This is the same as is marked on the meter shown in Figure A.1(a).

Step 3 In Figure A.1(a), the fs is marked by the arrow pointing to it.

Step 4 The major divisions, 0, 0.2, 0.4, 0.6, 0.8, and 1.0, are marked on the scale in Figure A.1(a). In between these divisions are 0.1, 0.3, 0.5, 0.7, and 0.9, which are not marked on the actual scale; however, these are noted in Figure A.1(a).

Step 5 The major divisions shown in Figure A.1(b) are 0.6 and 0.8. Sometimes we can use divisions closer to the pointer. In Figure A.1(b) we can actually use 0.7 (which we have to remember because it isn't marked) and 0.8.

Step 6 In Figure A.1(c) we have 5 spaces (or 4 lines) between 0.7 and 0.8; therefore we divide the difference between major divisions by the number of spaces

$$\frac{0.8 - 0.7}{5} = \frac{0.1}{5} = 0.02$$

Step 7 The actual meter reading is 0.74 A.

In Figure A.1(c) the pointer is exactly on the second line between major divisions; therefore the reading is 0.74. If the pointer were in between the short lines, then you would have to estimate where it was. If it were halfway between 0.74 and 0.76, then the reading would be 0.75. If it were three-fourths of the way, then the reading would be 0.755.

Example A.2 The meter shown in Figure A.2 has a full-scale value of 5 V. What is the meter reading?

Solution

Step 1 There is only one scale. It is shown in Figure A.2(a).

Step 2 The full-scale reading given is 5 V. This is the same as that marked on the meter shown in Figure A.2(a).

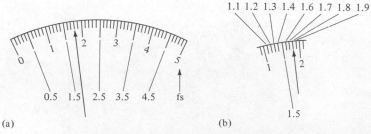

(a)   (b)

Figure A.2   Meter scale for Example A.2

Step 3 In Figure A.2(a), the fs is marked by the arrow pointing to it.

Step 4 The major divisions 0, 1, 2, 3, 4, and 5 are marked on the scale in Figure A.2(a). The divisions 0.5, 1.5, 2.5, 3.5, and 4.5 are marked below the scale. These divisions are not marked on the actual meter.

Step 5 The major divisions on each side of the pointer are 1 and 2.

Step 6 In Figure A.2(b) there are 10 spaces between 1 and 2. Therefore each space is equal to $1/10 = 0.1$ V.

Step 7 The actual meter reading is 1.8 V.

Figure A.3 Meter for Problem A.1, 1.0 A full scale

Problem A.1 The meter shown in Figure A.3 has a full-scale value of 1.0 A. What is the meter reading?

Problem A.2 The meter shown in Figure A.4 has a full-scale value of 1.0 A. What is the meter reading?

Figure A.4 Meter for Problem A.2, 1.0 A full scale

Problem A.3 The meter shown in Figure A.5 has a full-scale value of 5 V. What is the meter reading?

Problem A.4 The meter shown in Figure A.6 has a full-scale value of 3 V. What is the meter reading?

Problem A.5 The meter shown in Figure A.7 has a full-scale value of 15 V. What is the meter reading?

Objective 2 Read any single-scale meter that has a multiple range switch.

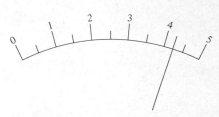

Figure A.5 Meter for Problem A.3, 5 V full scale

Example A.3 The meter shown in Figure A.1 has a full-scale value of 10 A. What is the meter reading?

Solution

Step 2 The full-scale reading given is 10 A. This is ten times that marked on the meter shown in Figure A.1(a).

Step 3 In Figure A.1(a), we mentally note that the 1.0 fs is really 10.

Step 4 Since the fs is 10 A, the major divisions now are equal to 0, 2, 4, 6, 8, and 10.

Step 5 The major divisions on each side of the pointer are 7 and 8.

Step 6 The value per space is 0.2.

Step 7 The actual reading is 7.4 A.

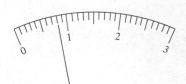

Figure A.6 Meter for Problem A.4, 3 V full scale

Problem A.6 The range switch for the meter scale shown in Figure A.2(a) is set on 50 V. List the major scale divisions and the value for each space. What is the meter reading?

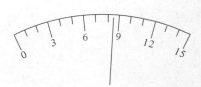

Figure A.7 Meter for Problem A.5, 15 V full scale

Figure A.8 Meter for Problem A.8, 50 V full scale

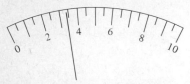

Figure A.9 Meter for Problem A.9, 100 mA full scale

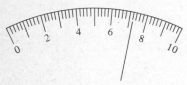

Figure A.10 Meter for Problem A.10, 1.0 A full scale

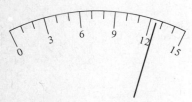

Figure A.11 Meter for Problem A.11, 1.5 V full scale

Problem A.7   The range switch for the meter scale shown in Figure A.2(a) is set on 0.5 V. List the major scale divisions and the value for each space. What is the meter reading?

Problem A.8   The range switch for the meter scale shown in Figure A.8 is set on 50 V (fs). What is the meter reading?

Problem A.9   The range switch for the meter scale shown in Figure A.9 is set on 100 mA. What is the meter reading?

Problem A.10   The range switch for the meter scale shown in Figure A.10 is set on 1.0 A. What is the meter reading?

Problem A.11   The range switch for the meter scale shown in Figure A.11 is set on 1.5 V. What is the meter reading?

Objective 3   Read any multiscale voltmeter or ammeter.

The meter face shown in Figure A.12(a) is an example of a multiscale meter. This meter measures resistance (ohms), dc

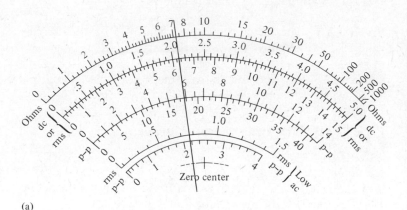

(a)

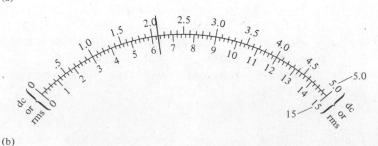

(b)

Figure A.12   Meter face for Examples A.4 and A.5

voltage and ac voltage (rms and peak to peak, p–p). The meter has two switches. The *function selector* switch tells you what you are measuring, and the *range selector* switch gives the range for the desired voltage or resistance measurement. The meter face shown in Figure A.13(a) is also an example of a multiscale meter. This meter measures resistance, dc and ac voltage, and dc current.

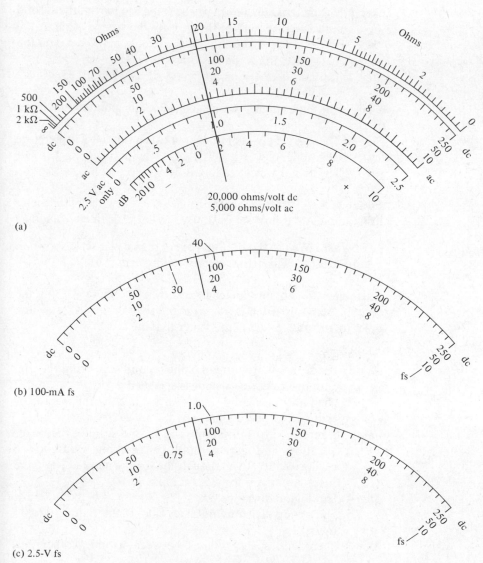

(a)

(b) 100-mA fs

(c) 2.5-V fs

Figure A.13   Meter face for Examples A.6 and A.7

Example A.4 For the meter face shown in Figure A.12(a), the function selector switch is set to + dc volts, and the range selector switch is set to 50 V. What is the meter reading?

Solution

Step 1 The scale we are to read is the second scale from the top, since the function selector is specified as dc volts.

Step 2 The full-scale reading is 50 V, since the range selector switch is set on 50 V.

Step 3 The meter scale we are to use is the top part of the second scale. Figure A.12(b) shows only the second scale or the dc volt scale. Even though the scale is marked 5.0 on the meter face, the fs for this example is 50.

Step 4 The major divisions are 5, 10, 15, 20, 25, 30, 35, 40, 45, and 50 V.

Step 5 The major divisions on each side of the pointer are 20 and 25.

Step 6 There are 5 spaces (4 lines) between 20 and 25; therefore each space is worth 1 V.

Step 7 The actual meter reading is 21 V.

Example A.5 Assume that the range selector switch of the meter in Example A.4 is now set on 150 V. What is the meter reading?

Solution We use the bottom part of the dc volt scale shown in Figure A.12(b). The actual meter reading is now 63 V.

Example A.6 For the meter face shown in Figure A.13(a), the function selector switch is set to dc and the range selector switch is set to 100 mA. What is the meter reading?

Solution

Step 1 The scale we are to read is the second scale from the top, since the function selector is specified as dc.

Step 2 The full-scale reading is 100 mA, since the range selector switch is set on 100 mA.

Step 3 Figure A.13(b) shows only the second scale of the meter face shown in Figure A.13(a). The markings we are to use are the 0 to 10. Even though the scale is marked 10 on the meter face, the fs for this example is 100.

Step 4 The major divisions are 0, 20, 40, 60, 80, and 100 mA.

Step 5 The major divisions on each side of the pointer are 30 and 40.

Step 6 There are 5 spaces (4 lines) between 30 and 40; therefore each space is worth 2 mA.

Step 7 The actual meter reading is 36 mA.

Example A.7 Assume that the range selector switch of the meter of Example A.6 is now set on 2.5V. What is the meter reading?

Solution The markings on the second scale of the meter face shown in Figure A.13(a) that we now use are 0 to 250. Refer to Figure A.13(c), and remember that fs is 2.5. The actual meter reading is now 0.9 V.

Problem A.12 For the meter face shown in Figure A.14, the function selector switch is set to +dc volts and the range selector switch is set to 15 V. What is the meter reading?

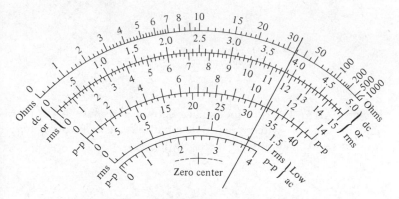

Figure A.14 Meter face for Problem A.12, + dc, 15 V

Problem A.13 For the meter face shown in Figure A.15, the function selector switch is set to +dc volts and the range selector switch is set to 500 V. What is the meter reading?

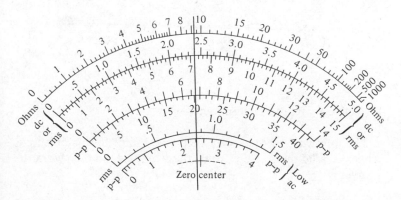

Figure A.15 Meter face for Problem A.13, + dc, 500 V

Problem A.14 For the meter face shown in Figure A.16, the function selector switch is set to dc and the range selector switch is set to 5 V. What is the meter reading?

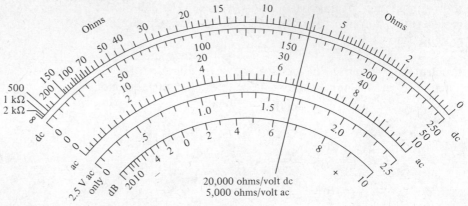

Figure A.16 Meter face for Problem A.14, dc, 5 V

Problem A.15 For the meter face shown in Figure A.17, the function selector switch is set to dc and the range selector switch is set to 1 mA. What is the meter reading?

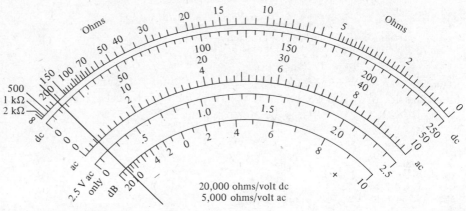

Figure A.17 Meter face for Problem A.15, dc, 1 mA

Objective 4 Read any ohmmeter.

The following six steps will help you complete this objective. After you read the six steps, study the examples and then work the problems.

1. Find the ohmmeter scale. On multiscale meters it is usually the top scale.

2. Determine the multiplier. The range switch tells you the multiplier to be used.
3. Locate the major divisions on each side of the pointer.
4. Count the number of spaces between the major divisions, then determine the value per space.
5. Determine the scale reading.
6. Multiply the scale reading by the multiplier for the actual ohmmeter reading.

The ohmmeter scales shown in Figure A.18 and A.19 are different from both voltage and current scales. There is no full-scale reading associated with ohms. The maximum ohms is infinity (∞). The ohm scale on the meter shown in Figure A.18 has 0 on the left-hand side and ∞ on the right-hand side. The ohm scale in Figure A.19 is just the reverse; the 0 is on the right-hand side and the ∞ on the left-hand side.

Example A.8   For the meter face shown in Figure A.18(a), the function selector switch is set to ohms and the range selector switch is set to $R \times 10$ kΩ. What is the meter reading?

Solution
Step 1 The scale we are to read is the top scale, since the function selector specifies ohms.

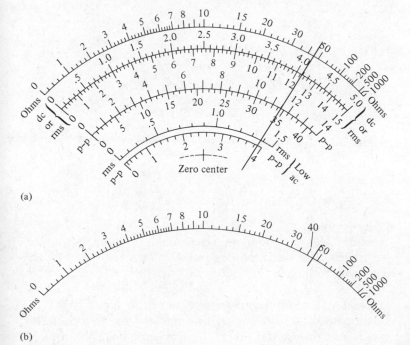

(a)

(b)

Figure A.18   Meter face for Example A.8

Step 2 The multiplier is 10 kΩ or 10,000.
Step 3 Figure A.18(b) shows only the top scale of the meter face shown in Figure A.18(a). The major divisions on each side of the pointer are 40 and 50.
Step 4 The pointer is actually on the line between 40 and 50.
Step 5 The scale reading is 45.
Step 6 The ohmmeter reading is 45 × 10 kΩ = 450 kΩ.

Example A.9   For the meter face shown in Figure A.19(a), the function selector switch is set to ohms and the range selector switch is set to *R* × 100. What is the meter reading?

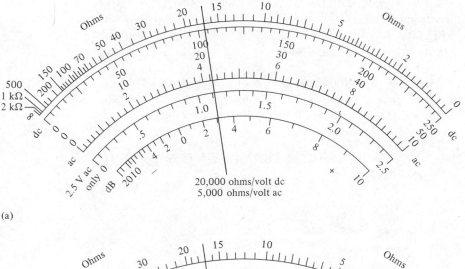

(a)

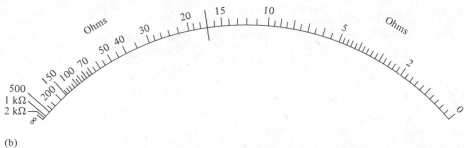

(b)

Figure A.19   Meter face for Example A.9

Solution   Note that this ohms scale has zero on the right and ∞ on the left. The scale reading [see Figure A.19(b)] is 17. The ohmmeter reading is 17 × 100 = 1700 Ω.

Problem A.16   For the meter face shown in Figure A.20, the function selector switch is set to ohms and the range selector switch is set to *R* × 100. What is the meter reading?

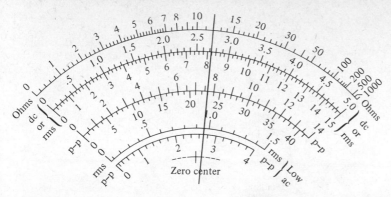

Problem A.17 For the meter face shown in Figure A.21, the function selector switch is set to ohms and the range selector switch is set to *R* × 100 kΩ. What is the meter reading?

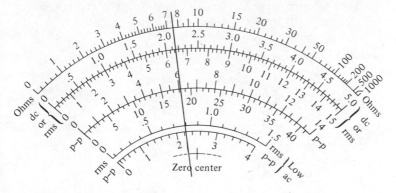

Figure A.21 Meter face for
Problem A.17, ohms, *R* × 100 kΩ

Problem A.18 For the meter face shown in Figure A.22, the function selector switch is set to ohms and the range selector switch is set to *R* × 10,000. What is the meter reading?

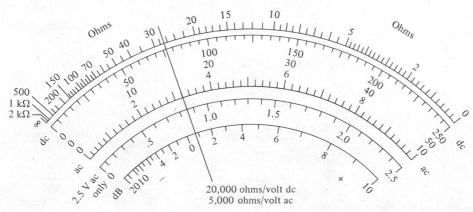

Figure A.22 Meter face for Problem A.18, ohms, *R* × 10,000

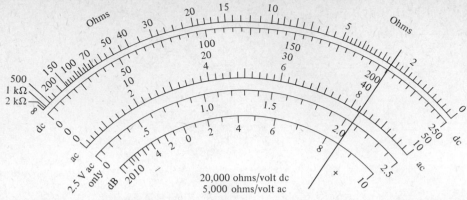

Figure A.23   Meter face for Problem A.19, ohms, $R \times 1000$

Problem A.19   For the meter face shown in Figure A.23, the function selector switch is set to ohms and the range selector switch is set to $R \times 1000$. What is the meter reading?

Answers

A.1. 0.86 A	A.2. 0.14 A	A.3. 4.25 V	A.4. 0.80 V
A.5. 8.5 V	A.6. 18 V	A.7. 0.18V	A.8. 27 V
A.9. 34 mA	A.10. 0.72 A	A.11. 1.25 V	A.12. 11.6 V
A.13. 240 V	A.14. 3.2 V	A.15. 0.05 mA	A.16. 1200 Ω
A.17. 750 kΩ	A.18. 270 kΩ	A.19. 2.6 kΩ	

Trigonometric Tables: Sine, Cosine, and Tangent   Appendix B

Degs.	Function	0.0°	0.1°	0.2°	0.3°	0.4°	0.5°	0.6°	0.7°	0.8°	0.9°
0	sin	0.0000	0.0017	0.0035	0.0052	0.0070	0.0087	0.0105	0.0122	0.0140	0.0157
	cos	1.0000	1.0000	1.0000	1.0000	1.0000	1.0000	0.9999	0.9999	0.9999	0.9999
	tan	0.0000	0.0017	0.0035	0.0052	0.0070	0.0087	0.0105	0.0122	0.0140	0.0157
1	sin	0.0175	0.0192	0.0209	0.0227	0.0244	0.0262	0.0279	0.0297	0.0314	0.0332
	cos	0.9998	0.9998	0.9998	0.9997	0.9997	0.9997	0.9996	0.9996	0.9995	0.9995
	tan	0.0175	0.0192	0.0209	0.0227	0.0244	0.0262	0.0279	0.0297	0.0314	0.0332
2	sin	0.0349	0.0366	0.0384	0.0401	0.0419	0.0436	0.0454	0.0471	0.0488	0.0506
	cos	0.9994	0.9993	0.9993	0.9992	0.9991	0.9990	0.9990	0.9989	0.9988	0.9987
	tan	0.0349	0.0367	0.0384	0.0402	0.0419	0.0437	0.0454	0.0472	0.0489	0.0507
3	sin	0.0523	0.0541	0.0558	0.0576	0.0593	0.0610	0.0628	0.0645	0.0663	0.0680
	cos	0.9986	0.9985	0.9984	0.9983	0.9982	0.9981	0.9980	0.9979	0.9978	0.9977
	tan	0.0524	0.0542	0.0559	0.0577	0.0594	0.0612	0.0629	0.0647	0.0664	0.0682
4	sin	0.0698	0.0715	0.0732	0.0750	0.0767	0.0785	0.0802	0.0819	0.0837	0.0854
	cos	0.9976	0.9974	0.9973	0.9972	0.9971	0.9969	0.9968	0.9966	0.9965	0.9963
	tan	0.0699	0.0717	0.0734	0.0752	0.0769	0.0787	0.0805	0.0822	0.0840	0.0857
5	sin	0.0872	0.0889	0.0906	0.0924	0.0941	0.0958	0.0976	0.0993	0.1011	0.1028
	cos	0.9962	0.9960	0.9959	0.9957	0.9956	0.9954	0.9952	0.9951	0.9949	0.9947
	tan	0.0875	0.0892	0.0910	0.0928	0.0945	0.0963	0.0981	0.0998	0.1016	0.1033
6	sin	0.1045	0.1063	0.1080	0.1097	0.1115	0.1132	0.1149	0.1167	0.1184	0.1201
	cos	0.9945	0.9943	0.9942	0.9940	0.9938	0.9936	0.9934	0.9932	0.9930	0.9928
	tan	0.1051	0.1069	0.1086	0.1104	0.1122	0.1139	0.1157	0.1175	0.1192	0.1210
7	sin	0.1219	0.1236	0.1253	0.1271	0.1288	0.1305	0.1323	0.1340	0.1357	0.1374
	cos	0.9925	0.9923	0.9921	0.9919	0.9917	0.9914	0.9912	0.9910	0.9907	0.9905
	tan	0.1228	0.1246	0.1263	0.1281	0.1299	0.1317	0.1334	0.1352	0.1370	0.1388
8	sin	0.1392	0.1409	0.1426	0.1444	0.1461	0.1478	0.1495	0.1513	0.1530	0.1547
	cos	0.9903	0.9900	0.9898	0.9895	0.9893	0.9890	0.9888	0.9885	0.9882	0.9880
	tan	0.1405	0.1423	0.1441	0.1459	0.1477	0.1495	0.1512	0.1530	0.1548	0.1566
9	sin	0.1564	0.1582	0.1599	0.1616	0.1633	0.1650	0.1668	0.1685	0.1702	0.1719
	cos	0.9877	0.9874	0.9871	0.9869	0.9866	0.9863	0.9860	0.9857	0.9854	0.9851
	tan	0.1584	0.1602	0.1620	0.1638	0.1655	0.1673	0.1691	0.1709	0.1727	0.1745
10	sin	0.1736	0.1754	0.1771	0.1788	0.1805	0.1822	0.1840	0.1857	0.1874	0.1891
	cos	0.9848	0.9845	0.9842	0.9839	0.9836	0.9833	0.9829	0.9826	0.9823	0.9820
	tan	0.1763	0.1781	0.1799	0.1817	0.1835	0.1853	0.1871	0.1890	0.1908	0.1926
11	sin	0.1908	0.1925	0.1942	0.1959	0.1977	0.1994	0.2011	0.2028	0.2045	0.2062
	cos	0.9816	0.9813	0.9810	0.9806	0.9803	0.9799	0.9796	0.9792	0.9789	0.9785
	tan	0.1944	0.1962	0.1980	0.1998	0.2016	0.2035	0.2053	0.2071	0.2089	0.2107
12	sin	0.2079	0.2096	0.2113	0.2130	0.2147	0.2164	0.2181	0.2198	0.2215	0.2232
	cos	0.9781	0.9778	0.9774	0.9770	0.9767	0.9763	0.9759	0.9755	0.9751	0.9748
	tan	0.2126	0.2144	0.2162	0.2180	0.2199	0.2217	0.2235	0.2254	0.2272	0.2290
13	sin	0.2250	0.2267	0.2284	0.2300	0.2318	0.2334	0.2351	0.2368	0.2385	0.2402
	cos	0.9744	0.9740	0.9736	0.9732	0.9728	0.9724	0.9720	0.9715	0.9711	0.9707
	tan	0.2309	0.2327	0.2345	0.2364	0.2382	0.2401	0.2419	0.2438	0.2456	0.2475
14	sin	0.2419	0.2436	0.2453	0.2470	0.2487	0.2504	0.2521	0.2538	0.2554	0.2571
	cos	0.9703	0.9699	0.9694	0.9690	0.9686	0.9681	0.9677	0.9673	0.9668	0.9664
	tan	0.2493	0.2512	0.2530	0.2549	0.2568	0.2586	0.2605	0.2623	0.2642	0.2661
Degs.	Function	0′	6′	12′	18′	24′	30′	36′	42′	48′	54′

Natural Sines, Cosines, and Tangents

Degs.	Function	0.0°	0.1°	0.2°	0.3°	0.4°	0.5°	0.6°	0.7°	0.8°	0.9°
15	sin	0.2588	0.2605	0.2622	0.2639	0.2656	0.2672	0.2689	0.2706	0.2723	0.2740
	cos	0.9659	0.9655	0.9650	0.9646	0.9641	0.9636	0.9632	0.9627	0.9622	0.9617
	tan	0.2679	0.2698	0.2717	0.2736	0.2754	0.2773	0.2792	0.2811	0.2830	0.2849
16	sin	0.2756	0.2773	0.2790	0.2807	0.2823	0.2840	0.2857	0.2874	0.2890	0.2907
	cos	0.9613	0.9608	0.9603	0.9598	0.9593	0.9588	0.9583	0.9578	0.9573	0.9568
	tan	0.2867	0.2886	0.2905	0.2924	0.2943	0.2962	0.2981	0.3000	0.3019	0.3038
17	sin	0.2924	0.2940	0.2957	0.2974	0.2990	0.3007	0.3024	0.3040	0.3057	0.3074
	cos	0.9563	0.9558	0.9553	0.9548	0.9542	0.9537	0.9532	0.9527	0.9521	0.9516
	tan	0.3057	0.3076	0.3096	0.3115	0.3134	0.3153	0.3172	0.3191	0.3211	0.3230
18	sin	0.3090	0.3107	0.3123	0.3140	0.3156	0.3173	0.3190	0.3206	0.3223	0.3239
	cos	0.9511	0.9505	0.9500	0.9494	0.9489	0.9483	0.9478	0.9472	0.9466	0.9461
	tan	0.3249	0.3269	0.3288	0.3307	0.3327	0.3346	0.3365	0.3385	0.3404	0.3424
19	sin	0.3256	0.3272	0.3289	0.3305	0.3322	0.3338	0.3355	0.3371	0.3387	0.3404
	cos	0.9455	0.9449	0.9444	0.9438	0.9432	0.9426	0.9421	0.9415	0.9409	0.9403
	tan	0.3443	0.3463	0.3482	0.3502	0.3522	0.3541	0.3561	0.3581	0.3600	0.3620
20	sin	0.3420	0.3437	0.3453	0.3469	0.3486	0.3502	0.3518	0.3535	0.3551	0.3567
	cos	0.9397	0.9391	0.9385	0.9379	0.9373	0.9367	0.9361	0.9354	0.9348	0.9342
	tan	0.3640	0.3659	0.3679	0.3699	0.3719	0.3739	0.3759	0.3779	0.3799	0.3819
21	sin	0.3584	0.3600	0.3616	0.3633	0.3649	0.3665	0.3681	0.3697	0.3714	0.3730
	cos	0.9336	0.9330	0.9323	0.9317	0.9311	0.9304	0.9298	0.9291	0.9285	0.9278
	tan	0.3839	0.3859	0.3879	0.3899	0.3919	0.3939	0.3959	0.3979	0.4000	0.4020
22	sin	0.3746	0.3762	0.3778	0.3795	0.3811	0.3827	0.3843	0.3859	0.3875	0.3891
	cos	0.9272	0.9265	0.9259	0.9252	0.9245	0.9239	0.9232	0.9225	0.9219	0.9212
	tan	0.4040	0.4061	0.4081	0.4101	0.4122	0.4142	0.4163	0.4183	0.4204	0.4224
23	sin	0.3907	0.3923	0.3939	0.3955	0.3971	0.3987	0.4003	0.4019	0.4035	0.4051
	cos	0.9205	0.9198	0.9191	0.9184	0.9178	0.9171	0.9164	0.9157	0.9150	0.9143
	tan	0.4245	0.4265	0.4286	0.4307	0.4327	0.4348	0.4369	0.4390	0.4411	0.4431
24	sin	0.4067	0.4083	0.4099	0.4115	0.4131	0.4147	0.4163	0.4179	0.4195	0.4210
	cos	0.9135	0.9128	0.9121	0.9114	0.9107	0.9100	0.9092	0.9085	0.9078	0.9070
	tan	0.4452	0.4473	0.4494	0.4515	0.4536	0.4557	0.4578	0.4599	0.4621	0.4642
25	sin	0.4226	0.4242	0.4258	0.4274	0.4289	0.4305	0.4321	0.4337	0.4352	0.4368
	cos	0.9063	0.9056	0.9048	0.9041	0.9033	0.9026	0.9018	0.9011	0.9003	0.8996
	tan	0.4663	0.4684	0.4706	0.4727	0.4748	0.4770	0.4791	0.4813	0.4834	0.4856
26	sin	0.4384	0.4399	0.4415	0.4431	0.4446	0.4462	0.4478	0.4493	0.4509	0.4524
	cos	0.8988	0.8980	0.8973	0.8965	0.8957	0.8949	0.8942	0.8934	0.8926	0.8918
	tan	0.4877	0.4899	0.4921	0.4942	0.4964	0.4986	0.5008	0.5029	0.5051	0.5073
27	sin	0.4540	0.4555	0.4571	0.4586	0.4602	0.4617	0.4633	0.4648	0.4664	0.4679
	cos	0.8910	0.8902	0.8894	0.8886	0.8878	0.8870	0.8862	0.8854	0.8846	0.8838
	tan	0.5095	0.5117	0.5139	0.5161	0.5184	0.5206	0.5228	0.5250	0.5272	0.5295
28	sin	0.4695	0.4710	0.4726	0.4741	0.4756	0.4772	0.4787	0.4802	0.4818	0.4833
	cos	0.8829	0.8821	0.8813	0.8805	0.8796	0.8788	0.8780	0.8771	0.8763	0.8755
	tan	0.5317	0.5340	0.5362	0.5384	0.5407	0.5430	0.5452	0.5475	0.5498	0.5520
29	sin	0.4848	0.4863	0.4879	0.4894	0.4909	0.4924	0.4939	0.4955	0.4970	0.4985
	cos	0.8746	0.8738	0.8729	0.8721	0.8712	0.8704	0.8695	0.8686	0.8678	0.8669
	tan	0.5543	0.5566	0.5589	0.5612	0.5635	0.5658	0.5681	0.5704	0.5727	0.5750
Degs.	Function	0′	6′	12′	18′	24′	30′	36′	42′	48′	54′

Natural Sines, Cosines, and Tangents

30°–44.9°

Degs.	Function	0.0°	0.1°	0.2°	0.3°	0.4°	0.5°	0.6°	0.7°	0.8°	0.9°
30	sin	0.5000	0.5015	0.5030	0.5045	0.5060	0.5075	0.5090	0.5105	0.5120	0.5135
	cos	0.8660	0.8652	0.8643	0.8634	0.8625	0.8616	0.8607	0.8599	0.8590	0.8581
	tan	0.5774	0.5797	0.5820	0.5844	0.5867	0.5890	0.5914	0.5938	0.5961	0.5985
31	sin	0.5150	0.5165	0.5180	0.5195	0.5210	0.5225	0.5240	0.5255	0.5270	0.5284
	cos	0.8572	0.8563	0.8554	0.8545	0.8536	0.8526	0.8517	0.8508	0.8499	0.8490
	tan	0.6009	0.6032	0.6056	0.6080	0.6104	0.6128	0.6152	0.6176	0.6200	0.6224
32	sin	0.5299	0.5314	0.5329	0.5344	0.5358	0.5373	0.5388	0.5402	0.5417	0.5432
	cos	0.8480	0.8471	0.8462	0.8453	0.8443	0.8434	0.8425	0.8415	0.8406	0.8396
	tan	0.6249	0.6273	0.6297	0.6322	0.6346	0.6371	0.6395	0.6420	0.6445	0.6469
33	sin	0.5446	0.5461	0.5476	0.5490	0.5505	0.5519	0.5534	0.5548	0.5563	0.5577
	cos	0.8387	0.8377	0.8368	0.8358	0.8348	0.8339	0.8329	0.8320	0.8310	0.8300
	tan	0.6494	0.6519	0.6544	0.6569	0.6594	0.6619	0.6644	0.6669	0.6694	0.6720
34	sin	0.5592	0.5606	0.5621	0.5635	0.5650	0.5664	0.5678	0.5693	0.5707	0.5721
	cos	0.8290	0.8281	0.8271	0.8261	0.8251	0.8241	0.8231	0.8221	0.8211	0.8202
	tan	0.6745	0.6771	0.6796	0.6822	0.6847	0.6873	0.6899	0.6924	0.6950	0.6976
35	sin	0.5736	0.5750	0.5764	0.5779	0.5793	0.5807	0.5821	0.5835	0.5850	0.5864
	cos	0.8192	0.8181	0.8171	0.8161	0.8151	0.8141	0.8131	0.8121	0.8111	0.8100
	tan	0.7002	0.7028	0.7054	0.7080	0.7107	0.7133	0.7159	0.7186	0.7212	0.7239
36	sin	0.5878	0.5892	0.5906	0.5920	0.5934	0.5948	0.5962	0.5976	0.5990	0.6004
	cos	0.8090	0.8080	0.8070	0.8059	0.8049	0.8039	0.8028	0.8018	0.8007	0.7997
	tan	0.7265	0.7292	0.7319	0.7346	0.7373	0.7400	0.7427	0.7454	0.7481	0.7508
37	sin	0.6018	0.6032	0.6046	0.6060	0.6074	0.6088	0.6101	0.6115	0.6129	0.6143
	cos	0.7986	0.7976	0.7965	0.7955	0.7944	0.7934	0.7923	0.7912	0.7902	0.7891
	tan	0.7536	0.7563	0.7590	0.7618	0.7646	0.7673	0.7701	0.7729	0.7757	0.7785
38	sin	0.6157	0.6170	0.6184	0.6198	0.6211	0.6225	0.6239	0.6252	0.6266	0.6280
	cos	0.7880	0.7869	0.7859	0.7848	0.7837	0.7826	0.7815	0.7804	0.7793	0.7782
	tan	0.7813	0.7841	0.7869	0.7898	0.7926	0.7954	0.7983	0.8012	0.8040	0.8069
39	sin	0.6293	0.6307	0.6320	0.6334	0.6347	0.6361	0.6374	0.6388	0.6401	0.6414
	cos	0.7771	0.7760	0.7749	0.7738	0.7727	0.7716	0.7705	0.7694	0.7683	0.7672
	tan	0.8098	0.8127	0.8156	0.8185	0.8214	0.8243	0.8273	0.8302	0.8332	0.8361
40	sin	0.6428	0.6441	0.6455	0.6468	0.6481	0.6494	0.6508	0.6521	0.6534	0.6547
	cos	0.7660	0.7649	0.7638	0.7627	0.7615	0.7604	0.7593	0.7581	0.7570	0.7559
	tan	0.8391	0.8421	0.8451	0.8481	0.8511	0.8541	0.8571	0.8601	0.8632	0.8662
41	sin	0.6561	0.6574	0.6587	0.6600	0.6613	0.6626	0.6639	0.6652	0.6665	0.6678
	cos	0.7547	0.7536	0.7524	0.7513	0.7501	0.7490	0.7478	0.7466	0.7455	0.7443
	tan	0.8693	0.8724	0.8754	0.8785	0.8816	0.8847	0.8878	0.8910	0.8941	0.8972
42	sin	0.6691	0.6704	0.6717	0.6730	0.6743	0.6756	0.6769	0.6782	0.6794	0.6807
	cos	0.7431	0.7420	0.7408	0.7396	0.7385	0.7373	0.7361	0.7349	0.7337	0.7325
	tan	0.9004	0.9036	0.9067	0.9099	0.9131	0.9163	0.9195	0.9228	0.9260	0.9293
43	sin	0.6820	0.6833	0.6845	0.6858	0.6871	0.6884	0.6896	0.6909	0.6921	0.6934
	cos	0.7314	0.7302	0.7290	0.7278	0.7266	0.7254	0.7242	0.7230	0.7218	0.7206
	tan	0.9325	0.9358	0.9391	0.9424	0.9457	0.9490	0.9523	0.9556	0.9590	0.9623
44	sin	0.6947	0.6959	0.6972	0.6984	0.6997	0.7009	0.7022	0.7034	0.7046	0.7059
	cos	0.7193	0.7181	0.7169	0.7157	0.7145	0.7133	0.7120	0.7108	0.7096	0.7083
	tan	0.9657	0.9691	0.9725	0.9759	0.9793	0.9827	0.9861	0.9896	0.9930	0.9965
Degs.	Function	0′	6′	12′	18′	24′	30′	36′	42′	48′	54′

Natural Sines, Cosines, and Tangents

Degs.	Function	0.0°	0.1°	0.2°	0.3°	0.4°	0.5°	0.6°	0.7°	0.8°	0.9°
45	sin	0.7071	0.7083	0.7096	0.7108	0.7120	0.7133	0.7145	0.7157	0.7169	0.7181
	cos	0.7071	0.7059	0.7046	0.7034	0.7022	0.7009	0.6997	0.6984	0.6972	0.6959
	tan	1.0000	1.0035	1.0070	1.0105	1.0141	1.0176	1.0212	1.0247	1.0283	1.0319
46	sin	0.7193	0.7206	0.7218	0.7230	0.7242	0.7254	0.7266	0.7278	0.7290	0.7302
	cos	0.6947	0.6934	0.6921	0.6909	0.6896	0.6884	0.6871	0.6858	0.6845	0.6833
	tan	1.0355	1.0392	1.0428	1.0464	1.0501	1.0538	1.0575	1.0612	1.0649	1.0686
47	sin	0.7314	0.7325	0.7337	0.7349	0.7361	0.7373	0.7385	0.7396	0.7408	0.7420
	cos	0.6820	0.6807	0.6794	0.6782	0.6769	0.6756	0.6743	0.6730	0.6717	0.6704
	tan	1.0724	1.0761	1.0799	1.0837	1.0875	1.0913	1.0951	1.0990	1.1028	1.1067
48	sin	0.7431	0.7443	0.7455	0.7466	0.7478	0.7490	0.7501	0.7513	0.7524	0.7536
	cos	0.6691	0.6678	0.6665	0.6652	0.6639	0.6626	0.6613	0.6600	0.6587	0.6574
	tan	1.1106	1.1145	1.1184	1.1224	1.1263	1.1303	1.1343	1.1383	1.1423	1.1463
49	sin	0.7547	0.7559	0.7570	0.7581	0.7593	0.7604	0.7615	0.7627	0.7638	0.7649
	cos	0.6561	0.6547	0.6534	0.6521	0.6508	0.6494	0.6481	0.6468	0.6455	0.6441
	tan	1.1504	1.1544	1.1585	1.1626	1.1667	1.1708	1.1750	1.1792	1.1833	1.1875
50	sin	0.7660	0.7672	0.7683	0.7694	0.7705	0.7716	0.7727	0.7738	0.7749	0.7760
	cos	0.6428	0.6414	0.6401	0.6388	0.6374	0.6361	0.6347	0.6334	0.6320	0.6307
	tan	1.1918	1.1960	1.2002	1.2045	1.2088	1.2131	1.2174	1.2218	1.2261	1.2305
51	sin	0.7771	0.7782	0.7793	0.7804	0.7815	0.7826	0.7837	0.7848	0.7859	0.7869
	cos	0.6293	0.6280	0.6266	0.6252	0.6239	0.6225	0.6211	0.6198	0.6184	0.6170
	tan	1.2349	1.2393	1.2437	1.2482	1.2527	1.2572	1.2617	1.2662	1.2708	1.2753
52	sin	0.7880	0.7891	0.7902	0.7912	0.7923	0.7934	0.7944	0.7955	0.7965	0.7976
	cos	0.6157	0.6143	0.6129	0.6115	0.6101	0.6088	0.6074	0.6060	0.6046	0.6032
	tan	1.2799	1.2846	1.2892	1.2938	1.2985	1.3032	1.3079	1.3127	1.3175	1.3222
53	sin	0.7986	0.7997	0.8007	0.8018	0.8028	0.8039	0.8049	0.8059	0.8070	0.8080
	cos	0.6018	0.6004	0.5990	0.5976	0.5962	0.5948	0.5934	0.5920	0.5906	0.5892
	tan	1.3270	1.3319	1.3367	1.3416	1.3465	1.3514	1.3564	1.3613	1.3663	1.3713
54	sin	0.8090	0.8100	0.8111	0.8121	0.8131	0.8141	0.8151	0.8161	0.8171	0.8181
	cos	0.5878	0.5864	0.5850	0.5835	0.5821	0.5807	0.5793	0.5779	0.5764	0.5750
	tan	1.3764	1.3814	1.3865	1.3916	1.3968	1.4019	1.4071	1.4124	1.4176	1.4229
55	sin	0.8192	0.8202	0.8211	0.8221	0.8231	0.8241	0.8251	0.8261	0.8271	0.8281
	cos	0.5736	0.5721	0.5707	0.5693	0.5678	0.5664	0.5650	0.5635	0.5621	0.5606
	tan	1.4281	1.4335	1.4388	1.4442	1.4496	1.4550	1.4605	1.4659	1.4715	1.4770
56	sin	0.8290	0.8300	0.8310	0.8320	0.8329	0.8339	0.8348	0.8358	0.8368	0.8377
	cos	0.5592	0.5577	0.5563	0.5548	0.5534	0.5519	0.5505	0.5490	0.5476	0.5461
	tan	1.4826	1.4882	1.4938	1.4994	1.5051	1.5108	1.5166	1.5224	1.5282	1.5340
57	sin	0.8387	0.8396	0.8406	0.8415	0.8425	0.8434	0.8443	0.8453	0.8462	0.8471
	cos	0.5446	0.5432	0.5417	0.5402	0.5388	0.5373	0.5358	0.5344	0.5329	0.5314
	tan	1.5399	1.5458	1.5517	1.5577	1.5637	1.5697	1.5757	1.5818	1.5880	1.5941
58	sin	0.8480	0.8490	0.8499	0.8508	0.8517	0.8526	0.8536	0.8545	0.8554	0.8563
	cos	0.5299	0.5284	0.5270	0.5255	0.5240	0.5225	0.5210	0.5195	0.5180	0.5165
	tan	1.6003	1.6066	1.6128	1.6191	1.6255	1.6319	1.6383	1.6447	1.6512	1.6577
59	sin	0.8572	0.8581	0.8590	0.8599	0.8607	0.8616	0.8625	0.8634	0.8643	0.8652
	cos	0.5150	0.5135	0.5120	0.5105	0.5090	0.5075	0.5060	0.5045	0.5030	0.5015
	tan	1.6643	1.6709	1.6775	1.6842	1.6909	1.6977	1.7045	1.7113	1.7182	1.7251
Degs.	Function	0′	6′	12′	18′	24′	30′	36′	42′	48′	54′

Natural Sines, Cosines, and Tangents

60°–74.9°

Degs.	Function	0.0°	0.1°	0.2°	0.3°	0.4°	0.5°	0.6°	0.7°	0.8°	0.9°
60	sin	0.8660	0.8669	0.8678	0.8686	0.8695	0.8704	0.8712	0.8721	0.8729	0.8738
	cos	0.5000	0.4985	0.4970	0.4955	0.4939	0.4924	0.4909	0.4894	0.4879	0.4863
	tan	1.7321	1.7391	1.7461	1.7532	1.7603	1.7675	1.7747	1.7820	1.7893	1.7966
61	sin	0.8746	0.8755	0.8763	0.8771	0.8780	0.8788	0.8796	0.8805	0.8813	0.8821
	cos	0.4848	0.4833	0.4818	0.4802	0.4787	0.4772	0.4756	0.4741	0.4726	0.4710
	tan	1.8040	1.8115	1.8190	1.8265	1.8341	1.8418	1.8495	1.8572	1.8650	1.8728
62	sin	0.8829	0.8838	0.8846	0.8854	0.8862	0.8870	0.8878	0.8886	0.8894	0.8902
	cos	0.4695	0.4679	0.4664	0.4648	0.4633	0.4617	0.4602	0.4586	0.4571	0.4555
	tan	1.8807	1.8887	1.8967	1.9047	1.9128	1.9210	1.9292	1.9375	1.9458	1.9542
63	sin	0.8910	0.8918	0.8926	0.8934	0.8942	0.8949	0.8957	0.8965	0.8973	0.8980
	cos	0.4540	0.4524	0.4509	0.4493	0.4478	0.4462	0.4446	0.4431	0.4415	0.4399
	tan	1.9626	1.9711	1.9797	1.9883	1.9970	2.0057	2.0145	2.0233	2.0323	2.0413
64	sin	0.8988	0.8996	0.9003	0.9011	0.9018	0.9026	0.9033	0.9041	0.9048	0.9056
	cos	0.4384	0.4368	0.4352	0.4337	0.4321	0.4305	0.4289	0.4274	0.4258	0.4242
	tan	2.0503	2.0594	2.0686	2.0778	2.0872	2.0965	2.1060	2.1155	2.1251	2.1348
65	sin	0.9063	0.9070	0.9078	0.9085	0.9092	0.9100	0.9107	0.9114	0.9121	0.9128
	cos	0.4226	0.4210	0.4195	0.4179	0.4163	0.4147	0.4131	0.4115	0.4099	0.4083
	tan	2.1445	2.1543	2.1642	2.1742	2.1842	2.1943	2.2045	2.2148	2.2251	2.2355
66	sin	0.9135	0.9143	0.9150	0.9157	0.9164	0.9171	0.9178	0.9184	0.9191	0.9198
	cos	0.4067	0.4051	0.4035	0.4019	0.4003	0.3987	0.3971	0.3955	0.3939	0.3923
	tan	2.2460	2.2566	2.2673	2.2781	2.2889	2.2998	2.3109	2.3220	2.3332	2.3445
67	sin	0.9205	0.9212	0.9219	0.9225	0.9232	0.9239	0.9245	0.9252	0.9259	0.9265
	cos	0.3907	0.3891	0.3875	0.3859	0.3843	0.3827	0.3811	0.3795	0.3778	0.3762
	tan	2.3559	2.3673	2.3789	2.3906	2.4023	2.4142	2.4262	2.4383	2.4504	2.4627
68	sin	0.9272	0.9278	0.9285	0.9291	0.9298	0.9304	0.9311	0.9317	0.9323	0.9330
	cos	0.3746	0.3730	0.3714	0.3697	0.3681	0.3665	0.3649	0.3633	0.3616	0.3600
	tan	2.4751	2.4876	2.5002	2.5129	2.5257	2.5386	2.5517	2.5649	2.5782	2.5916
69	sin	0.9336	0.9342	0.9348	0.9354	0.9361	0.9367	0.9373	0.9379	0.9385	0.9391
	cos	0.3584	0.3567	0.3551	0.3535	0.3518	0.3502	0.3486	0.3469	0.3453	0.3437
	tan	2.6051	2.6187	2.6325	2.6464	2.6605	2.6746	2.6889	2.7034	2.7179	2.7326
70	sin	0.9397	0.9403	0.9409	0.9415	0.9421	0.9426	0.9432	0.9438	0.9444	0.9449
	cos	0.3420	0.3404	0.3387	0.3371	0.3355	0.3338	0.3322	0.3305	0.3289	0.3272
	tan	2.7475	2.7625	2.7776	2.7929	2.8083	2.8239	2.8397	2.8556	2.8716	2.8878
71	sin	0.9455	0.9461	0.9466	0.9472	0.9478	0.9483	0.9489	0.9494	0.9500	0.9505
	cos	0.3256	0.3239	0.3223	0.3206	0.3190	0.3173	0.3156	0.3140	0.3123	0.3107
	tan	2.9042	2.9208	2.9375	2.9544	2.9714	2.9887	3.0061	3.0237	3.0415	3.0595
72	sin	0.9511	0.9516	0.9521	0.9527	0.9532	0.9537	0.9542	0.9548	0.9553	0.9558
	cos	0.3090	0.3074	0.3057	0.3040	0.3024	0.3007	0.2990	0.2974	0.2957	0.2940
	tan	3.0777	3.0961	3.1146	3.1334	3.1524	3.1716	3.1910	3.2106	3.2305	3.2506
73	sin	0.9563	0.9568	0.9573	0.9578	0.9583	0.9588	0.9593	0.9598	0.9603	0.9608
	cos	0.2924	0.2907	0.2890	0.2874	0.2857	0.2840	0.2823	0.2807	0.2790	0.2773
	tan	3.2709	3.2914	3.3122	3.3332	3.3544	3.3759	3.3977	3.4197	3.4420	3.4646
74	sin	0.9613	0.9617	0.9622	0.9627	0.9632	0.9636	0.9641	0.9646	0.9650	0.9655
	cos	0.2756	0.2740	0.2723	0.2706	0.2689	0.2672	0.2656	0.2639	0.2622	0.2605
	tan	3.4874	3.5105	3.5339	3.5576	3.5816	3.6059	3.6305	3.6554	3.6806	3.7062
Degs.	Function	0'	6'	12'	18'	24'	30'	36'	42'	48'	54'

Natural Sines, Cosines, and Tangents

75°–89.9°

Degs.	Function	0.0°	0.1°	0.2°	0.3°	0.4°	0.5°	0.6°	0.7°	0.8°	0.9°
75	sin	0.9659	0.9664	0.9668	0.9673	0.9677	0.9681	0.9686	0.9690	0.9694	0.9699
	cos	0.2588	0.2571	0.2554	0.2538	0.2521	0.2504	0.2487	0.2470	0.2453	0.2436
	tan	3.7321	3.7583	3.7848	3.8118	3.8391	3.8667	3.8947	3.9232	3.9520	3.9812
76	sin	0.9703	0.9707	0.9711	0.9715	0.9720	0.9724	0.9728	0.9732	0.9736	0.9740
	cos	0.2419	0.2402	0.2385	0.2368	0.2351	0.2334	0.2317	0.2300	0.2284	0.2267
	tan	4.0108	4.0408	4.0713	4.1022	4.1335	4.1653	4.1976	4.2303	4.2635	4.2972
77	sin	0.9744	0.9748	0.9751	0.9755	0.9759	0.9763	0.9767	0.9770	0.9774	0.9778
	cos	0.2250	0.2232	0.2215	0.2198	0.2181	0.2164	0.2147	0.2130	0.2113	0.2096
	tan	4.3315	4.3662	4.4015	4.4374	4.4737	4.5107	4.5483	4.5864	4.6252	4.6646
78	sin	0.9781	0.9785	0.9789	0.9792	0.9796	0.9799	0.9803	0.9806	0.9810	0.9813
	cos	0.2079	0.2062	0.2045	0.2028	0.2011	0.1994	0.1977	0.1959	0.1942	0.1925
	tan	4.7046	4.7453	4.7867	4.8288	4.8716	4.9152	4.9594	5.0045	5.0504	5.0970
79	sin	0.9816	0.9820	0.9823	0.9826	0.9829	0.9833	0.9836	0.9839	0.9842	0.9845
	cos	0.1908	0.1891	0.1874	0.1857	0.1840	0.1822	0.1805	0.1788	0.1771	0.1754
	tan	5.1446	5.1929	5.2422	5.2924	5.3435	5.3955	5.4486	5.5026	5.5578	5.6140
80	sin	0.9848	0.9851	0.9854	0.9857	0.9860	0.9863	0.9866	0.9869	0.9871	0.9874
	cos	0.1736	0.1719	0.1702	0.1685	0.1668	0.1650	0.1633	0.1616	0.1599	0.1582
	tan	5.6713	5.7297	5.7894	5.8502	5.9124	5.9758	6.0405	6.1066	6.1742	6.2432
81	sin	0.9877	0.9880	0.9882	0.9885	0.9888	0.9890	0.9893	0.9895	0.9898	0.9900
	cos	0.1564	0.1547	0.1530	0.1513	0.1495	0.1478	0.1461	0.1444	0.1426	0.1409
	tan	6.3138	6.3859	6.4596	6.5350	6.6122	6.6912	6.7720	6.8548	6.9395	7.0264
82	sin	0.9903	0.9905	0.9907	0.9910	0.9912	0.9914	0.9917	0.9919	0.9921	0.9923
	cos	0.1392	0.1374	0.1357	0.1340	0.1323	0.1305	0.1288	0.1271	0.1253	0.1236
	tan	7.1154	7.2066	7.3002	7.3962	7.4947	7.5958	7.6996	7.8062	7.9158	8.0285
83	sin	0.9925	0.9928	0.9930	0.9932	0.9934	0.9936	0.9938	0.9940	0.9942	0.9943
	cos	0.1219	0.1201	0.1184	0.1167	0.1149	0.1132	0.1115	0.1097	0.1080	0.1063
	tan	8.1443	8.2636	8.3863	8.5126	8.6427	8.7769	8.9152	9.0579	9.2052	9.3572
84	sin	0.9945	0.9947	0.9949	0.9951	0.9952	0.9954	0.9956	0.9957	0.9959	0.9960
	cos	0.1045	0.1028	0.1011	0.0993	0.0976	0.0958	0.0941	0.0924	0.0906	0.0889
	tan	9.5144	9.6768	9.8448	10.02	10.20	10.39	10.58	10.78	10.99	11.20
85	sin	0.9962	0.9963	0.9965	0.9966	0.9968	0.9969	0.9971	0.9972	0.9973	0.9974
	cos	0.0872	0.0854	0.0837	0.0819	0.0802	0.0785	0.0767	0.0750	0.0732	0.0715
	tan	11.43	11.66	11.91	12.16	12.43	12.71	13.00	13.30	13.62	13.95
86	sin	0.9976	0.9977	0.9978	0.9979	0.9980	0.9981	0.9982	0.9983	0.9984	0.9985
	cos	0.0698	0.0680	0.0663	0.0645	0.0628	0.0610	0.0593	0.0576	0.0558	0.0541
	tan	14.30	14.67	15.06	15.46	15.89	16.35	16.83	17.34	17.89	18.46
87	sin	0.9986	0.9987	0.9988	0.9989	0.9990	0.9990	0.9991	0.9992	0.9993	0.9993
	cos	0.0523	0.0506	0.0488	0.0471	0.0454	0.0436	0.0419	0.0401	0.0384	0.0366
	tan	19.08	19.74	20.45	21.20	22.02	22.90	23.86	24.90	26.03	27.27
88	sin	0.9994	0.9995	0.9995	0.9996	0.9996	0.9997	0.9997	0.9997	0.9998	0.9998
	cos	0.0349	0.0332	0.0314	0.0297	0.0279	0.0262	0.0244	0.0227	0.0209	0.0192
	tan	28.64	30.14	31.82	33.69	35.80	38.19	40.92	44.07	47.74	52.08
89	sin	0.9998	0.9999	0.9999	0.9999	0.9999	1.000	1.000	1.000	1.000	1.000
	cos	0.0175	0.0157	0.0140	0.0122	0.0105	0.0087	0.0070	0.0052	0.0035	0.0017
	tan	57.29	63.66	71.62	81.85	95.49	114.6	143.2	191.0	286.5	573.0
Degs.	Function	0′	6′	12′	18′	24′	30′	36′	42′	48′	54′

Exponential Table: Values of ϵ^x and ϵ^{-x} — Appendix C

x	Function	0.00	0.01	0.02	0.03	0.04	0.05	0.06	0.07	0.08	0.09
0.0	ϵ^x	1.0000	1.0101	1.0202	1.0305	1.0408	1.0513	1.0618	1.0725	1.0833	1.0942
	ϵ^{-x}	1.0000	0.9900	0.9802	0.9704	0.9608	0.9512	0.9418	0.9324	0.9231	0.9139
0.1	ϵ^x	1.1052	1.1163	1.1275	1.1388	1.1503	1.1618	1.1735	1.1853	1.1972	1.2093
	ϵ^{-x}	0.9048	0.8958	0.8869	0.8781	0.8694	0.8607	0.8521	0.8437	0.8353	0.8270
0.2	ϵ^x	1.2214	1.2337	1.2461	1.2586	1.2712	1.2840	1.2969	1.3100	1.3231	1.3364
	ϵ^{-x}	0.8187	0.8106	0.8025	0.7945	0.7866	0.7788	0.7711	0.7634	0.7558	0.7483
0.3	ϵ^x	1.3499	1.3634	1.3771	1.3910	1.4049	1.4191	1.4333	1.4477	1.4623	1.4770
	ϵ^{-x}	0.7408	0.7334	0.7261	0.7189	0.7118	0.7047	0.6977	0.6907	0.6839	0.6771
0.4	ϵ^x	1.4918	1.5068	1.5220	1.5373	1.5527	1.5683	1.5841	1.6000	1.6161	1.6323
	ϵ^{-x}	0.6703	0.6637	0.6570	0.6505	0.6440	0.6376	0.6313	0.6250	0.6188	0.6126
0.5	ϵ^x	1.6487	1.6653	1.6820	1.6989	1.7160	1.7333	1.7507	1.7683	1.7860	1.8040
	ϵ^{-x}	0.6065	0.6005	0.5945	0.5886	0.5827	0.5769	0.5712	0.5655	0.5599	0.5543
0.6	ϵ^x	1.8221	1.8404	1.8589	1.8776	1.8965	1.9155	1.9348	1.9542	1.9739	1.9939
	ϵ^{-x}	0.5488	0.5434	0.5379	0.5326	0.5273	0.5220	0.5169	0.5117	0.5066	0.5017
0.7	ϵ^x	2.0138	2.0340	2.0544	2.0751	2.0959	2.1170	2.1383	2.1598	2.1815	2.2034
	ϵ^{-x}	0.4966	0.4916	0.4868	0.4819	0.4771	0.4724	0.4677	0.4630	0.4584	0.4538
0.8	ϵ^x	2.2255	2.2479	2.2705	2.2933	2.3164	2.3396	2.3632	2.3869	2.4109	2.4351
	ϵ^{-x}	0.4493	0.4449	0.4404	0.4360	0.4317	0.4274	0.4232	0.4190	0.4148	0.4107
0.9	ϵ^x	2.4596	2.4843	2.5093	2.5345	2.5600	2.5857	2.6117	2.6379	2.6645	2.6912
	ϵ^{-x}	0.4066	0.4025	0.3985	0.3946	0.3906	0.3867	0.3829	0.3791	0.3753	0.3716
1.0	ϵ^x	2.7183	2.7456	2.7732	2.8011	2.8292	2.8577	2.8864	2.9154	2.9447	2.9743
	ϵ^{-x}	0.3679	0.3642	0.3606	0.3570	0.3535	0.3499	0.3465	0.3430	0.3396	0.3362
1.1	ϵ^x	3.0042	3.0344	3.0649	3.0957	3.1268	3.1582	3.1899	3.2220	3.2544	3.2871
	ϵ^{-x}	0.3329	0.3296	0.3263	0.3230	0.3198	0.3166	0.3135	0.3104	0.3073	0.3042
1.2	ϵ^x	3.3201	3.3535	3.3872	3.4212	3.4556	3.4903	3.5254	3.5609	3.5966	3.6328
	ϵ^{-x}	0.3012	0.2982	0.2952	0.2923	0.2894	0.2865	0.2837	0.2808	0.2780	0.2753
1.3	ϵ^x	3.6693	3.7062	3.7434	3.7810	3.8190	3.8574	3.8962	3.9354	3.9749	4.0149
	ϵ^{-x}	0.2725	0.2698	0.2671	0.2645	0.2618	0.2592	0.2567	0.2541	0.2516	0.2491
1.4	ϵ^x	4.0552	4.0960	4.1371	4.1787	4.2207	4.2631	4.3060	4.3492	4.3929	4.4371
	ϵ^{-x}	0.2466	0.2441	0.2417	0.2393	0.2369	0.2346	0.2322	0.2299	0.2276	0.2254
1.5	ϵ^x	4.4817	4.5267	4.5722	4.6182	4.6646	4.7115	4.7588	4.8066	4.8550	4.9037
	ϵ^{-x}	0.2231	0.2209	0.2187	0.2165	0.2144	0.2122	0.2101	0.2080	0.2060	0.2039
1.6	ϵ^x	4.9530	5.0028	5.0531	5.1039	5.1552	5.2070	5.2593	5.3122	5.3656	5.4195
	ϵ^{-x}	0.2019	0.1999	0.1979	0.1959	0.1940	0.1920	0.1901	0.1882	0.1864	0.1845
1.7	ϵ^x	5.4739	5.5290	5.5845	5.6407	5.6973	5.7546	5.8124	5.8709	5.9299	5.9895
	ϵ^{-x}	0.1827	0.1809	0.1791	0.1773	0.1755	0.1738	0.1720	0.1703	0.1686	0.1670
1.8	ϵ^x	6.0496	6.1104	6.1719	6.2339	6.2965	6.3598	6.4237	6.4883	6.5535	6.6194
	ϵ^{-x}	0.1653	0.1637	0.1620	0.1604	0.1588	0.1572	0.1557	0.1541	0.1526	0.1511
1.9	ϵ^x	6.6859	6.7531	6.8210	6.8895	6.9588	7.0287	7.0993	7.1707	7.2427	7.3155
	ϵ^{-x}	0.1496	0.1481	0.1466	0.1451	0.1437	0.1423	0.1409	0.1395	0.1381	0.1367

Values of ϵ^x and ϵ^{-x}

2.00–3.99

x	Function	0.00	0.01	0.02	0.03	0.04	0.05	0.06	0.07	0.08	0.09
2.0	ϵ^x	7.3891	7.4633	7.5383	7.6141	7.6906	7.7679	7.8460	7.9248	8.0045	8.0849
	ϵ^{-x}	0.1353	0.1340	0.1327	0.1313	0.1300	0.1287	0.1275	0.1262	0.1249	0.1237
2.1	ϵ^x	8.1662	8.2482	8.3311	8.4149	8.4994	8.5849	8.6711	8.7583	8.8463	8.9352
	ϵ^{-x}	0.1225	0.1212	0.1200	0.1188	0.1177	0.1165	0.1153	0.1142	0.1130	0.1119
2.2	ϵ^x	9.0250	9.1157	9.2073	9.2999	9.3933	9.4877	9.5831	9.6794	9.7767	9.8749
	ϵ^{-x}	0.1108	0.1097	0.1086	0.1075	0.1065	0.1054	0.1044	0.1033	0.1023	0.1013
2.3	ϵ^x	9.9742	10.074	10.176	10.278	10.381	10.486	10.591	10.697	10.805	10.913
	ϵ^{-x}	0.1003	0.0993	0.0983	0.0973	0.0963	0.0954	0.0944	0.0935	0.0926	0.0916
2.4	ϵ^x	11.023	11.134	11.246	11.359	11.473	11.588	11.705	11.822	11.941	12.061
	ϵ^{-x}	0.0907	0.0898	0.0889	0.0880	0.0872	0.0863	0.0854	0.0846	0.0837	0.0829
2.5	ϵ^x	12.182	12.305	12.429	12.554	12.680	12.807	12.936	13.066	13.197	13.330
	ϵ^{-x}	0.0821	0.0813	0.0805	0.0797	0.0789	0.0781	0.0773	0.0765	0.0758	0.0750
2.6	ϵ^x	13.464	13.599	13.736	13.874	14.013	14.154	14.296	14.440	14.585	14.732
	ϵ^{-x}	0.0743	0.0735	0.0728	0.0721	0.0714	0.0707	0.0699	0.0693	0.0686	0.0679
2.7	ϵ^x	14.880	15.029	15.180	15.333	15.487	15.643	15.800	15.959	16.119	16.281
	ϵ^{-x}	0.0672	0.0665	0.0659	0.0652	0.0646	0.0639	0.0633	0.0627	0.0620	0.0614
2.8	ϵ^x	16.445	16.610	16.777	16.945	17.116	17.288	17.462	17.637	17.814	17.993
	ϵ^{-x}	0.0608	0.0602	0.0596	0.0590	0.0584	0.0578	0.0573	0.0567	0.0561	0.0556
2.9	ϵ^x	18.174	18.357	18.541	18.728	18.916	19.106	19.298	19.492	19.688	19.886
	ϵ^{-x}	0.0550	0.0545	0.0539	0.0534	0.0529	0.0523	0.0518	0.0513	0.0508	0.0503
3.0	ϵ^x	20.086	20.287	20.491	20.697	20.905	21.115	21.328	21.542	21.758	21.977
	ϵ^{-x}	0.0498	0.0493	0.0488	0.0483	0.0478	0.0474	0.0469	0.0464	0.0460	0.0455
3.1	ϵ^x	22.198	22.421	22.646	22.874	23.104	23.336	23.571	23.807	24.047	24.288
	ϵ^{-x}	0.0450	0.0446	0.0442	0.0437	0.0433	0.0429	0.0424	0.0420	0.0416	0.0412
3.2	ϵ^x	24.533	24.779	25.028	25.280	25.534	25.790	26.050	26.311	26.576	26.843
	ϵ^{-x}	0.0408	0.0404	0.0400	0.0396	0.0392	0.0388	0.0384	0.0380	0.0376	0.0373
3.3	ϵ^x	27.113	27.385	27.660	27.938	28.219	28.503	28.789	29.079	29.371	29.666
	ϵ^{-x}	0.0369	0.0365	0.0362	0.0358	0.0354	0.0351	0.0347	0.0344	0.0340	0.0337
3.4	ϵ^x	29.964	30.265	30.569	30.877	31.187	31.500	31.817	32.137	32.460	32.786
	ϵ^{-x}	0.0334	0.0330	0.0327	0.0324	0.0321	0.0317	0.0314	0.0311	0.0308	0.0305
3.5	ϵ^x	33.115	33.448	33.784	34.124	34.467	34.813	35.163	35.517	35.874	36.234
	ϵ^{-x}	0.0302	0.0299	0.0296	0.0293	0.0290	0.0287	0.0284	0.0282	0.0279	0.0276
3.6	ϵ^x	36.598	36.966	37.338	37.713	38.092	38.475	38.861	39.252	39.646	40.045
	ϵ^{-x}	0.0273	0.0271	0.0268	0.0265	0.0263	0.0260	0.0257	0.0255	0.0252	0.0250
3.7	ϵ^x	40.447	40.854	41.264	41.679	42.098	42.521	42.948	43.380	43.816	44.256
	ϵ^{-x}	0.0247	0.0245	0.0242	0.0240	0.0238	0.0235	0.0233	0.0231	0.0228	0.0226
3.8	ϵ^x	44.701	45.150	45.604	46.063	46.525	46.993	47.465	47.942	48.424	48.911
	ϵ^{-x}	0.0224	0.0221	0.0219	0.0217	0.0215	0.0213	0.0211	0.0209	0.0207	0.0204
3.9	ϵ^x	49.402	49.899	50.400	50.907	51.419	51.935	52.457	52.985	53.517	54.055
	ϵ^{-x}	0.0202	0.0200	0.0198	0.0196	0.0195	0.0193	0.0191	0.0189	0.0187	0.0185

Values of ϵ^x and ϵ^{-x}

4.00–5.99

x	Function	0.00	0.01	0.02	0.03	0.04	0.05	0.06	0.07	0.08	0.09
4.0	ϵ^x	54.598	55.147	55.701	56.261	56.826	57.397	57.974	58.557	59.145	59.740
	ϵ^{-x}	0.0183	0.0181	0.0180	0.0178	0.0176	0.0174	0.0172	0.0171	0.0169	0.0167
4.1	ϵ^x	60.340	60.947	61.559	62.178	62.803	63.434	64.072	64.715	65.366	66.023
	ϵ^{-x}	0.0166	0.0164	0.0162	0.0161	0.0159	0.0158	0.0156	0.0155	0.0153	0.0151
4.2	ϵ^x	66.686	67.357	68.033	68.717	69.408	70.105	70.810	71.522	72.240	72.966
	ϵ^{-x}	0.0150	0.0148	0.0147	0.0146	0.0144	0.0143	0.0141	0.0140	0.0138	0.0137
4.3	ϵ^x	73.700	74.440	75.189	75.944	76.708	77.478	78.257	79.044	79.838	80.640
	ϵ^{-x}	0.0136	0.0134	0.0133	0.0132	0.0130	0.0129	0.0128	0.0127	0.0125	0.0124
4.4	ϵ^x	81.451	82.269	83.096	83.931	84.775	85.627	86.488	87.357	88.235	89.121
	ϵ^{-x}	0.0123	0.0122	0.0120	0.0119	0.0118	0.0117	0.0116	0.0114	0.0113	0.0112
4.5	ϵ^x	90.017	90.922	91.836	92.759	93.691	94.632	95.583	96.544	97.514	98.494
	ϵ^{-x}	0.0111	0.0110	0.0109	0.0108	0.0107	0.0106	0.0105	0.0104	0.0103	0.0102
4.6	ϵ^x	99.484	100.48	101.49	102.51	103.54	104.58	105.64	106.70	107.77	108.85
	ϵ^{-x}	0.0101	0.0100	0.0099	0.0098	0.0097	0.0096	0.0095	0.0094	0.0093	0.0092
4.7	ϵ^x	109.95	111.05	112.17	113.30	114.43	115.58	116.75	117.92	119.10	120.30
	ϵ^{-x}	0.0091	0.0090	0.0089	0.0088	0.0087	0.0087	0.0086	0.0085	0.0084	0.0083
4.8	ϵ^x	121.51	122.73	123.97	125.21	126.47	127.74	129.02	130.32	131.63	132.95
	ϵ^{-x}	0.0082	0.0081	0.0081	0.0080	0.0079	0.0078	0.0078	0.0077	0.0076	0.0075
4.9	ϵ^x	134.29	135.64	137.00	138.38	139.77	141.17	142.59	144.03	145.47	146.94
	ϵ^{-x}	0.0074	0.0074	0.0073	0.0072	0.0072	0.0071	0.0070	0.0069	0.0069	0.0068
5.0	ϵ^x	148.41	149.90	151.41	152.93	154.47	156.02	157.59	159.17	160.77	162.39
	ϵ^{-x}	0.0067	0.0067	0.0066	0.0065	0.0065	0.0064	0.0063	0.0063	0.0062	0.0062
5.1	ϵ	164.02	165.67	167.34	169.02	170.72	172.43	174.16	175.91	177.68	179.47
	ϵ^{-x}	0.0061	0.0060	0.0060	0.0059	0.0059	0.0058	0.0057	0.0057	0.0056	0.0056
5.2	ϵ^x	181.27	183.09	184.93	186.79	188.67	190.57	192.48	194.42	196.37	198.34
	ϵ^{-x}	0.0055	0.0055	0.0054	0.0054	0.0053	0.0052	0.0052	0.0051	0.0051	0.0050
5.3	ϵ^x	200.34	202.35	204.38	206.44	208.51	210.61	212.72	214.86	217.02	219.20
	ϵ^{-x}	0.0050	0.0049	0.0049	0.0048	0.0048	0.0047	0.0047	0.0047	0.0046	0.0046
5.4	ϵ^x	221.41	223.63	225.88	228.15	230.44	232.76	235.10	237.46	239.85	242.26
	ϵ^{-x}	0.0045	0.0045	0.0044	0.0044	0.0043	0.0043	0.0043	0.0042	0.0042	0.0041
5.5	ϵ^x	244.69	247.15	249.64	252.14	254.68	257.24	259.82	262.43	265.07	267.74
	ϵ^{-x}	0.0041	0.0040	0.0040	0.0040	0.0039	0.0039	0.0038	0.0038	0.0038	0.0037
5.6	ϵ^x	270.43	273.14	275.89	278.66	281.46	284.29	287.15	290.03	292.95	295.89
	ϵ^{-x}	0.0037	0.0037	0.0036	0.0036	0.0036	0.0035	0.0035	0.0034	0.0034	0.0034
5.7	ϵ^x	298.87	301.87	304.90	307.97	311.06	314.19	317.35	320.54	323.76	327.01
	ϵ^{-x}	0.0033	0.0033	0.0033	0.0032	0.0032	0.0032	0.0032	0.0031	0.0031	0.0031
5.8	ϵ^x	330.30	333.62	336.97	340.36	343.78	347.23	350.72	354.25	357.81	361.41
	ϵ^{-x}	0.0030	0.0030	0.0030	0.0029	0.0029	0.0029	0.0029	0.0028	0.0028	0.0028
5.9	ϵ^x	365.04	368.71	372.41	376.15	379.93	383.75	387.61	391.51	395.44	399.41
	ϵ^{-x}	0.0027	0.0027	0.0027	0.0027	0.0026	0.0026	0.0026	0.0026	0.0025	0.0025

Appendix D   Mathematical Formulas

C = circumference
D = diameter ($D = 2r$)
r = radius
A = area

Circle

$$C = \pi D = 2\pi r$$

$$A = \pi r^2 = \frac{\pi D^2}{4}$$

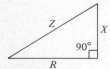

Right triangle

$$A = \tfrac{1}{2}RX, \qquad Z = \sqrt{R^2 + X^2}$$
$$R = \sqrt{Z^2 - X^2}, \qquad X = \sqrt{Z^2 - R^2}$$

Square

$$A = a^2$$
$$d = a\sqrt{2}$$

Rectangle

$$A = ab, \quad d = \sqrt{a^2 + b^2}$$

Trigonometry functions

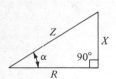

$$\text{sine (sin) } \alpha = \frac{X}{Z}$$

$$\text{cosine (cos) } \alpha = \frac{R}{Z}$$

$$\text{tangent (tan) } \alpha = \frac{\text{sine } \alpha}{\text{cosine } \alpha} = \frac{X}{R}$$

$$\text{cotangent (cot) } \alpha = \frac{R}{X}$$

$$\sin^2 \alpha + \cos^2 \alpha = 1$$
$$\sin 2\alpha = 2 \sin \alpha \cos \alpha$$
$$\cos 2\alpha = 2 \cos^2 \alpha - 1 = \cos^2 \alpha - \sin^2 \alpha$$
$$\sin^2 \alpha = \tfrac{1}{2}(1 - \cos 2\alpha)$$
$$\sin (\alpha \pm \beta) = \sin \alpha \cos \beta \pm \cos \alpha \sin \beta$$

Appendix E

Name	Letter Capital	Lower-case	Use in Text
alpha	A	α	α for angles
beta	B	β	
gamma	Γ	γ	
delta	Δ	δ, ∂	Δ for small change in value
epsilon	E	ε	
zeta	Z	ζ	
eta	H	η	η for efficiency
theta	Θ	θ	
iota	I	ι	
kappa	K	κ	κ for dielectric constant
lambda	Λ	λ	
mu	M	μ	μ for prefix micro or permeability
nu	N	ν	
xi	Ξ	ξ	
omicron	O	o	
pi	Π	π	π a constant = 3.1416
rho	P	ρ	ρ for resistivity
sigma	Σ	σ	Σ for summation
tau	T	τ	τ for time constant
upsilon	Υ	υ	
phi	Φ	φ	φ for angles and magnetic flux
chi	X	χ	
psi	Ψ	ψ	
omega	Ω	ω	Ω for ohms; ω for angular velocity

Appendix F Solutions to Test in Unit 3

2. $E = IR = (0.05)(22,000)$
   $= 1100 \text{ V}$

3. $R = \dfrac{E}{I} = \dfrac{12}{0.002} = 6000 \ \Omega$

4. $I = \dfrac{E}{R} = \dfrac{15}{5000} = 0.003 \text{ A}$

5. $I_T = I_1 + I_2$
   $= 2.75 \text{ A} + 0.935 \text{ A}$
   $= 3.685 \text{ A}$

6. $I_2 = I_T - I_1$
   $= 0.082 \text{ A} - 0.017 \text{ A}$
   $= 0.065 \text{ A}$

7. $P = I^2 R$
   $= (0.04)(0.04)(5100)$
   $= 8.16 \text{ W}$

8. $I = \sqrt{\dfrac{P}{R}} = \sqrt{\dfrac{1.8}{50,000}}$
   $= 0.006 \text{ A}$

9. $E = \sqrt{PR} = \sqrt{(0.2)(720)}$
   $= 12 \text{ V}$

10. $V = \dfrac{P}{I}$

11. $C = \dfrac{1}{2\pi f X_C}$

12. $\rho = \dfrac{RA}{l}$

13. $E = \sqrt{PR}$

14. $R_{eq} = \dfrac{R_5 R_6}{R_5 + R_6}$

15. $R_{eq} = \dfrac{(3300)(10,000)}{13,300}$
    $= 2481 \ \Omega$

Answers to Selected Problems   Appendix G

(Answers are given for most of the *odd*-numbered problems.)

Unit 1

1. See text

2. See Figure 1.12 and text

Unit 2

1. 10,000; 100,000; 1,000,000;
   10; 100

3. 300; 92.8; 7500; 412.3

5. 10.09; 5.76; 0.00235

7. 2.02×10^3; 8.25×10^1;
   6.24×10^{18}; 1.0×10^4;
   3.26×10^0; 6.43×10^8;
   4.7×10^4; 2.37×10^{-1};
   7.35×10^{-2}; 4.99×10^{-8};
   9.3×10^{-5}; 1.2×10^{-9}

9. See text

11. 10.37×10^3; 200×10^{-5};
    99.913×10^1

13. See text

15. $1.5 \times 10^6 \ \Omega = 1.5 \ M\Omega$
    $108 \times 10^6 \ Hz = 108 \ MHz$
    $33 \times 10^3 \ \Omega = 33 \ k\Omega$
    $8.2 \times 10^3 \ \Omega = 8.2 \ k\Omega$
    $12 \times 10^0 \ V = 12 \ V$
    $1.42 \times 10^6 \ Hz = 1.42 \ MHz$
    $2 \times 10^0 \ A = 2 \ A$
    $12 \times 10^{-3} \ A = 12 \ mA$
    $430 \times 10^{-9} \ A = 430 \ nA$
    $150 \times 10^{-3} \ V = 150 \ mV$

    $70 \times 10^{-9} \ s = 70 \ ns$
    $6 \times 10^{-6} \ A = 6 \ \mu A$
    $5 \times 10^{-12} \ s = 5 \ ps$
    $5.6 \times 10^3 \ \Omega = 5.6 \ k\Omega$
    $120 \times 10^0 \ V = 120 \ V$
    $3 \times 10^{-3} \ V = 3 \ mV$
    $390 \times 10^3 \ \Omega = 390 \ k\Omega$
    $80 \times 10^9 \ Hz = 80 \ GHz$
    $10 \times 10^3 \ \Omega = 10 \ k\Omega$
    $95 \times 10^6 \ Hz = 95 \ MHz$

17. 21 mA; 973 μs;
    4999.9 pF $\approx$ 5000 pF;
    76.3 kΩ; 514 MHz

19. $6,200,000 = 6.2 \ M\Omega \pm 5\%$
    $82,000 = 82 \ k\Omega \pm 20\%$
    $9100 \ \Omega \pm 5\%$; $180 \ \Omega \pm 10\%$

21. brown, black, green, none;
    brown, brown, yellow, gold;
    orange, white, red, silver;
    violet, green, black, gold

Unit 3

1–32. Answers given in text

33. $I = \dfrac{Q}{t}$; $t = \dfrac{Q}{I}$

35. $E = \dfrac{P}{I}$; $I = \dfrac{P}{E}$

37. $E = IR$

39. $W = Pt$; $t = \dfrac{W}{P}$

41. $A = \dfrac{\rho l}{R}$; $l = \dfrac{RA}{\rho}$

43. $R = \dfrac{V^2}{P}$

45. $E = \sqrt{PR}$ 47. $d = \sqrt{\dfrac{\rho l}{R}}$

49. $R_{eq} = \dfrac{12}{7}$

Unit 4

1. See text   3. $9\,\Omega$
5. 1 mA 7. 2 V
9. $20\text{ k}\Omega$ 11. 1.5 V
13. 2 mA 15. $4\text{ k}\Omega$
17. $386\text{ k}\Omega$

Unit 5

1. See text   3. $2.404 \times 10^{-16}\text{ C}$
5. 5 A 7. 2 A
9. 14 A; no   11. 9 V
13. 3 V; 7 V; 10 V 15. 277.8 mA

Unit 6

1. $15\,\Omega$ 3. $15\,\Omega$
5. $4\,\Omega$ 7. $10.37\,\Omega$
9. 1.01 ft 11. $6.516 \times 10^{-3}\,\Omega$
13. $1.645 \times 10^{-8}\,\Omega\text{-m}$;   15. 1.82 ft
    $1.724 \times 10^{-8}\,\Omega\text{-m}$;   17. $49.75\,\Omega$
    $2.443 \times 10^{-8}\,\Omega\text{-m}$;   19. $256\,\mu\text{S}$
    $2.83 \times 10^{-8}\,\Omega\text{-m}$;   21. 15 W
    $5.485 \times 10^{-8}\,\Omega\text{-m}$;   23. 12.13 mA
    $6.980 \times 10^{-8}\,\Omega\text{-m}$;   25. 7.9 W
    $7.811 \times 10^{-8}\,\Omega\text{-m}$;   27. 3.84 kWh
    $9.640 \times 10^{-8}\,\Omega\text{-m}$;
    $1.147 \times 10^{-7}\,\Omega\text{-m}$;
    $4.504 \times 10^{-7}\,\Omega\text{-m}$;
    $4.903 \times 10^{-7}\,\Omega\text{-m}$;
    $1.097 \times 10^{-6}\,\Omega\text{-m}$;

Unit 7

1. See Figure 7.10   5. $V_2 = 12\text{ V}$, $V_1 = 8\text{ V}$
3(b) $R_T = 15\text{ k}\Omega$, $E = 15\text{ V}$
  (c) $I = 1\text{ mA}$,   7. $R_2 = 20\text{ k}\Omega$,
  (d) $V_1 = 5\text{ V}$, $V_2 = 10\text{ V}$ $R_1 = 10\text{ k}\Omega$
  (e) $V_T = 15\text{ V}$ 9. $R_3 = 2\text{ k}\Omega$
11. $V_3 = 10\text{ V}$, $V_1 = 40\text{ V}$ 13. $V_2 = 6\text{ V}$
   $R_1 = 8\text{ k}\Omega$, $R_2 = 5\text{ k}\Omega$ 15. $V_1 = 4\text{ V}$, $V_3 = 10\text{ V}$
                                                         $V_1 + V_2 = 10\text{ V}$

17. $R_T = \infty\ \Omega$, $V_{out} = 0$ V
19. $V_L = 11.76$ V
21. I_L increases, V_L decreases
23. $R_{int} = 12.5\ \Omega$
25. $R_L = 0.5\ \Omega$, $R_L = 300\ \Omega$
27. $R_S = 1\ k\Omega$

Unit 8

1. See Figure 8.16(a)
3(a) $I_2 = 100$ mA,
  (b) $I_1 = 200$ mA,
  (c) $I_T = 50$ mA
5. $I_1 = 30$ mA, $I_2 = 22$ mA,
   $I_3 = 17$ mA
9(a) $I_1 = 300$ mA,
   $I_2 = 200$ mA,
   $I_T = 500$ mA
  (b) $I_1 = 200$ mA,
   $I_2 = 50$ mA,
   $I_T = 250$ mA
  (c) $I_1 = I_2 = 20$ mA,
   $I_T = 40$ mA
  (d) $I_1 = 3$ mA, $I_2 = 2$ mA,
   $I_T = 5$ mA
  (e) $I_1 = 0.8$ mA,
   $I_2 = 0.2$ mA,
   $I_T = 1.0$ mA
  (f) $I_1 = I_2 = 200\ \mu A$,
   $I_T = 400\ \mu A$
25. $R_{BD} = 0\ \Omega$, $R_T = 15\ k\Omega$
   $V_{AB} = 7$ V, $V_{BD} = 0$ V,
   $V_{BC} = 0$ V
   $V_{CD} = 0$ V, $V_{DE} = 14$ V
27. 6.8-$k\Omega$ resistor is shorted

7(a) $R_{eq} = 60\ \Omega$,
  (b) $R_{eq} = 80\ \Omega$,
  (c) $R_{eq} = 250\ \Omega$,
  (d) $R_{eq} = 2.4\ k\Omega$,
  (e) $R_{eq} = 16\ k\Omega$,
  (f) $R_{eq} = 50\ k\Omega$
11. $R_{eq} = R_T = 2.5\ k\Omega$
13. $R_{eq} = 2642\ \Omega$
15. $I_3 = 5$ mA
17. $I_1 = 2$ mA, $V_1 = 8$ V,
   $V_2 = 8$ V
   $I_2 = 0.667$ mA,
   $I_3 = 1.333$ mA
   $P_1 = 16$ mW,
   $P_2 = 5.33$ mW,
   $P_3 = 10.66$ mW
19. $R_T = 13\ \Omega$
21. $R_T = 10\ k\Omega$, $I_1 = 2$ mA
23. *Open*: $R_T = 4\ k\Omega$,
   $I_{source} = 2$ mA,
   $V_{source} = 8$ V,
   Shorted: $R_T = 0\ \Omega$,
   $I_{source} = \infty$ mA,
   $V_{source} = 0$ V

Unit 9

1. $I = 80$ mA
5. $I_L = 52$ mA, $V_L = 4.16$ V
9. $I_{40\ \Omega} = 80$ mA
11. $I_{40\ \Omega} = 80$ mA
13. $V_{out} = 1$ V

3. $V_L = 8$ V, $I_L = 0.364$ mA
7. $V_{10\ k\Omega} = 10$ V,
   $I_{10\ k\Omega} = 1$ mA,
   $I_{2\ k\Omega} = 2$ mA
15. $I_{meter} = 46.4\ \mu A$, to the
   right

Unit 10

1(a) $I = 0.667$ mA,
  (b) $V_{a-ground} = 50$ V
  (c) $V_{10\ k\Omega} = 6.67$ V,
  (d) $V_{b-ground} = 26.67$ V

3. $I = 5$ mA, $V = 30$ V
5. $I = 7$ mA, $V = 28$ V
7. $I_m = 2$ mA

Unit 11

1. See text
3. Voltage
5. $C_2 = 133$ pF
7. 0.0163 μF
9. $C_T = 70$ μF
   $Q_{10\ \mu F} = 500$ μC
   $Q_{20\ \mu F} = 1000$ μC
   $Q_{40\ \mu F} = 2000$ μC
11. $C_T = 0.01$ μF
   $V_{0.015\ \mu F} = 10$ V,
   $V_{2000\ pF} = 15$ V,
   $V_{0.04\ \mu F} = 3$ V,
   $V_{0.01\ \mu F} = 12$ V
   $Q_{0.015\ \mu F} = 0.15$ μC

Unit 12

1. See text
3. See text
5. $\mu = 6 \times 10^{-5}$ Wb/At-in.
7. $I = 1.346$ A

Unit 13

1. See text
3. $L \cong 355$ μH
5. $L = 20$ mH
7. $L_T = 20$ mH
9. $W_m = 24$ μJ

Unit 14

1. $e^{-1} = 0.368$, $e^{-2} = 0.135$
   $e^{-3} = 0.05$, $e^{-4} = 0.018$
3. $y = 0.135$
5. $x = 1.609$
7(a) 0 V, (b) 15.54 V,
  (c) 50 s
11. When:
   $t = 8$ ms, $i = 5.19$ mA,
9. $x = 1$, $v_C = 8.96$ V
   $v_L = 4.06$ V
   $x = 2$, $v_C = 15.94$ V
   $t = 12$ ms, $i = 5.70$ mA,
   $x = 3$, $v_C = 18.51$ V
   $v_L = 1.49$ V
   $x = 4$, $v_C = 19.45$ V
   $t = 20$ ms, $i = 6.0$ mA,
   After 1 s, $v_C = 8.96$ V
   $v_L = 0$ V
13. No waveforms are provided

Unit 15

1. See text
5. -12.43 V
3(a) 9.06 V, (b) 1.736 V
7(a) 5.88 V, (b) -10 V
  (c) -5.74 V, (d) -8.09 V
  (c) 0 V, (d) 9.54 V
9. 325.22 V_{p-p}
11(a) 1414 V, (b) 1000 V
  (c) 2828 V, (d) 500 Hz
  (e) 2 ms, (f) 0 V

Unit 16

1. Short phasor B_m is leading by 90°
5. See text
7. 2.5 kΩ

3(a) $0, 2\pi$, (b) $\pi, 3\pi$

(c) $\dfrac{\pi}{2}, \dfrac{3\pi}{2}, \dfrac{5\pi}{2}$, (d) $5\,\omega$,

(e) $-5\,\omega$

15. 2.4 Hz

9. 16.79 H

11. $-j2.652$ kΩ

13. $+j5.278$ kΩ

17. $X_L = 0\ \Omega;\ X_C = \infty\ \Omega$

Unit 17

1(a) $4 + j6$, (b) $5 - j2$
(c) $-3 + j3$, (d) $-7 - j4$
7. $2.09 - j19.89$
11. $50\,\underline{/+100^\circ}$
15. $5 + j100$
19. $7.61 + j13.71$
23. $45\,\underline{/+80^\circ}$
27. $51 + j33$
31. $-0.47 + j3.88$
35. 3.61 mA $\underline{/56.31^\circ}$
39. $I = 5$ mA $\underline{/0^\circ}$

3. $18.91 + j6.51$
5. $-16.77 + j10.89$
9. $50\,\underline{/+28^\circ}$
13. $50\,\underline{/+345^\circ}$ or $50\,\underline{/-15^\circ}$
17. $22.32 + j18.66$
21. $-13.74 + j28.04$
25. $4428\,\underline{/+35^\circ}$
29. $28 - j36$
33. $14.14\,\underline{/315^\circ}$
37. $\mathbf{V}_R = 16$ V $\underline{/0^\circ}$,
 $\mathbf{V}_L = j3.02$ V
41. $\mathbf{I}_R = 2$ mA $\underline{/0^\circ}$,
 $\mathbf{I}_L = 5$ mA $\underline{/-90^\circ}$,
 $\mathbf{I}_C = 10$ mA $\underline{/+90^\circ}$

43. 4.47 kΩ $\underline{/-63.43}$

Unit 18

1. See text
5. For $R = 5$ kΩ:
 $X_L = 5.28$ kΩ
 $\mathbf{Z} = 7.27$ kΩ $\underline{/46.6^\circ}$,
 $\mathbf{I} = 2.75$ mA $\underline{/0^\circ}$
 $\mathbf{V}_R = 13.75$ V $\underline{/0^\circ}$
 $\mathbf{V}_L = 14.51$ V $\underline{/90^\circ}$
 $P = 37.81$ mW,
 $VAR_L = 39.9$ mvar,
 $VA = 55$ mVA
 For $R = 10$ kΩ:
 $X_L = 5.28$ kΩ
 $\mathbf{Z} = 11.31$ kΩ $\underline{/27.8^\circ}$,
 $\mathbf{I} = 1.77$ mA $\underline{/0^\circ}$
 $\mathbf{V}_R = 17.7$ V $\underline{/0^\circ}$,
 $\mathbf{V}_L = 9.34$ V $\underline{/90^\circ}$
 $P = 31.28$ mW,
 $VAR_L = 16.53$ mvar,
 $VA = 35.4$ mVA

3. See text
7(a) $\mathbf{I} = 3.54$ mA $\underline{/0^\circ}$
 (b) $X_L = 2.64$ kΩ,
 (c) $L = 7$ H,
 (d) $\mathbf{Z} = 5.65$ kΩ $\underline{/27.8^\circ}$

Unit 19

1. See text

3(a) $\mathbf{I} = 2.129$ mA $\underline{/0°}$
   (b) $X_C = 26.52$ kΩ,
   (c) $C = 0.1$ μF,
   (d) $X_L = 5.2$ kΩ
   (e) $L = 14$ H

Unit 20

1. 2.65 kΩ, 5.66 mA $\underline{/+90°}$
   1.5 mA $\underline{/0°}$, 5.86 mA $\underline{/+75.2°}$
   2.56 kΩ $\underline{/-75.2°}$
   5.86 mA $\underline{/+75.2°}$
   2.65 kΩ, 5.66 mA $\underline{/+90°}$
   0.45 mA $\underline{/0°}$,
   5.86 mA $\underline{/+85.4°}$
   2.64 kΩ $\underline{/-85.4°}$,
   5.68 mA $\underline{/+85.4°}$
   2.65 kΩ, 0.566 mA $\underline{/+90°}$
   3.0 mA $\underline{/0°}$,
   3.05 mA $\underline{/+10.7°}$
   4.91 kΩ $\underline{/-10.7°}$
   3.05 mA $\underline{/+10.7°}$
   26.5 kΩ, 0.566 mA $\underline{/+90°}$
   1.5 mA $\underline{/0°}$, 1.6 mA $\underline{/+20.7°}$
   9.36 kΩ $\underline{/-20.7°}$,
   1.6 mA $\underline{/+20.7°}$
   26.5 kΩ, 0.566 mA $\underline{/+90°}$
   0.45 mA $\underline{/0°}$,
   0.72 mA $\underline{/+51.5°}$
   20.66 kΩ $\underline{/-51.5°}$,
   0.73 mA $\underline{/+51.5°}$

3. $\mathbf{I}_R = 5$ mA $\underline{/0°}$
   $\mathbf{I}_L = 15$ mA $\underline{/-90°}$
   $\mathbf{I}_C = 3$ mA $\underline{/+90°}$
5. See text
7. $\mathbf{G} = 1$ mS $\underline{/0°}$
   $B_L = 0.331$ mS
   $\mathbf{Y} = 1.053$ mS $\underline{/-18.3°}$
   $\mathbf{I}_T = 21.07$ mA $\underline{/-18.3°}$
   $\mathbf{Z}_{||} = 950$ Ω $\underline{/+18.3°}$

Unit 21

1. $P = 28.8$ W, $P_m = 57.6$ W

5. See text

9. $I = 560$ A
   $C = 3444$ μF
11. $C = 28.4$ μF

3. $P = 0$ W,
   $Q = 46.17$ var,
   $VA = 46.17$ VA
7(a) Lagging,
   (b) $P = 16$ W
   (c) $VA = 20$ VA,
   (d) PF = 0.8,
   (e) $Q = 12$ var,   (f) 0.6

Unit 22

1. $\mathbf{Z}_T = 24.1\ k\Omega + j12.19\ k\Omega$
   $= 27\ k\Omega\ \underline{/26.8°}$
   $I = 4.44\ mA,$
   $\mathbf{V}_1 = 88.9\ V\ \underline{/+60°},$
   $\mathbf{V}_2 = 66.6\ V\ \underline{/-20°}$
3. $\mathbf{Z}_T = 11.1\ k\Omega\ \underline{/+13.2°}$
5. $\mathbf{V}_1 = 88.9\ V\ \underline{/+60°},$
   $\mathbf{V}_2 = 66.6\ V\ \underline{/-20°}$
7. $\mathbf{I}_1 = 12\ mA\ \underline{/-20°}$
9. $R_s = 198\ \Omega,\ X_s = 1980\ \Omega$
11. $R_p = 2\ k\Omega,\ X_p = 2\ k\Omega$
15. $\mathbf{Z}_{ab} = 1.32\ \Omega\ \underline{/-88.5°}$
13. $\mathbf{Z}_{ab} = 22.4\ k\Omega\ \underline{/-26.5°},$
    $\mathbf{Z}_{cd} = 5\ \Omega\ \underline{/0°}$

Unit 23

1. See text
3. $L = 25.3\ \mu H$
5. $BW = 20\ kHz,$
   $f_1 = 1190\ kHz,$
   $f_2 = 1210\ kHz$
7. See text
9. *Load removed*:
   $Z_{||} = 50\ k\Omega,\ Q = 50,$
   $BW = 15.92\ kHz$
   With load:
   $Z_{||} = 40\ k\Omega,\ Q = 40,$
   $BW = 19.9\ kHz$

Unit 24

1. See text
3. 6:1
5. $P = 20.7\ kW,$
   $I_p = 11.25\ A$
7. $Z_p = 160\ k\Omega$
9. $I_p = 745.8\ mA$
11. In phase
13. $R_e = 4\ \Omega,\ X_e = 4.47\ \Omega$
    $R_c = 4\ \Omega,\ X_m = 9.17\ \Omega$
15. See text
17. See text

Unit 25

1. See text
3. $E_p = 2400\ V$
5. $E_L = E_p = 190.5\ V$
7. $I_p = 12\ A,\ I_L = 20.78\ A,$
   $P_T = 8640\ W$
9. See text
11. $\mathbf{I}_A = -3.5\ A + j2.6\ A,$
    $\mathbf{I}_B = 5\ A + j5.2\ A,$
    $\mathbf{I}_C = -1.5\ A - j7.8\ A,$
    $P_T = 1980\ W$

Unit 26

1(a) Odd, 12; (b) Even, 8;   3. Odd
  (c) Even, 4

5.   1.1518   1.1522   7. $V_{\text{eff}} = 0.6149$ V
       0.9619   0.9611 $P = 54$ mW
       0.9609   0.9621   9. $V_{\text{eff}} = 80.078$ V
       1.1528   1.1512
       0.0041   0.0061
      -1.1508 -1.1532
      -0.9629 -0.9601
      -0.9598 -0.9630
      -1.1539 -1.1502
      -0.0081 -0.0122

Unit 27

1. See text   3. $R_{\text{sh}} = 222$ Ω
5. $I_{\text{m}} = 3.922$ mA versus 4 mA   7. $R_{\text{m}} = 1.5$ kΩ
                                      0.6 mA with $R_{\text{ext}} = 1$ kΩ
                                      0.2 mA with $R_{\text{ext}} = 6$ kΩ

9. Yes, meter deflects half of   11. See text
   full scale ($I = 25$ μA)

Appendix H Answers to Tests

Unit 1

1 through 3: See text

4. Add voltmeter across battery to Figure 1.13

Unit 2

Part 1
1. 314
2. 72,860
3. 12
4. 1,207,000
5. 840
6. 0.328
7. 1.008
8. 0.07325
9. 0.000004
10. 0.000000001

Part 2
1. $1.5 \times 10^6 \ \Omega$
2. $3 \times 10^{-3} \ A$
3. $1.2 \times 10^2 \ V$
4. $7 \times 10^{-6} \ A$
5. $9.5 \times 10^1 \ V$
6. $3.9 \times 10^3 \ \Omega$
7. $4.0 \times 10^{-2} \ V$
8. $6.0 \times 10^1 \ Hz$
9. $1 \times 10^4 \ \Omega$
10. $9 \times 10^0 \ V$

Part 3
1. 3.7×10^4
2. 27.55×10^4
3. 12.33×10^2
4. 52.0×10^{-4}
5. 3.12×10^7
6. 8.61
7. 15.73×10^{-3}
8. 5.8×10^1
9. 5.61×10^2
10. 2.12×10^{-4}

Part 4
1. $0.035 \ V = 35,000 \ \mu V$; m
2. $1.5 \ MV = 1,500,000 \ V$; M
3. $0.75 \ mA = 750 \ \mu A$; μ
4. $0.0075 \ A = 7500 \ \mu A$; m
5. $1500 \ k\Omega = 1.5 \ M\Omega$; M
6. $39,000 \ \Omega = 0.039 \ M\Omega$; k
7. $5 \ GHz = 5,000,000 \ kHz$; G
8. $2 \ kHz = 0.002 \ MHz$; k
9. $95 \ ms = 95,000 \ \mu s$; m
10. $0.006 \ \mu s = 6 \ ns$; n

Part 5
1. 560 kΩ; 672 kΩ; 448 kΩ
2. 27 kΩ; 29.7 kΩ; 24.3 kΩ
3. 2400 Ω; 2520 Ω; 2280 Ω
4. 4.3 Ω; 4.515 Ω; 4.085 Ω
5. 300 kΩ; 315 kΩ; 285 kΩ
6. 15 kΩ; 18 kΩ; 12 kΩ
7. red, black, red, gold
8. orange, orange, black, silver
9. white, brown, yellow, gold
10. brown, red, orange, none
11. green, blue, green, silver
12. yellow, violet, silver, gold

Unit 3

See Appendix F

Unit 4

1. $I = 0.3 \ A$
3. $E = 10 \ V$
5. $R = 50 \ \Omega$
2. $R = 15 \ \Omega$
4. $I = 300 \ \mu A$
6. $I = 10 \ mA$

7. $E = 100$ mV
9. $I_1 = 5$ mA

8. $V = 20$ V
10. $V_C = 15$ V

Unit 5

1. 6.24×10^{18} electrons
3. $Q = 1.5 \times 10^{-2}$ C;
   $I = 15$ mA
5. $Q = 0.2$ C
7. $V = 12$ V

2. $Q = 1.6 \times 10^{-18}$ C
4. $I = 5$ A
6. 4 V
8. 13.90 h
9. $W = 3$ J

Unit 6

1. $R = 333\ \Omega$

2. $R = \dfrac{\rho l}{A}$

3. $R_2 = 10\ \Omega$
5. $2.83 \times 10^{-8}\ \Omega$-meters
7. $G = 12.2\ \mu$S
9. $P = 1W$
11. $R = 14.4\ \Omega; I = 8.33$ A
13. $I = 44.7$ mA

4. $R = 170\ \Omega$
6. $R = 0.787\ \Omega$
8. See text
10. 26.88¢
12. $E = 50$ V
14. $\eta = 74.6\%$

Unit 7

1(b) $R_T = 60$ kΩ,
  (c) $I = 100\ \mu$A
  (d) $V_1 = 1$ V, $V_2 = 2$ V,
    $V_3 = 3$ V
3. $R = 2\ \Omega; P = 18$ W
5. $V_2 = 20$ V
6. $R_1 = 30\ \Omega$, $R_2 = 20\ \Omega$
8. $I = 400$ mA, $R_1 = 15\ \Omega$
10. $R_1 = 8$ kΩ
12. $V = 24$ V
14. $R_{int} = 2\ \Omega$
15. $R_L = 100\ \Omega$
16. $E = 690$ V

2(a) Series, (b) $R_T = 5$ kΩ
  (c) $I = 3$ mA,
  (d) $V_1 = 12$ V; $V_2 = 3$ V
  (e) $V_T = 15$ V
4. $V_2 = 5$ V, $V_1 = 12$ V,
   $E = 9$ V
7. $R_1 = 4.5$ kΩ, $R_2 = 3$ kΩ
9. $R_2 = 50$ kΩ, $R_3 = 25$ kΩ
11. $V_2 = 25$ V; $V_1 = 5$ V
   $V_1 = 12$ V, $V_2 = 8$ V
   $V_1 = 3$ V, $V_2 = 27$ V
13(a) $V_{out} = 0$ V,
  (b) $V_{out} = 40$ V,
  (c) $V_{out} = 40$ V,
  (d) $V_{out} = 0$ V

Unit 8

1(b) $V_{8\ k\Omega} = 16$ V,
  (c) $R_{eq} = 1.6$ kΩ,
  (d) $I = 10$ mA,
  (e) $I_{2\ k\Omega} = 8$ mA

2. $R_{eq} = 5$ kΩ
3. $V_2 = 15$ V
4. $I_1 = 700$ mA
5. $R_{eq} = 7.5$ kΩ

6. $R_1 = 10 \text{ k}\Omega$

7. $I_2 = 1.5 \text{ mA}$

8. $R_2 = 2 \text{ k}\Omega, \quad R_3 = 5 \text{ k}\Omega$

9. $R_T = 20 \text{ k}\Omega, \quad V_2 = 16 \text{ V}$

10. $I_2 = 10 \text{ mA}, \quad I_1 = 30 \text{ mA}$
    $R_3 = 2 \text{ k}\Omega, \quad R_2 = 3 \text{ k}\Omega,$
    $R_1 = 2 \text{ k}\Omega$

11. $R_T = 40 \text{ k}\Omega$
    $I_1 = 0.5 \text{ mA}, \quad I_2 = 0.3 \text{ mA},$
    $I_3 = 0.2 \text{ mA}$
    $I_4 = 0.1 \text{ mA}, \quad I_5 = 0.1 \text{ mA}$
    $P_1 = 2 \text{ mW},$
    $P_6 = 0.16 \text{ mW}$
    $V_4 = 4 \text{ V}$

12(a) $V_{AB} = 24 \text{ V};$
    $V_{BC} = V_{CD} = V_{BD} = 0 \text{ V}$

(b) $V_{AB} = 0 \text{ V}; V_{BC} = 14.4 \text{ V}$
    $V_{CD} = 9.6 \text{ V}; V_{BD} = 24 \text{ V}$

(c) $V_{AB} = 15.72 \text{ V};$
    $V_{BC} = 0 \text{ V}$
    $V_{CD} = V_{BD} = 8.28 \text{ V}$

(d) $V_{AB} = 6.86 \text{ V}; V_{BC} = 0 \text{ V}$
    $V_{CD} = V_{BD} = 17.14 \text{ V}$

(e) $V_{AB} = 24 \text{ V};$
    $V_{BC} = V_{CD} = V_{BD} = 0 \text{ V}$

Unit 9

1. $I_L = 5 \text{ mA}; V_L = 7.5 \text{ V}$

2. $I_L = 0.5 \text{ mA}; V_L = 1 \text{ V}$

3. $I_{3 \text{ k}\Omega} = 3.75 \text{ mA};$
    $V_{3 \text{ k}\Omega} = 11.25 \text{ V}$

4. $R_L = 3 \text{ k}\Omega; P_L = 3 \text{ mW}$

5. $I_L = 5 \text{ mA}$

6. $I_{2 \text{ k}\Omega} = 500 \text{ } \mu\text{A}$

7. $I_L = 5 \text{ mA}$

8. $V_{\text{out}} = 7 \text{ V}$

9. $I_B = 10 \text{ } \mu\text{A}$

10. $I_m = 1.49 \text{ mA}$

11(a) $I_{2.7 \text{ k}\Omega} = 0.851 \text{ mA}$

(b) $I_{5 \text{ k}\Omega} = 0.701 \text{ mA}$

Unit 10

1. $I = -0.1 \text{ mA},$
    $V_{10 \text{ k}\Omega} = -1 \text{ V},$
    $V_{a-b} = 7 \text{ V}$

2. $I_1 = 2.5 \text{ mA}, I_2 = 2.5 \text{ mA}$
    $V_{\text{meter}} = 0 \text{ V}$

3. and 4. $I = 2.33 \text{ mA},$
    $V = 14 \text{ V}$

5. $I_{1 \text{ k}\Omega} = 4 \text{ mA to right}$
    $V_{1 \text{ k}\Omega} = 4 \text{ V}$

6. $I_{2 \text{ k}\Omega} = 0 \text{ mA}$

7. $I = 4 \text{ mA}, V = 20 \text{ V}$

Unit 11

1. Changes in voltage

2. $C = 0.05 \text{ } \mu\text{F}$

3. $C_T = 0.06 \text{ } \mu\text{F}$

4. $Q = 0.5 \text{ } \mu\text{C}$

5. $C_T = 0.014 \text{ } \mu\text{F}$

6. $C_T = 24 \text{ } \mu\text{F}$

7. $Q_T = 480 \text{ pC}$

8. $V_{60 \text{ pF}} = 8 \text{ V}$

9. $C_T = 12 \text{ pF},$
    $V_{30 \text{ pF}} = 40 \text{ V}$

10. 8 mJ

Unit 12

1. through 3. See text
5. $I = 96$ mA
7. $\mu \cong 1.25 \times 10^{-3}$
9. $I \cong 300$ mA

4. $\mathscr{R} = 400 \times 10^3$ At/Wb
6. $\mathscr{R} = 2.39 \times 10^9$ At/Wb
8. $I \cong 843.8$ A
10. $I \cong 1.93$ A

Unit 13

1. Changes in current
3. $L = 50$ mH
5. $L = 10$ mH
7. $L = 24$ mH
9. $W_m = 75$ μJ

2. $L = 400$ mH
4. $L = 2.5$ H
6. $L = 400$ mH
8. $L \cong 32.9$ μH

Unit 14

1. $x = 0, y = 1$
   $x = 1, y = 0.368$
   $x = 2, y = 0.135$
   $x = 5, y = 0$
6(a) $v_C = 59.02$ V,
 (b) $v_C = 21.71$ V

2. $v_C = 40.65$ V
3. $v_C = 9.89$ V
4. $v_C = 16.59$ V
5. $t = 41.7$ ms, $i = 38$ mA
7. At $t = 0.6$ s, $v_C = 45.12$ V
   At $t = 20$ s, $v_C = 6.11$ V

Unit 15

1(a) See text
 (b) $E_m = 50$ V
 (c) $E = 35.35$ V
 (d) $E_{p-p} = 100$ V p–p
 (e) $f = 100$ Hz,
 (f) $T = 10$ ms
 (g) $e = -25$ V

2. $\theta = 37°$ and $143°$
3. $e = -29.39$ V
4(a) 400 V p–p
 (b) 0 V,   (c) 141.4 V
5. $I_m = 32.14$ A
6. $V_{HWA} = 127.6$ V

Unit 16

1. See text
3. $+j2$ kΩ
5. $+j0$
7. $L = 126.78$ μH

2. $-j10$ kΩ
4. $-j2.653$ kΩ
6. $f = 79.62$ Hz

Unit 17

1. See text
2(a) $-43.3 + j25$
 (b) $+35 - j60.62$
 (c) $-21.21 - j21.21$,
 (d) $-34.64 + j20$

3(a) $15 \, \underline{/-53.13°}$
 (b) $5.66 \, \underline{/+225°}$
 (c) $5 \, \underline{/+53°}$
 (d) $4.47 \, \underline{/+116.57°}$

4. $7 + j2$
6. $40 \underline{/+45°}$
8. $6.45 + j25.66$
10. $0.122 - j0.902$

5. $5 + j5$
7. $3 \underline{/+25°}$
9. $-6 - j90$

Unit 18

1. $\mathbf{V}_L = 22.36$ V $\underline{/90°}$
2. $X_L = 5.278$ kΩ
   $\mathbf{Z} = 22.62$ kΩ $\underline{/+13.49°}$
   $\mathbf{I} = 0.884$ mA $\underline{/0°}$
   $\mathbf{V}_R = 19.45$ V $\underline{/0°}$
   $\mathbf{V}_L = 4.66$ V $\underline{/+90°}$
   $P = 17.19$ mW
   $VAR = 4.128$ mvar
   $VA = 17.68$ mVA

3. $\mathbf{I} = 2.113$ mA $\underline{/0°}$
   $X_C = 26.536$ kΩ
   $C = 0.1$ μF
   $\mathbf{Z} = 37.86$ kΩ $\underline{/-44.5°}$
4. $\mathbf{Z} = 15$ kΩ $\underline{/-36.9°}$
   $\mathbf{I} = 4$ mA $\underline{/0°}$
   $\mathbf{V}_R = 48$ V $\underline{/0°}$
   $\mathbf{V}_C = 36$ V $\underline{/-90°}$
5. $L = 60.78$ H

Unit 19

1. $\mathbf{I} = 4.84$ mA $\underline{/0°}$
   $X_C = 2645$ Ω
   $C = 1.0$ μF
   $X_L = 5.28$ kΩ
   $L = 14$ H
3. $\mathbf{Z} = 5$ kΩ $\underline{/-53.1°}$
   $\mathbf{I} = 20$ mA $\underline{/0°}$
   $\mathbf{V}_R = 60$ V $\underline{/0°}$
   $\mathbf{V}_L = 120$ V $\underline{/+90°}$
   $\mathbf{V}_C = 200$ V $\underline{/-90°}$

2. $\mathbf{I} = 16.04$ mA $\underline{/0°}$
   $X_C = 7.54$ kΩ
   $C = 0.351$ μF
   $X_L = 5.3$ kΩ
   $L = 14$ H
   $VAR_L = 1364.8$ mvar
   $VAR_C = 1939.8$ mvar
   $VA = 962.4$ mVA
4. $X_L = 5.278$ kΩ
   $X_C = 2.65$ kΩ
   $\mathbf{Z} = 10.34$ kΩ $\underline{/+14.72°}$
   $\mathbf{I} = 4.84$ mA $\underline{/0°}$
   $\mathbf{V}_R = 48.4$ V $\underline{/0°}$
   $\mathbf{V}_L = 25.55$ V $\underline{/+90°}$
   $\mathbf{V}_C = 12.83$ V $\underline{/-90°}$

Unit 20

1. $X_C = 26.525$ kΩ
   $\mathbf{Z} = 9.36$ kΩ $\underline{/-20.66°}$
   $\mathbf{I}_R = 5$ mA $\underline{/0°}$
   $\mathbf{I}_C = 1.89$ mA $\underline{/+90°}$
   $\mathbf{I}_T = 5.35$ mA $\underline{/+20.71°}$

2. $\mathbf{I}_T = 25$ mA $\underline{/-53.1°}$
3. $\mathbf{Z}_{||} = 7.07$ kΩ $\underline{/+45°}$
   $\mathbf{Y} = 0.141$ mS $\underline{/-45°}$
4. $\mathbf{I}_T = 13$ mA $\underline{/-22.62°}$
   $\mathbf{Z}_{||} = 5.54$ kΩ $\underline{/22.62°}$
   $\mathbf{Y} = 0.181$ mS $\underline{/-22.62°}$

Unit 21

1. $P = 1.108$ kW
   $P_m = 2.215$ kW

2. $P = 72$ W, $Q = 0$
   $P_{in} = 72$ VA

3. $P = 0$, $Q = 3.77$ var
   $P_{in} = 3.77$ VA
5. See text
7. $P = 770$ mW, PF $= 0.164$
   Lagging
9. PF $= 0.741$, $\phi = 42.2°$
   $VA = 1080$ VA
11. PF $= 0.8$, $P = 8$ kW
    $VA = 10$ kVA
13. $C = 23.28$ μF
15. $C = 235$ μF

4. $P = 0$, $Q = 0.314$ var
   $P_{in} = 0.314$ VA
6. PF $= 0.56$, $\phi = 56.25°$
8. $I = 227.3$ A
   $VA = 50$ kVA
10. $P = 41.4$ kW
    $VA = 46$ kVA, $\phi = 25.84°$
12. PF $= 0.141$,
    $\mathbf{I}_T = 227.3$ A $\underline{/0°}$
    $\mathbf{I}_C = 1590.9$ A $\underline{/90°}$
    $\mathbf{I}_L = 1607$ A $\underline{/-81.9°}$
14. $I_m = 2$ A, $I_C = 1.05$ A
    $I_L = 1.7$ A
16. $Q = 139.2$ var, $C = 25.6$ μF

Unit 22

1. $\mathbf{Z}_T = 31.3$ kΩ $\underline{/-26.57°}$,
   $\mathbf{I} = 3.83$ mA $\underline{/0°}$
   $\mathbf{V}_1 = 54.21$ V $\underline{/+45°}$
   $\mathbf{V}_2 = 114.9$ V $\underline{/-53.1°}$
3. $\mathbf{Z}_T = 674$ Ω $\underline{/+14.24°}$
5. $\mathbf{V}_2 = 115$ V $\underline{/-53.1°}$
6. $\mathbf{V}_L = 26.47$ V $\underline{/+90°}$
7. $\mathbf{I}_2 = 4.0$ mA $\underline{/53.1°}$
8. $\mathbf{I}_3 = 3.376$ mA $\underline{/+32°}$
9. $R_s = 31.13$ Ω
   $X_s = 498.05$ Ω
11. $R_p = 3$ kΩ, $X_p = 1.5$ kΩ
13. $R_p = 6.07$ kΩ
    $X_p = -j9.229$ kΩ
15. $\mathbf{V}_{oc} = 70.72$ V $\underline{/+45°}$
    $\mathbf{Z}_{oc} = 20 + j40$
    $\mathbf{I}_L = 1.961$ A $\underline{/-11.3°}$
    $\mathbf{V}_L = 19.61$ V $\underline{/-101.3°}$

2. $\mathbf{Z}_T = 13.55$ kΩ $\underline{/+18.4°}$
   $\mathbf{I}_T = 8.86$ mA $\underline{/-18.4°}$
   $\mathbf{I}_1 = 8.49$ mA $\underline{/-45°}$
   $\mathbf{I}_2 = 4$ mA $\underline{/53.13°}$
4. $\mathbf{Z}_T = 2.026$ kΩ $\underline{/-7.2°}$
   $\mathbf{I}_T = 9.872$ mA $\underline{/7.2°}$
   $\mathbf{I}_1 = 5.128$ mA $\underline{/0°}$
   $\mathbf{I}_2 = 1.908$ mA $\underline{/-17.4°}$
   $\mathbf{I}_3 = 3.374$ mA $\underline{/+32.5°}$
10. $R_s = 297$ Ω
    $X_s = 29.7$ Ω
12. $R_p = 10$ kΩ, $X_p = 30$ kΩ
14. $\mathbf{Z}_{ab} = 15$ Ω $\underline{/0°}$
    $\mathbf{Z}_{cd} = 10 - j20$
16. $\mathbf{Z}_{in} = 150$ Ω $\underline{/+36.87°}$

Unit 23

1. $f_0 = 39.808$ kHz
   $Z_s = 125$ Ω
   $I = 200$ mA, $V_R = 25$ V
   $V_L = V_C = 2$ kV
4. $f_0 = 796$ kHz
   BW $= 6$ kHz, $Q \cong 132.7$
   $R \cong 9.42$ Ω
6. $Z_{||} = 15$ kΩ, $Q = 15$
   BW $= 53.07$ kHz

2. $L = 26.4$ μH
3. $f_0 = 9.952$ kHz
   $Q = 50$, BW $= 199$ Hz
5. $Z_{||} = 20$ kΩ, $Q = 20$
   BW $= 39.8$ kHz
7. $V_{out} = 40$ V
8. $R_L = 400$ kΩ

Unit 24

1. See text
3. 2.67:1
5. $I_p = 25$ mA, $I_s = 6.52$ A

7. $Z = 1$ kΩ
9. $I_p = 3.61$ A
11. $P_{sl} = 37.8$ W
13. $R_c = 0.25\ \Omega$, $X_m = 2.59\ \Omega$

2. $E = 6$ V
4. $I = 35.56$ mA
6. $V = 267$ mV, $I = 66.7$ mA
   $Z_p = 5625\ \Omega$
8(a) 19, (b) $V = 6.32$ V
  (c) $Z = 1805\ \Omega$
10. Out of phase
12. $R_e = 0.202\ \Omega$, $X_e = 0.36\ \Omega$
14. $E_s = 80$ V, $I_p = 6.67$ A
   $I_s = 10$ A, $I_{BC} = 3.33$ A

Unit 25

1. No waveform given
3. $E_L = E_p = 220$ V
5(a) 120 V, (b) 6 A
  (c) 6 A, (d) 2.16 kW
  (e) 0
7. $\mathbf{I_A} = -10.39 + j6$
    $= 12$ A $\underline{/150°}$
  $\mathbf{I_B} = -j8 = 8$ A $\underline{/270°}$
  $\mathbf{I_C} = 8.66 + j5$
    $= 10$ A $\underline{/+30°}$
  $\mathbf{I_N} = -1.73 + j3$
    $= 3.46$ A $\underline{/+120°}$
  $P_T = 3600$ W

2. $E = 2400$ V
4(a) 3 A, (b) 5.2 A
  (c) 1080 W
6. $-40 + j17.32,\ 35 + j8.66$
   $5 - j25.98,\ P_T = 18$ kW

Unit 26

1. Even, period $= 8$
  Odd, period $= 4$
  Odd, period $= 8$
3. Even
  10 ms, $A = 60.92$
  20 ms, $A = 50.96$
  40 ms, $A = -60.82$
  60 ms, $A = -61.01$
  80 ms, $A = 50.66$
  100 ms, $A = +120$

2. Even
  dc amplitude $= E$
  Fundamental amplitude
    $= 2E$
  Second harmonic $= 2E$
  Third harmonic $= 2E$
4. $V_{eff} = 0.9656$ V,
  $P = 93$ mW
5. $V_{out} = 110 + 10 \sin 62.8t$
  $V_{eff} = 110.23$ V

Unit 27

1. $V = 50$ mV
3. $R_1 = 40\ \Omega$, $R_2 = 9\ \Omega$
  $R_3 = 1\ \Omega$

2. $R \cong 0.505\ \Omega$
4. 2.985 mA versus 3 mA
  9.375 mA versus 11.11 mA

5. $R_s = 950\ \Omega$, $R_T = 1\ \text{k}\Omega$
   $R_s = 2.45\ \text{k}\Omega$, $R_T = 2.5\ \text{k}\Omega$
   $R_s = 9.95\ \text{k}\Omega$, $R_T = 10\ \text{k}\Omega$

6. 3.43 V versus 8 V
   9.84 V versus 10 V

7. $R_m = 6\ \text{k}\Omega$, $\ 0\ \Omega$ at 1 mA
   1.5 kΩ at 0.8 mA
   6 kΩ at 0.5 mA

8. Yes, 30 μA or three-fifths
   of scale

9. Yes, 40 μA or four-fifths
   of scale

10. See text

11. See text

Index

ABCDEFGHIJ-VB-8210/79